The Porphyrin Handbook

Volumes in This Set

The Porphyrin Handbook

Volume 20 / Phthalocyanines: Structural Characterization

Editors

Karl M. Kadish
Department of Chemistry
University of Houston
Houston, Texas

Kevin M. Smith
Department of Chemistry
Louisiana State University
Baton Rouge, Louisiana

Roger Guilard
Faculté des Sciences Gabriel
Université de Bourgogne
Dijon, France

ACADEMIC PRESS

An imprint of Elsevier Science

Amsterdam · Boston · London · New York · Oxford · Paris
San Diego · San Francisco · Singapore · Sydney · Tokyo

Academic Press
An imprint of Elsevier Science
525 B Street, Suite 1900
San Diego, California 92101-4495
USA

First edition 2003
ISBN: **0-12-393220-3 (Set of Volumes 11–20)**
0-12-393221-1 (Volume 11)
0-12-393222-X (Volume 12)
0-12-393223-8 (Volume 13)
0-12-393224-6 (Volume 14)
0-12-393225-4 (Volume 15)
0-12-393226-2 (Volume 16)
0-12-393227-0 (Volume 17)
0-12-393228-9 (Volume 18)
0-12-393229-7 (Volume 19)
0-12-393230-0 (Volume 20)

∞ The paper used in this publication meets the requirements of ANSI/NISO Z39.48-1992 (Permanence of Paper).

Printed in The Netherlands.

Contents of Volume 20

Preface

Volumes 1–10 of *The Porphyrin Handbook* were first published in late 1999 and met with resounding international acclaim. After only two months on the shelves, our publisher, Academic Press, was awarded the 1999 Chemistry award by the Professional and Scholarly Publishing Division of the American Association of Publishers, Inc.; this honor recognized *The Porphyrin Handbook* as the "best chemistry book of the year." At that time we, the three editors, began receiving calls and requests from our colleagues and our Academic Press publisher to "expand the work and fill in the gaps" so as to assemble the best ever up-to-date compendium of every possible aspect of the field of porphyrins, other tetrapyrroles, and related macrocycles.

Shortly after publication of the first ten volumes of *The Porphyrin Handbook*, the First International Conference on Porphyrins and Phthalocyanines (ICPP-1) took place in Dijon, France (June 2000), hosted by one of the three editors of the Handbook. At that major international event, the Society of Porphyrins and Phthalocyanines (SPP) was created, and in early 2002 our fledgling SPP took over publication of the *Journal of Porphyrins and Phthalocyanines* (JPP) from John Wiley & Sons, with one of the three Handbook editors serving as Editor-in-Chief. At ICPP-2 in Kyoto, Japan (July 2002), it became abundantly clear that *The Porphyrin Handbook*, Volumes 1–10, was serving well our community of research scientists. This applied equally to our society (SPP) and journal (JPP), which are incrementally enhancing the international visibility of our field. Our new series of Handbook volumes (11–20) and the continuation of our popular international conference series (ICPP-3, to be held in New Orleans, USA, in 2004) each in their own way will contribute further to the scientific enrichment of the porphyrin and phthalocyanine research areas.

However, even before publication of Volumes 1–10, we were becoming aware of the great interest and excitement about the treatise from our real audience, the scientists, clinicians, and engineers working in our field – "you won't be covering biology or biosynthesis, let's have Volume 11 soon" – "what about phthalocyanines? They're (tetraazatetrabenzo)porphyrins too!" So, in very short order the three of us met, put together a plan for continuation, and obtained the publisher's approval and blessings for ten more volumes; this last action was no mean accomplishment since consolidations were taking place in the publishing sector, and the new set of volumes would be published by Elsevier.

Our sequel, in ten more volumes (11–20), builds upon the first published set of works by extending its authoritative treatise of the tetrapyrroles. We have moved on from synthesis, chemistry, spectroscopy and applications of porphyrins to address, in depth, many of the sub-fields not covered in the first treatise (Volumes 1–10). To be sure, there were still some loose ends, but we believe that our plan was fairly comprehensive. New topics this time include biology and medical implications of porphyrin systems, the biosynthesis of porphyrins, chlorophylls and vitamin B_{12}. Other areas covered include clinical and biochemical aspects of genetically transmitted or drug induced diseases associated with errors in heme metabolism, as well as the transformation of hemes into bile pigments, the organic synthesis of bilins, recent work on phytochrome and the pathways of degradation of chlorophyll in senescent plants. In addition, Volumes 11–20 address every aspect of the synthesis, chemistry, structure and spectroscopy of phthalocyanines.

Our 61 chapters (coincidentally the same number of chapters as in the 1–10 set of volumes) are once again written by internationally recognized experts and we were particularly gratified by two organizational aspects; firstly, we received almost no rejections of our requests for chapters, and secondly, almost all of our contracted authors provided us with chapters in a timely manner which, we believe, are as good in quality, or even better, than those in the first ten volumes. Indeed, once again we believe that our contributors have produced research articles that will be dominant in their specific areas for another fifteen years.

We sincerely hope that our readers will appreciate this new treatise as much as they liked our first venture. The tetrapyrrole field continues to expand into new areas and to gather new recruits and converts. We think that having the research field available at ones' fingertips (even if it comprises twenty volumes) is a wonderful thing, and this can only serve to facilitate future expansion while at the same time recording our field's rich history.

Karl M. Kadish, *Houston*
Kevin M. Smith, *Baton Rouge*
Roger Guilard, *Dijon*

Contributors to Volumes 1–20

Takuzo Aida
Department of Chemistry and Biotechnology
Graduate School of Engineering
The University of Tokyo
Hongo, Bunkyo-ku, Tokyo 113-8656, Japan
aida@macro.t.u-tokyo.ac.jp
Chapters 23, 42

Muhammad Akhtar
School of Biological Sciences
University of Southampton
Bassett Crescent East
Southampton SO16 7PX, UK
m.akhtar@soton.ac.uk
Chapter 71

Karl E. Anderson
Departments of Preventive Medicine and
Community Health Internal Medicine,
and Pharmacology and Toxicology
The University of Texas Medical Branch at Galveston
Texas 77555-1109, USA
kanderso@utmb.edu
Chapter 94

John Arnold
Department of Chemistry
University of California, Berkeley
Berkeley, California 94720-1460, USA
arnold@socrates.Berkeley.edu
Chapter 17

Karine Auclair
Department of Chemistry
McGill University
801 Sherbrooke Street West
Montréal, Québec, Canada H3A 2K6
karine.auclair@mcgill.ca
Chapter 75

Nick Bampos
University Chemical Laboratory
Cambridge University
Cambridge, CB2 1EW, UK
Chapter 15

Lucia Banci
CERM and Department of Chemistry
University of Florence
50019 Sesto Fiorentino (Florence), Italy
banci@cerm.unifi.it
Chapter 39

Jean-Michel Barbe
Laboratoire d'Ingénierie Moléculaire
pour la Séparation et les Applications des Gaz
LIMSAG (UMR 5633)
Université de Bourgogne
Faculté des Sciences "Gabriel"
6, Bd Gabriel 21000 Dijon, France
jean-michel.barbe@u-bourgogne.fr
Chapters 19, 116

Pierrette Battioni
Université René Descartes
Laboratoire de Chimie et Biochimie
Pharmacologiques et Toxicologiques UMR 8061
75270 Paris, France
battioni@biomedicale.univ-paris5.fr
Chapter 26

Carl E. Bauer
Department of Biology
Indiana University
Bloomington, Indiana 47405, USA
cbauer@bio.indiana.edu
Chapter 80

Marc Bénard
Laboratoire de Chimie Quantique UMR 7551
Université Louis Pasteur
67000 Strasbourg, France
benard@quantix.u-strasbg.fr
Chapter 48

Ehud Ben-Hur
Consultant in Photomedicine
160 West End Avenue # 24P, New York, NY 10023, USA
ehudbenhur@yahoo.com
Chapter 117

Jean Bernadou
Laboratoire de Chimie de Coordination du CNRS
31077 Toulouse, France
bernadou@lcc-toulouse.fr
Chapter 31

Ivano Bertini
CERM and Department of Chemistry
University of Florence
50019 Sesto Fiorentino (Florence), Italy
bertini@cerm.unifl.it
Chapter 39

David W. Bollivar
Department of Biology
Illinois Wesleyan University
P.O. Box 2900, Bloomington, Illinois 61702-2900, USA
dbolliva@iwu.edu
Chapter 78

Arnaud Bondon
UMR 6509, Organométallique et Catalyse,
Chimie et Électrochimie Moléculaires
Université de Rennes I
35042 Rennes, France
arnaud.bondon@univ.rennes1.fr
Chapter 38

Sylvia S. Bottomley
Department of Medicine
University of Oklahoma College of Medicine and
Department of Veterans Affairs Medical Center
Hematology-Oncology Section (111J)
921 N.E. 13th Street
Oklahoma City
Oklahoma 73104-5007, USA
sylvia-bottomley@ouhsc.edu
Chapter 85

Roman Boulatov
Department of Chemistry
Stanford University
Stanford, California 94305-5080, USA
boulatov@stanford.edu
Chapter 62

Marcel Bouvet
Laboratoire de Chimie Inorganique et Matériaux Moléculaires
CNRS – UMR7071
Université Pierre et Marie Curie
4, place Jussieu, Case Courrier 42
75252 Paris cedex 05, France
marcel.bouvet@cspci.fr
Chapter 118

Johann W. Buchler
Institut für Anorganische Chemie
Darmstadt University of Technology
D-64287 Darmstadt, Germany
dg7m@hrzpub.tu-darmstadt.de
jwbuchler@t-online.de
Chapter 20

Henry J. Callot
Faculté de Chimie
Université Louis Pasteur
F-67000 Strasbourg, France
callot@chimie.u-strasbg.fr
Chapter 7

Isabelle Chambrier
Wolfson Materials & Catalysis Centre
School of Chemical Sciences and Pharmacy
University of East Anglia
Norwich NR4 7TJ, UK
i.fernandes@uea.ac.uk
Chapter 108

Jean-Claude Chambron
Laboratoire de Chimie Organo-Minérale
Université Louis Pasteur
Institut Le Bel
67070 Strasbourg, France
jean-claude.chambron@u-bourgogne.fr
Chapter 40

Wai-Shun Chan
Progenitor Cell Therapy, L.L.C.
Park 80 West, Plaza II, Suite 200
Saddle Brook, New Jersey 07663, USA
wschan@progenitorcell.net
Chapter 117

Jennifer Cheek
School of Medicine
University of South Carolina
Columbia, South Carolina 29208, USA
Chapter 53

Lin Cheng
Department of Chemistry and Biochemistry
University of Oklahoma
Norman, Oklahoma 73019, USA
Chapter 33

Jun-Hong Chou
School of Chemical Sciences
University of Illinois at Urbana-Champaign
Urbana, Illinois 61801, USA
Chapter 41

Zöe Clyde-Watson
University Chemical Laboratory
Cambridge University
Cambridge, CB2 1EW, UK
Chapter 15

James P. Collman
Department of Chemistry
Stanford University
Stanford, California 94305-5080, USA
jpc@stanford.edu
Chapter 62

Michael J. Cook
Wolfson Materials & Catalysis Centre
School of Chemical Sciences and Pharmacy
University of East Anglia
Norwich NR4 7TJ, UK
m.cook@uea.ac.uk
Chapter 108

Anne Corrigall
Lennox Eales Porphyria Laboratories
Department of Medicine
University of Cape Town Medical School
Observatory, South Africa 7925
anne@uctgsh1.uct.ac.za
Chapter 89

Timothy M. Cox
Department of Medicine
University of Cambridge, Box 157
Addenbrooke's Hospital, Cambridge CB2 2QQ, UK
jbg20@medschl.cam.ac.uk
Chapter 90

Claudia Crestini
Dipartimento di Scienze e Tecnologie Chimiche
Tor Vergata University
Via Della Ricerca scientifica
00133, Rome, Italy
crestini@uniroma2.it
Chapter 66

Roman S. Czernuszewicz
Department of Chemistry
University of Houston
Houston, Texas 77204-5003, USA
roman@uh.edu
Chapter 52

Harry A. Dailey
Biomedical and Health Sciences Institute
Department of Microbiology, and
Department of Biochemistry & Molecular Biology
A220 Life Sciences Building
University of Georgia
Athens, Georgia 30602-7229, USA
hdailey@uga.edu
Chapter 72

Tamara A. Dailey
Biomedical and Health Sciences Institute
Department of Microbiology, and
Department of Biochemistry & Molecular Biology
A220 Life Sciences Building
University of Georgia
Athens, Georgia 30602-7229, USA
tdailey@uga.edu
Chapter 72

Scott L. Darling
University Chemical Laboratory
Cambridge University
Cambridge, CB2 1EW, UK
Chapter 15

John Dawson
Department of Chemistry and Biochemistry
University of South Carolina
Columbia, South Carolina 29208, USA
dawson@psc.sc.edu
Chapter 53

Gema de la Torre
Departamento de Química Orgánica
Universidad Autónoma de Madrid
28049 Cantoblanco, Madrid, Spain
gema.delatorre@uam.es
Chapter 99

Felix W. M. de Rooij
Department of Internal Medicine, Erasmus MC
University Medical Center Rotterdam
P.B. 2040, 3000 CA Rotterdam
The Netherlands
f.derooij@erasmusmc.nl
Chapters 93, 95

Jeffery S. de Ropp
NMR Facility
University of California, Davis
Davis, California 95616, USA
jsderopp@ucdavis.edu
Chapter 37

Hubert de Verneuil
Laboratoire de Pathologie
Moléculaire et Thérapie Génique, INSERM E0217
Université V Segalen Bordeaux 2
146 rue Léo Saignat
33076 Bordeaux cédex, France
verneuil@u-bordeaux2.fr
Chapter 87

Jean-Charles Deybach
Centre Français des Porphyries, INSERM U409
178 rue des Renouillers
92701 Colombes Cedex, France;
and Faculté de Médecine Xavier Bichat
Université Paris VII
16 rue Henri Huchard
BP416, 75870 Paris Cedex 18, France
jc.deybach@wanadoo.fr
Chapter 86

Danilo Dini
Institut für Organische Chemie
Universität Tübingen
72076 Tübingen, Germany
danilo.dini@uni-tuebingen.de
Chapters 107, 114

Maria Pia Donzello
Dipartimento di Chimica
Università degli Studi di Roma "La Sapienza"
P. le Aldo Moro 5, I-00185 Roma, Italy
mariapia.donzello@uniroma1.it
Chapter 112

Annie Edixhoven
Department of Internal Medicine, Erasmus MC
University Medical Center Rotterdam
P.B. 2040, 3000 CA Rotterdam
The Netherlands
a.edixhoven@erasmusmc.nl
Chapter 93

G. H. Elder
Department of Medical Biochemistry
University of Wales College of Medicine
Heath Park, Cardiff CF14 4XN, UK
elder@cardiff.ac.uk
Chapter 88

Michael Klaus Engel
Dainippon Ink and Chemicals, Inc.
Central Research Laboratories
631 Sakado, Sakura-shi
Chiba-ken 285-8668, Japan
michael-engel@ma.dic.co.jp
Chapter 122

Christoph Erben
Lucent Technologies—Bell Laboratories
Murray Hill, New Jersey 07974, USA
chris_erben@hotmail.com
Chapter 12

Claudio Ercolani
Dipartimento di Chimica
Università degli Studi di Roma "La Sapienza"
P. le Aldo Moro 5, I-00185 Roma, Italy
claudio.ercolani@uniroma1.it
Chapters 101, 112

Peter Erk
Performance Chemicals Research
GVP/C – J 550, BASF AG
67056 Ludwigshafen/Rh, Germany
peter.erk@basf-ag.de
Chapter 119

Jean Fischer
Laboratoire de Chimie Organométallique et de Catalyse
Institut le Bel
Université Louis Pasteur de Strasbourg
4, rue Blaise Pascal, 67070 Strasbourg, France
fischer@chimie.u-strasb.fr
Chapter 105

Steven R. Flom
Code 5613, Naval Research Laboratory
Washington, DC 20375-5338, USA
flom@nrl.navy.mil
Chapter 121

Carlo Floriani
Institut de Chimie Minérale et Analytique, BCH
Université de Lausanne
Lausanne CH-1015, Switzerland
Chapters 24, 25

Rita Floriani-Moro
Institut de Chimie Minérale et Analytique, BCH
Université de Lausanne
Lausanne CH-1015, Switzerland
Chapters 24, 25

Barbara Floris
Dipartimento di Scienze e Tecnologie Chimiche
Università di Roma "Tor Vergata"
Via della Ricerca Scientifica
I-00133, Roma, Italy
floris@uniroma2.it
Chapter 112

Nicole Frankenberg
Section of Molecular and Cellular Biology
University of California, Davis
One Shields Avenue
Davis, California 95616, USA
nfrankenberg@tu-bs.de
Chapter 83

Yuichi Fujita
Graduate School of Bioagricultural Sciences
Nagoya University
Nagoya 464-8601, Japan
fujita@nuagr1.agr.nagoya-u.ac.jp
Chapter 80

Shunichi Fukuzumi
Department of Material and Life Sciences
Graduate School of Engineering
Osaka University
Suita, Osaka 565-0871, Japan
fukuzumi@chem.eng.osaka-u.ac.jp
Chapter 56

Philip A. Gale
Department of Chemistry
University of Southampton
Southampton, S017 1BJ, UK
philip.gale@soton.ac.uk
Chapter 45

Andreas Gebauer
Department of Chemistry
University of California at Berkeley
Berkeley, California 94720, USA
andreas@socrates.berkeley.edu
Chapters 8, 9

Cécile Ged
Laboratoire de Pathologie
Moléculaire et Thérapie Génique
INSERM E0217
Université V Segalen Bordeaux 2
146 rue Léo Saignat
33076 Bordeaux cédex, France
cecile.ged@u-bordeaux2.fr
Chapter 87

Abhik Ghosh
Institute of Chemistry, Faculty of Science
University of Trømsø
Trømsø, Norway;
and San Diego Supercomputer Center
University of California at San Diego
La Jolla, California 92093-0505, USA
Abhik@chem.iut.no
Chapter 47

Avram Gold
Department of Environmental Science and Engineering
The University of North Carolina at Chapel Hill
Chapel Hill, North Carolina 27599-7400, USA
agold@sophia.sph.unc.edu
Chapter 29

Albert Gossauer
Department of Chemistry
University of Fribourg
Route du Musée 9
CH-1700 Fribourg, Switzerland
albert.gossauer@unifr.ch
Chapter 84

Vincent Gotte
European Synchrotron Radiation Facility
F-38043 Grenoble, France
Chapter 49

José Goulon
European Synchrotron Radiation Facility
F-38043 Grenoble, France
goulon@esrf.fr
Chapter 49

Chantal Goulon-Ginet
European Synchrotron Radiation Facility
F-38043 Grenoble, France; *and*
Faculté de Pharmacie, Université de Grenoble I
F-38700 La Tronche, France
cgoulon@esrf.fr
Chapter 49

Harry B. Gray
Beckman Institute, MC 139-74
California Institute of Technology
Pasadena, California 91125, USA
hbgray@caltech.edu
Chapter 63

Bernhard Grimm
Institute of Biology/Plant Physiology
Humboldt University
Unter den Linden 6
D 10099 Berlin, Germany
bernhard.grimm@rz.hu.berlin.de
Chapter 69

John T. Groves
Department of Chemistry
Princeton University
Princeton, New Jersey 08544-1009, USA
jtgroves@princeton.edu
Chapter 27

Lena Gueletti
Department of Chemistry
University of Houston
Houston, Texas 77204-5003, USA
Chapter 59

Roger Guilard
Laboratoire d'Ingénierie Moléculaire
pour la Séparation et les Applications des Gaz
LIMSAG (UMR 5633)
Université de Bourgogne
Faculté des Sciences "Gabriel"
6, Bd Gabriel 21000 Dijon, France
roger.guilard@u-bourgogne.fr
Chapters 19, 21, 116

Devens Gust
Department of Chemistry and Biochemistry
Arizona State University
Tempe, Arizona 85287-1604, USA
gust@asu.edu
Chapter 57

Peter Hambright
Department of Chemistry
Howard University
Washington, D.C. 20059, USA
whambright@howard.edu
Chapter 18

Michael Hanack
Institut für Organische Chemie
Universität Tübingen
72076 Tübingen, Germany
hanack@uni-tuebingen.de
Chapters 107, 114

Mats Hansson
Department of Biochemistry
Lund University
S-22100 Lund, Sweden
mats.hansson@biokem.lu.se
Chapter 77

Pierre D. Harvey
Département de Chimie
Université de Sherbrooke
Sherbrooke, P.Q., Canada, J1K 2R1
pierre.harvey@USherbrooke.ca
Chapter 113

Fred M. Hawkridge
Department of Chemistry
Virginia Commonwealth University
Richmond, Virginia 23284, USA
fmhawkri@vcu.edu
Chapter 58

Joanne C. Hawley
University Chemical Laboratory
Cambridge University
Cambridge, CB2 1EW, UK
Chapter 15

Takashi Hayashi
Department of Chemistry and Biochemistry
Graduate School of Engineering
Kyushu University
Higashi-ku Fukuoka 812-8581, Japan
thayatcm@mbox.nc.kyushu-u.ac.jp
Chapter 46

Valérie Heitz
Laboratoire de Chimie Organo-Minérale
Université Louis Pasteur
Institut Le Bel
67070 Strasbourg, France
heitz@chimie.u-strasbg.fr
Chapter 40

Heidi Hengelsberg
Marketing Pigments
EVP/M – J 550, BASF AG
67056 Ludwigshafen/Rh, Germany
heidi.hengelsberg@basf-ag.de
Chapter 119

Richard J. Hift
Lennox Eales Porphyria Laboratories
Department of Medicine
University of Cape Town Medical School
Observatory, South Africa 7925
rjh@uctgsh1.uct.ac.za
Chapters 89, 91, 96

Shohei Inoue
Department of Industrial Chemistry
Faculty of Engineering
Science University of Tokyo
Kagurazaka, Shinjuku-ku
Tokyo 162-8601, Japan
Chapter 42

Kazuyuki Ishii
Department of Chemistry
Graduate School of Science
Tohoku University
Sendai 980-8578, Japan
k-ishii@mail.cc.tohoku.ac.jp
Chapter 102

N. I. Jaeger
Institut für Angewandte und Physikalische Chemie
Fachbereich 2 (Biologie/Chemie)
Universität Bremen
Postfach 33 04 40
D-28334 Bremen, Germany
jse@uni-bremen.de
Chapter 106

Laurent Jaquinod
Department of Chemistry
University of California, Davis
Davis, California 95616, USA
jaquinod@indigo.ucdavis.edu
Chapter 5

Dong-Lin Jiang
Department of Chemistry and Biotechnology
Graduate School of Engineering
The University of Tokyo
Hongo, Bunk-yo, Tokyo 113-8656, Japan
jiang@macro.t.u-tokyo.ac.jp
Chapter 23

Karl M. Kadish
Department of Chemistry
University of Houston
Houston, Texas 77204-5003, USA
kkadish@uh.edu
Chapters 12, 21, 55, 59, 116

Hee-Joon Kim
University Chemical Laboratory
Cambridge University
Cambridge, CB2 1EW, UK
Chapter 15

Mutsumi Kimura
Department of Functional Polymer Science
Faculty of Textile Science and Technology
Shinshu University
Ueda 386-8567, Japan
mkimura@giptc.shinshu-u.ac.jp
Chapter 120

James R. Kincaid
Chemistry Department
Marquette University
Milwaukee, Wisconsin 53233, USA
kincaid@vms.csd.mu.edu
Chapter 51

Nagao Kobayashi
Department of Chemistry, Graduate School of Science
Tohoku University
Sendai 980-8578, Japan
nagaok@mail.cc.tohoku.ac.jp
Chapters 13, 100, 102

Margaret E. Kosal
School of Chemical Sciences
University of Illinois at Urbana-Champaign
Urbana, Illinois 61801, USA
Chapter 41

Bernhard Kräutler
Institute of Organic Chemistry
University of Innsbruck
Innrain 52a, A-6020 Innsbruck, Austria
bernhard.kraeutler@uibk.ac.at
Chapters 68, 82

Yasuhisa Kuroda
Department of Polymer Science
Kyoto Institute of Technology
Matsugasaki, Kyoto 606 0962, Japan
ykuroda@ipc.kit.ac.jp
Chapter 46

J. Clark Lagarias
Section of Molecular and Cellular Biology
University of California, Davis
One Shields Avenue
Davis, California 95616, USA
jclagarias@ucdavis.edu
Chapter 83

Gerd N. La Mar
Department of Chemistry
University of California, Davis
Davis, California 95616, USA
lamar@indigo.ucdavis.edu
Chapter 37

Timothy D. Lash
Department of Chemistry
Illinois State University
Normal, Illinois 61794-4160, USA
tdlash@ilstu.edu
Chapter 10

Lechosław Latos-Grażyński
Department of Chemistry
University of Wrocław
Wrocław 50 383, Poland
llg@wchuwr.chem.uni.wroc.pl
Chapter 14

Claude Lecomte
Laboratoire de Cristallographie et Modélisation des Matériaux Minéraux et Biologiques URA CNRS 809
Université Henri Poincaré-Nancy 1
54506 Vandoeuvre-lés-Nancy, France
lecomte@lcm3b.u-nancy.fr
Chapter 48

Jinbo Lee
Department of Chemistry
Princeton University
Princeton, New Jersey 08544-1009, USA
Chapter 27

Paul Le Maux
Laboratoire de Chimie Organométallique et Biologique
UMR CNRS 6509, Institut de Chimie
Université de Rennes 1
35042 Rennes cedex, France
paul.lemaux@univ-rennes1.fr
Chapter 65

Maurice L'Her
Université de Bretagne Occidentale
6521/Faculté des Sciences
B.P. 809, 29285 Brest Cedex, France
maurice.lher@univ-brest.fr
Chapter 104

Jonathan S. Lindsey
Department of Chemistry
North Carolina State University
Raleigh, North Carolina 27695-8204, USA
jlindsey@ncsu.edu
Chapter 2

Claudio Luchinat
CERM and Department of Soil Science and Plant Nutrition
University of Florence
50019 Sesto Fiorentino (Florence), Italy
luchinat@cerm.unifi.it
Chapter 39

John Mack
Department of Chemistry
University of Western Ontario, London
Ontario, Canada, N6A 5B7
mack@uwo.ca
Chapter 103

Estelle M. Maes
Department of Chemistry
University of Houston
Houston, Texas 77204-5003, USA
emaes@uh.edu
Chapter 52

Chi Ching Mak
University Chemical Laboratory
Cambridge University
Cambridge, CB2 1EW, UK
Chapter 15

Tadeusz Malinski
Department of Chemistry and Biochemistry
Ohio University
Athens, Ohio 45701, USA
malinski@ohio.edu
Chapter 44

Daniel Mansuy
Université René Descartes
Laboratoire de Chimie et Biochimie
Pharmacologiques et Toxicologiques
UMR 8061, 75270 Paris, France
daniel.mansuy@biomedicale.univ-paris5.fr
Chapter 26

Jean-Claude Marchon
Laboratoire de Chimie Inorganique et Biologique
Département de Recherche Fondamentale sur la Matière Condensée
CEA-Grenoble, 38054 Grenoble, France
jcmarchon@cea.fr
Chapter 64

Pavel Martásek
Department of Pediatrics, First Faculty of Medicine
Charles University
12109 Prague, Czech Republic
martasek@egap.cz
Chapter 34

Bettie Sue S. Masters
Department of Biochemistry
The University of Texas Health Science Center at San Antonio
San Antonio, Texas 78284-7760, USA
masters@uthscsa.edu
Chapter 34

Neil B. McKeown
Department of Chemistry
University of Manchester
Manchester M13 9PL, UK
neil.mckeown@man.ac.uk
Chapter 98

Craig J. Medforth
Department of Chemistry
University of California, Davis
Davis, California 95616, USA
medforth@indigo.ucdavis.edu
Chapter 35

Peter N. Meissner
Lennox Eales Porphyria Laboratories
Department of Medicine
University of Cape Town Medical School
Observatory, South Africa 7925
pete@uctgsh1.uct.ac.za
Chapters 89, 91, 96

Bernard Meunier
Laboratoire de Chimie de Coordination du CNRS
31077 Toulouse, France
bmeunier@lcc.toulouse.fr
Chapter 31

Tadashi Mizutani
Department of Synthetic Chemistry and Biological Chemistry
Graduate School of Engineering
Kyoto University
Yoshida, Sakyo-ku, Kyoto 606 8501, Japan
mizutani@sbchem.kyoto-u.ac.jp
Chapter 46

Tatsushi Mogi
Yoshida ATP System Project, ERATO
Japan Science and Technology Corporation
Green Hills Suzukakedai 1F, 5800-3 Nagatsuta, Midori-ku
Yokohama 226-0026;
and Department of Biological Sciences
Graduate School of Science
University of Tokyo
7-3-1 Hongo, Bunkyo-ku, Tokyo 113-0033, Japan
tmogi@res.titech.ac.jp
Chapter 74

Michael R. Moore
National Research Centre for Environmental Toxicology
(EnTox), The University of Queensland and
Queensland Health Scientific Services
39 Kessels Road, Coopers Plains, Brisbane 4108
Queensland, Australia
m.moore@uq.edu.au
Chapter 96

Thomas A. Moore
Department of Chemistry and Biochemistry
Arizona State University
Tempe, Arizona 85287-1604, USA
Tom.Moore@asu.edu
Chapter 57

François Moreau-Gaudry
Laboratoire de Pathologie Moléculaire et Thérapie Génique
INSERM E0217
Université V Segalen Bordeaux 2
146 rue Léo Saignat, 33076 Bordeaux cédex
France
francois.moreau-gaudry@u-bordeaux2.fr
Chapter 87

Hari Singh Nalwa
Hitachi Research Laboratory
Hitachi Ltd.
Hitachi City, Ibaraki 319 1292, Japan
Chapter 41

Dennis K.P. Ng
Department of Chemistry
The Chinese University of Hong Kong
Shatin, N.T., Hong Kong
dkpn@cuhk.edu.hk
Chapter 20

Jun Nishimoto
Analytical Research Center for Experimental Sciences
Saga University
Saga 840-8502, Japan
nishimoj@cc.saga-u.ac.jp
Chapter 60

Roeland J. M. Nolte
Department of Organic Chemistry, NSR Center
The University of Nijmegen
Toernooiveld 1
NL-6525 ED Nijmegen, The Netherlands
nolte@sci.kun.nl
Chapter 115

Rubén Ocampo
Faculté de Chimie
Université Louis Pasteur
67008 Strasbourg, France
ocampo@chimie.u-strasbg.fr
Chapter 7

T. Oekermann
Graduate School of Engineering
Environmental and Renewable Energy Systems (ERES) Division
Gifu University
Yanagido 1-1, Gifu 501-1193, Japan
torsten@apchem.gifu-u.ac.jp
Chapter 106

Hisanobu Ogoshi
Fukui National College of Technology
Geshi, Sabae
Fukui 916 8507, Japan
ogoshi@fukui-nct.ac.jp
Chapter 46

Paul R. Ortiz de Montellano
Department of Pharmaceutical Chemistry
School of Pharmacy
University of California
San Francisco, California 94143-0446, USA
ortiz@cgl.ucsf.edu
Chapter 75

Sigrid Ostermann
Institute of Organic Chemistry
University of Innsbruck, Innrain 52a
A-6020 Innsbruck, Austria
sigrid.ostermann@uibk.ac.at
Chapter 68

Ravindra K. Pandey
Photodynamic Therapy Center and
Department of Nuclear Medicine
Roswell Park Cancer Institute
Buffalo, New York 14263-0001, USA
ravindra.pandey@roswellpark.org
Chapter 43

Roberto Paolesse
Dipartimento di Scienze e Tecnologie Chimiche
Università di Roma "Tor Vergata"
00133 Roma, Italy
roberto.paolesse@uniroma2.it
Chapter 11

Annig Pondaven
Université de Bretagne Occidentale
6521/Faculté des Sciences
B.P. 809, 29285 Brest Cedex, France
annig.pondaven@univ-brest.fr
Chapter 104

Thomas L. Poulos
Department of Molecular Biology and Biochemistry
University of California, Irvine
Irvine, California 92697-3900, USA
poulos@uci.edu
Chapter 32

Geneviève Pratviel
Laboratoire de Chimie de Coordination du CNRS
31077 Toulouse, France
pratviel@lcc-toulouse.fr
Chapter 31

Hervé Puy
Centre Français des Porphyries, INSERM U409
178 rue des Renouillers
92701 Colombes Cedex, France;
and Faculté de Médecine
Paris Ile de France Ouest
104 boulevard Raymond Poincaré, 92380 Garches, France
herve.puy@apr.ap-hop-paris.fr
Chapter 86

J. Martin E. Quirke
Department of Chemistry
Florida International University
Miami, Florida 33199, USA
quirke@servms.fiu.edu
Chapter 54

Stephen W. Ragsdale
Department of Biochemistry
Beadle Center, 19th and Vine Streets
University of Nebraska
Lincoln, Nebraska 68588-0664, USA
sragsdale1@unl.edu
Chapter 67

Neal A. Rakow
School of Chemical Sciences
University of Illinois at Urbana-Champaign
Urbana, Illinois 61801, USA
Chapter 41

C. S. Raman
Department of Molecular Biology and Biochemistry
University of California, Irvine 92697-3900
Irvine, California, USA
raman@bragg.bio.uct.edu
Chapter 34

René Ramasseul
Laboratoire de Chimie Inorganique et Biologique
Département de Recherche Fondamentale sur la Matière Condensée
CEA-Grenoble
38054 Grenoble, France
ramasse@drfmc.ceng.cea.fr
Chapter 64

J. Graham Rankin
Department of Chemistry
Marshal University
Huntington, West Virginia 25701, USA
rankin@mu.edu
Chapter 52

George B. Richter-Addo
Department of Chemistry and Biochemistry
University of Oklahoma
Norman, Oklahoma 73019, USA
grichteraddo@ou.edu
Chapter 33

Anne Robert
Laboratoire de Chimie de Coordination du CNRS
31077 Toulouse, France
arobert@lcc-toulouse.fr
Chapter 31

M. Salomé Rodríguez-Morgade
Departamento de Química Orgánica
Universidad Autónoma de Madrid
28049 Cantoblanco, Madrid, Spain
salome.rodriguez@uam.es
Chapter 99

Charles A. Roessner
Center for Biological NMR, Department of Chemistry
Texas A&M University
College Station, Texas 77843, USA
c-roessner@tamu.edu
Chapter 76

Marie-Madeleine Rohmer
Laboratoire de Chimie Quantique UMR 7551
Université Louis Pasteur
67000 Strasbourg, France
rohmer@quantix.u-strasbg.fr
Chapter 48

Alan E. Rowan
Department of Organic Chemistry, NSR Center
The University of Nijmegen
Toernooiveld 1
NL-6525 ED Nijmegen, The Netherlands
rowan@sci.kun.nl
Chapter 115

Guy Royal
Department of Chemistry
University of Houston
Houston, Texas 77204-5003, USA
guy.royal@ujf-grenoble.fr
Chapter 55, 59

W. Rüdiger
Department Biologie I, Botanik
Universität München
Menzingerstrasse 67
D-80638, München, Germany
ruediger@botanik.biologie.uni-muenchen.de
Chapter 79

Jeremy K. M. Sanders
University Chemical Laboratory
Cambridge University
Cambridge, CB2 1EW, UK
Chapters 15, 22

Patricio J. Santander
Center for Biological NMR, Department of Chemistry
Texas A&M University
College Station, Texas 77843, USA
santander@tamu.edu
Chapter 76

James D. Satterlee
Department of Chemistry
Washington State University
Pullman, Washington 99164, USA
hemeteam@wsu.edu
Chapter 37

Jean-Pierre Sauvage
Laboratoire de Chimie Organo-Minérale
Université Louis Pasteur
Institut Le Bel
67070 Strasbourg, France
sauvage@chimie.u-strasbg.fr
Chapter 40

W. Robert Scheidt
Department of Chemistry and Biochemistry
University of Notre Dame
Notre Dame, Indiana 46556-5670, USA
scheidt.1@nd.edu
Chapter 16

D. Schlettwein
Physikalische Chemie 1, Fachbereich 9 (Chemie)
Universität Oldenburg
Postfach 2503
D-26111 Oldenburg, Germany
derck.schlettwein@uni-oldenburg.de
Chapter 106

Günter Schnurpfeil
Universität Bremen
Institut für Organische und Makromolekulare Chemie
P.O. Box 33 04 40
28334 Bremen, Germany
gschnur@chemie.uni-bremen.de
Chapter 110

A. Ian Scott
Center for Biological NMR, Department of Chemistry
Texas A&M University
College Station, Texas 77843, USA
aiscott@tamu.edu
Chapter 76

Mathias O. Senge
Institut für Chemie, Organische Chemie
Freie Universität Berlin
D-14195 Berlin, Germany
mosenge@chemie.fu-berlin.de
Chapters 6, 61

Jonathan L. Sessler
Department of Chemistry and Biochemistry
Institute for Cellular and Molecular Biology
The University of Texas at Austin
Austin, Texas 78712-0165, USA
sessler@mail.utexas.edu
Chapters 8, 9, 45

Kirill Shalyaev
Department of Chemistry
Princeton University
Princeton, New Jersey 08544-1009, USA
Chapter 27

Wesley M. Sharman
CIHR Group in the Radiation Sciences
Faculty of Medicine
Université de Sherbrooke
Sherbrooke, Québec, Canada, J1H 5N4
wesleysharman@hotmail.com
Chapter 97

John A. Shelnutt
Biomolecular Materials and Interfaces Department
Sandia National Laboratories
1001 University Boulevard SE, Albuquerque
New Mexico 87106, USA; *and*
Department of Chemistry
University of Georgia
Athens, Georgia 30602-2556, USA
jasheln@unm.edu
Chapter 50

Hirofusa Shirai
Department of Functional Polymer Science
Faculty of Textile Science and Technology
Shinshu University
Ueda 386-8567, Japan
hshirai@giptc.shinshu-u.ac.jp
Chapter 120

Peter M. Shoolingin-Jordan
School of Biological Sciences
University of Southampton
Bassett Crescent East
Southampton, SO16 7PX, UK
pmsj@soton.ac.uk
Chapter 70

Gérard Simonneaux
Organométallique et Catalyse, Chimie et Électrochimie Moléculaires
Université de Rennes I
35042 Rennes, France
gerard.simonneaux@univ-rennes1.fr
Chapters 38, 65

Andrew G. Smith
MRC Toxicology Unit, Hodgkin Building
University of Leicester
Lancaster Road, Leicester LE1 9HN, UK
ags5@le.ac.uk
Chapter 92

Kevin M. Smith
Department of Chemistry
Louisiana State University
130 David Boyd Hall
Baton Rouge, Louisiana 70803-2755, USA
kmsmith@lsu.edu
Chapters 1, 3, 81

Arthur W. Snow
Chemistry Division, Code 6123
Naval Research Laboratory
4555 Overlook Avenue, SW
Washington, DC 20375, USA
snow@ccf.nrl.navy.mil
Chapter 109

Christine Stern
Laboratoire d'Ingénierie Moléculaire pour la Séparation et les Applications des Gaz LIMSAG (UMR 5633)
Université de Bourgogne
Faculté des Sciences "Gabriel"
6, Bd Gabriel 21000 Dijon, France
christine.stern@u-bourgogne.fr
Chapter 116

Martin J. Stillman
Department of Chemistry
University of Western Ontario, London
Ontario, Canada, N6A 5B7
martin.stillman@uwo.ca
Chapter 103

Pavel A. Stuzhin
Department of Organic Chemistry
Ivanovo State University of Chemistry and Technology
153460 Ivanovo, Russian Federation
stuzhin@isuct.ru
Chapter 101

Christopher J. Sunderland
Department of Chemistry
Stanford University
Stanford, California 94305-5080, USA
sunland@stanford.edu
Chapter 62

Kenneth S. Suslick
School of Chemical Sciences
University of Illinois at Urbana-Champaign
Urbana, Illinois 61801, USA
ksuslick@uiuc.edu
Chapters 28, 41

Alain Tabard
Laboratoire d'Ingénierie Moléculaire pour la Séparation et les Applications des Gaz LIMSAG (UMR 5633)
Université de Bourgogne
Faculté des Sciences "Gabriel"
6, Bd Gabriel 21000 Dijon, France
atabard@u-bourgogne.fr
Chapter 21

Masaaki Tabata
Department of Chemistry
Faculty of Science and Engineering
Saga University
Saga 890-8502, Japan
tabatam@cc.saga-u.ac.jp
Chapter 60

Pietro Tagliatesta
Dipartimento di Scienze e Tecnologie Chimiche
Tor Vergata University
Via Della Ricerca scientifica
00133, Rome, Italy
pietro.tagliatesta@uniroma2.it
Chapter 66

Isao Taniguchi
Department of Applied Chemistry
Kumamoto University
Kumamoto 860, Japan
taniguch@gpo.kumamoto-u.ac.jp
Chapter 58

James Terner
Department of Chemistry
Virginia Commonwealth University
Richmond, Virginia 23284-2006, USA
jterner@saturn.vcu.edu
Chapter 29

Pall Thordarson
Department of Organic Chemistry, NSR Center
The University of Nijmegen
Toernooiveld 1
NL-6525 ED Nijmegen, The Netherlands
pall@sci.kun.nl
Chapter 115

Russell Timkovich
Department of Chemistry
University of Alabama
Tuscaloosa, Alabama 35487-0336, USA
rtimkovi@bama.ua.edu
Chapter 73

Tomás Torres
Departamento de Química Orgánica
Universidad Autónoma de Madrid
28049 Cantoblanco, Madrid, Spain
tomas.torres@uam.es
Chapter 99

Alfred X. Trautwein
Institut für Physik
Universität zu Lübeck
D-23538 Lübeck, Germany
trautwein@physik.uni-luebeck.de
Chapter 29

Paola Turano
CERM and Department of Chemistry
University of Florence
50019 Sesto Florentino (Florence), Italy
turano@cerm.unifi.it
Chapter 39

Eric Van Caemelbecke
Department of Chemistry
University of Houston
Houston, Texas 77204-5003, USA
Chapters 21, 55, 59

Johan E. van Lier
CIHR Group in the Radiation Sciences
Faculty of Medicine
Université de Sherbrooke
Sherbrooke, Québec, Canada, J1H 5N4
jvanlier@courrier.usherb.ca
Chapter 97

Maria da Graça H. Vicente
Department of Chemistry
Louisiana State University
Baton Rouge, Louisiana 70803-2755, USA
vicente@isu.edu
Chapter 4

Emanuel Vogel
Institut für Organische Chemie
Universität zu Köln
50939 Köln, Germany
Chapter 8

F. Ann Walker
Department of Chemistry
University of Arizona
Tucson, Arizona 85721-0041, USA
awalker@u.arizona.edu
Chapter 36

Michael Wark
Institute for Physical Chemistry and Electrochemistry
University Hanover
Callin Str. 3-3A
D-30167 Hanover, Germany
michael.wark@pci.uni-hannover.de
Chapter 111

Yoshihito Watanabe
Institute for Molecular Science
Myodaiji, Okazaki 444-8585, Japan
yoshi@ims.ac.jp
Chapter 30

Simon J. Webb
University Chemical Laboratory
Cambridge University
Cambridge, CB2 1EW, UK
Chapter 15

Steven J. Weghorn
Tonah Products, Inc.
Milton, Wisconsin 53563, USA
sweg@jvlnet.com
Chapter 9

Raymond Weiss
Laboratoire de Cristallochimie et de Chimie Structurale UMR 7513
Institut Le Bel
Université Louis Pasteur
67070 Strasbourg, France
weiss@chimie.u-strasbg.fr
Chapter 29
and
Laboratoire de Chimie Supramoléculaire
Institut de Science et d'Ingénierie Supramoléculaires
Université Louis Pasteur de Strasbourg
8, rue Gaspard Monge, F-67000 Strasbourg, France
weiss@chimie.u-strasbg.fr
Chapter 105

Stefan Will
Institut für Organische Chemie
Universität zu Köln
50939 Köln, Germany
Chapter 12

Robert D. Willows
Department of Biological Sciences
Macquarie University
North Ryde, 2109 Australia
rwillows@rna.bio.mq.edu.au
Chapter 77

J. H. P. Wilson
Department of Internal Medicine
Erasmus MC
University Medical Center Rotterdam
P.B. 2040, 3000 CA Rotterdam
The Netherlands
j.wilson@erasmusmc.nl
Chapters 93, 95

Jay R. Winkler
Beckman Institute, MC 139-74
California Institute of Technology
Pasadena, California 91125, USA
winklerj@caltech.edu
Chapter 63

Dieter Wöhrle
Universität Bremen
Institut für Organische und Makromolekulare Chemie
P.O. Box 33 04 40
28334 Bremen, Germany
woehrle@chemie.uni-bremen.de
Chapter 110

Gang Zheng
Department of Nuclear Medicine
Roswell Park Cancer Insitute
Buffalo, New York 14263-001, USA
pdtctr@sc3101.med.buffalo.edu
Chapter 43

Contents of Volumes 1–20

Single-Crystal Structures of Phthalocyanine Complexes and Related Macrocycles

122

MICHAEL KLAUS ENGEL
Dainippon Ink and Chemicals, Inc., Central Research Laboratories, 631 Sakado, Sakura-shi, Chiba-ken 285-8668, Japan

I. Introduction

A. NOTATIONS AND REMARKS

The review consists of several sections; first, a discussion of the molecular and crystal structures of phthalocyanine and related complexes followed by a database section. The latter contains several Appendices; in Appendices 1–6, basic crystallographic data of all known phthalocyanine and phthalocyanine-related structures are compiled; Appendix 7 is a cross-reference table of structures and structure code arranged by central atoms; Appendices 8 and 9 are cross-reference tables of structure codes and chemical structures. Drawings of all molecular structures are compiled in Appendices 10–23 and

ISBN 0-12-393220-3

the atomic coordinates of several structures, not included in the Cambridge Structural Database (CSD), can be found in Appendices 24–41.

Crystal structures within the text are referenced in the following way: e.g., PcH_2 (PHTHCY01,[1] App. 10:19), where "PHTHCY01" represents the structure code used by the CSD and "App. 10:19" is the location of a drawing of the molecular structure in Appendix 10. Structures with "CCDCxxxxxx" code (where x represents a number) are structures which are under processing at the Cambridge Crystallographic Data Centre and which have not yet obtained a final code. Most, but not all, of the structures of this review are available in the v5.23 (April 2002) release of the CSD. "#YYYY-xx" is the code for structures which are not contained in the Cambridge Structural Database, where "YYYY" represents the year of their publication and "xx" a number.

If a molecule exists in several polymorphic modifications, the nomenclatures of the original publications are used, e.g., I-PcTiO[2] or C-PcTiO.[3] If the nomenclature has changed over time, the latest nomenclature is used, e.g., "PcH_2[1]" became "β-PcH_2" after several other polymorphs had been discovered. The structural drawings in this review were redrawn from the atomic coordinates of the original publications, from entries of the Cambridge Structural Database or from personal communications obtained from their authors. Table 1 gives an alphabetical list of abbreviations used within this review.

B. HISTORY OF PHTHALOCYANINE CRYSTALLOGRAPHICAL RESEARCH

We owe the discovery of phthalocyanines to an accident, which occurred at the right time and the right place. A blue colored compound was detected in a reaction flask where only colorless materials were expected.[4] Normally, such an impurity is simply discarded, however, this accident occurred in a dye company (Scottish Dyes, Ltd., Grangemouth, Scotland) and the blue product attracted immediate interest. Initial research showed the material's potential as an exceptionally stable pigment and after the first patent had been applied for, the job of conducting further research on this promising material was placed in the hands of Sir Patrick Reginald Linstead[5] from the Imperial College (London, U.K.). Linstead and coworkers determined the molecular structure of phthalocyanine through the oxidative decomposition of the macrocycle and a thorough investigation of the resulting fragments.[6–11] During their investigation, Linstead and coworkers were able to grow the first single-crystals of phthalocyanines by low-pressure sublimation in a CO_2 current at temperatures in excess 500°C. This enormous thermal stability had previously been remarked upon by earlier discoverers who accidentally synthesized copper phthalocyanine, copper octamethylphthalocyanine, and copper naphthalocyanine but failed to fully characterize and value these materials.[12] The phthalocyanine single-crystals were then handed to J. Monteath Robertson from the Davy Faraday Research Laboratory of the Royal Institution in London, U.K. who conducted an intensive crystallographic investigation on these materials. Firstly, Robertson, Linstead, and Dent jointly determined the molecular weight of the phthalocyanine complexes using a combination of both physical (density) and crystallographic data.[13] Through direct methods, Robertson then determined the two-dimensional crystal structures of metal-free phthalocyanine, PcH_2 (PHTHCY01,[1] App. 10:19), nickel phthalocyanine, PcNi (NIPHTC,[14] App. 11:57), and platinum phthalocyanine, PcPt (PTPOCY02,[15] App. 11:70). This crystal data provided many researchers with the groundwork to conduct their own research on phthalocyanine crystal structures and was a structural basis for the calculation or prediction of their physical properties. Between Robertson's initial research and the next full determination of a phthalocyanine structure lies a lengthy gap of some 30 years, a period where researchers relied exclusively on the data of Robertson. Due to the lack of calculation resources, it took until the mid 1960s before truly three-dimensional crystal structures of phthalocyanines could be established. Nowadays, single-crystal structural analysis has become an essential analytical tool, a technique used almost as frequently and easily as NMR or IR spectroscopy. This can be clearly seen by the steadily growing number of publications of crystal structures within the literature. Figure 1 gives an overview of structural reports and crystal structures by year, between 1935 and 2001. There are three periods which stand out after the initial research of Robertson; the first started in the mid 1960s when computer-controlled X-ray diffraction equipment became available and a strong interest in porphyrin structural chemistry was directed towards the phthalocyanine field, resulting in a steady number of published crystal structures and reports. Period two started around 1980 when the metallic conductivity found in organic materials deepened the interest in one-dimensional phthalocyanines and resulted in an increase in phthalocyanine research and crystal structures.

Table 1. Alphabetical List of Abbreviations and Definitions

Band	a structural arrangement where the molecules are one-dimensionally arranged and where the normal to the macrocycle is (approximately) perpendicular to the band axis and where no overlap can be found
Bis-phthalocyanine	Pc-M-Pc complexes, sometimes also called double-decker molecules[54]
1-BrN	1-bromonaphthalene
C_{α}	isoindole α-carbon atom (see Figure 2a)
ClB	chlorobenzene
1-ClN	1-chloronaphthalene
CN1	phthalocyanine complexes with central atom of coordination number 1 (see Figure 5a)
CN4	phthalocyanine complexes with central atom of coordination number 4 (see Figure 5b)
CN5	phthalocyanine complexes with central atom of coordination number 5 (see Figure 5d,e,m,n)
CN5*	phthalocyanine complexes with two central atoms of coordination number 5, located on opposite sides of the phthalocyanine macrocycle (see Figure 3c)
CN6*cis*	phthalocyanine complexes with central atom of coordination number 6 and two axial ligands located on one side of the phthalocyanine macrocycle (see Figure 5f,p)
CN6*tr*	phthalocyanine complexes with central atom of coordination number 6 and two axial ligands located on opposite sides of the phthalocyanine macrocycle (see Figure 5c,o)
CN7*cis*	phthalocyanine complexes with central atom of coordination number 7 and three axial ligands located on one side of the phthalocyanine macrocycle (see Figure 5g,q)
CN8*cis*	phthalocyanine complexes with central atom of coordination number 8 and four axial ligands located on one side of the phthalocyanine macrocycle (see Figure 5h,r)
CN8*bis*	phthalocyanine complexes with central atom of coordination number 8 sandwiched between two phthalocyanine ligands (see Figure 5i–l)
Column	a structural arrangement where the molecules are one-dimensionally arranged and in a view perpendicular to the macrocycle neighboring molecules overlap
DCB	dichlorobenzene
DMF	dimethylformamide
DMSO	dimethylsulfoxide
Isostructural, isomorphous, isomorphic	indicate structures which are very similar
Layer	a two-dimensional analog of a band
Mac	macrocycle
MATBP	monoazatetrabenzoporphyrin
mepy	methylpyridine
Pc	phthalocyanine, phthalocyaninato
PcH_2	metal-free phthalocyanine
PcM	metal phthalocyanine
N_{br}	bridging azamethine nitrogen atom (see Figure 2a)
Nc	naphthalocyanine, naphthalocyaninato
N_i	isoindole nitrogen atom (see Figure 2a)
OEP	octaethylporphyrin
oop	out-of-plane
PAZ	porphyrazine
PN	phthalonitrile
PNP	*bis*(triphenylphosphine)iminium
Sheet	a two-dimensional structure similar to a column or a stack
Stack	a structural arrangement where the molecules are one-dimensionally arranged and in a view perpendicular to the macrocycle neighboring molecules overlap
TATBP	triazatetrabenzoporphyrin
TBP	tetrabenzoporphyrin
THF	tetrahydrofuran
TPP	tetraphenylporphyrin
TPyP	tetrapyridinoporphyrin
Tris-phthalocyanine	Pc-M-Pc-M-Pc complexes, sometimes also called triple-decker molecules[54]

Period three then begun in the early 1990s when the group of Heiner Homborg (>20% of all phthalocyanine structures) and also the group of Kubiak and Janczak (7%) started publishing their results.

Apart from its uses in the pigment industries,[16,17] phthalocyanine compounds have become widely used as dyes,[17] and more recently, they have found application as photosensitive compounds in charge-generation materials for laser beam printers and photocopiers,[18–20] and as molecular materials for transistor applications.[21] However, only the pigment and photoconductor industries have invested a deep interest in the research on

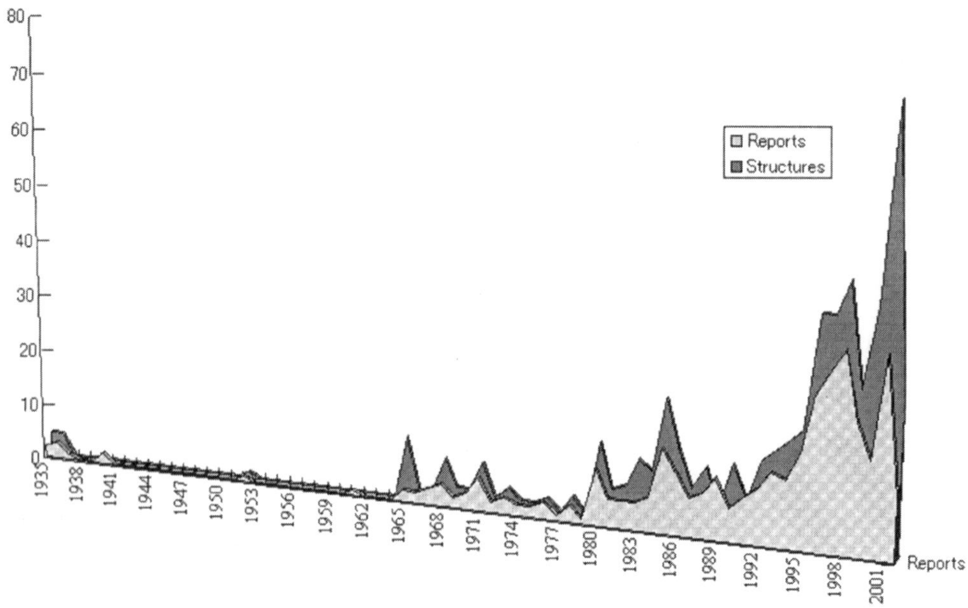

Figure 1. Graphical overview of published structural reports and phthalocyanine crystal structures.

crystal structures of phthalocyanines. In both applications, the compounds are used as solid-state materials where the property (color, photosensitivity) is related to the crystal structure, i.e., the arrangement of adjacent molecules towards each other. Different arrangements of the molecules lead to materials with different physical properties and polymorphic modifications were therefore referred to as physical isomers. Great interest nowadays lies in the prediction of crystal structures from the molecular structure and calculations of physical properties therefrom.[22] However, despite initial success stories, it has only recently become possible to use crystal structure prediction as a common research tool.

In general, however, interest in crystal structures of phthalocyanines has been decidedly limited and until 1997, when two independent reviews were published,[23,24] very little work had been undertaken to classify and compare the crystal structures. Since 1997, the number of crystal structures has greatly increased and the need to update the 1997 review is paramount. This work will give an overview of all published crystal structures, briefly discuss molecular structures of the phthalocyanine complexes and attempt to identify common arrangement patterns which are seen among the different phthalocyanine structures. It will also point to some unknown or lesser-known structures, which have been published in doctoral theses but never gained entrance into the peer-reviewed literature. Several of these structures have been reprinted with kind permission of the authors (Appendices 24–41). Due to the vast amount of data, it is not possible to report all findings in great detail; therefore, this review focuses more on the neutral phthalocyanine complexes and treats only briefly ionic phthalocyanines, peripherically substituted phthalocyanines, and phthalocyanines with central metals carrying large axial ligands.

C. BASICS OF PHTHALOCYANINE STRUCTURES

To clarify structural relationships between the phthalocyanine complexes, a new classification scheme using the molecular shape was introduced.[23] The shape is intrinsically related to the central atom(s), the number of additional ligands bound to the central atom and the ligand geometry around the central metal. The arrangement of molecules in the crystal can be qualitatively predicted from the shape of the molecule. If a phthalocyanine cocrystallizes with solvent molecules or counter ions, however, then the arrangement pattern is adversely affected, making the resulting crystal structures difficult to predict.

The basic structure of a phthalocyanine molecule (Figure 2a) consists of four isoindole units, circularly linked via azamethine bridges (N_{br}). The resemblance with porphyrins has been subject of many studies and contributed greatly to the interest in phthalocyanine molecules. Comparisons of the main structural features of porphyrins and phthalocyanines have been published elsewhere.[25,26] One common structural feature is the cavity within the center of the molecule, which is able to accommodate one or two central atom(s). Principally, a central atom only coordinates to the phthalocyanine ligand in a limited number of ways (Figure 3), either a) inside the molecular plane of the macrocycle or b–c) out-of-plane coordinated, i.e., it has a certain distance to the plane of the macrocycle.

With the exception of metal-free phthalocyanine complexes, only one atom is coordinated inside the central cavity, e.g., Co in cobalt phthalocyanine, PcCo (COPTCY01,[27,28] App. 11:4). It is located exactly in the center of the phthalocyanine molecule and coordinates to the two sets of centrosymmetrically-related isoindole nitrogen atoms either by covalent or coordinate bonds (Figure 2a).

In the case of metal-free phthalocyanine, PcH_2, the two central atoms are hydrogens, which are small enough to be both coordinated inside the central cavity (Figure 4). The hydrogen atoms are either coordinated a) to one set of centrosymmetrically related isoindole nitrogen atoms (localized hydrogen atoms) or b) to both sets (disordered half-hydrogen atoms which means that every isoindole nitrogen atom is coordinated to 50% of one of the inner hydrogen atoms). There has been considerable discussion within the literature[31–40] to determine whether this disorder is dynamic, i.e., if it is a molecular property where the hydrogen atoms are steadily changing positions, or if the disorder is static, i.e., where molecules with localized hydrogen atoms are disordered and X-ray diffraction only sees a statistical disorder. The current thinking implies this disorder is a molecular property. It has been suggested[23] that the hydrogen atoms are localized, i.e., positionally ordered, providing the molecule has two distinct molecular axes. The hydrogen atoms are then expected to occupy positions on the axis that provides adequate space to accommodate them. With increasing difference between the two axes, a parameter often equal to the increasing deformation of the macrocycle, the activation energy between the two states of such a dynamical equilibrium will increase and localization will become more probable. The first phthalocyanine structures to be identified which showed localized hydrogen atoms were two octasubstituted

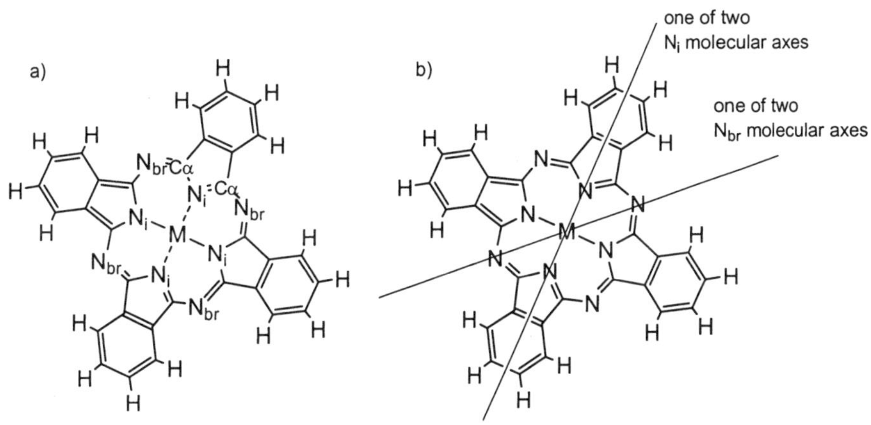

Figure 2. (a) The basic molecular structure of metal-free phthalocyanine ($M{=}H_2$) and metallophthalocyanine; N_{br}, N_i, and C_α denotes bridging azamethine nitrogen atoms, isoindole nitrogen atoms and isoindole α-carbon atoms, respectively. Plain lines indicate covalent bonds while broken lines represent dative bonds. (b) The location of one of the two N_i molecular axes and of one of the two N_{br} molecular axes; the other two axes are perpendicular to the axes shown.

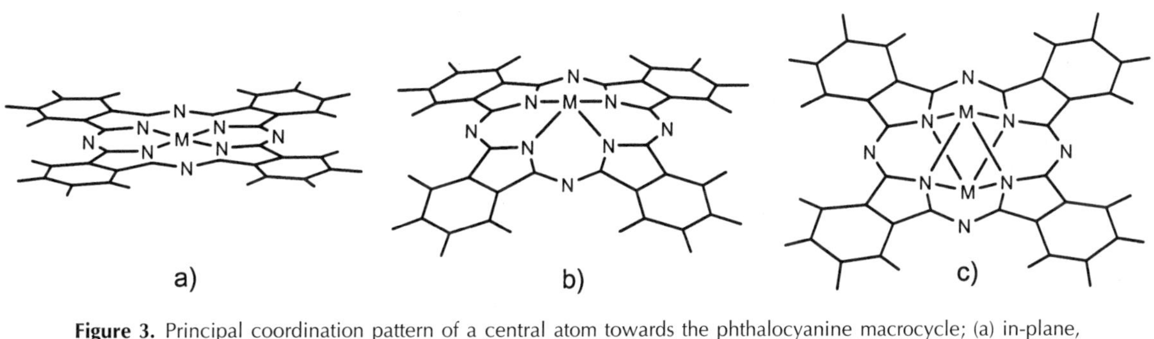

Figure 3. Principal coordination pattern of a central atom towards the phthalocyanine macrocycle; (a) in-plane, (b) out-of-plane (oop) coordinated, and (c) two oop coordinated atoms on opposite sides of a (μ-coordinated) phthalocyanine macrocycle.

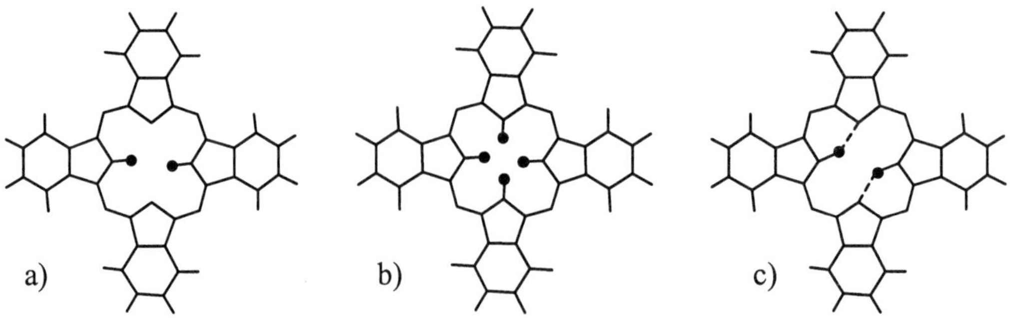

Figure 4. Three basic arrangement patterns of the inner hydrogen atoms in metal-free phthalocyanine; (a) localized, as found in PHTHCY07,[29,30] (b) disordered, so-called "half-hydrogen" atoms as found in #1971-03[31] and (c) intramolecularly hydrogen bonded hydrogen atoms as seen in Figure 8. The inner hydrogen atoms are represented as filled circles.

phthalocyanines (SUKSUM,[41] App. 10:4–6, and SOZBUE,[42] App. 10:10–12) and a mixed phthalocyanine–porphyrazine macrocycle known as norphthalocyanine (WETJUA,[43] WETJUA10,[44] App. 22:46–47). More recently, a further structure, derived from the parent-unsubstituted phthalocyanine has been identified with localized hydrogen atoms, and its structural analysis confirms the assumption of different axes (PHTHCY07,[29] App. 10:22). If the axes are not different, i.e., the theoretical locations of the hydrogen atoms of the two axes are very near to each other, then also substituted phthalocyanine complexes with deformed macrocycles show "half-hydrogen" atoms (WOKDAB,[45] App. 10:1–3).

In the case of a single central atom, coordination within the inner cavity depends on the size of the ion.[46] From geometrical considerations, it was calculated that the center-N_i radius with minimized strain of the relatively rigid macrocycle is *ca.* 1.90 Å.[47] If the ion is smaller or larger than the optimum size, then insertion of the ion into the cavity introduces conformational stress, which the macrocycle can partially reduce by moving the isoindole nitrogen atoms closer to or further away from the center accordingly. However, if the ion is too large, and is unable to enter the cavity it rests "atop" or "out-of-plane coordinated," a situation that is sometimes referred to as extra-coordination.[48] The macrocycle adapts to this special bonding situation by deformation (see Section III: Phthalocyanine complexes with an out-of-plane coordinated central atom). A special case of this coordination is the μ-coordination (Figure 3c) where the phthalocyanine ligand coordinates to two out-of-plane coordinated metals located on opposite sides of the macrocycle, as in $Pc(M\{L\})_2$ [e.g., $PcTl_2$ (HEBHOL,[49] App. 13:9) or $Pc(Na\text{-}OP\phi_3)_2$ (QASKAW,[50] App. 13:7) or in the *tris*-phthalocyanines. In the latter case, both symmetrical and unsymmetrical coordination of the central metals to the isoindole nitrogen atoms of the μ-coordinating macrocycle is known, symmetrical in the case of Pc_3Bi_2 (DIQPIC[51] and DIQPIC01,[52] App. 18:5–6 and Figure 5k), where each Bi atom coordinates symmetrically to the four isoindole nitrogen atoms of the inner phthalocyanine, and unsymmetrical in the case of Pc_3In_2 (PIFDEN,[53] App. 18:37–38 and Figure 5l), where each In atom coordinates to two of the four isoindole nitrogen atoms more strongly than to the remaining two. In the case of symmetrical μ-coordination, the inner macrocycle remains mostly planar while in the case of an unsymmetrical μ-coordination, it becomes more deformed in a similar way to the out-of-plane coordinated complexes.

An overview of the molecular structures of phthalocyanine complexes (Figure 5) shows that they basically derive from two types of phthalocyanine structures, either from molecules with an approximately planar macrocycle or from molecules with a bent macrocycle where the macrocycle takes the appearance of a saucer.[54] Recently, the saucer shape of several phthalocyanine complexes has been made visible by STM investigations.[55,56]

Of those molecules, which adopt a planar macrocycle configuration, there exists a sub-classification according to the ligand stereochemistry of the central atom (Table 2). Complexes where the central cavity contains two hydrogen atoms coordinated to the isoindole nitrogen atoms are referred to as CN1 complexes (Figure 5a), those containing one metal as CN4 complexes (Figure 5b) or those containing one metal with two axial ligands perpendicular to the phthalocyanine ring located on opposite sides are referred to as CN6*tr* complexes (Figure 5c). Dimers (Figure 5o) or larger oligomers can be derived from the case shown in Figure 5c when one or more axial ligand(s) is(are) μ-coordinated to two metal-phthalocyanine units, $X\text{-}PcM\text{-}[Y\text{-}PcM]_{n\geq 1}\text{-}Z$.

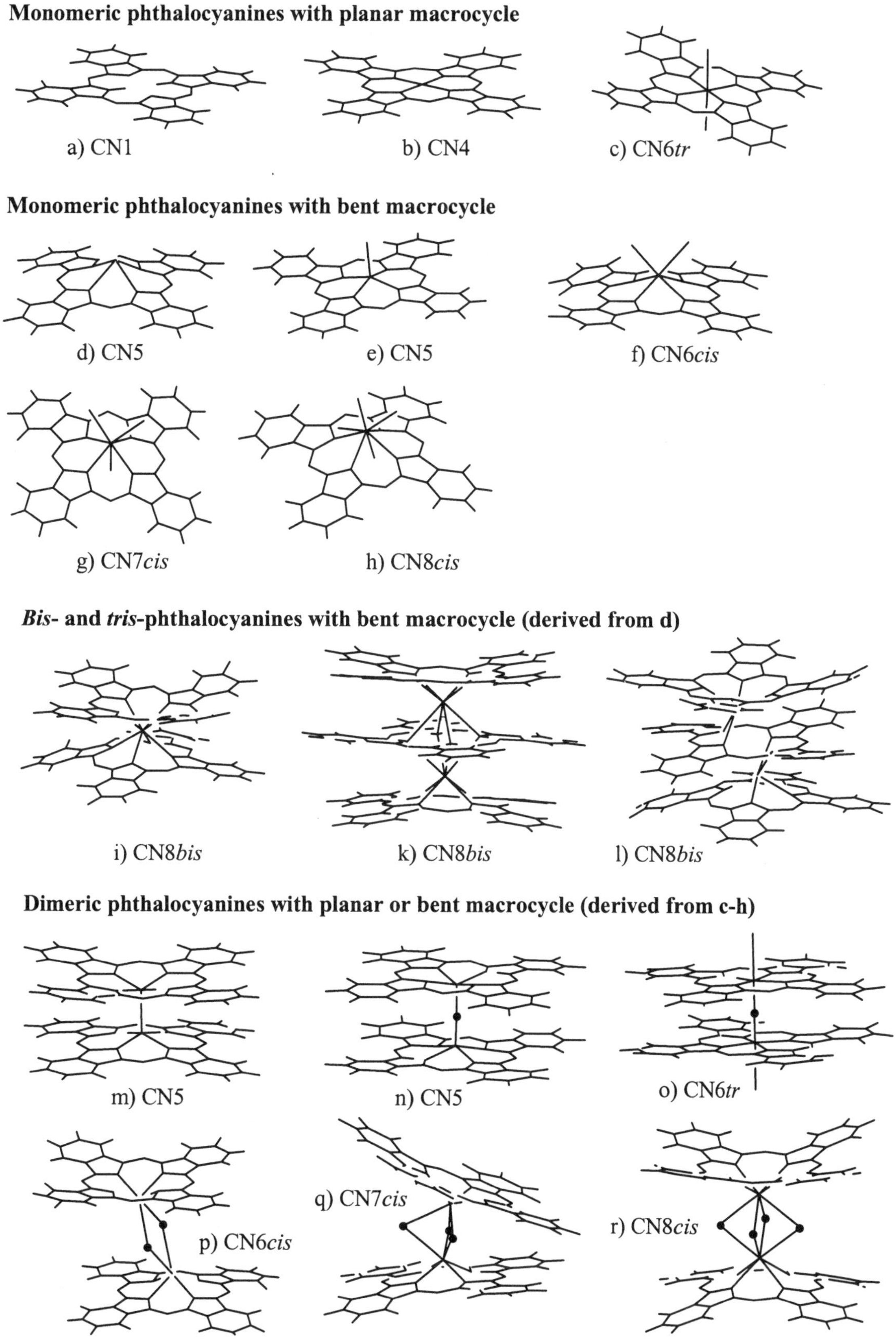

Figure 5. Overview of basic structures of phthalocyanine complexes, μ-coordinating ligands are represented as filled circles.

Table 2. Stereochemistry of the Metal Atoms in Metallophthalocyanines

Coordination number	Oxidation state of the central atom	Coordination geometry[a–d]		Symmetry[a]	Example [Ref.]
4	2	Square planar	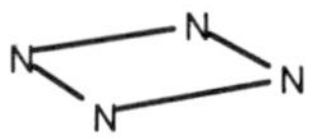	D_{4h}	PcCu[76]
5	2,3,4,5	Square pyramidal	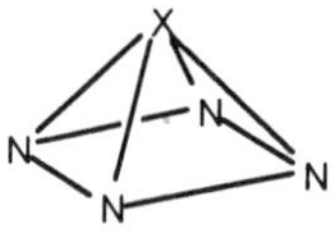	C_{4v}	PcSn,[31,142,143] PcAlCl,[154–156] PcTiO,[2] PcReN[e]
6*tr*	4,5,6	Octahedral	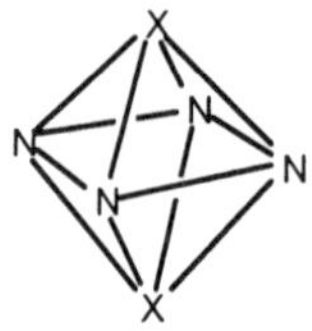	O_h	$PcSnCl_2$,[116,117] PcMoOCl,[124] $PcMoO_2$[f]
6*cis*	4	Trigonal prism	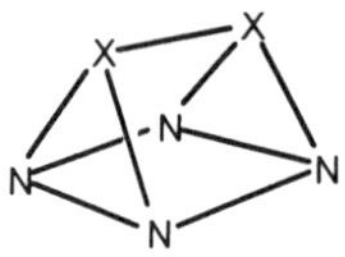	D_{3h}	$PcTiCl_2$,[158] $PcMoO_2$[f]
7*cis*	5	Capped trigonal prism	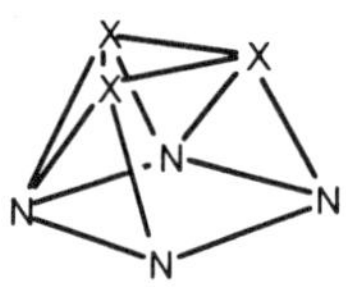	C_{2v}	$PcNbCl_3$[124,163]
8*cis*/8*bis*	3,4	Cube	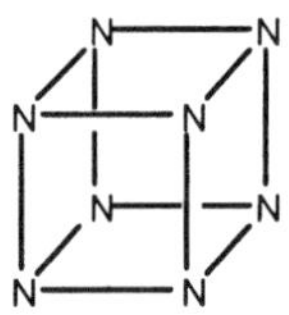	O_h	$[Pc_2Nd][NBu_4]$[197]
8*cis*/8*bis*	3,4	Square antiprism	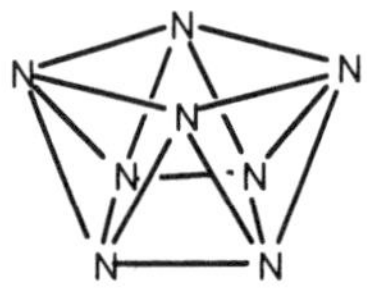	D_{4d}	Pc_2Lu,[159] Pc_2Zr[215]

[a]The geometries and symmetries are only approximate.
[b]No example of coordination number 7 with capped octahedral geometry (C_{3v}) is known.
[c]No examples of coordination number 8 with the geometry of a bicapped trigonal prism (C_{2v}, C_{3v}) are known.
[d]"N" and "X" denote the isoindole-nitrogen and an axial ligand, respectively.
[e]No absolute structure is known, but preliminary results indicate a rhenium atom with square-pyramidal coordination geometry.[146,147]
[f]No absolute structure is known, but similar porphyrin complexes indicate two structural possibilities, a linear O=Mo=O fragment, similar to O=U=O in superphthalocyanine,[260] and a peroxo group, as in $OEPTiO_2$ (OEP = octaethylporphyrin).[453]

Among the molecules with bent macrocycles, a similar sub-classification is equally useful. In these classes, the bent macrocycle is always exclusively on one side of the central atom with the additional axial ligands on the other side. They are therefore also referred to as *cis* complexes and also include the case of a single axial ligand, PcM-X (CN5 complex, Figure 5e) and the special case PcM (Figure 5d), where an electron-pair of the central atom is considered to be the axial ligand. Other *cis* complexes comprise the case where there exists two additional ligands, *cis*-$PcMX_2$ (CN6*cis* complex, Figure 5f), three additional ligands, *cis*-$PcMX_3$ (CN7*cis* complex, Figure 5g), or four additional ligands, $PcMX_4$ (CN8*cis* complex, Figure 5h). Two monodentate axial ligands can equally be replaced by a bidentate ligand. Special cases include the *bis*-phthalocyanines (CN8*bis* complex, Figure 5i), where two bent macrocycles coordinate to the same central atom and *tris*-phthalocyanines (Figure 5k,l), where the terminal macrocycles are bent while the μ-coordinating middle macrocycle is much less deformed. Similar to the dimers mentioned above, dimers derived from complexes with bent macrocycles are known where two PcM are joined via a single bond, as in PcM-MPc (Figure 5m) or via a μ-coordinating axial ligand such as PcM-X-MPc (Figure 5n). Metallophthalocyanines joined via more than one bond are also known, selected examples of which are shown in Figure 5p–r.

II. Phthalocyanine Complexes with In-plane Coordinated Central Atoms

As shown in Figure 5a–c, phthalocyanine complexes with in-plane coordinated central atoms and planar phthalocyanine macrocycle can be subclassified in CN1, CN4, and CN6*tr* complexes. Due to their structural similarities, CN1 and CN4 complexes will be discussed together.

A. CN1 AND CN4 COMPLEXES

1. Neutral Phthalocyanines

a. Introduction and Molecular Structures

The periodic table of CN1 and CN4 complexes with the number of structures given next to each element is shown in Figure 6; elements in parentheses indicate that only unit-cell data but no atomic coordinates are known. A compilation of the basic crystallographic data of the structures is given in Appendix 1 and their molecular structures are shown in Appendices 10 and 11.

Robertson's first paper concerning the molecular and crystal structures of the phthalocyanines[57] already contains most of today's knowledge on the structural features of CN1 and CN4 complexes: space group $P2_1/a$ points to a center of symmetry in the molecule; the dimensional similarity of the unit-cells of PcH_2, PcNi, and PcCu points to a molecule with a large rigid framework which coordinates the metal within it; and the different dimensions of the unit-cell of PcPt point to a change of orientation of the molecule, rather than to a different molecule as the volume of the unit-cell is identical or very similar to that of the others. Robertson concluded that the phthalocyanine molecules are large, square, and probably flat, stacked along the crystallographic *b*-axis with molecules of identical orientation recurring at intervals of 4.7 or 3.8 Å with the molecules inclined to the ac plane.[57] He also concluded that a projection along the crystallographic *b* axis should show a complete picture of the molecule and all of its component atoms without individual atoms obscured by overlap. Still, Robertson also identified some remaining difficulties with regards to the fine structure, e.g., the location of the o-quinoid ring and the central hydrogen atoms. These topics were also covered by Dent *et al.* in a separate publication.[11] Robertson, Linstead, and coworkers discussed the possibility of the inner hydrogen atoms being located outside of the molecular plane, giving rise to the possible existence of both *cis*- and *trans*-isomers.[11,57] Robertson added that only the latter would be strictly compatible with the center of symmetry induced by the space group $P2_1/a$ and we now know that the only unsubstituted metal-free phthalocyanine with localized hydrogen atoms exists in the *trans*-form.[29] In the case of a metal occupying the central cavity of the molecule, the metal atom and the four surrounding isoindole nitrogen atoms are necessarily coplanar.

The first quantitative structural determination of a phthalocyanine complex, the metal-free phthalocyanine, PcH_2 (PHTHCY01,[1] App. 10:19), was reported in 1936 by Robertson. This was a substantial achievement as it was the first direct X-ray analysis of an organic compound. The obtained structure depicted a

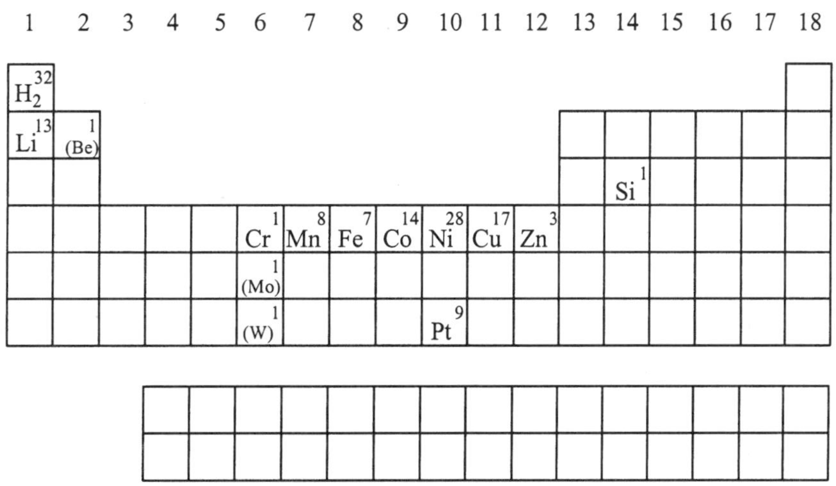

Figure 6. Periodic table of CN1 and CN4 complexes (Figure 5a,b) with number of structures given next to each element. Elements in parentheses indicate that only unit-cell data but no atomic coordinates are known.

two-dimensional projection along the *b*-axis onto the ac plane providing the *x*- and *z*-atomic coordinates. To calculate the third dimension, Robertson used the assumption that the benzene rings in the phthalocyanine macrocycle were regular hexagons with a C–C bond length of 1.39 Å. Nowadays we know that this assumption is not entirely true; however, the exact data is within the error margin of Robertson's data. Comparing the actual size of a benzene C–C bond distance with the theoretical value allowed Robertson to calculate the inclination of the molecule towards the ac-plane and to calculate the *y*-coordinates of the atoms. Robertson then calculated the projection of the crystal structure along the *c*-axis and compared the calculated results with the experimental data. As both sets of data were consistent, he concluded that the assumption of a planar molecule must be valid. Robertson also noticed the departure of the metal-free phthalocyanine from true tetragonal symmetry and he brought the possibility of intramolecular hydrogen bonds into the discussion believing that such bonding might help to explain the distortion. The distance between the two nearest isoindole nitrogen atoms is approximately 2.65 Å, a distance for which hydrogen bonding is acceptable.

In the same year, Linstead and Robertson compiled the first review on PcM crystal structures.[58] This review contains a collection of unit-cell data for the macrocycles PcH_2 and PcM, where M represents Be^{II}, Mn^{II}, Fe^{II}, Co^{II}, Ni^{II}, Cu^{II}, or Pt^{II}. With the exception of PcPt, Linstead and coworkers observed that all structures were isomorphous. They also remarked that the monoclinic crystals of PcZn and (the anhydrous) PcMg were identical. This review also mentioned the fact that while the stereochemistry of planar 4-coordinate bivalent metals was extremely rare, the authors were the first to report such an example in the form of a metal phthalocyanine complex. Additionally, they were also the first to report the stereochemistry of the metal-coordinated isoindole nitrogen. The authors noted some small deviations in the case of PcBe, but unfortunately, until now, no one has reinvestigated its structure. It would be very interesting to know if the small beryllium ion is located exactly in the center of the macrocycle or if it is unsymmetrically coordinated to two isoindole nitrogen atoms. Another possibility would be an out-of-plane coordinated beryllium atom similar to the case found recently with magnesium in the anhydrous PcMg (XETZIF01-153733,[59] App. 12:29–31, and XETZIF,[60,61] App. 12:32). It is somewhat surprising that several authors report that PcMg is isomorphous with PcH_2 and PcNi,[58,62] and yet recent structures have found quite different unit-cell parameters. It would be a very interesting case if PcMg were found to possess structures with an Mg ion coordinated either in-plane or out-of-plane.

A second of the early structures, which needs to be reinvestigated, is PcNi (NIPHTC,[14] App. 11:57). As in the metal-free phthalocyanine structure, the *y*-coordinates of the atoms of the PcNi structure have been calculated. A comparison of Robertson's structure of metal-free phthalocyanine and those of others shows that there are many differences in both the bond lengths and bond angles. We can therefore expect that the accuracy of Robertson's PcNi structure is somewhat limited. Robertson pointed out some differences of the inner macrocycle of the nickel phthalocyanine compared to that of the metal-free macrocycle, one of which was a shift of the four isoindole nitrogen atoms towards the nickel atom; there were also different C–N bond lengths in the 16-membered inner ring. However, he had already remarked that these differences were not much greater than the limit of uncertainty imposed by the incomplete resolution of the various atoms. He also wrote that PcNi is more truly tetragonal than the metal-free compound but was unable to explain the one distortion present in both PcNi and the metal-free compound alike. Robertson observed that the nickel atom is surrounded by six nitrogen atoms, four of which are located at the corners of a square, 1.83 Å from the nickel atom and bound to it by covalent (or coordinate) links, while the two other nitrogen atoms lie vertically above and below the nickel atom at a distance of 3.38 Å, thus forming a distorted octahedron. A more recent example of such an octahedral coordination is shown in Figure 7 for the case of β-PcCr (#1999-01).[63] The nomenclature "β" had been attributed retrospectively to these polymorphs by comparison with the nomenclature introduced by Susich.[64,65]

In 1939, Barret *et al.*[66] reported preliminary X-ray investigations of needle-shaped single-crystals of tetrabenzotriazaporphyrin and its copper derivative. The results indicated that the structures would essentially be identical with those of the corresponding phthalocyanine molecules. Robertson, who performed the X-ray investigations, assumed that the molecules display a statistical center of symmetry in the crystal due to a random distribution of the molecules with respect to the unique –CH=N– group. Polymorphic structures probably exist as both plate-like and needle-like crystals were obtained following crystallization the material from chloronaphthalene. While needle-type crystals usually indicate β-type structures, plate-like structures were found with α-type polymorphs.[64]

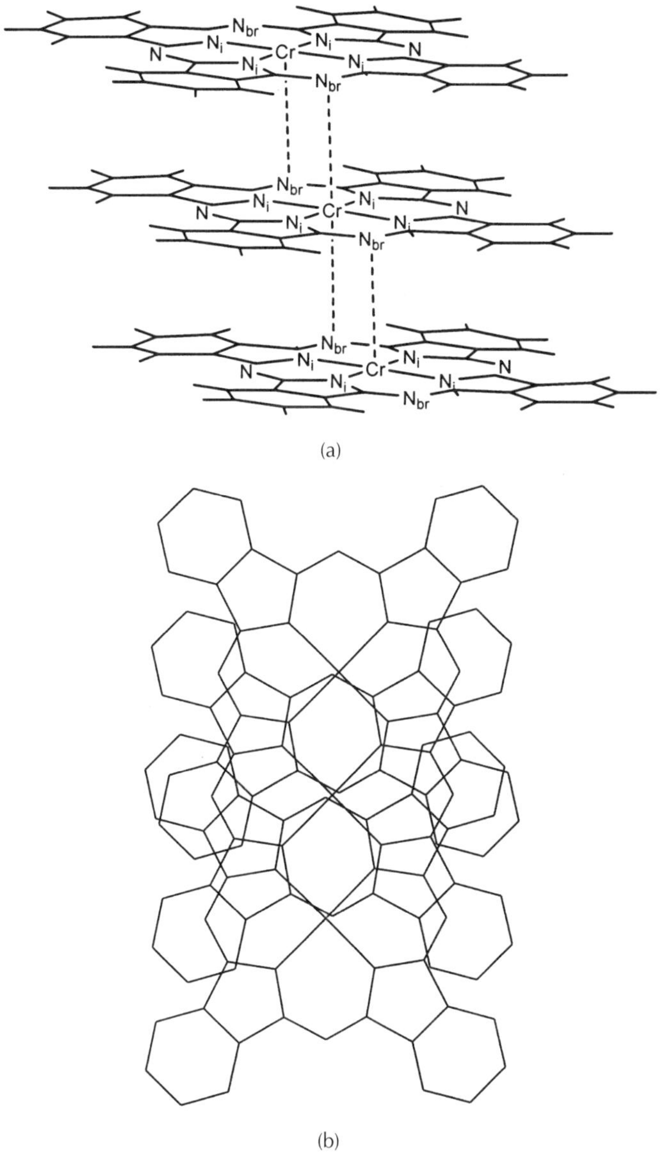

Figure 7. Distorted octahedral coordination around the central atom in PcCr (#1999-01[63]) as an example for β-polymorphs of CN4 complexes, (a) Cr–N_i bonds are drawn in plain lines while the Cr–N_{br} interactions with adjacent macrocycles are represented as broken lines, (b) is a view perpendicular to the phthalocyanine macrocycles.

In 1940, Robertson and Woodward then published the first structure of a polymorphic modification of a phthalocyanine complex, PcPt (PTPOCY02,[15] App. 11:70). In accordance with the structures of PcH_2 and PcNi, it was necessary to calculate the *y*-coordinates of the PcPt structure. The obtained bond lengths and bond angles have to be regarded with caution as Robertson pointed at the large margin of error in some of the atomic coordinates. Changes in the crystal structure of PcPt are due to a different arrangement of the molecules in the solid state. Generally, the molecules are inclined at a much smaller angle to the ac-plane than those found in other reported structures. When looking along a normal projection onto the molecules, it can be seen that in no case the atoms of adjacent molecules are directly above each other. The molecular structure, however, is identical with that of previously reported structures.

Also in 1940, Woodward compared X-ray studies of porphyrin complexes including phthalocyanine and phthalocyanine-type complexes.[67] In this study, the author also included preliminary data on the crystal structures of tetrabenzomonoazaporphyrin and tetrabenzoporphyrin. Woodward assumed the inclination angle between the molecular plane and the ac-plane to be around 60° and, as a consequence, a different orientation of the molecules is expected. These assumptions have since been confirmed by another report.[68]

In 1945, a photographic Fourier synthesis of the X-ray data of Robertson's PcH_2 was undertaken by Huggins.[69] From the photograph shown, it can be assumed that electron density is evident along the smaller side of the isoindole nitrogen rectangle attributable to inner hydrogen atoms.

In 1948 and 1950, unpublished results, dating back from research conducted in the early 1930s, about the polymorphic forms of phthalocyanine were published.[64,65] The author was able to identify and characterize two polymorphic structures of copper phthalocyanine (α,β) and three polymorphs of metal-free phthalocyanine (α,β,γ). Susich also reported the isolation of plate-like single-crystals of α-PcCu, a photograph of these crystals is shown.[64] Today's nomenclature of α- and β-polymorphs of phthalocyanine complexes originates from Susich's research.

In 1952, Donohue rejected the possibility of intramolecular hydrogen bonds in metal-free phthalocyanine as impossible, citing the large differences between the two bond angles N_i-H...N_i as the main reason for his opposition.[33] However, Donohue only considered linear bonding N_i-H...N_i (similar to Figure 4c) and failed to take into account the possible existence of "curved" bonding, a situation, where the hydrogen atoms lie approximately on lines connecting the centrosymmetrically related isoindole nitrogen atoms (Figure 4a), and where the N_i-H...N_i geometry is similar to the B–H–B bonds found in boranes.

In 1961, Fleischer reinvestigated the 2D X-ray collated by Robertson to revisit the inner hydrogen dilemma.[34] Fleischer refined the data using a much smaller temperature factor than Robertson and obtained an electron-density map that clearly shows inner hydrogen atoms connecting two neighboring isoindole nitrogen atoms (Figure 8). The map appears strinkingly similar to the results of the photographic Fourier synthesis obtained by Huggins.[69] Appendix 24 contains the atomic coordinates of this structure.

While Robertson reviewed his own research on phthalocyanine structures,[70] Moser and Thomas[71] and Lever[72] published detailed reports on further aspects of crystallographic research on phthalocyanines (electron microscopy of thin films, powder diffraction results). In 1965, Fielding and Stephenson reported the isolation of crystals of several β-polymorphs, 5–6% more dense and with different unit-cell dimensions, than those previously reported by Robertson.[73] Unfortunately, they failed to remeasure the full 3D crystal structures.

In the same year, an important paper concerning the polymorphism of phthalocyanines was published by Honigmann *et al.*[74] The authors' primary interest concerned the unstable forms of copper phthalocyanines, macrocycles that are of great interest to the pigment industry. In the following, the nomenclature of Honigmann *et al.* is used. To obtain detailed structural information of the unstable forms, they searched for isomorphous and stable forms of different phthalocyanine complexes, which could be investigated by X-ray diffraction. Honigmann *et al.* found that the stable modification of αI-4-Cl_4PcCu is isomorphous with the unstable form of αI-PcCu, while the stable α-PcPt is isomorphous with the unstable αII-PcCu (previously named γ-PcCu). From these stable forms, they were able to grow single-crystals and derive important information about the space groups and the arrangement of the molecules in space. Unfortunately, the obtained single-crystals were too small to perform a full structural determination on them. The publication undertook a structural comparison of β-PcCu (at the time this paper went to publication, no full structure was known, but it was known to be isomorphous with β-PcNi), the different PcPt structures of Robertson and Honigmann, 4-Cl_1PcCu and 4-Cl_4PcCu (see Section II.A.1.b for a detailed description of the arrangements).

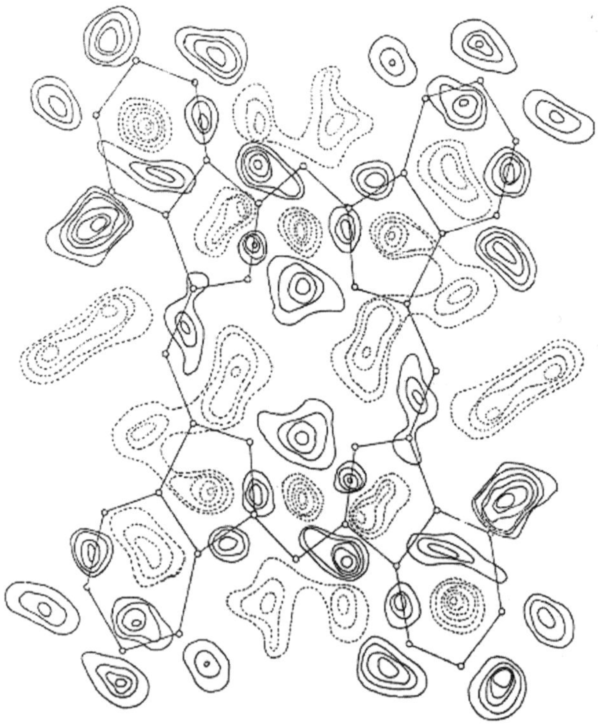

Figure 8. Electron density map of metal-free phthalocyanine, PcH_2. Refinement of the original data[1] by Fleischer. Reproduced with permission from the Ph.D. thesis of Everly B. Fleischer.[34]

In 1967, Suito *et al.*[75] also published a review about the PcM structures. The authors introduced their own electron microscopy research about phthalocyanines and reported it in context with the X-ray research in the literature.

In 1968, Brown then published his research on β-PcCu (CUPOCY10,[76] App. 11:16), α-PcPt (PTPOCY10,[77] App. 11:71), and γ-PcPt (PTPOCY11,[77] App. 11:72). The crystal structures reported by Brown are the first known 3D structures of metal phthalocyanines. They showed that the structure of β-PcCu, apart from small differences, is identical with that of β-PcNi. It appears as if the copper atom pushes one set of the isoindole nitrogen atoms further away from the center than the other set. This results in two different sets of Cu–N_i bond lengths and bond angles C_α–N_i–C_α, where the angle around the former nitrogen atoms is significantly larger. No explanation for this distortion is given in the original publications, however. It is purported here that this effect might result from the localization of covalent and coordinative bonds in the central cavity, due to small differences in the molecular axes, a similar effect as found in the localization of hydrogen atoms in metal-free phthalocyanines.

Brown also reinvestigated PcPt crystals from the original batch of Linstead, provided by a former student of Linstead, and found two types of crystals with differing structures.[77] In his paper, Brown questioned the crystal structure space group reported by Robertson and concluded that Robertson's $P2_1/a$ was not a true representation of the unit-cell. Brown also claimed[77] that his α-polymorph structure is a better description of the polymorph found 28 years earlier by Robertson. However, a comparison of the spatial orientation of the molecules in neighboring columns in Robertson's structure and the α- and γ-polymorphs of Brown reveals that the γ-polymorph of Brown is identical with Robertson's structure while the α-polymorph is identical with the α-PcPt described qualitatively by Honigmann *et al.*[74] (see also "Note Added in Proof").

Research on the high-pressure crystal structures of metal-free and nickel phthalocyanine was first reported by Kirk in 1968[78] and then in 1969.[79] He observed that below 7 kbar, the unit-cell data of nickel phthalocyanine appears to indicate a β-polymorph configuration. Around 7 kbar, a phase transition takes place and the structure appears to change to α-type configuration. Similarly, an α-type structure of metal-free phthalocyanine has been obtained under high pressure.

In 1969, Hoskins *et al.* reported a neutron diffraction study of single-crystals of metal-free phthalocyanine, PHTHCY.[38] They found the inner hydrogen atoms in the molecular plane of the phthalocyanine macrocycle, and showed them to be disordered, so that on average, one half of a hydrogen atom ("half-hydrogen") is associated with each isoindole nitrogen atom. These so-called half-hydrogen atoms lie very close to the connecting lines between the centrosymmetrically-related isoindole nitrogen atoms. More details about this research, including the full crystal structure, were later published in the Ph.D. thesis of Mason, PcH_2 (#1971-03,[31] App. 10:18), the atomic coordinates can be found in Appendix 26.

In 1970, E.C. Bissell reported the first 3D X-ray structure of metal-free phthalocyanine in his Ph.D. thesis, β-PcH_2 (#1970-02,[40] App. 10:20), a structure which had been presented at the Summer Meeting of the American Crystallographic Association in 1968.[80] The author was not able to locate the inner hydrogen atoms, but, instead, identified the existence of residual 4-lobed electron density. The four lobes are located at the same positions occupied by the half-hydrogen atoms in the neutron diffraction structure. A comparison of this structure and the structure reported by Robertson reveals several discrepancies in bond lengths and bond angles. The atomic coordinates can be found in Appendix 25. Bissell also discussed the possibility of statistical disorder vs. tautomeric exchange of protons. His argument states that the statistical disorder of molecules with localized, unshared hydrogen atoms should be reflected by a greater uncertainty in the location of the isoindole nitrogen atoms, due to differences in the nature of the isoindoles with or without hydrogen atoms attached. He concluded that the half-hydrogen/residual density effect is a molecular, rather than a crystallographic effect as there is no evidence for the uncertainty effect described above.

In 1972, Das and Chandhuri obtained a two-dimensional X-ray crystal structure of tetrabenzomonoazaporphyrin (BZAZPN,[68] App. 10:37). This structure has a statistical center of symmetry as the differences between the –CH= and the –N= groups have only a small effect on the scattering. The authors subsequently refined the (010) and (001) projection of tetrabenzomonoazaporphyrin and derived the atomic coordinates from the solutions. The molecules are observed to stack in columns along the *b*-axis, where the calculated value of the angle between the mean molecular plane and the (010) plane is 62° (60° as predicted by Woodward[67]). The authors noticed some distortions in the molecule, which they attributed to the limited accuracy of structures derived from two-dimensional analyses.

In 1976, Kirner *et al.* reported the full structures of PcFe (FEPHCY, App. 11:21), and PcMn (MNPHCY, App. 11:44).[81] The structures are very similar to that of β-PcCu and β-PcNi, displaying similar octahedral coordination around the metal (Figure 7). Kirner *et al.* found that one of the azamethine nitrogen atoms from each adjacent molecule deviates slightly from the mean plane of the phthalocyanine toward the central metal atom of the neighbor molecule. In 1977, Scheidt and Dow reported the structure of PcZn (PTHCZN,[82] App. 11:76) in which the zinc atom, despite its size, is located exactly inside the metal cavity and not out-of-plane coordinated as found for the Sn atom in PcSn. The structure has exceptionally short Zn–N bonds with distances less than 2 Å. A small radial expansion of the core has been attributed as the main reason, which allows the phthalocyanine to accommodate the Zn atom. The structure is very similar to other previously reported β-polymorphic structures. In 1978 and 1979, Figgis *et al.*[83] and Mason *et al.*[84] reported the crystal structures of PcMn (MNPHCY11, App. 11:41) and PcCo (COPTCY10, App. 11:7), respectively. These they found to be isomorphous with other β-polymorphic structures. The molecules have essentially D_{4h} symmetry but the authors found small but significant deviations from ligand planarity.

Mason *et al.* compared the structures of metal phthalocyanines with those of the metal-free analogs and remarked that the "half-hydrogen" atoms, with the larger of the two least-squares-determined occupancy factors, are associated with the isoindole nitrogen atoms positioned furthest away from the center of the cavity, indicating the presence of repulsive interaction between the inner hydrogen atoms due to overcrowding within the inner cavity.[84] The authors also re-investigated the metal-nitrogen axial contact distances ($M–N_{br}'$) and angles ($N_i–M–N_{br}'$) in the β-polymorphic structures and found that the further the phthalocyanine molecules are apart, the smaller the angle deviates from 90°. A decrease in bond distance is also related to a decrease from D_{4h}. Furthermore, an increase in a "bending-away" distortion of some of the benzene rings can be related to a decrease in intermolecular separation, i.e., the most planar macrocycle can be found in PcH_2 and PcCu, where the molecules are the furthest apart. The authors discuss the relationship between the metal spin state and the size of the inner cavity of the phthalocyanine ligand. In 1980, Figgis *et al.*[85] refined the structure of PcMn (MNPHCY21, App. 11:39, and #1980-01, App. 11:40), using the data reported by Mason *et al.*[84] The results of a neutron diffraction of PcCo at 4.3K (COPTCY01, App. 11:4), and a room temperature X-ray diffraction study of the same macrocycle (COPTCY10, App. 11:7) have been published by Williams *et al.*[27] The authors report significant and systematic differences between the bond lengths, with the 4.3K bond lengths being, in general, larger than the corresponding 295K bond lengths. They discuss this phenomenon in regard to a shift away from the true positions in the case of X-ray diffractions and also in a less pronounced thermal motion in the case of the 4.3K structures. The authors concluded that the positions derived by neutron diffraction are more realistic than those obtained by X-ray structural determination at room temperature. From a decrease in the $Co–N_{br}'$ intermolecular distance, they concluded that the molecules pack more closely at lower temperatures. Figgis *et al.* determined the structure of β-PcMn with neutron diffraction at 5.8K in 1980 (MNPHCY03,[86] App. 11:43). Similar to the case of PcCo, significant deviations between higher-temperature X-ray and low-temperature neutron diffraction studies are observed.

In the same year, the β-polymorphic structure of $PcSi_{0.17}$ was published, which according to the authors is better described as $PcSi*4H_2Pc*Pc$ (PHPRSI,[87] App. 11:74–75). This structure seems to be a β-PcH_2 containing some impurities. A two-dimensional crystal structure of hexadecachlorinated copper phthalocyanine, obtained by electron microscopy, was reported by Kirkland in 1982 and by Saito and Shiojiri two years later (#1982-01, App. 11:15).[88,89]

In 1985, Homborg and Teske investigated the structural characteristics of lithium phthalocyanines.[90] From powder diffraction studies, they concluded that PcHLi (COBGOP) and the monoclinic modification of PcLi (DIXNIH01) crystallized isomorphous with the β-polymorphs of phthalocyanine. They also uncovered a tetragonal modification (DIXNIH), whose full structure was published in 1986 by Sugimoto *et al.* (DIXNIH02,[91] App. 11:26). In this structure, the phthalocyanines are stacked linearly with the normal parallel to the stacking axis. The distance between adjacent phthalocyanine molecules was found to be 3.245 Å (Li...Li), where the molecules are staggered by 38.7° around the Li–Li axis. The PcLi molecular structure shows some distinct differences in bond lengths and angles compared to the α-, β-, or γ-polymorphs.[23] In 1987, Perovic used electron microscopy to gain more detailed information about x-PcH_2.[92] The author suggested this structure is basically monoclinic with partial orthorhombic structure. Unfortunately, the author was unable to derive more detailed information about the arrangement of the molecules or about the molecular structure itself.

It is known that partial chlorination of the phthalocyanine macrocycle stabilizes the α-forms. Similarly, sublimation of a 1:1 mixture of metal-free phthalocyanine and its 4-chloro derivative lead to single-crystals of the α-form, as shown by Yase *et al.* in 1988 (GAKWUK,[93] App. 10:27). The structure is isomorphous with that of α-PcPt reported by both Brown[77] and Honigmann *et al.*[74]

In 1989, Figgis *et al.* investigated the structure of PcCo using X-ray diffraction at 115K (COPTCY05,[94] App. 11:6). They showed that the molecular structure was almost 2/m in symmetry, with a two-fold axis along the Co–N(1) vector and an almost exact mirror plane through the Co–N(3) molecular normal plane. The intermolecular interactions between adjacent molecules mentioned above link Co and N(3). The deviations from 4/mmm symmetry are small but significant. In 1991, Reynolds *et al.* published a structure of PcCo obtained from neutron diffraction at 115K (COPTCY06,[95] App. 11:3). They were able to reconfirm the data of Figgis *et al.*,[94] and to show that the macrocycle adopted molecular bowed deformation.

In the 1990's, Janczak and Kubiak started synthesizing phthalocyanine complexes by solid-state synthesis. With this method, they obtained many phthalocyanine compounds in the form of single-crystals. In 1992, they published two papers on, what appears to be, metal-free phthalocyanines.[96,97] According to these publications, Kubiak and Janczak claimed to have obtained a precursor of the phthalocyanines (without hydrogen atoms in the center) in both β- (PHTHCY03,[96] App. 10:21) and α-modifications (PHTHCY04,[97] App. 10:16). They assumed this result from the fact that "the supplementary hydrogen atoms introduced into appropriate central positions of the phthalocyanine molecule ring were consequently rejected during the refining cycles." Unfortunately, their papers did not contain additional chemical and physical data, e.g., colour of the crystal, absorption and mass spectral data etc., which could validate their claim. The authors, meanwhile, refer to these materials as metal-free phthalocyanines.[98] The α-forms of PcH_2/PcM are less stable than the β-form and can be easily transformed into the latter. Usually, stabilization of the α-form is achieved either through the use of additives, or by exchanging some of the peripheral hydrogen atoms for halogens. Janczak and Kubiak obtained single crystals of the α-form directly from solid-state synthesis.[97] The residual electron density observed outside of the molecule, they attributed to metal inclusions (0.077 Bi atoms per unit-cell) from the starting materials. The authors infer that such inclusions might impart stability in the crystal, the structure of which is isomorphous with that of α-PcPt, reported earlier by Brown.[77]

Due to its success as a component in charge-generating materials, x-PcH_2, or τ-PcH_2, as it is also referred to, has been investigated by several groups, *x*-PcH_2[98,99] (PHTHCY05, App. 10:24), τ-PcH_2,[100] *x*-PcH_2[98,101] (PHTHCY06, App. 10:25), *x*-PcH_2[102] (#1997-03, App. 10:26). Information obtained from X-ray powder diffraction patterns was used to calculate the arrangement of the molecules of *x*-PcH_2 in the solid-state, and to derive a structure–property relationship to explain the different photosensitivity of the α-, β-, and *x*-polymorphs of metal-free phthalocyanines. The molecular structures were taken from the β-polymorph and held rigid during calculation. An analysis of the orientation of the molecules in neighboring columns shows, that the structures of PHTHCY06 and #1997-03 are, however, polymorphic modifications, see Section II.A.1.b.

Liu *et al.* calculated the crystal arrangement of PcCu in 1996 but only provided general information about the crystal structures.[103] In the same year, Morishige and Araki reported the crystal structures of 2,3-naphthalocyaninato nickel (RADPER, App. 19:2–3), copper (RADPIV, App. 19:1) and zinc (RADPOB, App. 19:4).[104] The authors report the molecule being essentially planar with only small deviations. The N(1) azamethine nitrogen atom is located furthest outside the mean plane of the molecule which the authors believe to be due to an interaction between the nitrogen atom and the central metal of a neighbor molecule, similar to the interaction found in β-polymorphs of metal phthalocyanines as shown in Figure 7 of this review. Figure 4 of the publication of Morishige and Araki shows, however, that this is an invalid claim and further investigation reveals it is more probably due to an interaction between the nitrogen atom and an aromatic hydrogen atom of a molecule in a neighboring column. Similar interactions can be found in crystal structures of *x*-PcH_2.

A year later, a single-crystal structure of β-PcH_2 (PHTHCY02, App. 10:23) was published by Zugenmaier *et al.*[102] The molecular and crystal structure of β-PcH_2 is isomorphous with those from earlier reports. From research on powders and thin films, Brinkmann *et al.* concluded that PcLi exists as three polymorphs.[105,106] In 1998, Ballirano *et al.* calculated the crystal structures of two polymorphs of

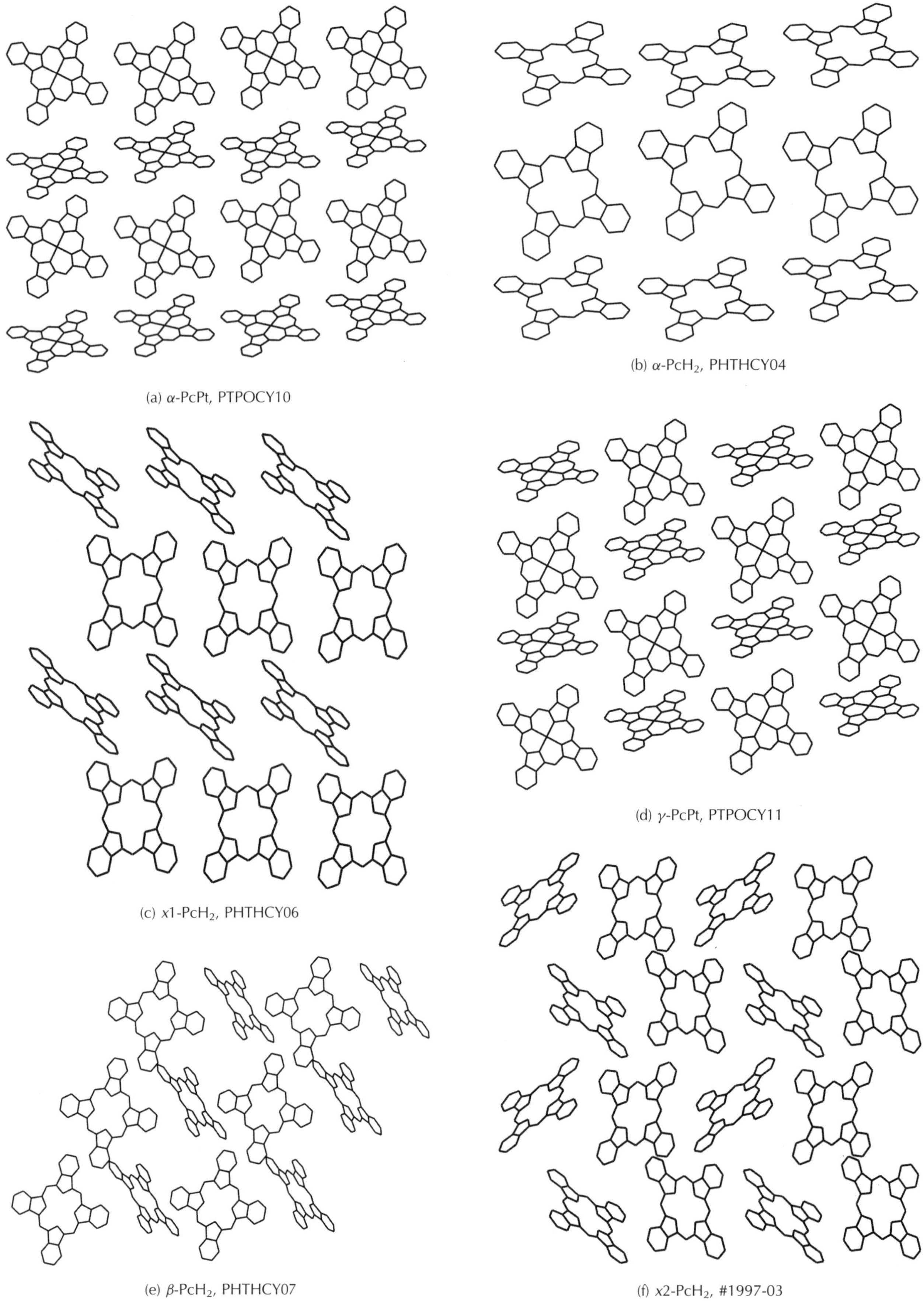

(a) α-PcPt, PTPOCY10

(b) α-PcH_2, PHTHCY04

(c) $x1$-PcH_2, PHTHCY06

(d) γ-PcPt, PTPOCY11

(e) β-PcH_2, PHTHCY07

(f) $x2$-PcH_2, #1997-03

Figure 9. Columnar arrangement of neutral CN4 complexes; (a–f) view along an axis tilted by 30° from the stacking axis to give a 3D impression of the arrangement, (g–i) view along the stacking axis.

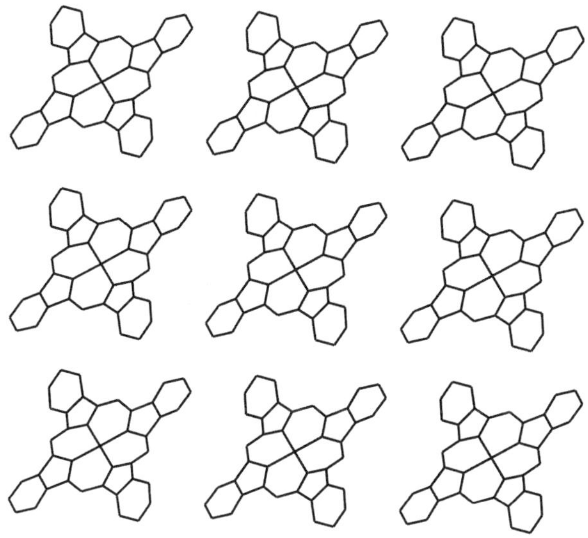

(g) α-PcCo, COPTCY07

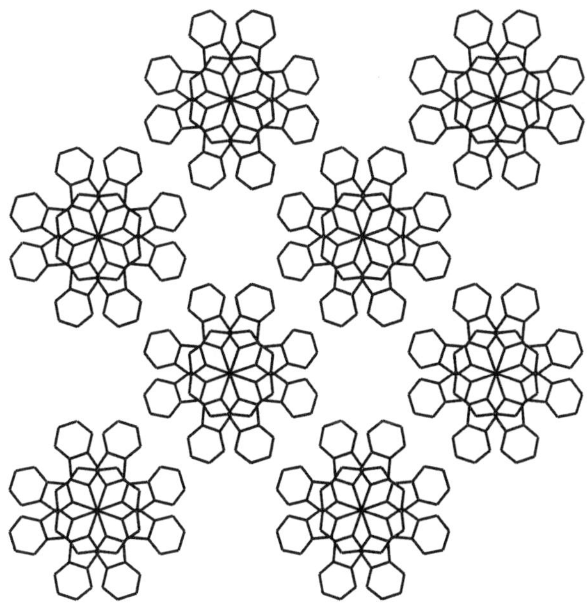

(h) m-PcLi, DIXNIH02

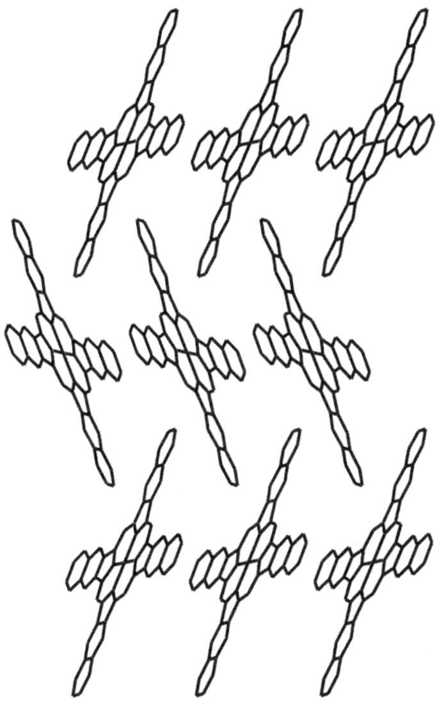

(i) 2,3-NcCu, RADPIV

Figure 9. Continued.

PcCo (α: COPTCY07, App. 11:1–2, β: COPTCY08, App. 11:5).[107] They found the molecular structures to be severely distorted. It is not known whether such distorted structures exist or if the distortion is due to the limited accuracy of the calculation method.

In 1999, Matsumoto *et al.* reported a single-crystal structure of β-PcH_2 (PHTHCY07, App. 10:22), which they found to be isomorphous with previously reported β-PcH_2 structures.[29] However, this time, the two internal hydrogen atoms were localized. A more precise structural analysis shows that the hydrogen atoms are located on the slightly more deformed molecular axis, as predicted by Engel in 1997.[23] The differences between the two axes are small. In 2000, Janczak published a further structure of α-PcH_2 (PHTHCY08,[98] App. 10:17). In contrast to his first "α-PcH_2" investigation in 1992,[97] this time, he was able to locate the inner hydrogen atoms as "half-hydrogen" atoms. Sievertsen and Homborg recently performed structural analysis of PcCr, and determined the crystal to be β-polymorphic (#1999-01, App. 11:13–14).[63,108] Their structure is identical with other β-polymorphic metallophthalocyanines reported in the literature.

b. Solid-State Arrangement

CN1 and CN4 complexes are observed to adopt two arrangement patterns. In the first case, the complexes stack one over the other, i.e., the normal to the molecular plane corresponds to the stacking axis. Examples include most of the cationic complexes and the monoclinic structure of PcLi, DIXNIH02[91] as shown in Figure 12a,b. A second type of stacking is known as slipped-stacking, where in projections perpendicular to the molecule, every molecule is offset with regards to its neighboring molecule, as shown in Figure 15a,b. In this kind of arrangement, the molecular planes are angled towards the stacking axis.

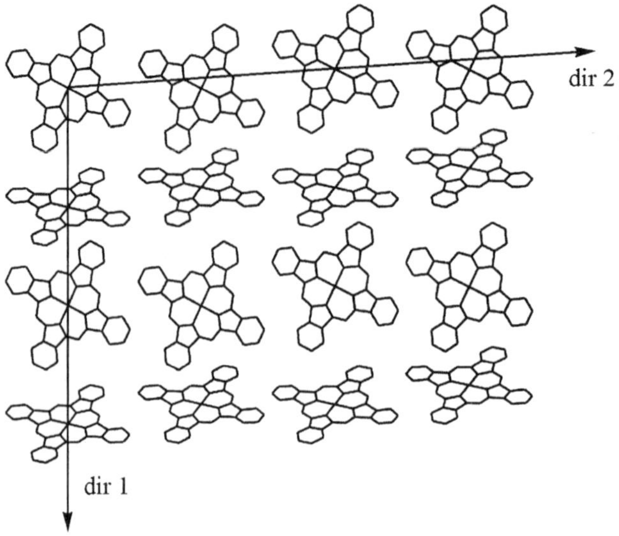

(a) α-PcPt, PTPOCY10

(c) View along dir2

(e) View along dir1

Figure 10. Structural comparison of a phthalocyanine α-polymorph (α-PcPt, PTPOCY10[77]) and a β-polymorph (β-PcH_2, PHTHCY07[29,30]); (a) and (b) are views along an axis tilted by 30° from the stacking axis, (c) and (d) are views perpendicular to the stacking axis along direction 2, and (e) and (f) along direction 1, respectively. The pattern shown in (c) and (d) are usually referred to as fish-bone pattern.

In general, the three-dimensional direction of the stacking axis towards the molecule is one of distinctive features of the different crystal structures. Other features include the distances between neighboring stacks, and the orientation of the molecules within these stacks towards each other. However, such an extensive comparison is beyond the boundaries of this review which will therefore focus only on a pictorial comparison of the main polymorphic structures.

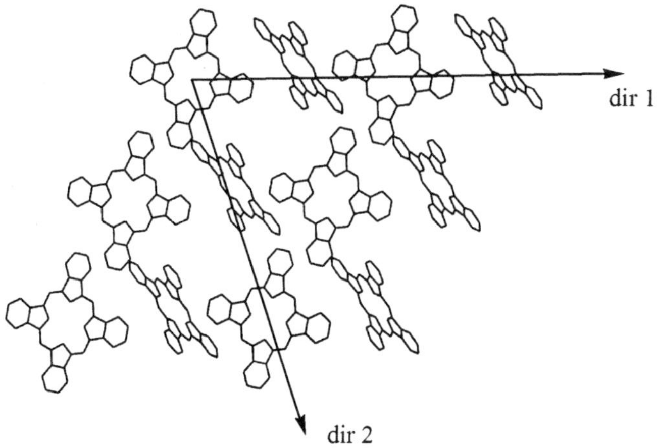

(b) β-PcH_2, PHTHCY07

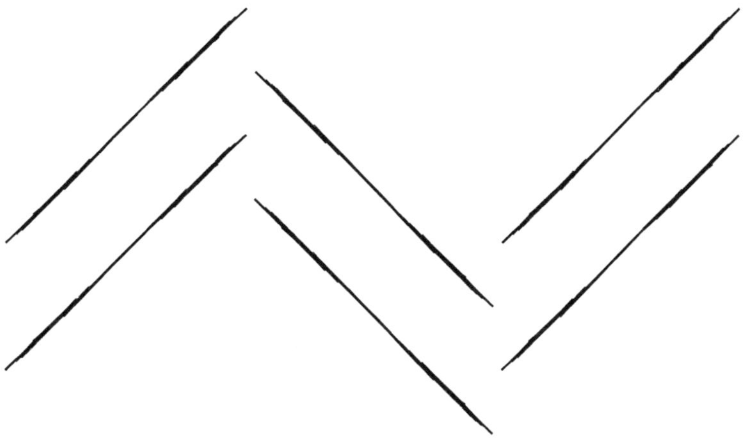

(d) View along dir2

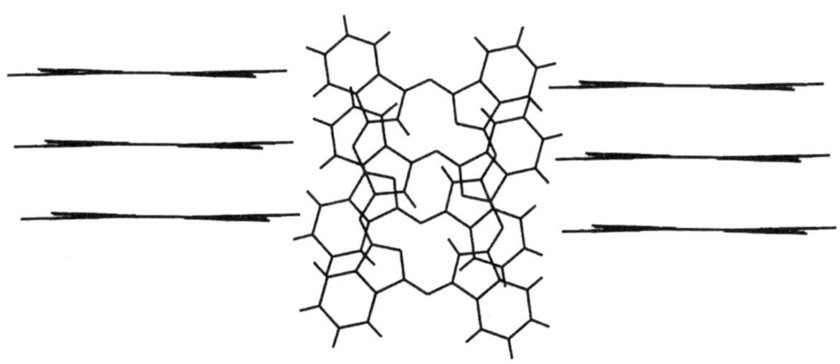

(f) View along dir1

Figure 10. Continued.

A comparison of CN4 crystal structures was first published by Honigmann *et al.* in 1965.[74] They divided the structures into two groups, the α-group, where the distance between neighboring molecules in a stack is approximately 3.8 Å, and the β-group, where the same distance is 4.7 Å. This distinction is essentially still valid, but new types of structures (x-PcH_2, PcLi), which do not fall into either category, have since been found. Before looking at these structures in detail, it is necessary to clarify some

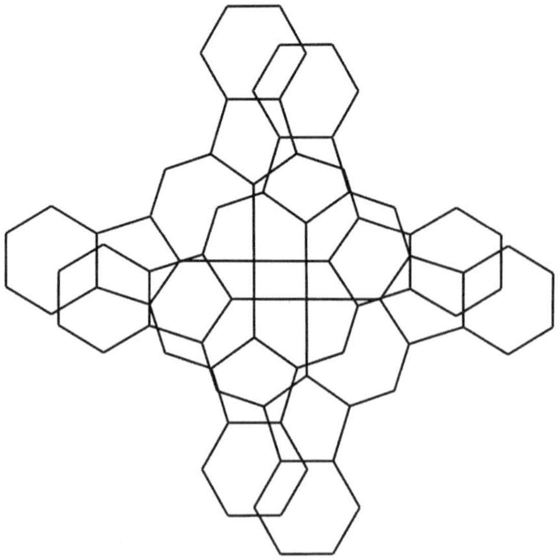

(a) α-PcPt, PTPOCY10

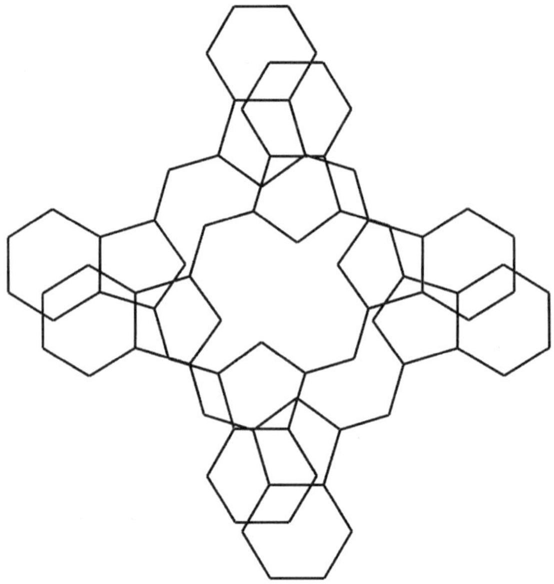

(b) α-PcH_2, PHTHCY04

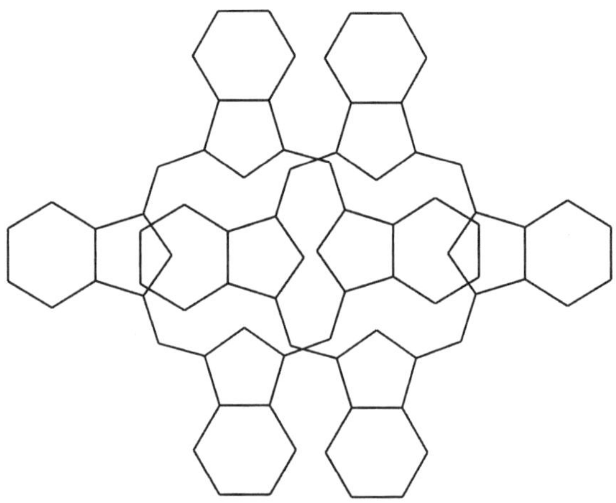

(c) x1-PcH_2, PHTHCY06

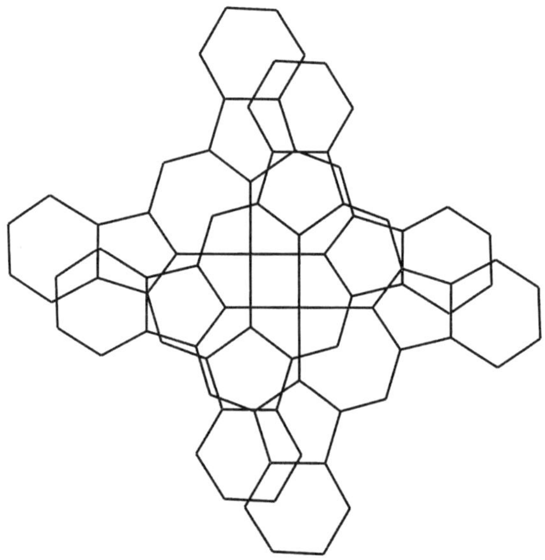

(d) γ-PcPt, PTPOCY11

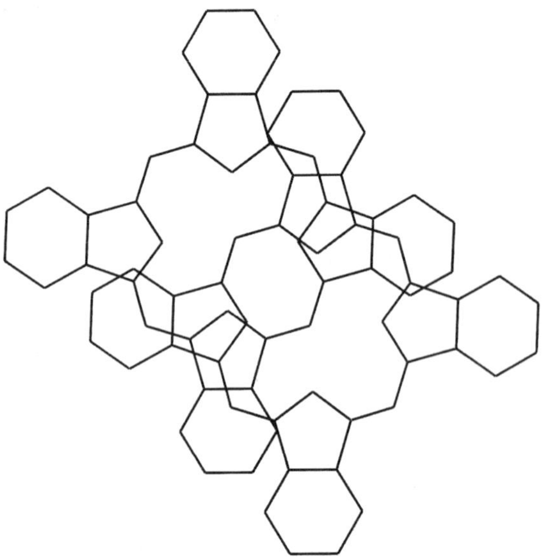

(e) β-PcH_2, PHTHCY07

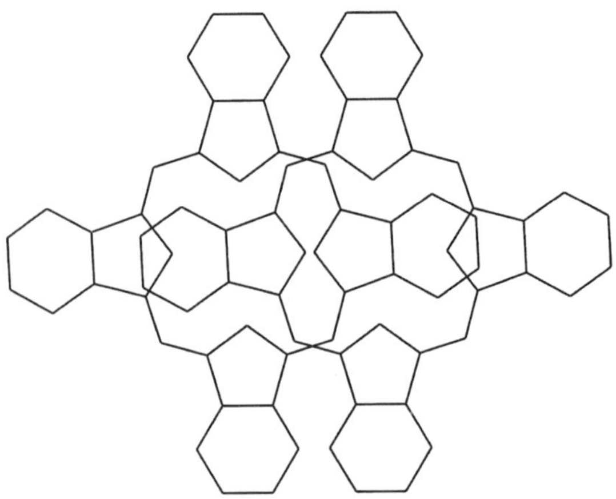

(f) x2-PcH_2, #1997-03

Figure 11. Overlapping geometries of two adjacent phthalocyanines of the columns shown in Figure 9.

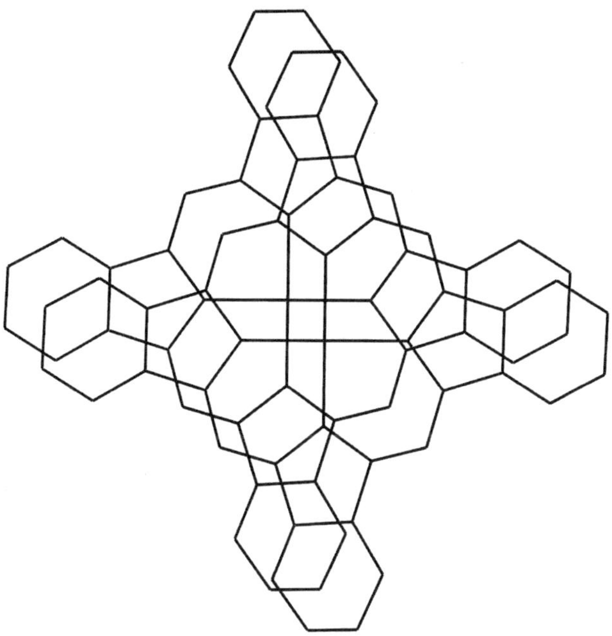

(g) α-PcCo, COPTCY07

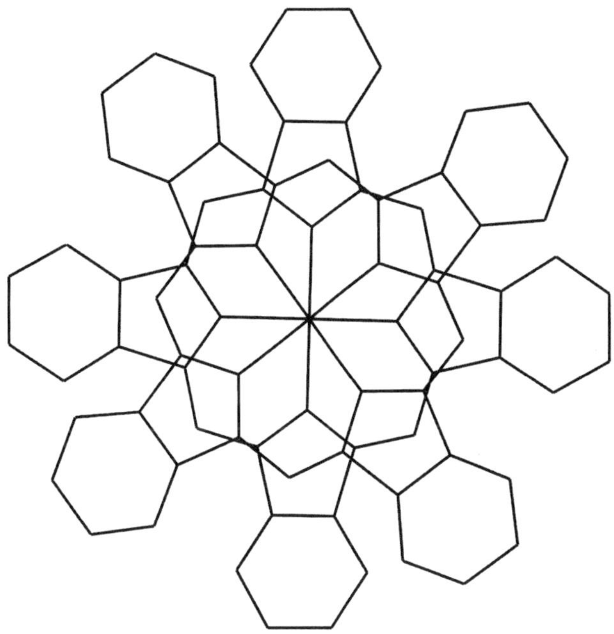

(h) m-PcLi, DIXNIH02

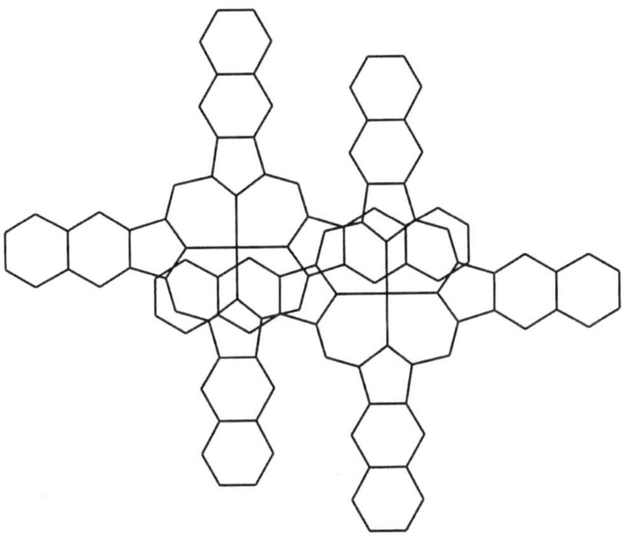

(i) 2,3-NcCu, RADPIV

Figure 11. Continued.

concepts: a *column* or a *stack* identifies a structural arrangement where the molecules are one-dimensionally arranged and in a view perpendicular to the macrocycle neighboring molecules overlap; a *band* represents an arrangement where the normal to the macrocycle is (approximately) perpendicular to the band axis and where no overlap can be found; a *sheet* is a two-dimensional structure similar to a column or a stack; and a *layer* is a two-dimensional analog of a band.

One common feature of all neutral CN1 and CN4 structures is the stacking arrangement of the molecules within the crystal. An overview of this arrangement for the different polymorphs is shown in Figure 9. The stacks are each represented by a molecule and are viewed at an angle of about 30° toward the stacking axis to give a three-dimensional impression of the arrangement. For simplicities sake, the structures have been divided into three main groups. In group 1 (Figure 9a–c,i), the common feature is that the spatial orientation of the molecules in neighboring columns along one crystallographic direction (in the drawings: horizontally), is identical, while in the vertical direction, they are at an angle to each other. In group 2, the molecules in neighboring stacks are inclined towards each other in both directions (Figure 9d–f), while in group 3, all molecules are parallel (Figure 9g–h).

A comparison of the arrangement geometries of the molecules in both α-polymorph (PTPOCY10) and β-polymorph (PHTHCY07) forms is shown in Figure 10. The arrangement along direction 1 is usually referred to as a fish-bone structure, where the molecules in neighboring stacks are related by a glide axis running parallel to the stacking axis and being located between the stacks (Figure 10c–d). It should be noted that the β-polymorph is reported to be stabilized by hydrogen bonding between the azamethine nitrogen atoms and aromatic hydrogen atoms along direction 1 in Figure 10b.[96] However, this interaction is also present

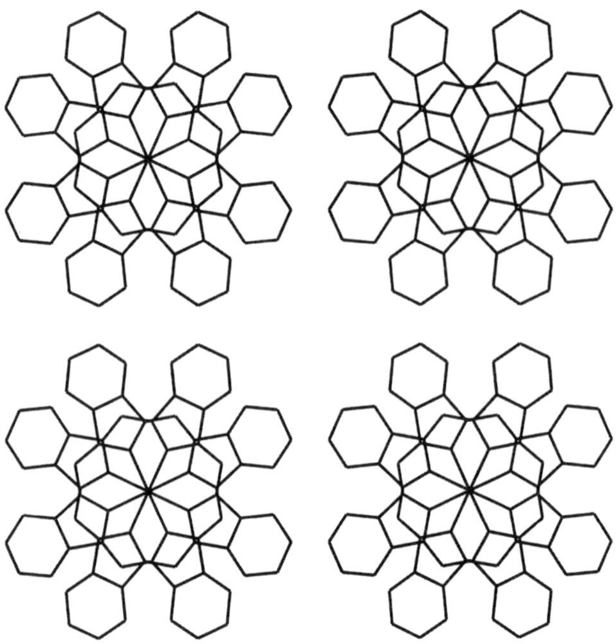

(a)

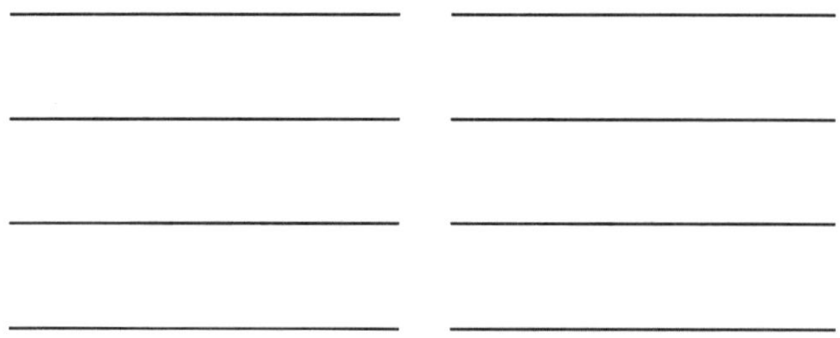

(b)

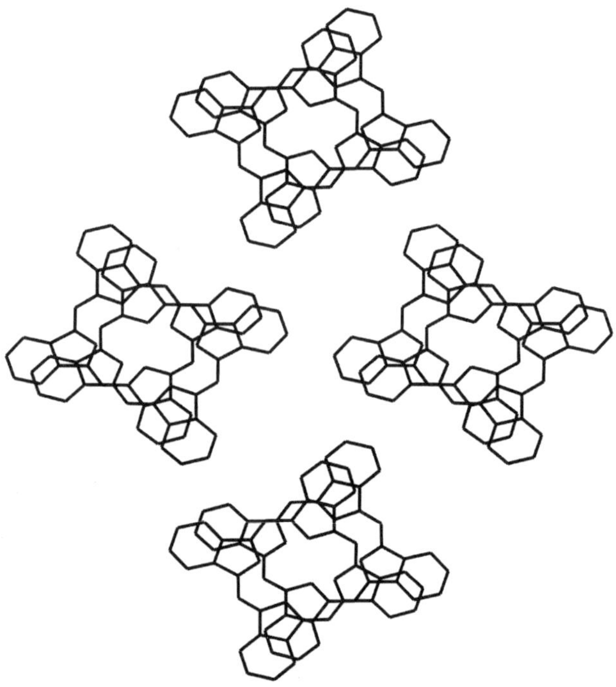

(c)

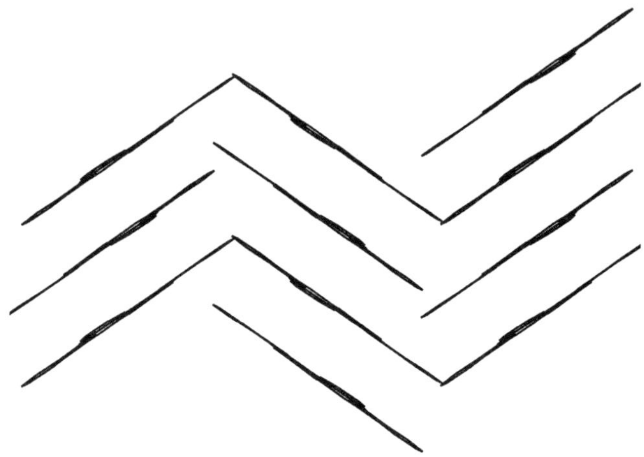

(d)

Figure 12. Arrangement pattern of cationic or radical CN4 complexes; (a) and (b) show perpendicular and side-on views on a phthalocyanine cation with non-integral charges, $[PcNi]^{+0.33}[(I_3)_{0.33}]^{-0.33}$ (NIPHTI[111]) while (c) and (d) show the corresponding views on a cation with integral charges, $[PcH_2]^{+1.0}[IBr_2]^{-1.0}$ (CCDC184514[114]).

in the α-structures; α: PTPOCY10 (N–H = 2.704 Å) and β: PHTHCY07 (N–H = 2.640 Å). An interaction that is present exclusively in β-polymorphs, is the interaction between the central metal and one of the bridging nitrogen atoms of each adjacent molecule, as shown in Figure 7. It is also possible that the β-polymorph of the metal-free phthalocyanine is similarly stabilized by an interaction between the hydrogen atoms in the center and the neighbor nitrogen atoms. The arrangement of the molecules in α-PcH_2 (PHTHCY04) is isostructural to that reported for α-PcPt (PTPOCY10), as is the arrangement of γ-PcPt (PTPOCY11) similar to β-PcH_2 (PHTHCY07), despite the differences in the angle of the molecules towards the stacking axis.

Three detailed reports concerning the crystal structure of x-PcH_2 have been documented in the literature.[99,101,102] While all the reports refer to one "x-polymorph," the drawings in Figure 9c,f show that these structures are actually two different polymorphs. In PHTHCY06, the molecules in the horizontal stacks are parallel, while those in #1997-03 are inclined to each other. Zugenmaier stated that PHTHCY05 is isomorphic with their structure #1997-03, while Janczak claimed that the unit-cell data of PHTHCY05 could be transformed into those of PHTHCY06.[98,102] A comparison of the stacking arrangements of both PHTHCY05 and PHTHCY06 shows that they are isostructural.

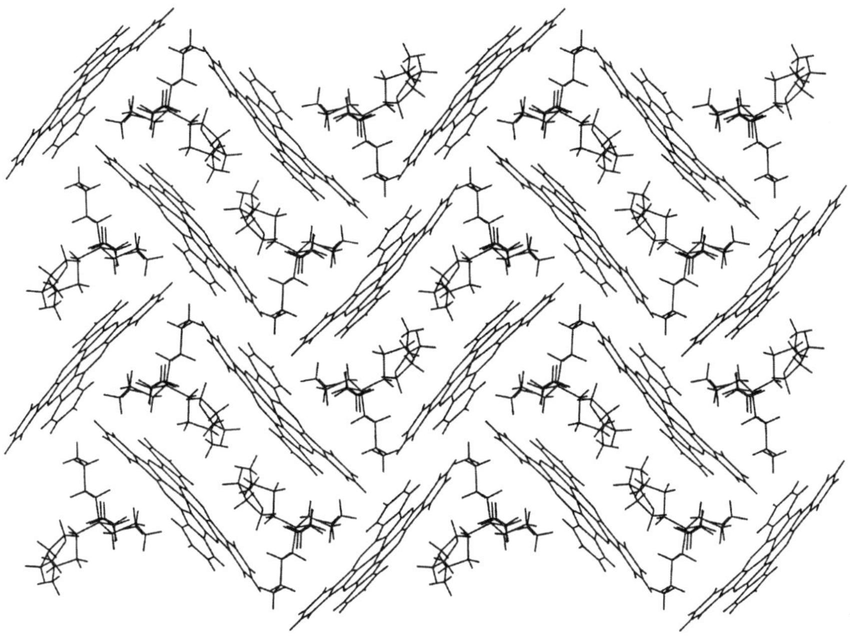

Figure 13. Arrangement pattern of a typical anionic CN4 complex, [PcLi][NBu_4][THF] (RESYAP[115]).

Table 3. Structural Description of Cationic CN1 and CN4 Complexes

Compound	Code [Ref.]	Figure	Structural description
PcLi	DIXNIH02[91]	11:26	Staggered by 38.7°, (Ni–Ni) 3.245 Å
[PcH_2][I_3]$_{1/3}$	DIJYEA[272]	–	Staggered by 40.0°, (center–center) 3.251 Å
[PcH_2][I_3]$_{1/3}$	LATLIB[273]	10:28	Isomorphous with [PcNi][I_3]$_{1/3}$
[$TBPH_2$][I_3]$_{1/3}$	?[454]	–	Isomorphous with [TBPNi][I_3]$_{1/3}$; the replacement of the nickel center with protons causes the coordinating isoindole-nitrogen atoms to move away from the center of the ring. Thus the angles around the nitrogen atoms are larger in the metal-free compound compared with the nickel compound. In the former compound the hole is expanded
[TBPCo][I_3]$_{1/3}$	?	?	?
[PcCo][I_3]$_{1/3}$	DIDTEP[112]	11:9	Staggered by 40°, (Co–Co) 3.123 Å
[PcCu][I_3]$_{1/3}$	FEGJOQ[113]	11:17	Staggered by 39.2°
[TATBPCu][I_3]$_{1/3}$	?[455]	–	Staggered by 39.6°
[PcFe][I_3]$_{1/3}$	DAPCEC[157]	11:22	–
[PcLi][I_3]$_{1/3}$	DIXMUS10[290]	11:36	Isomorphous with [PcNi][I_3]$_{1/3}$
[PcNi][I_3]$_{1/3}$	NIPHTI[111]	11:64	Staggered by 39.5°, (Ni–Ni) 3.244 Å; columns of PcNi are surrounded by four parallel tunnels containing symmetric polyiodide molecules, I_3^-; iodine chains are disordered with the thermal parameters along the chain direction twice that perpendicular to the chain, supercells of iodine chains
TBPNiI	?[456]		Staggered by 41°, (Ni–Ni) 3.217 Å
[PcNi][Br_3]$_{1/3}$	DOXCOI[300]	11:63	Staggered by 39.1°, (Ni–Ni) 3.24 Å
[PcH_2][IBr_2]	CCDC184514[114]	10:34	Columns of two crystallographically different molecules along [100]; column axes form hexagonal network; orientation of column parallel along [001], different orientation along [010]; column consists of slipped-stacked dimers, dimers overlap along N_i molecular axis
[PcH_2][IBr_2][Br][1-BrN]	CCDC184515[114]	10:35	Columns along [100], parallel but shifted; in column two different molecules overlap along N_{br} molecular axis
[PcCo][AsF_6]$_{0.5}$	VIYFEO[278]	11:8	Staggered by 41.7°; isomorphous with other [PcM][I_3]$_{1/3}$; AsF_6 anions are located in channels parallel to the PcCo columns, slightly displaced ($z = 0.095$ Å) from the PcCo layers. The site occupancy is 0.25, so that anions displaced from the mirror planes at $z = 0$ and $z = 0.5$ are alternately located along the *c* axis without three-dimensional order. In addition, the anions are positionally disordered between the two possible sites ($z = \pm 0.095$) in the one-dimensional channel along the phthalocyanine molecular column. Large temperature factors of the anion indicate that they are rotationally disordered

(*Continued*)

Table 3. Continued

Compound	Code [Ref.]	Figure	Structural description
$[PcCo_{0.25}Ni_{0.75}][AsF_6]_{0.5}$	CCDC156503[280]	11:11	Isomorphous to $[PcNi][AsF_6]_{0.5}$
$[PcCo_{0.45}Ni_{0.55}][AsF_6]_{0.5}$	CCDC156502[280]	11:12	Isomorphous to $[PcCo][AsF_6]_{0.5}$
$[PcNi][AsF_6]_{0.5}$	FIYSEL[274,292–294]		Spacegroup was first given as Pcc2 but was revised later, see KARDAI
$[PcNi][AsF_6]_{0.5}$	KARDAI[293,295]	11:58	No zigzag deformation of the PcNi stacks (as in $[PcNi][SbF_6]_{0.5}$); AsF_6 is disordered occupying one of the two possible sites alternately along the *c* axis
$[PcPt][AsF_6]_{0.5}$	[301]		?
$[PcNi][SbF_6]_{0.5}$	FANZUP, FANZUP10[292–295]	11:62	(Slightly zigzag) linear columns along the c axis; molecules staggered by 41°, (mol–mol) 3.22 Å; counterions in channels between the stacks; large-sized counterions introduce zigzag deformation of the stack; deviation of the Ni atom from the slide planes are 0.12 Å and 0.18 Å in the directions of *b*- and *a*-axes, respectively; (Ni–Ni) 3.23 Å; SbF_6 is ordered with large temperature factors
$[PcCo][ClO_4]$	?[457]	–	n.a.
$[PcCu]_3[OReO_3]_2$	#2002-01[302]	12:10–11	Sheets parallel (001), slipped-stacked bands parallel [−110]; Cu in the terminal macrocycles of the trimeric units 0.119 Å above the N_4 plane, in-plane in the central macrocycle
$[PcNi][(BF_4)_{0.33}]$	[296–298]		structurally similar to the iodinated metallophthalocyanines; full structure n.a.
$[PcNi][(ClO_4)_{0.42}]$	FUFJOF, FUFJOF01[297,299]		$T > 230$K, the unit-cell is P4/mcc with $a = 13.957$ Å, $c = 6.4672$ Å, and $Z = 2$; molecules are staggered by 39.5°, (Ni–Ni) 3.233 Å; ClO_4 sites are only 42% occupied and the tetrahedral ClO_4 ions appear to be rotationally disordered in such a way that partial oxygen atoms occupy the vertices of a cube with the Cl atom at the center. Also the Cl and O thermal parameters in the c-directions are *ca.* 20 times greater than in transverse directions, suggesting translational disorder along the stacking axis $T < 230$°C: P4/mnc, $a = 19.543$ Å, $c = 6.416$ Å (153K); transformation is due to rotational ordering of the ClO_4 ions
$[PcPt][(ClO_4)_{0.5}]$	JECGAZ[294,301]		Staggered by 40.7°, (Ni–Ni) 3.217 Å; ClO_4 ions are located at interlayer positions. The site occupancy of the ClO_4 ions is 0.5, so that the anions are positionally disordered. Due to the large size of the anions, the authors suggest that they are alternately located along the *c* axis and these ordered one-dimension columns are randomly arranged along the a and b axes. The large temperature factors indicate that the anions are also rotationally disordered.
$[PcH_2]_2[AsF_6]_3[1\text{-}ClN]$	?[274]		Isomorphous to $[TATBPCu]_3[ReO_4]_2[1\text{-}ClN]$ and $[TBPNi]_3[AsF_6]_2[1\text{-}ClN]$
$[PcH_2]_2[AsF_6]_3[1\text{-}ClN]$	CCDC168923[275]	10:31–32	Columns along [100], column consists of trimeric units, in unit staggered overlap, neighbor units overlap slipped-stacked along N_{br} molecular axis
$[TBPH_2][(AsF_6)_{0.67}]$	?[458]	–	Probably similar to $[PcH_2]_2[AsF_6]_3[1\text{-}ClN]$
$[PcCu]_3[AsF_6]_2[1\text{-}ClN]$	CCDC168920[275]	11:18–19	Similar to $[PcH_2]_2[AsF_6]_3[1\text{-}ClN]$
$[TBPNi]_3[AsF_6]_2[1\text{-}ClN]$	?[459]		Deformed one-dimensional stacking with a periodic unit composed of three macrocycles; staggering angle between molecules A and B and between B and C is 45°, while the overlapping between C and A′ (from the neighbor unit) is nearly eclipsed but the molecules are shifted along the N_{br}–N_{br} molecular axis. The columns along c are (pseudo-)hexagonally packed
$[PcNi][AsF_6]_{0.67}$	?[460]		Probably similar to $[PcH_2]_2[AsF_6]_3[1\text{-}ClN]$
$[PcH_2]_3[(SbF_6)_2][1\text{-}ClN]$	CCDC168921[275]	10:33	Similar to $[PcH_2]_2[AsF_6]_3[1\text{-}ClN]$
$[PcCu]_3[(SbF_6)_2][1\text{-}ClN]$	CCDC168919[275]	11:20	Similar to $[PcH_2]_2[AsF_6]_3[1\text{-}ClN]$
$[PcNi]_3[(SbF_6)_2][1\text{-}ClN]$	CCDC168922[275]	11:66	First reported but not realized in[292–295]
$[TATBPCu]_3[ReO_4]_2[1\text{-}ClN]$	[287]	–	Isostructural to $[TBPNi]_3[AsF_6]_2[1\text{-}ClN]$
$[tatbpCu]_3[ReO_4]_2[1\text{-}ClN]$	JECGON[287]		Probably similar to $[PcH_2]_2[AsF_6]_3[1\text{-}ClN]$
$[PcNi]_3[(ReO_4)_2][1\text{-}ClN]$	CCDC168918[275]	11:65	Similar to $[PcH_2]_2[AsF_6]_3[1\text{-}ClN]$

Another basic difference between the polymorphic structures lies in the interaction between adjacent molecules in a stack. From the different overlap geometries shown in Figure 11, it can be seen that all the *α*- and *γ*-structures (Figure 11a,b,d,g) overlap similarly along the N_{br} molecular axis of the phthalocyanine macrocycle. The *β*-polymorphic molecules also overlap along the N_{br} molecular axis, but the degree of overlap is less pronounced (Figure 11e). The *x*1- and *x*2-PcH_2 polymorphic molecules

Table 4. Structural Description of Anionic CN4 Complexes

Compound	Code [Ref.]	Figure	Structural description
[PcCo][NBu_4][Me_2CO]	NOYYIJ[279]	11:10	Bands along [100], honeycomb structure where four bands form a channel for the counterions
[PcFe][K.cntnd.222]	FEDXOB[288]	11:24–25	Bands along [001], honeycomb structure where four bands form a channel for the counterions
[PcFe][Li.cntnd.18c.6] [Li(THF).cntnd.18c6]	FEDXUH[288]	11:23	Bands along [001], honeycomb structure where six bands form a channel for the counterions
[PcLi][F][NBu_4]$_2$[0.5H_2O]	RESYET[115]	11:27–28	Bands along [001], layered structure of alternating counter-ion and phthalocyanines
[PcLi][NBu_4][THF]	RESYAP[115]	11:29–30	Bands along [100], honeycomb structure where four bands form a channel for the counterions
[PcLi][PNP][CH_2Cl_2]	ZEPKIO[289]	11:31–32	Bands along [010], honeycomb structure where four bands form a channel for the counterions
[PcLi][PNP][H_2O]	ZEPKOU[289]	11:33–35	Bands along [100], honeycomb structure where four bands form a channel for the counterions
[PcLi][Li(OPϕ_3)$_4$][EtOH,H_2O]	MAKHEL[136]	11:37–38	Bands along [001], honeycomb structure where four bands form a channel for the counterions

(Figure 11c,f), in contrast, overlap along the N_i molecular axis. The overlapping molecules in PcLi (Figure 11h) on the other hand, are not offset, but rather staggered, with an angle of rotation of 38.7°. 2,3-NcM molecules overlap along a direction intermediate to the N_i- and N_{br} molecular axes (Figure 11i). The structures of 2,3-NcM, M=Cu (RADPIV, App. 19:1), M=Ni (RADPER, App. 19:2–3), and M=Zn (RADPOB, App. 19:4), are basically similar to the *x*1-polymorph of metal-free phthalocyanine. The stacking angle of 62° in the columns of tetrabenzo-monoazaporphyrazine, BZAZPN, molecules leads to not-overlapping macrocycles.

2. Ionic Phthalocyanines

a. Cationic Phthalocyanines

In the 1970s, it was discovered that coordination compounds become electrically conductive, if the metal-ligand molecules are arrayed in close communication, occupy crystallographically similar environments, and if the metal-ligand complex adopts a nonintegral oxidation state.[109,110] In the search for organic room-temperature superconductors, various metal-ligand systems, e.g., metal phthalocyanines, were investigated. This renewed interest in phthalocyanine complexes lead to the discovery of a variety of new phthalocyanine crystal structures.[111–113]

While unsubstituted, neutral phthalocyanines crystallize in columns of slipped-stacked molecules,[23] the radical, cationic and anionic phthalocyanine complexes usually do not.[111] In the case of most cationic and radical CN1 or CN4 complexes, the phthalocyanine molecules stack linearly, such that the normal to the phthalocyanine ring corresponds to the stacking axis. Variations do occur due to the type of counterion, its location with respect to the phthalocyanine, and its temperature-dependant order. Details about the cationic and radical phthalocyanine complexes have been compiled in Table 3 and typical solid-state arrangements are shown in Figure 12. Recently however, structures of the first slipped-stacked, cationic phthalocyanine complexes have been published, [PcH_2][IBr_2] (CCDC184514, App. 10:34) and [PcH_2][IBr_2][Br][1-BrN] (CCDC184515, App. 10:35), respectively.[114] While the overlap between neighboring phthalocyanines in CCDC184515 is uniform along the N_{br} molecular axis, the molecules in CCDC184514 alternately overlap along the N_{br} and N_i molecular axes.

b. Anionic Phthalocyanines

In the case of anionic complexes, phthalocyanine macrocycles do not overlap. All known structures are very similar, consisting of a honeycomb structure with a phthalocyanine support and counterions and cocrystallizing solvents occupying the vacant holes. Details about the anionic phthalocyanine complexes are compiled in Table 4 and a typical arrangement is shown in Figure 13.[115]

B. CN6*tr* COMPLEXES

1. Neutral Phthalocyanines

a. Introduction and Molecular Structures

The periodic table of CN6*tr* complexes with the number of structures given next to each element is

Table 5. Molecular Deformation of the Macrocycles in the Crystal Structures of Si-, Ge-, Sn Phthalocyanines

Axial ligand	Si	Ge	Sn
Chlorine	Planar (VIKVEQ, CCDC173621)	Unknown	Deformed (#1972-01)
Bromine	Unknown	Unknown	Unknown
Iodine	Unknown	Deformed (RESFIE, GODLEQ)	Deformed (RESFOK, RESFOK01)
Hydroxy	Planar (QEKBEN, QEKBEN01)	Planar (#1985-01, #1985-02)	Unknown

Table 6. CN6*tr* Complexes with Halogen Axial Ligands

Group	Subgroup	Members	Structure code	Reference
Halogens	1	$PcCoCl_2$	KEHXEA	123
	1	$PcCrCl_2$	KEHWUB	123
	1	$PcFeCl_2$	KEHXAW	123
	1	$PcSiCl_2$	VIKVEQ	127
	1	$PcSiCl_2$	CCDC173621	128
	1	PcMoOCl	#1990-03	124
	1	PcWOCl	#1990-02	124
	2	$PcSnI_2$	RESFOK	119
	2	$PcGeI_2$	RESFIE	119
	3	$PcSnI_2$	RESFOK01	118
	3	$PcSnCl_2$	CPHCYT	116
			#1972-01	117
	4	PcGaF	BAMSEN,	121
			BAMSEN10	122

Table 7. CN6*tr* Complexes with Hydroxy Axial Ligands

Group	Subgroup	Members	Structure code	Reference
Hydroxy	1	α-$PcGe(OH)_2$	#1985-01	120
	2	β-$PcGe(OH)_2$	#1985-02	120
	2	II-$PcSi(OH)_2$	QEKBEN	126
	3	γ-$PcSi(OH)_2$	n.a.	394
	3	γ-$PcGe(OH)_2$	n.a.	125
	3	I-$PcSi(OH)_2$	QEKBEN01	126

Table 8. Hydrogen Bonding Distances in CN6*tr* Crystal Structures

Complex	Intracolumn/Å < 3.5Å	Intercolumn/Å < 3.5Å
$PcSiCl_2$ (CCDC173621)	2.883, 3.106	2.836, 2.914
$PcSnI_2$ (RESFOK)	3.122	3.146
$PcSnI_2$ (RESFOK01)	3.070, 3.228	3.149
α-$PcGe(OH)_2$ (#1985-01)	2.621, 2.706; O-H...N 2.980	2.568
β-$PcGe(OH)_2$ (#1985-02)	2.712, 2.721; O-H...N 3.044	2.592, 2.801

shown in Figure 14. A compilation of the basic crystallographic data of the structures is given in Appendix 1 and their molecular structures are shown in Appendix 15.

The molecular structure of a typical CN6*tr* complex is similar to that of a CN4 complex, consisting of a planar phthalocyanine with an atom occupying the central cavity. In addition, the central atom coordinates two axial ligands located on opposite sides of the macrocycle (Figure 5c). For a better structural comparison, it is necessary to divide the CN6*tr* complexes into two groups, (i) where the central atom coordinates

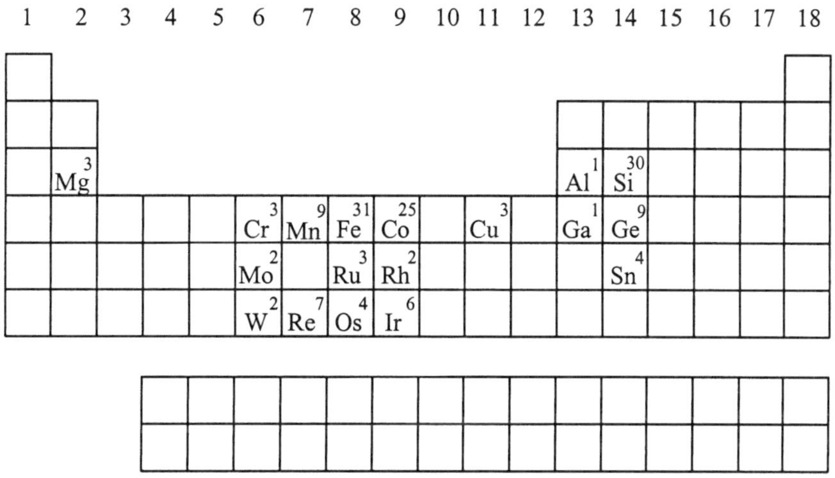

Figure 14. Periodic table of CN6*tr* complexes (Figure 5c,o) with number of structures given next to each element.

small axial ligands, e.g., halogens or hydroxyl groups and (ii), where the central atom carries either larger axial ligands, or where the phthalocyanine component is found to cocrystallize with counterions. In the following section, the main focus lies on the compounds in the first group.

The first CN6*tr* phthalocyanine structure containing a central metal atom bound by two small axial ligands, $PcSnCl_2$ (CPHCYT), was published in 1971.[116] The authors neglected to publish the atomic coordinates until a year later when they were subsequently published in the doctoral thesis of one of the authors (#1972-01,[117] App. 15:122–123), the atomic coordinates are contained in Appendix 35. It is said that the Sn(IV) ion is too large to fit within the central cavity of the phthalocyanine macrocycle.[116] To accommodate such a large ion, the phthalocyanine macrocycle would need to rearrange itself by forcing the isoindole nitrogen atoms further away from the center, while simultaneously rotating the opposing isoindole units around their C_α–C_α-axes (alternately up or down) out of the plane. After analyzing the arrangement of $PcSnCl_2$ molecules in Osborn's structure, Engel remarked[23] that the deformation could also be a crystal structure effect, facilitated or even induced by hydrogen bonding between the aromatic hydrogen atoms of the benzo rings and the chlorine atoms of neighboring molecules. More recent investigations show that the same distortion pattern can be found in the isomorphous structure of $PcSnI_2$ (RESFOK01,[118] App. 15:125, Figures 16e, 17c, 18d, and 19c). Differently deformed macrocycles have been found in a polymorphic structure of $PcSnI_2$ (RESFOK,[119] App. 15:124, Figures 16d, 17b, 18c, and 19b) and in $PcGeI_2$ (RESFIE,[119] App. 15:57) while planar macrocycles with no deformation have been found in α-$PcGe(OH)_2$ (#1985-01, App. 15:55, Figures 16f, 18e, and 19d) and β-$PcGe(OH)_2$ (#1985-02, App. 15:56, Figures 16g, 17d, 18f, and 19e), published by Krämer in 1985.[120] The atomic coordinates of the $PcGe(OH)_2$ structures can be found in Appendices 29 and 30, respectively. Table 5 gives an overview of the presence or absence of macrocycle deformations and Figures 17 and 19 show the deformation in relation to the crystal environment of selected Si, Ge, or Sn phthalocyanines. The presence of non-planar and planar macrocycles in Ge phthalocyanine complexes indicates that the crystal environment might indeed play a significant role in the deformation of the macrocycle. However, this assumption can only be validated when further crystal structures of, e.g., $PcSn(OH)_2$, $PcGeCl_2$, or $PcSiI_2$ become known.

In 1981 and 1985, the crystal structure of $[PcGaF]_n$ (BAMSEN,[121] BAMSEN10,[122] App. 15:54, Figure 20) was published. This structure contains an axial fluoro ligand μ-coordinated to two gallium phthalocyanines, forming a one-dimensional stack. The molecular structure is slightly deformed. In 1990, Moubaraki *et al.* published three isomorphic structures of $PcCrCl_2$ (KEHWUB, App. 15:24), $PcFeCl_2$ (KEHXAW, App. 15:38), and $PcCoCl_2$ (KEHXEA, App. 15:10).[123] In the same year, Gingl published the crystal structures of PcMoOCl (#1990-03, App. 15:78) and PcWOCl (#1990-02, App. 15:127) in his doctoral thesis.[124] The atomic coordinates of PcMoOCl and PcWOCl are contained in Appendices 32 and 36, respectively. In both structures, the central atom is statistically located in two possible locations outside

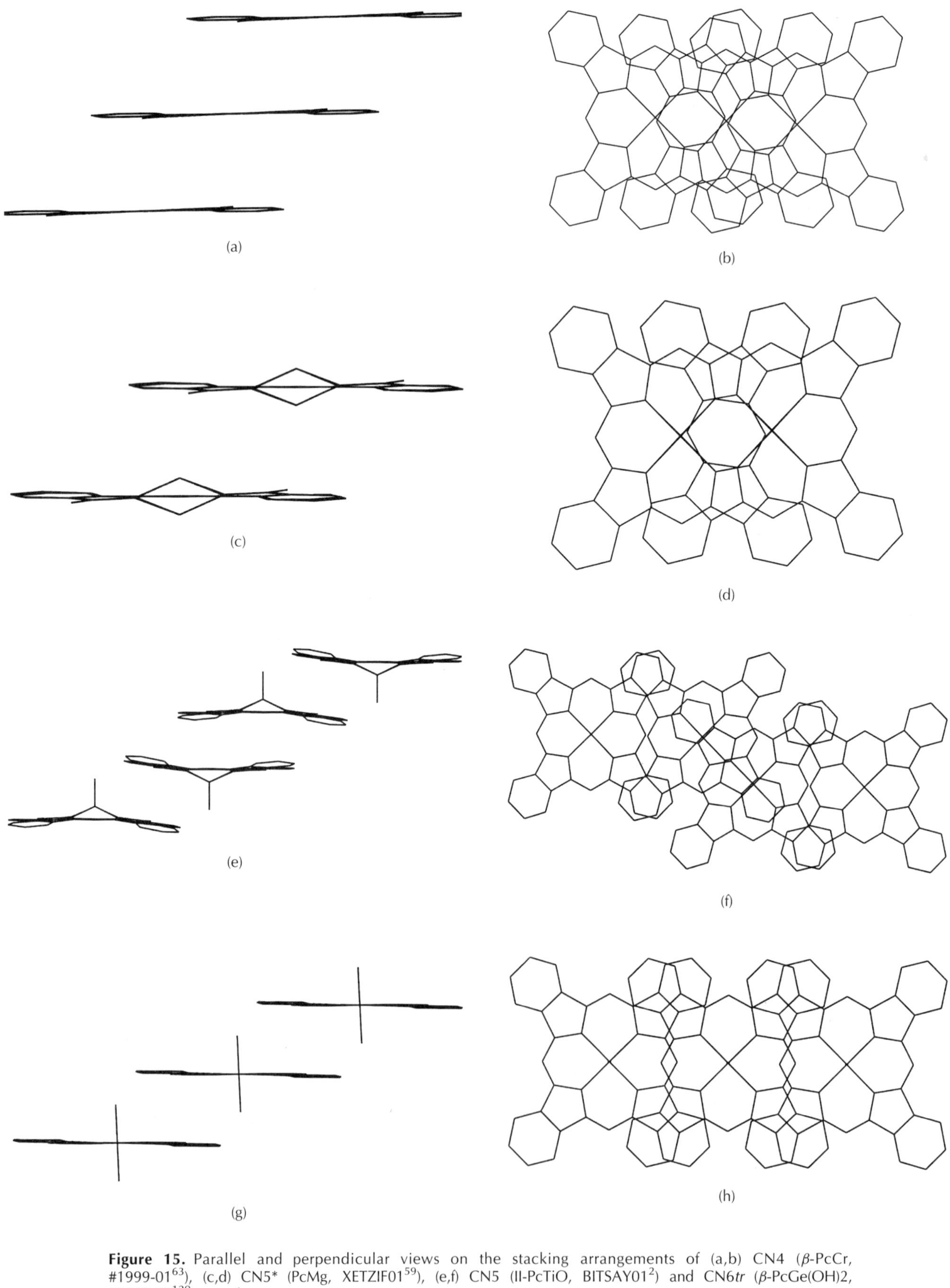

(a) (b) (c) (d) (e) (f) (g) (h)

Figure 15. Parallel and perpendicular views on the stacking arrangements of (a,b) CN4 (β-PcCr, #1999-01[63]), (c,d) CN5* (PcMg, XETZIF01[59]), (e,f) CN5 (II-PcTiO, BITSAY01[2]) and CN6*tr* (β-PcGe(OH)2, #1985-02[120]) complexes.

the macrocycle plane. These locations are on opposite sides of the phthalocyanine macrocycle, therefore, the central atom in average is occupying a position within it. Another structure where such a statistical distribution occurs is PcMg (XETZIF,[59,60] see Section III). The Gingl structures are isomorphic with $PcCrCl_2$, $PcFeCl_2$, and $PcCoCl_2$. Molecular structures of CN6*tr* complexes show only minor differences in bond lengths and bond angles,[23] and will therefore not be analyzed in greater detail. Generally, the main differences lie in the deformation pattern of the macrocycles and in the arrangement of the molecules in space.

Triclinic structures of an isomorphic series of $PcSnI_2$ (RESFOK, App. 15:124), and $PcGeI_2$ (RESFIE, App. 15:57) structures were reported in 1997.[119] In the same year, Kobayashi *et al.* investigated the structures of germanium and silicon phthalocyanine thin films by electron microscopy, and found triclinic polymorphs for both $PcGe(OH)_2$ and $PcSi(OH)_2$.[125] In 1999, Kojima *et al.* calculated the structures of two of the polymorphs of $PcSi(OH)_2$ using information obtained from powder diffraction pattern.[126] While one of the polymorphs, II-$PcSi(OH)_2$ (QEKBEN, App. 15:105), seems to be isomorphous with the structure found by Kobayashi *et al.*,[125] the other, I-$PcSi(OH)_2$ (QEKBEN01, App. 15:104), is isomorphous with the β-polymorph of $PcGe(OH)_2$.[120] Also in 1999, Janczak and Kubiak discovered a new polymorph of $PcSnI_2$ (RESFOK01, App. 15:125), which is isomorphous with the structure of $PcSnCl_2$.[118] In 2000, Kojima *et al.* published the structure of $PcSiCl_2$ (VIKVEQ, App. 15:116), which they obtained from micrometer-sized single-crystals.[127] The structure is isomorphous with the structures of $PcCrCl_2$, $PcFeCl_2$, and $PcCoCl_2$.[123] A year later, Silver *et al.* published their own structural analysis of $PcSiCl_2$ (CCDC173621, App. 15:117) reporting analogous results.[128]

b. Solid-State Arrangement

As shown in Figure 15a,b, CN4 complexes form slipped-stacked columns, in which neighboring molecules overlap strongly in direction along the columns. CN6*tr* complexes can also form slipped-stack columns, however, the presence of axial ligands prevents the complexes from stacking too closely, therefore reducing the degree of overlap (see Figure 15g,h). This different stacking behavior also results in smaller stacking angles of the molecules towards the stacking axes; while α-PcH_2 has an angle of *ca.* 65° and β-PcH_2 *ca.* 45°, those of α- and β-$PcGe(OH)_2$ are *ca.* 28° and the $PcMHal_2$ have angles between 23° and 19°. One can confer that the larger the axial ligand, the greater influence it has on the stacking arrangement, until further interaction is precluded, and overlap between the phthalocyanine moieties is prevented.

In the following section, the basic arrangement pattern for those CN6*tr* complexes with small axial ligands will be discussed. Analysis of the structural data of CN6*tr* complexes in Appendix 1 resulted in the identification of several groups with either halogen axial ligands (Table 6) or hydroxyl axial ligands (Table 7). One of the main differences between these groups is the interaction between their adjacent stacking molecules in a column, as shown in Figure 16. In such arrangements, strong hydrogen bonding exists between the "aromatic" hydrogen atoms and the axial ligands of the adjacent molecule. In addition, repulsive forces push the axial ligands away from the normal to the macrocycle plane (e.g., see Figure 17). Depending on the number of hydrogen bonds, the molecular arrangement is either unsymmetrical, as seen in subgroup 1 of the halogen group, Figure 16a–c, or symmetrical as in the other groups (Figure 16d–h). In the former case, each molecule on either side possesses a strong hydrogen bond, whereas in the symmetrical arrangement, two strong hydrogen bonds exist on each side. Where the axial ligand is a hydroxyl, an additional interaction between the hydrogen atom of this group and the neighboring azamethine nitrogen atom is expected, on account of the distance (*ca.* 3.0 Å) between the oxygen and the azamethine nitrogen atom. Hydrogen bonding distances for selected CN6*tr* complexes with small axial ligands have been compiled in Table 8. A second notable difference between the various CN6*tr* structures is the deformation of their macrocycles within the stacks (Figure 17). While those phthalocyanine macrocycles which have axial hydroxyl groups (e.g., Figure 17d), or are members of subgroup 1 of the halogen group (Figure 17a), are essentially planar, those in group 2–4 are not (Figure 17b–c).

In Figure 18, views along the stacking axes of these CN6*tr* complexes are shown. These views best describe the molecular arrangement of these complexes within the crystal. It can be seen that the spatial orientation of the molecules in the vertical columns is identical, while those drawn horizontally have a different orientation. The exception to this rule is $PcSnI_2$ (RESFOK) where the spatial orientation of all molecules is identical, similar to the case of α-PcCo (COPTCY07, Figure 9g). The 3D arrangement of the

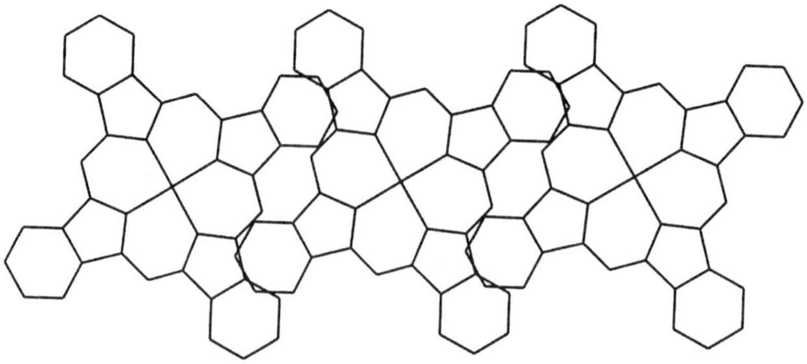

(a) $PcSiCl_2$, CCDC173621

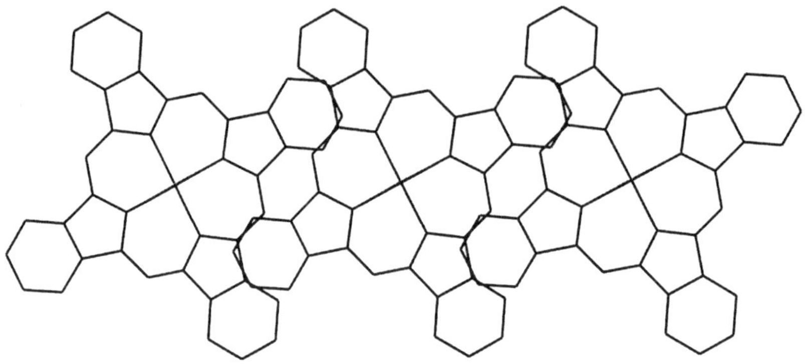

(b) PcMoOCl, #1990-03

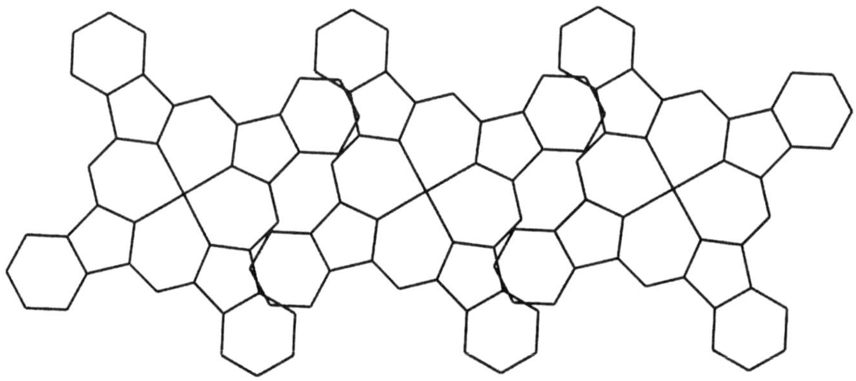

(c) $PcCoCl_2$, KEHXEA

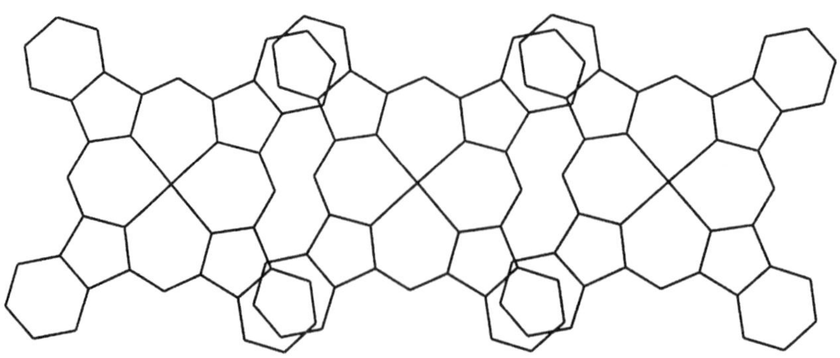

(d) $PcSnI_2$, RESFOK

Figure 16. Overlapping geometries of slipped-stacked, adjacent phthalocyanines in columnar stacks of CN6*tr* complexes.

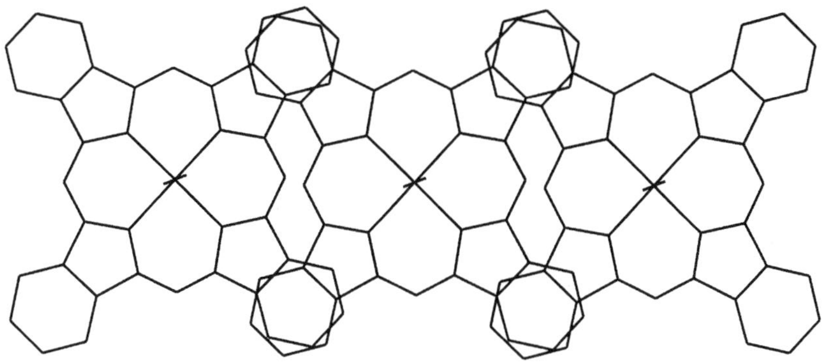

(e) $PcSnI_2$, RESFOK01

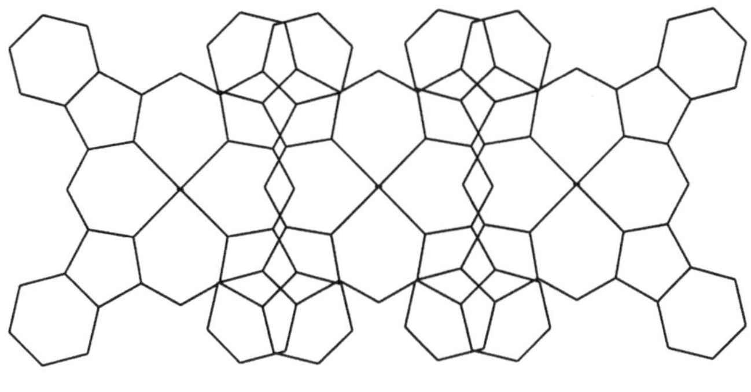

(f) α-$PcGe(OH)_2$, #1985-01

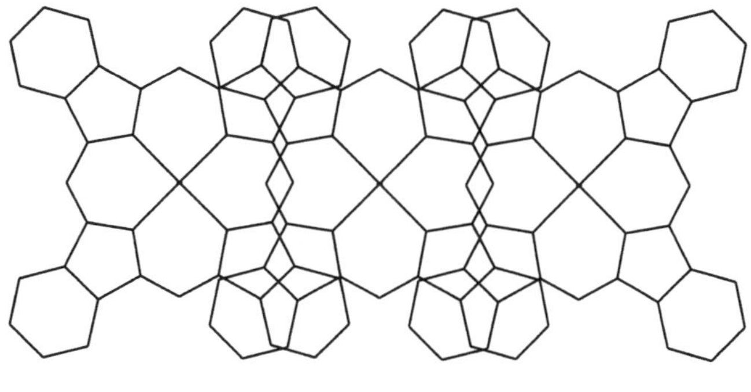

(g) β-$PcGe(OH)_2$, #1985-02

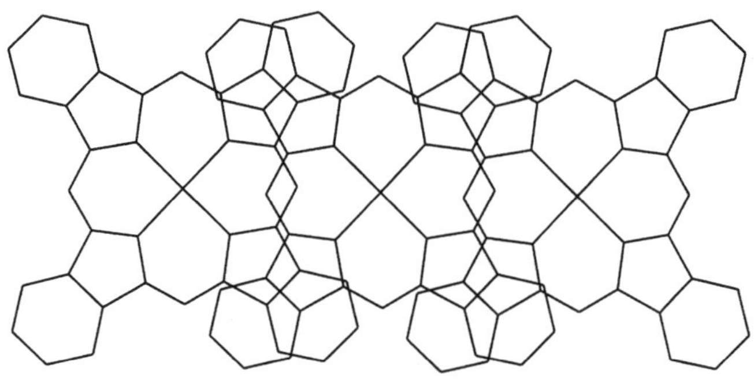

(h) I-$PcSi(OH)_2$, QEKBEN01

Figure 16. Continued.

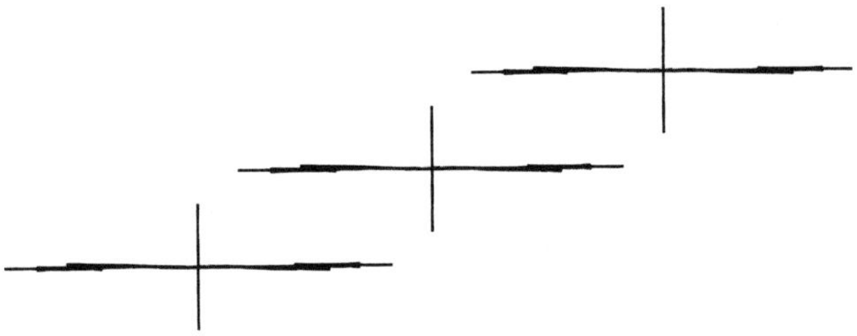

(a) $PcSiCl_2$, CCDC173621

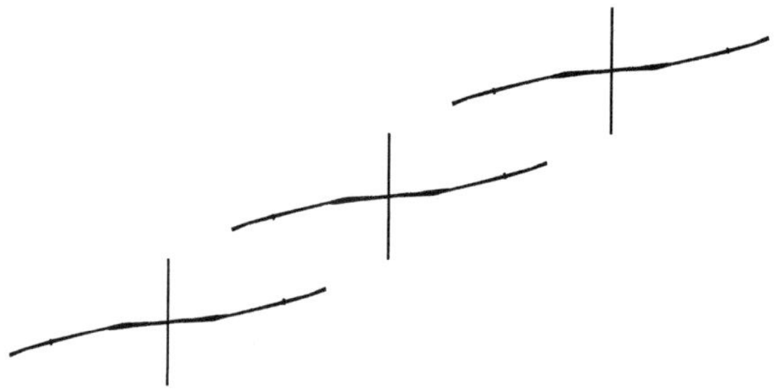

(b) $PcSnI_2$, RESFOK

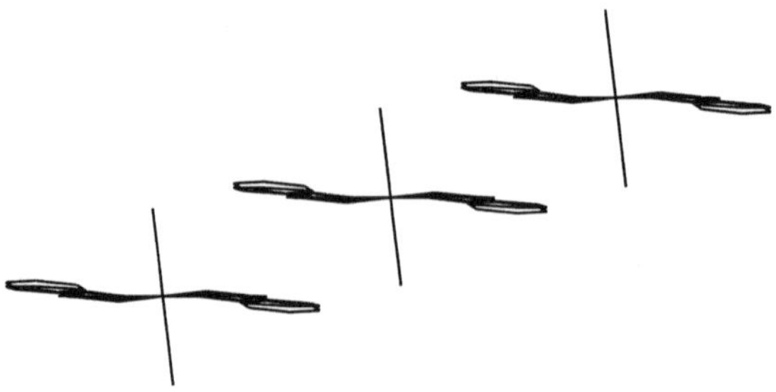

(c) $PcSnI_2$, RESFOK01

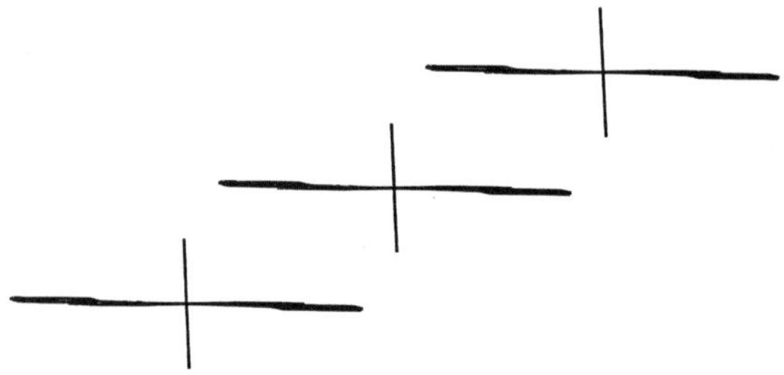

(d) β-$PcGe(OH)_2$, #1985-02

Figure 17. Planar and deformed macrocycles in columns of CN6*tr* complexes.

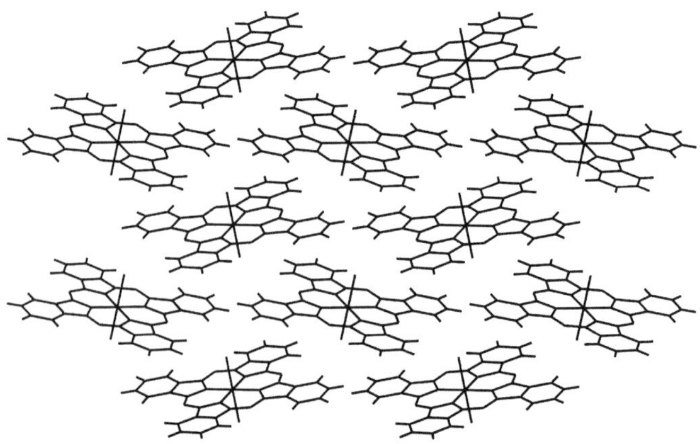

(a) $PcSiCl_2$, CCDC173621

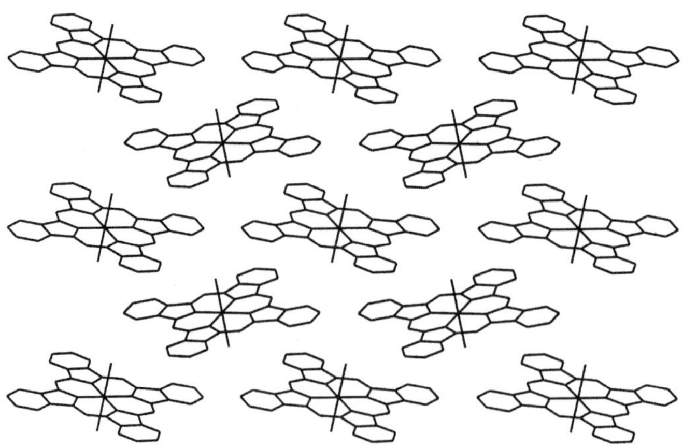

(b) $PcCoCl_2$, KEHXEA

Figure 18. Columnar arrangement of neutral CN6*tr* complexes, (a–c,e,f) views along the stacking axes, (d) view along an axis tilted by 6° from the stacking axis.

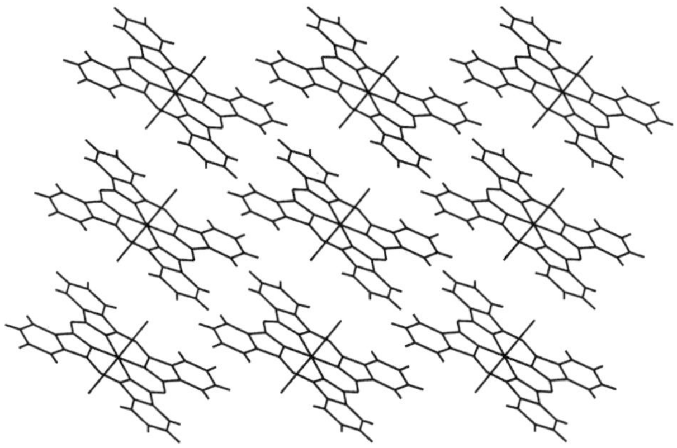

(c) $PcSnI_2$, RESFOK

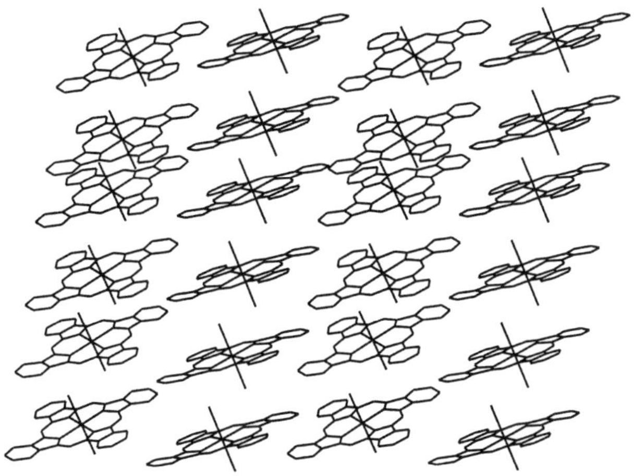

(d) $PcSnI_2$, RESFOK01

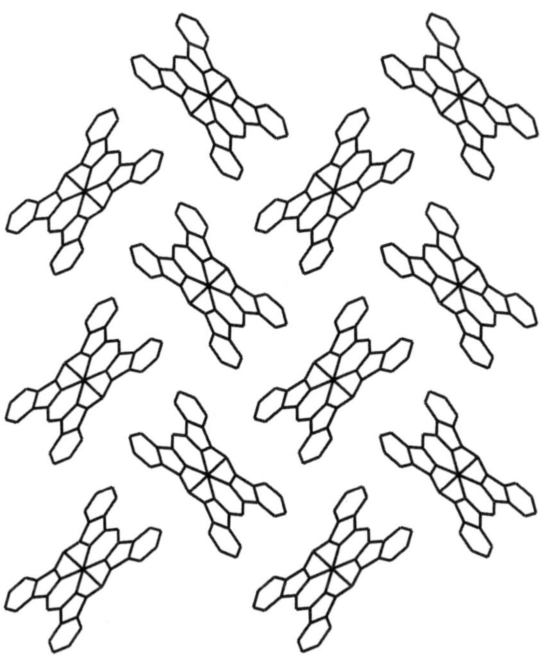

(e) α-$PcGe(OH)_2$, #1985-01

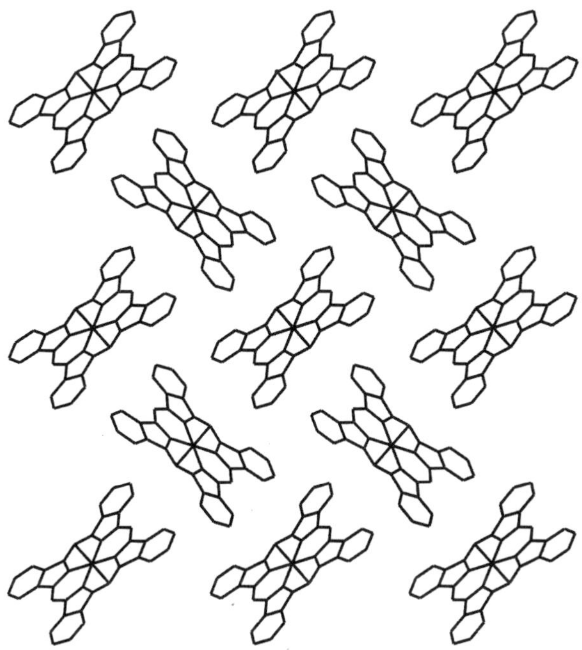

(f) β-$PcGe(OH)_2$, #1985-02

Figure 18. Continued.

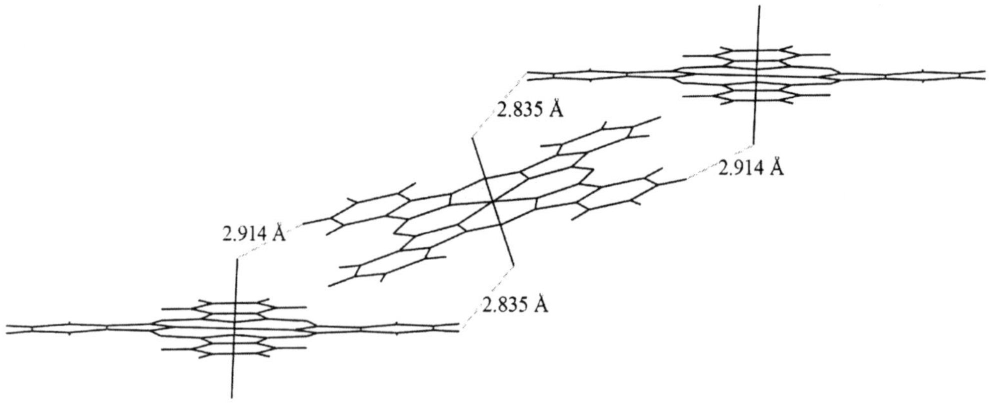

(a) $PcSiCl_2$, CCDC173621

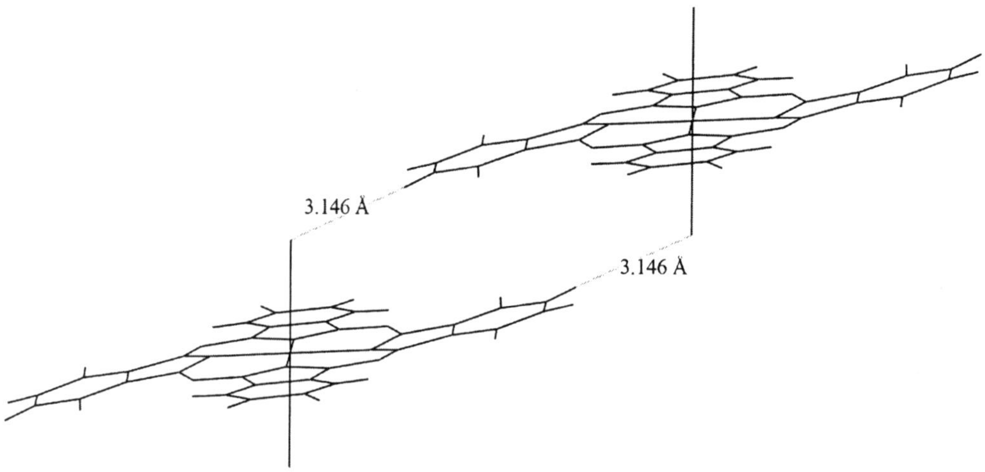

(b) $PcSnI_2$, RESFOK

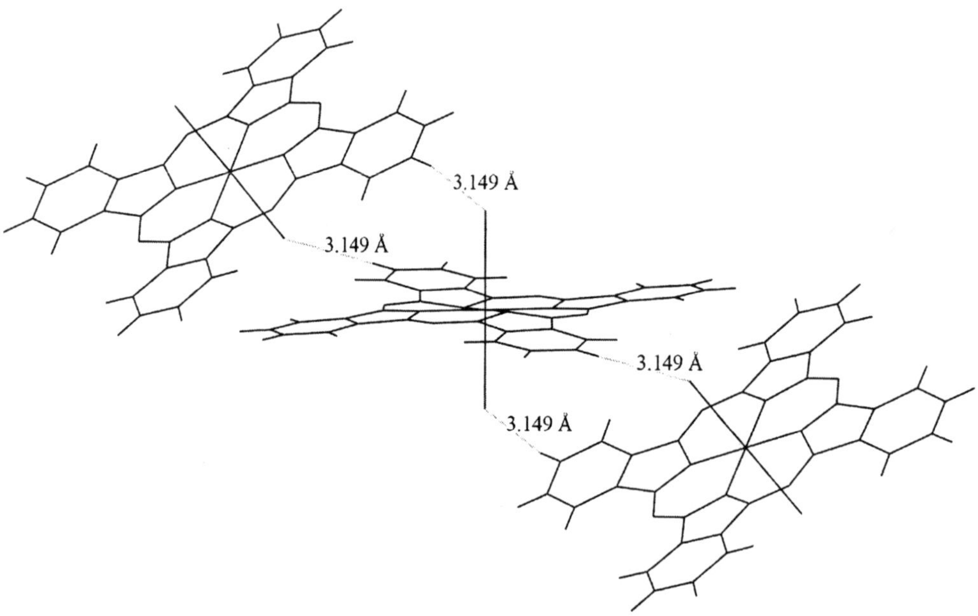

(c) $PcSnI_2$, RESFOK01

Figure 19. Intermolecular hydrogen bonding pattern in selected neutral CN6*tr* complexes.

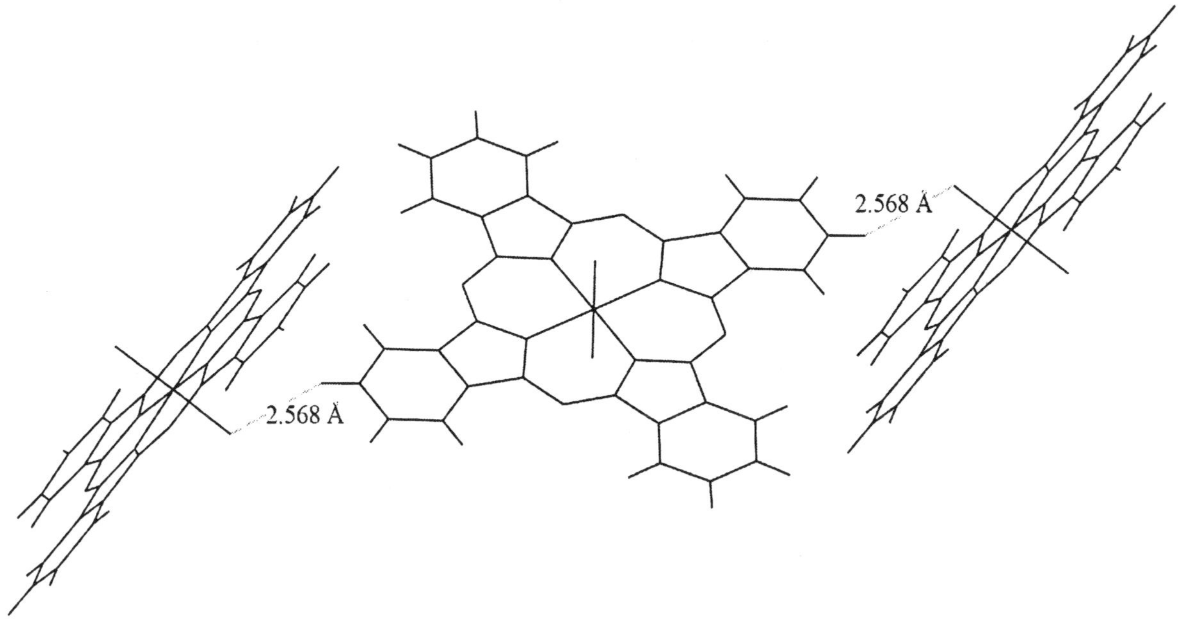

(d) α-$PcGe(OH)_2$, #1985-01

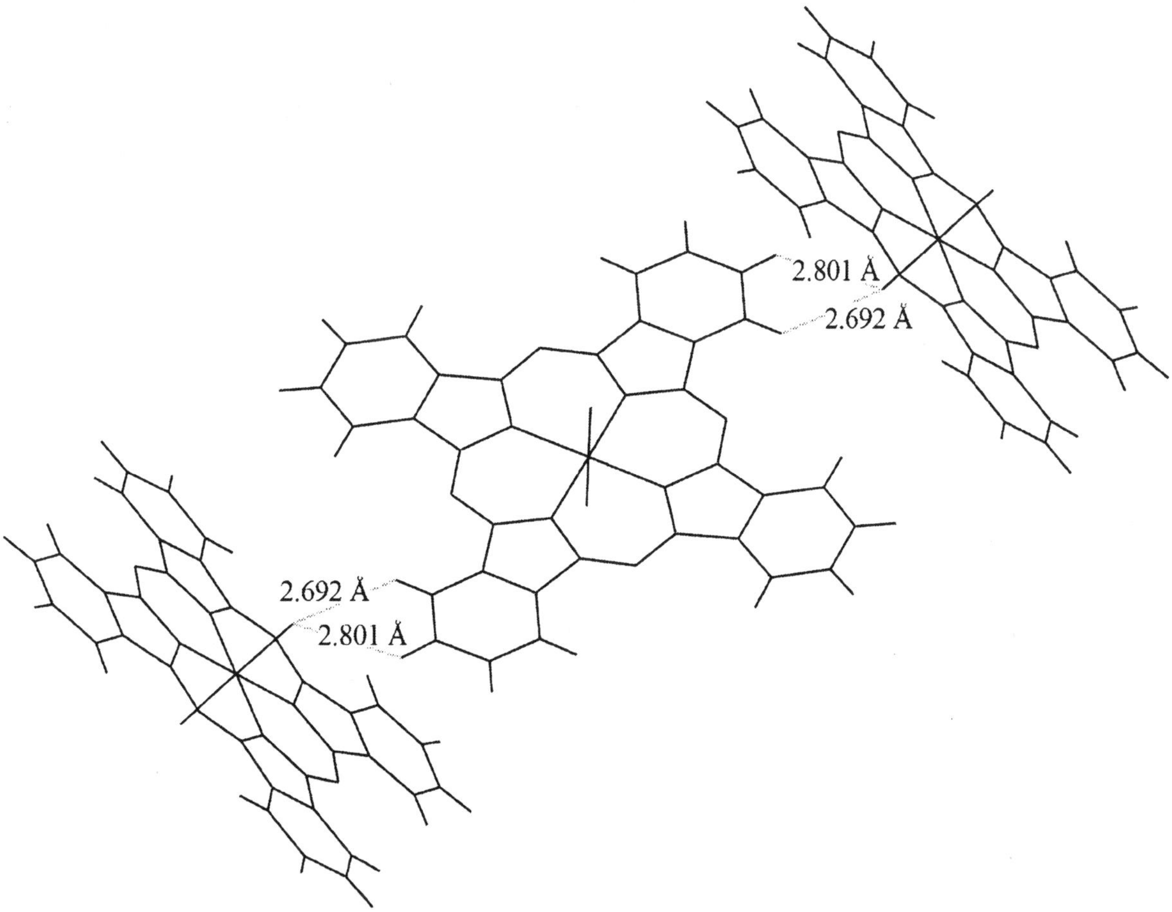

(e) β-$PcGe(OH)_2$, #1985-02

Figure 19. Continued.

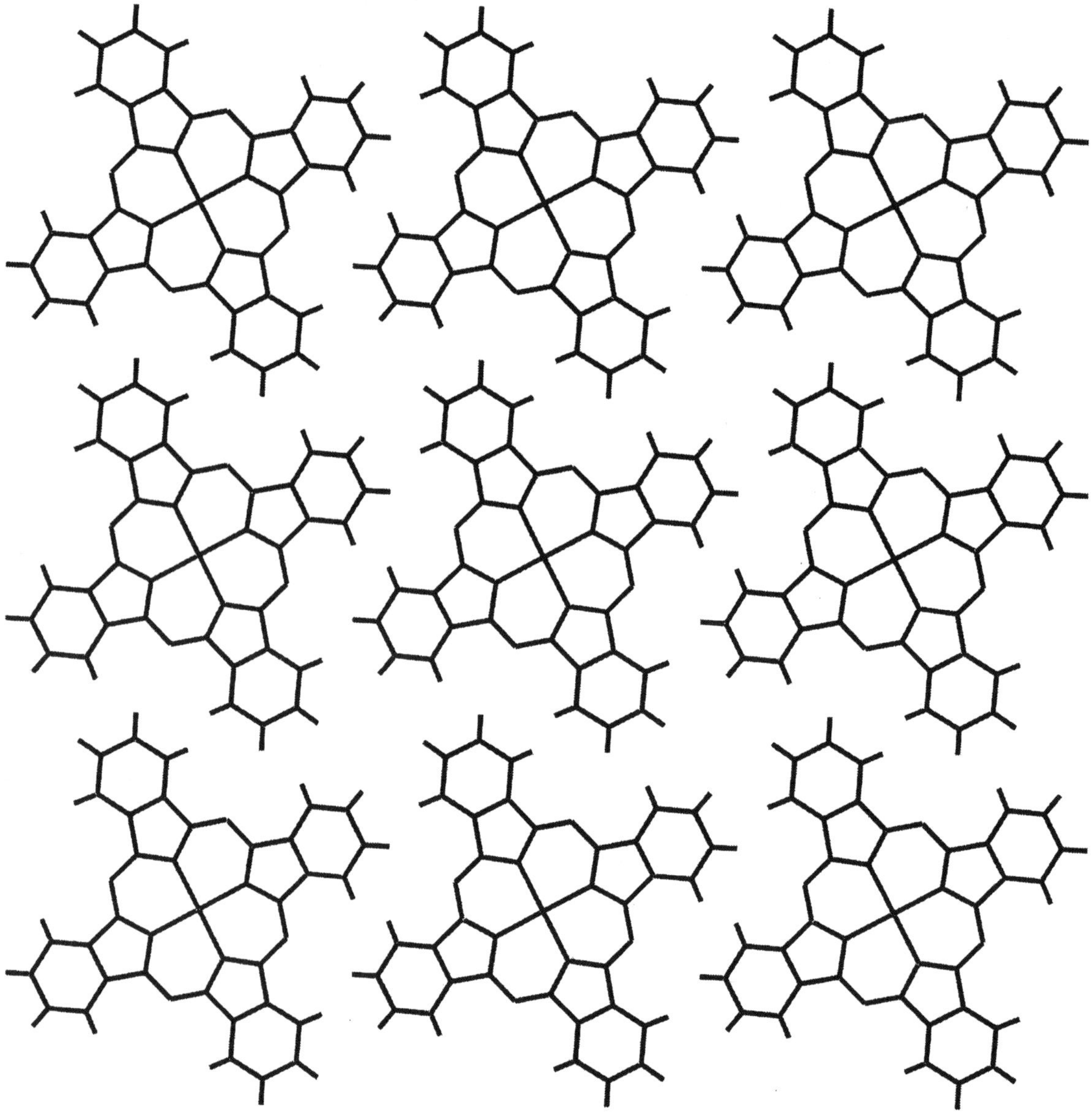

(a)

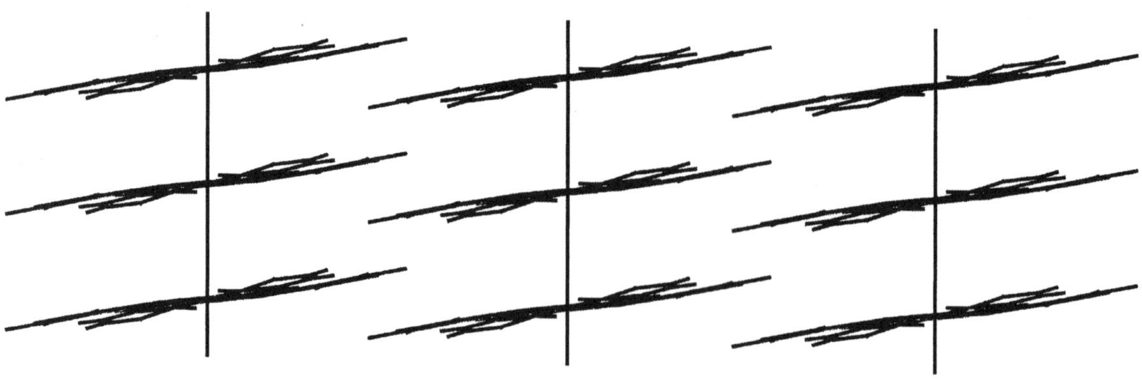

(b)

Figure 20. Crystal structure of $[PcGaF]_n$ (BAMSEN,[121] BAMSEN10[122]); (a) view along the stacking axis, (b) view perpendicular to the stacking axis.

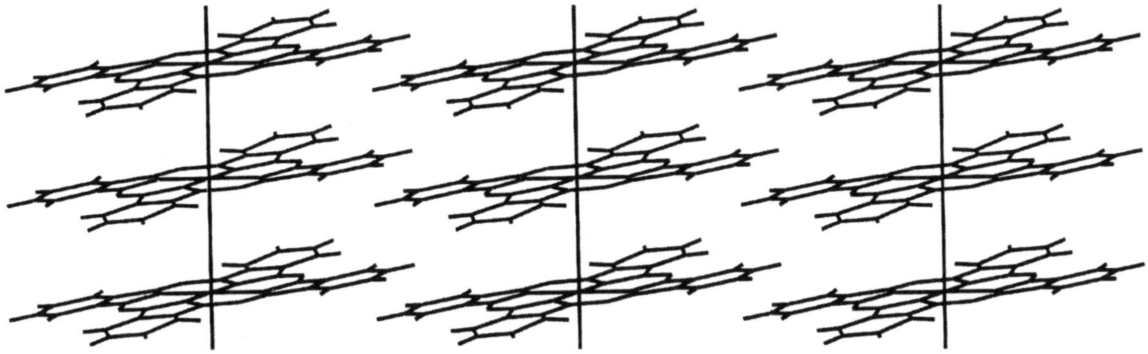

(c)

Figure 20. Continued.

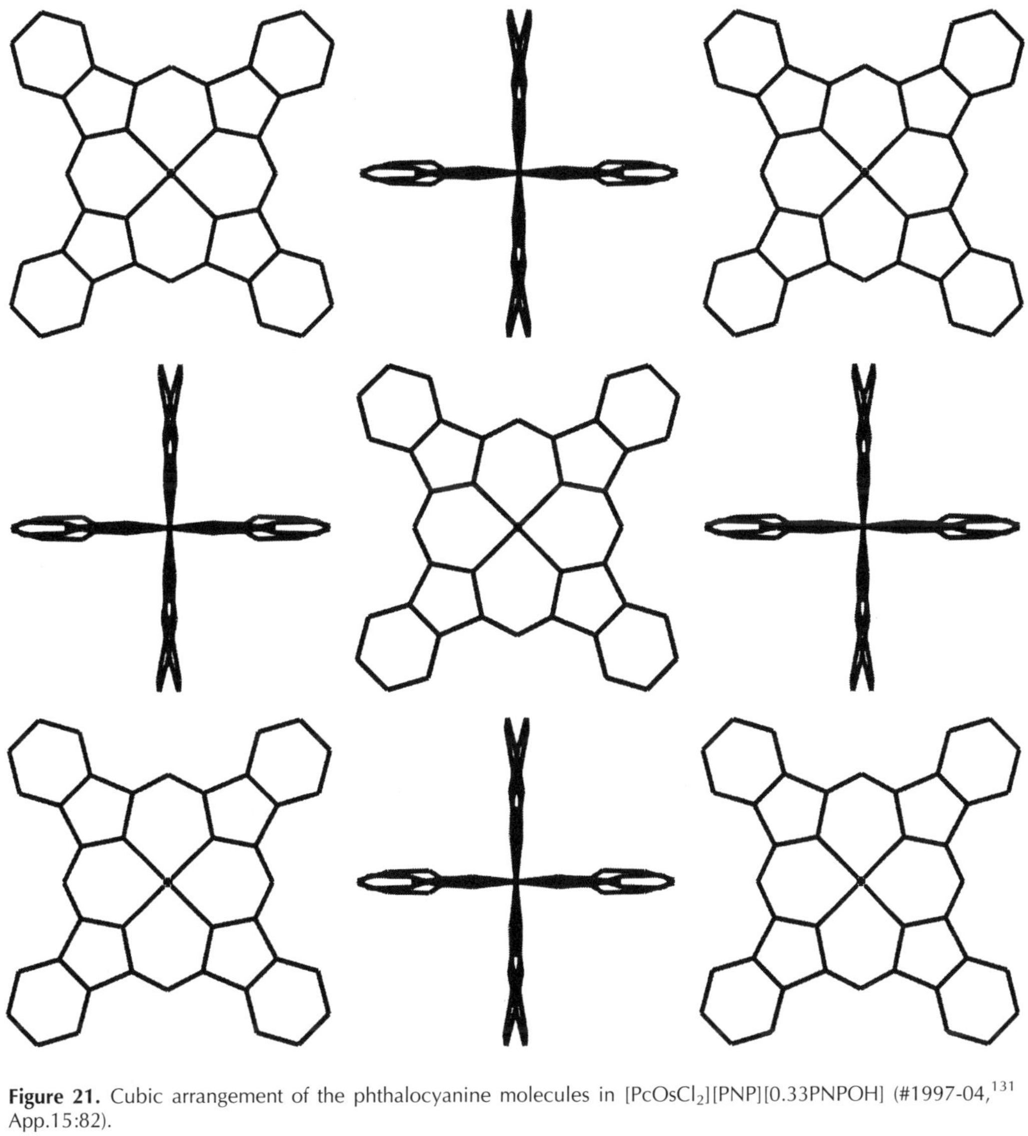

Figure 21. Cubic arrangement of the phthalocyanine molecules in $[PcOsCl_2][PNP][0.33PNPOH]$ (#1997-04,[131] App.15:82).

Table 9. Structural Description of [PcM(CN)$_2$] Complexes

Compound	Code [Ref.]	Figure	Structural description
[PcCo(CN)$_2$][PTMA] [MeCN] {113K}	NIRPAF[336]	15:13	Slipped-stacked columns along [010], counterions in channels between columns, neighboring columns along [001] are parallel while along [010] differently oriented; in column symmetrical overlap along N_{br} molecular axis, macrocycle is deformed along column
[PcCo(CN)$_2$][PTMA] [MeCN] {295K}	NIRDOH[336]	15:14	Isostructural to NIRPAF
[PcCo(CN)$_2$][2CHBr$_3$]	YAVPEQ[327,328]	15:3	Columns along [100], neighboring columns along [001] are parallel while along [100] differently oriented; in column unsymmetrical overlap along N_{br} molecular axis
[PcCo(CN)$_2$][PXX]$_2$	BAPZOH[130,337]	15:15	Slipped-stacked double-columns along [001], nearly symmetrical overlap in column, along [010] alternating parallel double-columns of phthalocyanine with columns of counterion; planes of counterions and phthalocyanine are nearly perpendicular oriented; along [100] orientation of columns is different; in column symmetrical overlap along N_{br} molecular axis
[PcFe(CN)$_2$][PNP]	DEFNUX[129,357]	15:45	Parallel columns along [001], alternating layers of phthalocyanine and counterion along [100]; in column overlap along N_i molecular axis
[PcCo(CN)$_2$][2DMSO]	DOVYUI[330]	15:5	Parallel columns along [001] form sheets parallel (010); counterions in channels between columns; in column overlap along N_{br} molecular axis; macrocycle deformation related to overlap along N_i molecular axis along [100]
[PcFe(CN)$_2$][PNP]$_2$[CH$_2$Cl$_2$]	DODXUP[358]	15:46	Isolated bands of not overlapping molecules along [001]; alternating layers of phthalocyanines and counterions along [010]
PcCo(CN)$_2$	DOVZAP[330]	15:9	Sheets parallel (100); sheet consists of parallel columns along [0–11] or [011], respectively, in columns, molecules are overlapping only via benzene rings and are tilted towards each other
[PcCo(CN)$_2$][Pϕ_4]$_{0.5}$	JAMRAQ[339,340]	15:19	Columns along [001], honeycomb structure, counterions in channels between four columns; in columns overlap along the N_{br} molecular axis, phthalocyanine macrocycle is deformed
[PcCo(CN)$_2$][2H$_2$O]	SILPEI[331,332]	15:6	Structure similar to NIRPAF
[PcCo(CN)$_2$][p-PPYNN]	SOPQET[338]	15:16	Double-bands along [100], along [010] alternating double-bands and counterions form sheets, spatial orientation of molecules in neighboring sheets is different, overlap in double-bands along N_i molecular axis
[PcCo(CN)$_2$][p-BPYNN]	SOPQIX[338]	15:17–18	Identical to SOPQET
[K][PcCo(CN)$_2$][5CH$_3$CN]	VENDAT10[331,332,335]	15:12	Parallel columns of 2 crystallographically independent molecules along [010], in column overlap along N_{br} molecular axis
[2,3-NcCo(CN)$_2$][2DCB]	WOHDUH[251]	19:7	Parallel columns along [001]
[2,3-NcCo(CN)$_2$][CH$_3$CN]	WOWDOB[251]	19:8	Parallel columns along [001]

Table 10. Hydrogen Bonding in MacM(CN)$_2$ Structures, Mac=Pc, Nc and M=Co, Fe

Complex	Orientation of columns (parallel to)	Intracolumnar hydrogen bonding/Å < 3.0Å	Intercolumnar hydrogen bonding/Å < 3.0Å
DEFNUX	[001]	N_i: 2.888, 2.912	n/a
DODXUP	n/a	n/a	n/a
VENDAT10	[010]	N_{br}: 2.400, 2.501	2.661, 2.818
SILPEI	[001]	N_{br}: 2.647, 2.647	n/a
ZODWAQ	[010]	N_{br}: 2.706	2.662
YAVPEQ	[100]	N_{br}: 2.391, 2.936	perp. 2.934
JAMRAQ	[001]	N_{br}: 2.650, 2.799	n/a
SOPQET(SOPQIX)	n/a	n/a	perp. 2.593
DOVZAP	[011]/[0–11]	1.752, 2.649, 2.709, 2.972	<=
DOVYUI	[001]	N_{br}: 2.497	2.663
BAPZOH	[001]	N_{br}: 2.649, 2.774	2.644, 2.774
NIRDOH	[010]	N_{br}: 2.723, 2.723	n/a
NIRPAF	[010]	N_{br}: 2.604, 2.604	n/a
WOHDUH	[001]	N_{br}: 2.666, 2.809	2.558
WOWDOB	[001]	N_{br}: 2.739, 2.813	2.675/2.798

Table 11. Structural Description of CN6*tr* Complexes (Monomers)

Compound	Code [Ref.]	Figure	Structural description
[PcAl(-ONO)$_2$][NBu$_4$]	ZUVYOE[324]	15:1	Sheets parallel [101], neighboring sheets by $\pm a/2$; in a sheet: columns along [001] with parallel axes, orientation alternating, counterion between columns; molecules in column overlapping along "N$_{br}$" molecular axis
[PcCo(4-mepy)$_2$][2mepy]	MPCOPC10[325,326]	15:2	Columns along [100] with parallel axes; alternating orientation of adjacent columns in [110], neighboring columns shifted by $\pm a/2$; alternating orientation of adjacent columns in [101], neighboring columns shifted by $\pm a/2$; solvent between adjacent columns in [110]; molecules in column not overlapping
[PcCo(SCN)(py)][CHCl$_3$]	FODLOZ[333]	15:7–8	Columns along [001] with parallel axes; in (100) sheet: alternating columns different orientation but the pyridine ligand always in one direction (chiral sheet), neighboring sheets related by inversion symmetry and shifted by $\pm b/2$; adjacent columns in [011] shifted by $\pm c/2$; in [101] sheet neighboring columns shifted by $\pm b/2$ and by a minute amount along c; in columns small overlap along intermediate axis between N$_i$ and N$_{br}$ molecular axes
[PcCr(-ONO)$_2$][PNP]	TOPVID[343]	15:25	Parallel columns along [100] with identical orientation, counterion in channels between four columns; molecules in columns overlap along N$_i$ molecular axis; bonded by hydrogen bonding between axial ligand and aromatic hydrogen atoms
PcCu(OReO$_3$)$_2$	QEMMOK[306]	15:26	Columns along [100], column axes form hexagonal lattice, adjacent columns in [110] sheet identical orientation; different orientation in neighboring sheets; in columns overlap along intermediate molecular axis, hydrogen bonding in column
PcCu(SbF$_6$)$_2$	QILWAJ[344]	15:27	Parallel columns along [100] with identical orientation, in columns overlap along N$_{br}$ molecular axis, hydrogen bonding in and between column(s)
[PcFe(4-mepy)$_2$][2mepy]	PTCYFE[345] PTCYFE01[326]	15:28	Isostructural to [PcCo(4-mepy)$_2$][2Mepy]
[PcFe(CO)(DMF)][DMF]	PHTHFE[346]	15:29	Isostructural to [PcCo(SCN)(py)][CHCl$_3$]
[PcFe(NC-Ru(Pϕ_3)$_2$Cp)$_2$][1.5CH$_2$Cl$_2$]	WOHVAQ[314]	15:30	Columns along [001], column axes form tetragonal lattice, along [010] and [100] adjacent columns alternating orientation, identical columns at $\pm a/2$ $\pm b/2$
[NC-(Pc)Fe-NC-Ru(Pϕ_3)$_2$Cp][2CHCl$_3$]	WOHTIW[314]	15:32	Double-columns along [010], along column no overlap, overlap between columns along N$_{br}$ molecular axis and N$_i$ molecular axis (on the side of the smaller axial ligand)
[PcFe(DMSO)$_2$][2DMSO]	SPCNFE[347]	15:31	Columns along [001] with tetragonal lattice of column axes; columns in sheets parallel to (010) have identical orientation; orientation of molecules in adjacent sheets is different; molecules are separated in columns due to large axial ligands
D_{4h}-1,2-NcFe(CN-cyclohex)$_2$	VACXOM[249]	19:9–10	Similar to PcCu(OReO$_3$)$_2$, columns along [001], very small overlap
tBu$_4$-D_{4h}-1,2-NcFe(CN-tBu)$_2$	VIWMOD[250]	19:11–14	Columns along [110], sheets parallel (001), molecules in a sheet parallel, alternating in adjacent sheets
PcFe(4-mepip)$_2$	XENLAD[351]	15:36–37	Similar to PcMg(py)$_2$
[PcFe(OMe)$_2$][PNP]$_2$ [1Me$_2$CO, 1.5MeOH]	NELCOW[360]	15:50	Isolated columns along [100], no overlap in columns
[PcFe(CN-Ru(Pϕ_3)$_2$Cp) (NC-Ru(Pϕ_3)$_2$Cp)][SbF$_6$]	WOHVEU[314]	15:51	Tetragonal lattice of double-columns along [010], no overlap between molecules in column, deformation of macrocycle in column is inversion-related to second column of double-column
[PcFe(NC-Fe(dppe)Cp)$_2$][SbF$_6$]	WOHTOC[314]	15:52	Columns parallel [010], sheets parallel (001), molecules in sheet identical orientation, different from neighbor sheet, overlap between molecules in columns along N$_{br}$ molecular axis

(*Continued*)

Table 11. Continued

Compound	Code [Ref.]	Figure	Structural description
[PcFe(NC-Ru(Pϕ_3)$_2$Cp)$_2$][SbF$_6$]	WOHTUI[314]	15:53	Columns along [100] and [010], sheets parallel (001), molecules in sheet identical orientation, different from neighbor sheet
[PcGeI$_2$][I$_2$]	GODLEQ[366]	15:58	Columns along [−101], I$_2$ connects molecules along this axis, [PcGeI-I$_2$]$_n$; slipped-stacked columns along [010], overlap along N$_{br}$ molecular axis, hydrogen bonding, 3.211 and 3.430 Å
[PcIr(Cl,-CC(O)C)][H$_2$O]	GUDZUA[367]	15:59	Columns along [100], sheets parallel (001), molecules in sheet identical orientation, different from neighbor sheet, in sheet alternating counterion, phthalocyanine columns, orientation of axial ligands along [001] chiral
[PcIr(py,I)][py]	RENJID[368]	15:60	Similar to [PcCo(SCN)(py)][CHCl$_3$]
[PcIr-(acido,N$_3$)][NBu$_4$] [0.25Et$_2$O,0.5CH$_2$Cl$_2$]	JAHSOA[370]	15:62	Double-bands along [001], overlapping along N$_{br}$ molecular axis; double-bands separated by layers of counterions
[PcIr(-NO$_2$)$_2$][PNP]	RILVOX[371]	15:63	Isolated columns along [100], overlap along N$_i$ molecular axis, hydrogen bonding
[PcIr(OH,OMe)][NBu$_4$]	LIQVAI[372]	15:64	Columns of two independent molecules along [100], layers parallel (010), alternating counterion layers, orientation in layers identical, in alternating layers different; overlap in column along N$_{br}$ molecular
PcMg(HOCH$_2$CH$_2$OMe)$_2$	QOPCON[373]	15:65	Columns along [100], sheets parallel (001), molecules in sheet identical orientation, different from neighbor sheet; no overlap in columns
[PcMg(OH$_2$)$_2$][NMP]$_2$	QOPCUT[373]	15:66	Columns along [100], sheets parallel (010), molecules in sheet identical orientation, different from neighbor sheet; strong overlap in column; hydrogen bonding
PcMg(py)$_2$	CCDC166120[374]	15:67	Parallel (shifted), slipped-stacked columns along [001] form sheets parallel (100); in column overlap along N$_{br}$ molecular axis
PcMn(NO)(OPϕ_3)	XEYREY[377]	15:71	Double-bands along [010], sheets parallel (001), molecules in sheet identical orientation, different from neighbor sheet; in double-bands strong overlap along N$_{br}$, hydrogen bonding, large axial ligands isolate double-bands
[PcMn(NO)(-NCO)][PNP]	XEYQOH[377]	15:74	n.a.
[PcMn(NO)(-ONO)][PNP]	XEYQIB[377]	15:75	n.a.
[PcMn(NO)$_2$][PNP]	XEYQUN[377]	15:76	Isolated columns along [100], overlap along N$_i$ molecular axis, hydrogen bonding
[PcMn(NCMe)$_2$][C$_4$(CN)$_6$]	1996-02[271]	15:77	Columns parallel [100], sheets parallel (001), alternating phthalocyanine counterions, orientation of all molecules identical, in sheet 2D overlap, strong hydrogen bonding
[PcMoBr$_2$][PNP]	RIKSIN[380]	15:79	Isolated columns along [100], overlap along N$_i$ molecular axis, hydrogen bonding
[PcOsCl$_2$][PNP][0.33PNPOH]	#1997-04[131]	15:82	Cubic structure, see Figure 21
[PcOsCl$_2$][PNP]	#1997-05[131]	15:83	Columns along [100], sheet parallel (001), orientation of all molecules identical, overlap in column along N$_i$ molecular axis, hydrogen bonding
PcOs(CO)(py)	PCPCOS[381]	15:80	Pseudo-hexagonal lattice of double-bands along [100], overlap along N$_{br}$
[PcOs(NO,C$_6$H$_{12}$O$_2$)][BF$_4$]	#1996-01[382]	15:81	Columns of two independent molecules along [100], layers parallel (010), molecules in sheet identical orientation, different from neighbor sheet; in column adjacent molecules point up or down, overlap along N$_{br}$ molecular axis, hydrogen bonding
PcRe(NO)(OPϕ_3)	XEYRIC[377]	15:84	Isostructural to PcMn(NO)(OPϕ_3)
PcRe(Pϕ_3)$_2$	NAPPAV[383]	15:85	Bands along [001], molecules are not overlapping; layers parallel (100), orientation of molecules in layers is identical, very small overlap along [100]
[PcReF$_2$][0.33(PNP)F][2H$_2$O]	GEQCOU[385]	15:88	Isostructural to [PcOsCl$_2$][PNP]
[PcRe(NO)$_2$][PNP]	XEYRAU[377]	15:89	Isostructural to [PcMn(NO)$_2$][PNP]

(*Continued*)

Table 11. Continued

Compound	Code [Ref.]	Figure	Structural description
[PcRe(NO)(OPϕ_3)][I_3][CH_2Cl_2]	XEYROI[377]	15:90	Double-bands along [100], sheets parallel (010), orientation of molecules in sheets identical, different in neighboring sheets, columns consist of alternately molecules pointing up or down, overlap of two molecules via the side of the smaller ligand along N_{br} molecular axis
[PcRh(OϕCl)$_2$][NBu_4]	1997-02[386]	15:92	Bands/columns along [100], layers/sheets parallel (001), orientation of molecules in sheets identical, different in neighboring sheets, small overlap along columns
[PcRu(-NO_2)$_2$][NBu_4]$_2$	RUJQES[388]	15:94	Layers parallel (001) consisting of bands along [100], molecular planes slightly inclined, alternating phthalocyanine-counterion layers, no pc–pc overlap
[PcRu(py,CN)][NBu_4]	GEQDIP[389]	15:95	Isolated molecules in sheets parallel (100)
[PcSi(-O-C(O)-Fc)$_2$][2$CHCl_3$]	NEYCEZ[392]	15:96	Columns along [100], orientation of neighboring columns is different, in column: small overlap along N_{br} molecular axis
[PcSi(OMe)$_2$][0.48H_2O]	VIKZAQ[127]	15:99	Columns along [001], orientation in all columns is identical, overlap along N_{br} molecular axis
[PcSiOSiEt$_3$,Me][0.5C_6H_{12}]	SIOPCY[449] #1971-01[270]	15:100	Double-layer parallel (010), consists of double-bands along [001] where always two molecules overlap via the side of the smallest axial ligand along the N_{br} molecular axis, solvent is located between two double-layers
PcSi(O-C(O)-$C_{10}H_{21}$)$_2$	CCDC173623[128]	15:102–103	Columns along [100], molecules in adjacent columns have different orientations, layers parallel [011] and [100]; overlap in column along N_{br} molecular axis
PcSi(OSiMe$_3$)$_2$	SIMPLC10[395]	15:106	Columns along [010], sheets parallel (100), in sheets chiral orientation of axial ligands, neighboring sheets have different orientation of axial ligands and phthalocyanine macrocycles, in columns no direct pc–pc overlap, overlap between neighboring columns in two sheets
PcSi(thff)$_2$	VILGOM[127]	15:107	Columns along [101], hydrogen bonding, overlap along N_i molecular axis
PcSi[OCH$_2$-C_6H_3((OCH$_2$CH$_2$O)$_3$Me)$_2$]$_2$	2001-02[396]	15:111–112	Columns along [001] in sheets parallel (010), overlap along N_{br} molecular axis
PcSi[OCH$_2$-C_6H_3(OCH$_2\phi$)$_2$]$_2$	NOLXOB[397]	15:113–114	Parallel bands along [001]
PcSi[Oepiglobulyl]$_2$	CAVZAA[398]	15:115	Nearly isolated molecules, small overlap in columns along [010]
PcSi[OSi(C_6H_{13})$_3$]$_2$	QUDQOV[391]	15:118–119	Nearly isolated molecules, small overlap in columns along [010]
PcSi[OSiMe$_2$(C_8H_{17})]$_2$	QUDQUB[391]	15:120–121	Nearly isolated molecules, small overlap in columns along [100]
PcSi[O(CO)(CH$_2$)$_2$ph(3,4-(OMe)$_2$)$_2$	ACUMOA[399]	15:110	Hexagonal array of slipped-stacked columns along [001]; in column overlap along N_{br} molecular axis
PcSi[O(CO)-(3-thienyl)]$_2$	ACUMIU[399]	15:109	Parallel columns along [001], alternating direction along [010]
PcSi[O(CO)ph-4tBu]$_2$	ACUMEQ[399]	15:108	Parallel columns along [100]; overlap along intermediate molecular axis
[PcWCl$_2$][PNP]	#2001-05[342]	15:126	Isolated columns along [100], in columns overlap along N_i molecular axis, hydrogen bonding; structure similar to [PcMo(Br)$_2$][PNP]

molecules in PcGe(OH)$_2$ structures closely resembles that of the x-PcH$_2$ molecules or that of the 2,3-NcM complexes (Figure 9c,f,i).

Figure 19 shows the hydrogen bonding pattern which exists between neighboring columns of different orientation. Similar interaction distances resulting from intra-columnar and inter-columnar hydrogen bonding indicates they are of similar strength (Table 8). Differences in these inter-columnar hydrogen bonding pattern is thought to be the main difference between the polymorphic structures of PcGe(OH)$_2$ (Figure 19d–e). The positions of the hydrogen atoms in all PcGe(OH)$_2$ structures have been calculated assuming a C–H bond length of 0.95 Å. In α-PcGe(OH)$_2$, interactions exist between aromatic β-hydrogen atoms and the axial ligand of the neighbor molecules, while in the

Table 12. Structural Description of CN6*tr* Complexes (Dimers and Polymers)

Compound	Code [Ref.]	Figure	Structural description
$[py\text{-}(Pc)Cr\text{-}O\text{-}Cr(Pc)\text{-}py][py]_2$	#2001-08[342]	15:23	Columns along [100]
$[R\text{-}(Pc)Fe\text{-}C\text{-}Fe(Pc)\text{-}R][3Me_2CO]$ {+M}	FUNGEA[348]	15:33	Bands along [010], sheets parallel (100), monomer between sheets
$[R\text{-}(Pc)Fe\text{-}O\text{-}Fe(Pc)\text{-}R][3Me_2CO]$ {+M}	KITPIM[349]	15:34	Isostructural to $[R\text{-}(Pc)Fe\text{-}C\text{-}Fe(Pc)\text{-}R][3Me_2CO]$ {+M}
Br-(Pc)Fe-N-Fe(Pc)-Br	VAFMIY[350]	15:35	Linear stacks along [001]
py-(Pc)Fe-C-Fe(Pc)-py	RERGEA[352]	15:39	Slipped-stacked columns along [100], network of double-columns with fish-bone structure
THF-(Pc)Fe-C-Fe(Pc)-THF	PAZREN[353]	15:40	Isostructural to py-(Pc)Fe-C-Fe(Pc)-py
THF-(Pc)Fe-C-Fe(tpp)-THF	PAZRAJ[353]	23:1	Isolated molecules
$[F\text{-}(Pc)Fe\text{-}C\text{-}Fe(Pc)\text{-}F][NBu_4]_2[3H_2O]$	NAPHIV[355]	15:43	Bands along [100], layers parallel (001); bands have different orientations
$[N_3\text{-}(Pc)Fe\text{-}N\text{-}Fe(Pc)\text{-}N_3][PNP]_2[I_2][_2Et_2O]$	NEKLUK[356]	15:44	Bands along [010], layers parallel (100)
$[H_2O\text{-}(Pc)Fe\text{-}N\text{-}Fe(tpp)\text{-}OH_2][I_5][_2THF]$	YUGKOA[361]	23:2	Isolated molecules in layers parallel (001)
$[THF\text{-}(Pc)Fe\text{-}N\text{-}Mn(TPP)\text{-}THF][I_5][2THF]$	#2001-01[362]	23:3	Isostructural to $[H_2O\text{-}(Pc)Fe\text{-}N\text{-}Fe(tpp)\text{-}OH_2][I_5][_2THF]$
py-(Pc)Ir-Ir(Pc)-py	TOCLIG[369]	15:61	Isostructural to py-(Pc)Fe-C-Fe(Pc)-py
(dppe)(Cp)Fe-(Pc)Mn-O-Mn(Pc)-NC-Fe(dppe)Cp	ROPTAR[375]	15:68	Layers parallel (100)
[py-(Pc)Mn-O-Mn(Pc)-py][2py]	PHCPMP10[378,379]	15:72–73	Isostructural to py-(Pc)Fe-C-Fe(Pc)-py
$[py\text{-}(Pc)Rh\text{-}Rh(Pc)\text{-}py][2C_6H_6]$	NEXSIS[387]	15:91	Layers parallel (100) with column-like arrangement of molecules along [010], columns with different arrangement along [001]
py-(Pc)Ru-C-Ru(Pc)-py	RERGIE[352]	15:93	Isostructural to py-(Pc)Fe-C-Fe(Pc)-py
$Me_2(tBu)SiO\text{-}(Pc)Si\text{-}O\text{-}Si(Pc)\text{-}OSi(tBu)Me_2$	DOHYAA[393]	15:101	Layers parallel (100) with bands along [010] and slipped-stacked columns along [001], adjacent columns have different orientation
$(Me_3SiO)_2MeSiO\text{-}((Pc)Si\text{-}O)_3\text{-}SiMe(OSiMe_3)_2$	1970-01[269]	15:97	Slipped-stacked columns along [100]
$[\text{-}Ga(Pc)F\text{-}]_n$	BAMSEN,[121] BAMSEN10[122]	15:54	Polymer chains along [100], see text
$[\text{-}(Pc)Mn\text{-}NC\text{-}C(CN)\text{-}CN\text{-}]_n$	PILBER[376]	15:69	Polymer chains along [101]
$[\text{-}\{(Pc)Mn\text{-}TCNE\}_3\text{-}]_n$ {β}	PILBAN[376]	15:70	Polymer chains along [111]
$[\text{-}(Pc)Si\text{-}(OSiMe_2\{CH_2\}_7SiMe_2O\text{-})]_n$	QUDQIP[390,391]	15:98	Polymer chains along [100]

β-polymorph, the interaction involves both α- and β-hydrogen atoms.

In $[PcGaF]_n$, the phthalocyanine complexes are stacked linearly in contrast to the slipped-stacking arrangement mentioned above. The axial ligands are μ-coordinated to two phthalocyanine complexes forming a 1D linear chain. These chains are themselves ordered in a 2D tetragonal network (Figure 20a). Each macrocycle has an angle toward the stacking axis and is somewhat deformed (Figure 20b).

2. Ionic Phthalocyanines and Neutral Phthalocyanines with Large Axial Ligands

The strong influence of additional molecules on the arrangement of basic phthalocyanine units can be best seen in the series of $[PcM(CN)_2]$[solvent or counterion]. The first such structure was published in 1985 by Küppers *et al.*[129] In 1989, Inabe and coworkers published the first of a large series of $[PcCo(CN)_2]$ and $[PcFe(CN)_2]$ crystal structures, an overview of their work was recently published.[130] In the vast majority of these structures, the phthalocyanines overlap in much the same way as those in the neutral CN6*tr* complexes, the molecules forming slipped-stacked columns and strong hydrogen bonding can be seen inside the structures (see Tables 9 and 10 for details). Similar to the effects counterions and solvent molecules have on a crystal structure, large axial ligands have an equally strong influence on the molecular arrangement (see Tables 11 and 12 for details). A special structure, worth mentioning, is the cubic arrangement of phthalocyanines in $[PcOsCl_2][PNP][0.33PNPOH]$ (#1997-04,[131] App. 15:82), shown in Figure 21.

III. Phthalocyanine Complexes with an Out-of-Plane Coordinated Central Atom

The periodic table of complexes with out-of-plane coordinated central atoms is shown in Figure 22, with the number of structures given next to each element. This chapter concentrates on the structural aspects of the CN5

(Figure 5d,e), CN6*cis* (Figure 5f), CN7*cis* (Figure 5g) and CN8*cis* (Figure 5h) complexes. The corresponding *bis*- and *tris*phthalocyanines (Figure 5i–l) will be discussed later in Section III.B. A compilation of the basic crystallographic data of the structures is given in Appendix 1 and their molecular structures are shown in Appendices 12–14, 16, and 17.

A. CN5, CN6*cis*, CN7*cis*, AND CN8*cis* COMPLEXES

1. Introduction and Molecular Structures

A typical phthalocyanine complex with an out-of-plane coordinated central atom, and one axial ligand is shown in Figure 23a. Hereafter, phthalocyanine complexes with out-of-plane coordinated central atoms will be referred to as "*cis* complexes" to emphasize that in these complexes all axial ligands are on the one side of the central atom and the phthalocyanine macrocycle on the other. Also, PcM with an out-of-plane coordinated central atom and PcM-X with only one axial ligand will be included into this notation because of their structural similarities with complexes where two or more axial ligands are on one side. In the case of CN5 complexes, the ligands surrounding the central atom form a square–pyramidal coordination, with the four isoindole nitrogen atoms defining the base. The ligand stereochemistry around the central atoms of the other

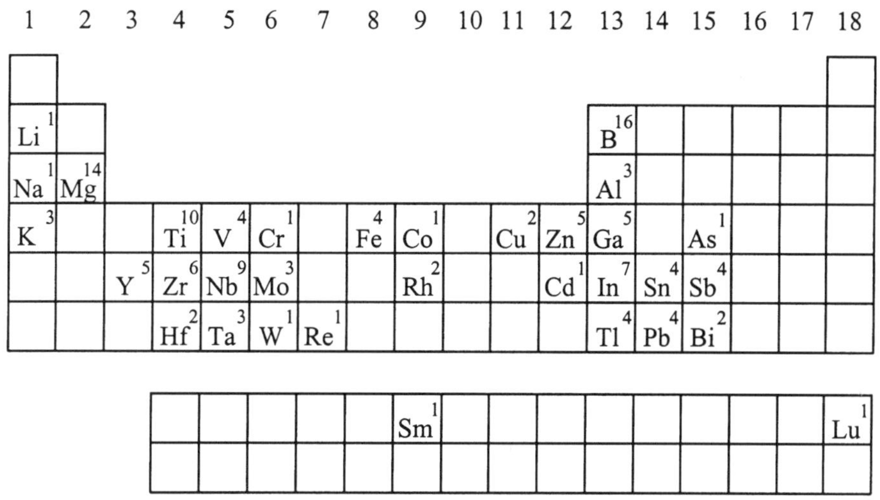

Figure 22. Periodic table of phthalocyanine complexes with out-of-plane coordinated central atoms with number of structures given next to each element (included are CN5, CN5*, CN6*cis*, CN7*cis*, and CN8*cis* complexes, excluded are *bis*- and *tris*-phthalocyanines).

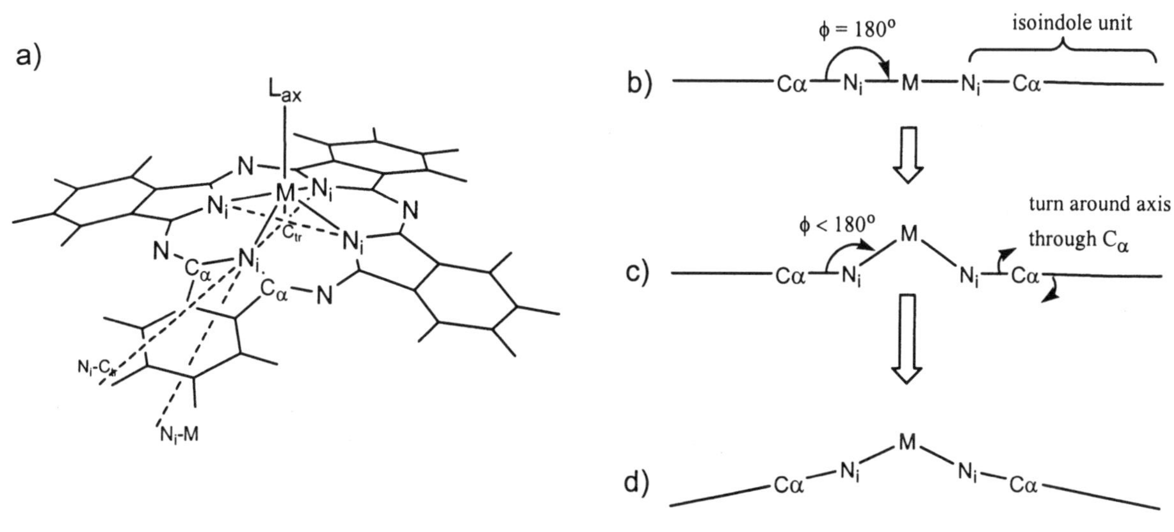

Figure 23. (a) The molecular structure of a phthalocyanine complex with an out-of-plane coordinated central atom M and an axial ligand L_{ax}. (b) The model of a complex with a small central atom that is in-plane coordinated. (c,d) The model of a complex with an oversized central atom which is coordinated out-of-plane, before and after rearrangement of the macrocycle into a saucer-shaped form.

cis complexes is shown in Table 2. In App. 12:28 or App. 12:39, e.g., it can be seen that these *cis* complexes have two different molecular sides, a "phthalocyanine-only" side, and a side, which is dominated by the central atom and the axial ligand. Due to the resemblance with optical lenses, Iyechika first adopted the terms "concave side" to represent the "phthalocyanine-only" side, and "convex side" to describe the "metal-axial-ligand side."[132]

In most of the simple metal phthalocyanines (PcM in Section II.A), the phthalocyanine molecule adopts a planar conformation (Figure 23b). In the case that a large ion is coordinated to the phthalocyanine ring, the ion is unable to enter the central cavity, but rather sits "atop" of the ring, located between 0.1 and 1.8 Å from the plane defined by the four N_i isoindole nitrogens (Figure 23c). This results in an angle (ϕ) of less than 180° between the plane of the isoindole unit and the N_i–M bond. Such a bonding situation introduces conformational stress into the phthalocyanine macrocycle, particularly for the sp^2 isoindole nitrogen atoms. The macrocycle responds to this situation by reorientating the isoindole units, rotating them around axes through their C_α atoms, as shown in Figure 23c,d), effectively reincreasing the angle between the isoindole unit and the N_i–M bond. The final angle results from an energetic compromise between a situation with an angle of 180° and interaction with neighbors, which flatten the molecule by repulsive interaction and decrease the angle. In the actual molecular structure, every quarter of the molecule can have different angles providing it is exposed to different crystalline environments. While phthalocyanine molecules in the monoclinic polymorph of PcPb (PBPHPY,[133] App. 12:37–38) adopt a saucer-like structure without any flattening, fully flattened macrocycles are found in the structures of, e.g., PcZnCl (DAMTEQ,[134] App. 12:63), [C_{2v}-1,2-NcCo-py][2py] (#2001-03,[135] App. 19:5–6) and [PcMg-OPϕ_3][THF] (MAKJAJ,[136] App. 12:28) (Figure 24). From Figure 24, it would appear as if the structure of [C_{2v}-1,2-NcCo-py][2py] [#2001-03] would be the only case in which the central cobalt atom is 5-coordinated and located in-plane rather than out-of-plane. However, a careful analysis of the environment around the cobalt reveals that it is actually more a CN6*tr* complex, where the sixth ligand is the azamethine nitrogen atom from an adjacent molecule (Co–N′: 3.448 Å), a case similar to the "hexacoordination" of the central atoms in the β-polymorphs of CN4 PcM complexes. Pronounced shuttlecock structures can also be found in the case of the subphthalocyanines

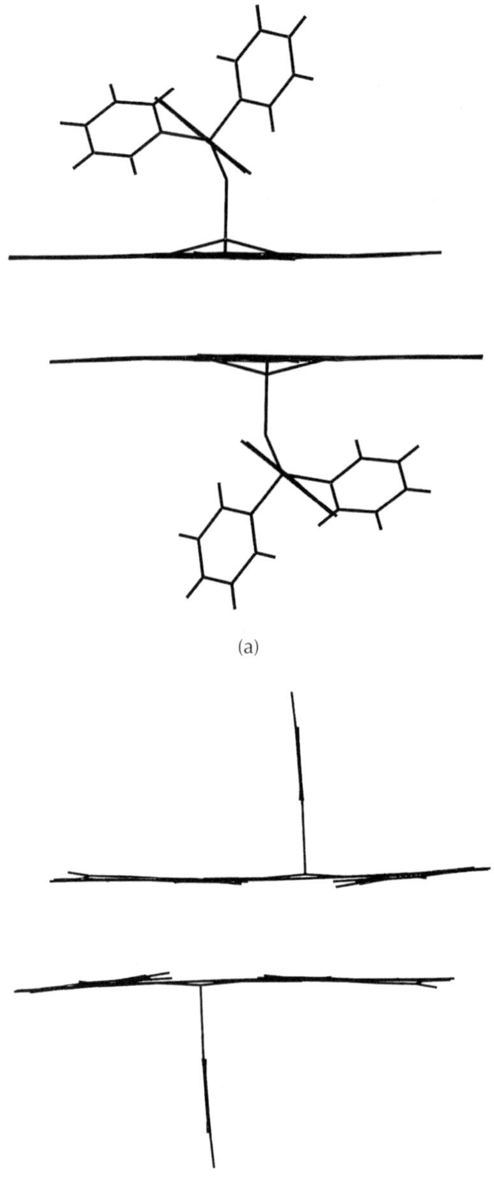

Figure 24. Side-on view of CN5 complexes with flattened macrocycles; flattening due to nearly full concave–concave overlap; (a) MAKJAJ[136] and (b) #2001-03.[135]

(App. 20:1–18),[137–141] where even the concave interaction with neighbors fails to fully flatten the macrocycles (see Section IV.D). The saucer-shaped form of PcPb and PcSn molecules is clearly visible in STM scans.[55,56] In these papers, the authors also report that the central metals can tunnel from on side of the macrocycle to the other.

Friedel *et al.* reported the first structure of a *cis*-complex in 1970, tin(II) phthalocyanine, PcSn (SNPHCY,[142] App. 12:49). Here, the Sn(II) ion is typically too large to be accommodated inside the central cavity, and is therefore out-of-plane coordinated, positioned 1.11 Å above the plane. A free electron pair on the tin atom takes the place of an axial ligand. The full crystal structure of PcSn was published one year later in a doctoral thesis of one of the authors;[31] the atomic coordinates can be found in Appendix 27. An identical structure was reported by Kubiak and Janczak in 1992 (SNPHCY01,[143] App. 12:50–51). PcSn adopts a saucer-shaped structure, which is partly flattened by a concave overlap with a neighbor. One year later, Kobayashi *et al.* reported the structure of PcZn(H_2N-C_6H_{13}) (ZNPTCY,[144] App. 12:60). The carbon atoms of the hexylamine ligand are disordered and could not be fully determined. The zinc atom is displaced by 0.480 Å from the plane. In 1971, Fischer *et al.* reported the structure of the monohydrated, dipyridyl magnesium phthalocyanine, PcMg-O($Hpy)_2$ (MGPCPH,[145] App. 12:23). The two pyridine ligands strongly coordinate the hydrogen atoms of the water molecule and can thus be considered as being part of one large axial ligand. The central magnesium atom is situated 0.496 Å out of the plane of the isoindole nitrogen atoms. In 1973, Ukei published the structure of the monoclinic polymorph of lead(II) phthalocyanine, PcPb (PBPHCY,[133] App. 12:37–38). This is the only known structure, with the exception of the Rietveld-calculated structure of C-PcTiO (BITSAY04,[3] App. 12:52), and the predicted structure of PcReN,[146,147] where a non-distorted, shuttlecock-shaped molecule can be found. In PBPHCY, the planes of the isoindole units deviate 14° from planarity and the lead atom is displaced by 0.906 Å. Ukei[148] also reported having obtained a triclinic polymorph of PcPb in 1976. Griffiths *et al.* reported the triclinic unit-cell and a structural model of the II polymorph of vanadyl phthalocyanine, PcVO (VPHTHC) in 1976.[149] They also proposed a model for the I-PcVO, which is similar to the monoclinic modification of PcPb. The full structure of PcVO was later published in 1980 by Ziolo *et al.* (VPHTHC,[150] App. 12:58). They found the vanadium positioned 0.575 Å above the plane. Przyborowski and Hamann reported the unit-cell data of another monoclinic form of PcPb in 1982.[151] In 1982 too, Iyechika *et al.*, reported the full structure of the triclinic polymorph of PcPb (PBPHCY01,[132] App. 12:39–40). The structure contains two crystallographically different molecules, both are very similar to the molecular structure of PcSn, but less flat. No explanation is given, why the displacement of the lead atom from the plane is 1.290 Å in the triclinic compared to the 0.906 Å of the monoclinic polymorph. In 1983, unit-cell data of another polymorph of PcPb was reported by Bluhm *et al.*[62] The crystal structures of the polymorphs of I-PcTiO (BITSAY, App. 12:53), and II-PcTiO (BITSAY01, App. 12:54) were published by Hiller *et al.* in 1982.[2] Both structures share many similarities and the titanium atoms are 0.70 (I) and 0.72 (II) Å above the molecular plane, respectively. II-PcTiO is isostructural with II-PcVO.[150]

The first structure of a *cis*-complex with two axial ligands, $PcNbCl_2$ (BEJTIT, App. 14:12–13), was published in 1982 by Ukei.[152] The Nb atom is displaced 0.98 Å from the central C_8N_8 plane, and the Nb–Cl bonds are oriented along one of the two N_{br} molecular axes. The four azamethine nitrogens and the two chlorine atoms form a trigonal prism around the Nb atom. The structure of molybdenyl phthalocyanine, PcMoO, was published by Börschel and Strähle in 1984 (CUTGIH,[153] App. 12:35). This complex is isostructural with the I-polymorph of PcTiO, with the Mo atom located 0.69 Å above the plane. The structures of PcGaCl and PcAlCl were reported by Wynne in 1984.[154] While Wynne successfully resolved the structure of PcGaCl (CUWNUD, App. 12:17), he was unable to fully determine the disordered structure of one of two crystallographically different PcAlCl (CUWPAL, App. 12:2) complexes. Fortunately, Hasegawa and Sato discovered an identical, but fully ordered structure of PcAlCl (#1995-01, App. 12:3) in 1995,[155,156] the atomic coordinates can be found in Appendix 28. The gallium atom in CUWNUD is positioned 0.439 Å above the plane, while the aluminum atom in the analogous PcAlCl complex is just 0.410 Å. A second aluminum phthalocyanine, PcAl-O-AlPc (DAYYOR, App. 12:4–5) has been reported by Wynne in 1985.[122] In this dimer, the two phthalocyanine molecules (with out-of-plane coordinated aluminum central atoms) are connected through a μ-coordinating oxygen axial ligand. The aluminum atoms sit 0.459 Å above the plane of the isoindole nitrogen atoms. The phthalocyanine rings are eclipsed and located 4.27 Å apart from each other. The structure of another dimer, $(PcFeCl)_2I_2$ (DAPCIG, App. 12:12–13), has been reported by Palmer *et al.* in 1985.[157] Neutral iodine connects two molecules of PcFeCl via their axial chloro ligands. The Fe atom lies 0.30 Å above the plane defined by the four coordinating isoindole nitrogen atoms.

The structure of PcZnCl (DAMTEQ, App. 12:63) was published by Mossoyan-Deneux *et al.* in 1985.[134] As previously mentioned, the molecular structure of PcZnCl macrocycle is planar due to full overlap with an adjacent molecule. The phthalocyanines are staggered by an angle of approximately 39° and the Zn atom is 0.59 Å outside the plane of the macrocycle. Another example of a *cis* complex with two axial ligands was reported in 1985 by Goedken *et al.*, $PcTiCl_2$ (DAGXUE,[158] App. 14:19–20). In accordance with the structure of $PcNbCl_2$, the two axial chloro ligands of $PcTiCl_2$ are oriented along one of the N_{br} molecular axes. The titanium atom is located 0.84 Å above the N_4 plane. The first example of a CN8*cis* complex is $PcLu(OAc,H_2O,H_2O)$ (DICCAT,[159] App. 17:4–5), where one of the groups coordinates as a bidentate ligand to the lutetium atom and the central atom is located 1.24 Å above the N_4 plane. The first example of a CN7*cis* complex, $PcTaCl_3$ (GIFZUQ, App. 16:20), was published by Gingl and Strähle.[124,160] They found the tantalum atom positioned 1.03 Å above the plane of the four isoindole nitrogen atoms. One of the three Ta–Cl bonds is oriented nearly parallel to one of the Ta–N_i bonds, while the other two are directed towards two azamethine nitrogen atoms. A comparison of several *cis* complexes shows that the distance between the central atom and the plane of the four isoindole atoms generally increases with increasing coordination number: CN8*cis* > CN7*cis* > CN6*cis* > CN5. The structure of an iron phthalocyanine anion with a methyl axial ligand was reported by Tahiri *et al.* in 1988.[161] The Fe atom in [PcFeMe][Na{18cr6}] (GEDKAB, App. 12:16) is only 0.10 Å above the isoindole nitrogen plane.

Two structures of a phthalocyanine dimer, with non-parallel phthalocyanine rings, have been published by Gingl and Strähle; $[\{PcNb\}_2O_2Br][Br_3]$, cocrystallized with either one molecule of phthalonitrile (SAMFAN, App. 16:17), or one molecule of 1-chloronaphthalene (SAMDUF, App. 16:15–16).[162] The phthalocyanine components of both structures are very similar. The coordination number of the Nb atoms is 7, which corresponds to a capped trigonal prism. A view parallel to the macrocycles of $[\{PcNb\}_2O_2Br][Br_3]$, given in App. 16:16, shows the Nb atoms positioned approximately 1.03 Å above the phthalocyanine rings. Gingl reported the structure of $PcNbCl_3$ (#1990-01, App. 16:11–12) in his doctoral thesis in 1990;[124] the atomic coordinates can be found in Appendix 37. A second, identical structure of $PcNbCl_3$ (NEXVAN, App. 16:9–10) was reported some years later by Cellucci *et al.*,[163] and found to be isostructural with $PcTaCl_3$ (GIFZUQ,[124,160] App. 16:20). In App. 16:10 and App. 16:12 it is visible that the crystal structures of $PcNbCl_3$ contain actually two chiral isomers, related by a crystallographic inversion center.

Bluhm *et al.*[164] calculated the structure of a further polymorph of PcTiO (IV-PcTiO: BITSAY03, App. 12:55), using a Rietveld analysis. The structure of the *Y*-polymorph (BITSAY02, App. 12:56) was obtained by Oka *et al.* in a similar fashion.[165] Both calculations used the molecular structure of II-PcTiO as a starting structure. In 1992, Kubiak and Janczak[143] reported the structure of PcSn (SNPHCY01, App. 12:50–51), and showed it to be identical to the same structure found by Friedel *et al.* in 1970.[31,142] The structure of $(H_{11}C_5)_8PcRh\text{-}Me$ (HASZEG, App. 12:46–48) was published by Chen *et al.*[166] where the phthalocyanine moiety of $(H_{11}C_5)_8PcRh\text{-}Me$ is almost free from distortion, unlike in most of the other substituted phthalocyanine complexes.[23] Only one isoindole unit is slightly twisted due to the interaction between two substituents. Okada *et al.*[3] used a Rietveld analysis to ascertain the structure of the C-polymorph of PcTiO (BITSAY04, App. 12:52). They found the structure to be similar to that reported for $[PcGaF]_n$ and the monoclinic polymorph of PcPb. Inami *et al.* used the same calculation method to determine the crystal structure of PcGa(OH), NALSOI.[167,168] Gorsch *et al.* reported the crystal structure of $[PcZrCl_3][PNP][2CH_2Cl_2]$ (YOKWEA,[169] App. 16:26–27), in which the Zr exhibits 7-coordination with the ligands forming a square base-capped trigonal prism, similar to the structures of $PcTaCl_3$ and $PcNbCl_3$, with the Zr lying 1.18 Å above the N_4 plane. Aßmann *et al.* reported that the NO_2 ligand in $[PcZn\text{-}ONO][PNP][H_2O]$ (ZUHCIO, App. 12:64) is coordinated to the zinc atom by one of its oxygen atoms, with the central Zn atom being displaced 0.469 Å out of the N_4 plane.[170] In 1998, Cellucci *et al.*[163] reported the structures of the $PcNbCl_3$ complex (NEXVAN, App. 16:9–10) and of the triclinic polymorph of $PcNbCl_2$ (BEJTIT01, App. 14:14). They found $PcNbCl_3$ to be isostructural with the structure of $PcTaCl_3$ (GIFZUQ,[124,160] App. 16:20), reported by Gingl and Strähle, and identical to another $PcNbCl_3$ structure, also previously reported by Gingl (#1990-01,[124] App. 16:11–12). The Nb atoms are displaced from the N_4 plane by 0.898 Å in BEJTIT01 (identical to BEJTIT) and by 1.039 Å in NEXVAN.

Chen *et al.* reported the structure of the dimeric $(H_{11}C_5)_8PcRh\text{-}H\text{-}RhPc(C_5H_{11})_8$ (TESFAY, App. 12:44–45).[171] The hydrogen atom between the rhodium atoms could not be resolved from the crystal structure, however its presence, the authors justified by the metal–metal separation of 3.347 Å being too long for a Rh–Rh single bond. The Rh atoms are not significantly displaced from the planes of the ligands, further support justifying the bridging hydrogen's existence.

The phthalocyanine ligands are planar and staggered by 35°. In the [PcMg-Cl][PNP][Me_2CO] complex, (TAYSIV, App. 12:34), Aßmann *et al.* reported that the central magnesium is displaced by 0.570 Å from the N_4 plane.[172] In [PcIn(-NCO)$_2$][PNP] (ZUFZOP,[173] App. 14:6–7), the two cyanate ligands coordinated to the In atom on the same side of the phthalocyanine ring. The two cyanate ligands are orientated along one of the N_{br} molecular axes, yet their respective conformation and their bonding to the In atom is structurally different. The distance of the In atom from the N_4 plane is 0.988 Å. Yamasaki *et al.* reported the crystal structures of II-PcGaCl (CUWNUD01, App. 12:18) and XI-PcGa(OH), NALSOI01, both derived from Rietveld analyses[174] (the latter structure could not be correctly displayed by software available to me, distances between adjacent molecules were in the range of single-bonds). In 1997, Kubiak and Ejsmont reported the crystal structure of a *cis*-7-coordinated zirconium phthalocyanine consisting of two iodo and one phthalonitrile ligand, $PcZrI_2$(PN) (NOFNUR, App. 16:22–23).[175,403] The phthalonitrile ligand (PN) is coordinated to the central Zr atom via one of its cyano groups. The Zr atom is displaced by 1.066 Å out of the N_4 plane. In the [PcIn(-O,-O′CO)][NBu_4] (RERGOK,[177] App. 14:8–9) structure, Schweiger *et al.* reported that the In atom is coordinated by a bidentate carbonate ligand. The In atom is located 0.903 Å above the N_4 plane. A *cis*-6-coordinated In phthalocyanine with two fluoro ligands was found by Schweiger *et al.* in 1998 [$PcInF_2$][NBu_4][H_2O] (NAPJIX, App. 14:4–5).[178] The distance between the In atom and the N_4 plane is 0.953 Å. The same authors also reported the structure of a mixed complex of *cis*-7-coordinated [$PcTaF_3$] and *cis*-8-coordinated [$PcTaF_4$]; [$(PcTaF_3)_{1-x}(PcTaF_4)_x$][$NBu_4$] (NAPRAX,[179] App. 16:28–29 and App. 17:1), in which the tantalum atom is located 1.234 Å above the N_4 plane. It is unknown why the structure does not show two different dislocation distances for the Ta atom. Schweiger *et al.* report that in [PcTl-I][CH_2Cl_2], (NELCIQ, App. 12:57), the Tl atom is coordinated to the four isoindole nitrogen atoms and an iodine, forming a square pyramidal arrangement,[180] in which the Tl atom is displaced by 0.959 Å from the N_4 plane. The structure of PcRe–RePc (NEKRAW, App. 12:41–42) was reported by Göldner *et al.* in 1998.[181] The structure resembles that of PcAl-O-AlPc, with the exception that the two metal atoms in NEKRAW are directly linked by a single bond. The distance between the two phthalocyanine moieties is 3.14 Å, compared to a much larger 4.27 Å in the case of $(PcAl)_2O$. The staggering angle is 0° and the Re atoms are located 0.426 Å outside the N_4 plane. Schweiger *et al.* also reported the formation of a *cis*-7-coordinated hafnium phthalocyanine [$PcHfF_3$][NBu_4] (POQYID,[182] App. 16:4–5), where the Hf atom is displaced by 1.218 Å from the N_4 plane. In 1998, Kubiak and Razik reported the first phthalocyanine dimer based on a CN6*cis*

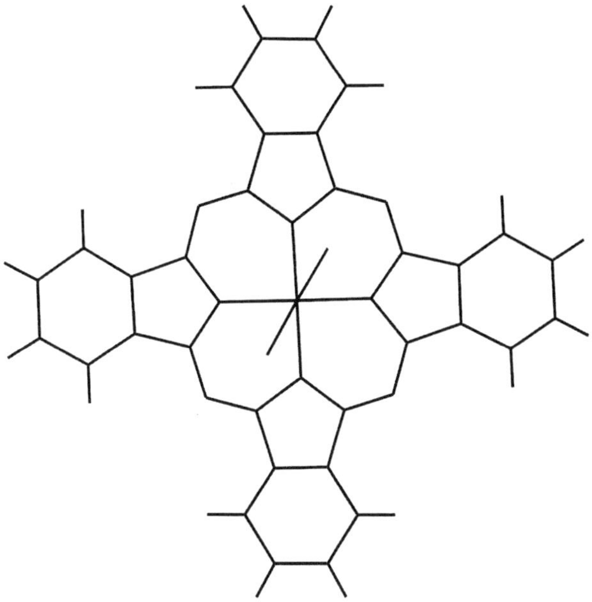

(a)

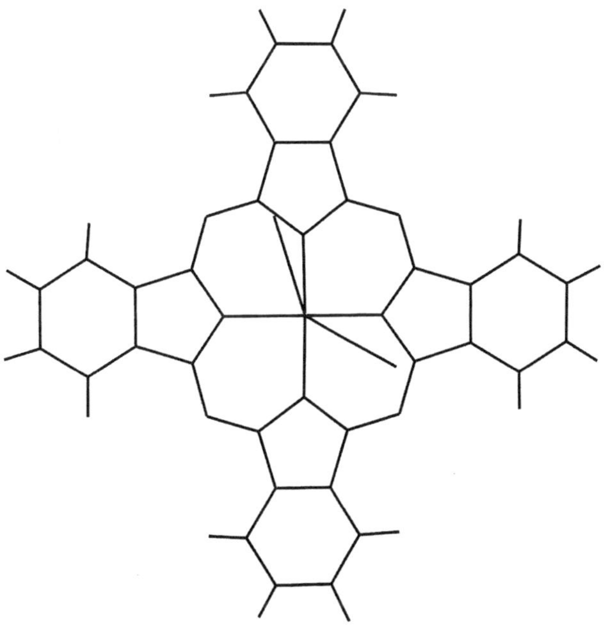

(b)

Figure 25. Perpendicular view on two CN6*cis* complexes; (a) $PcTiCl_2$ (DAGXUE[158]) with symmetrical orientation of the axial ligands towards the phthalocyanine ligand and one macrocycle of $[PcSbI]_2$ (NILVUZ[183]) with unsymmetrical orientation of the axial ligands.

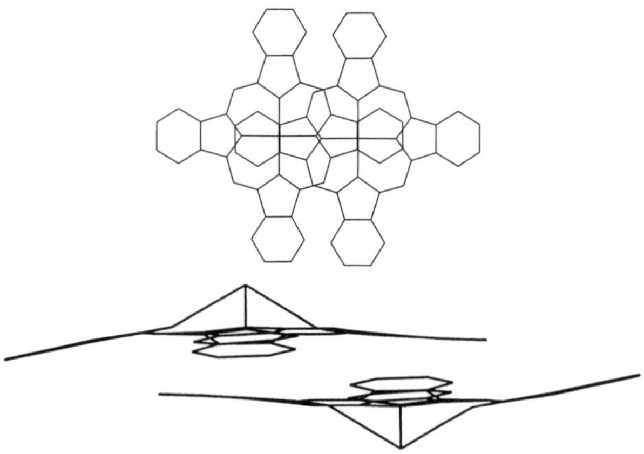

(a) PcSn, SNPHCY01

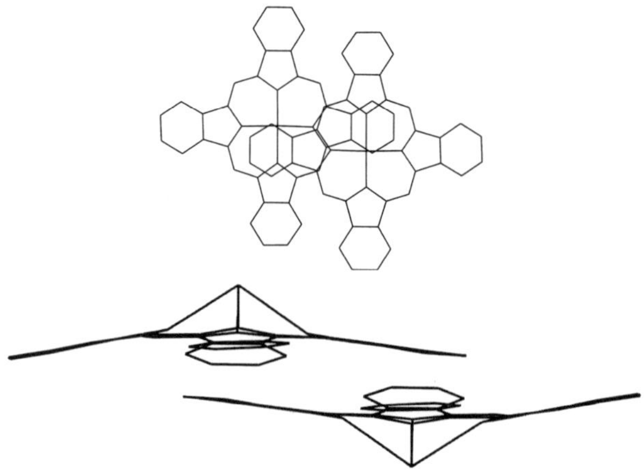

(b) triclinic-PcPb, PBPHCY01

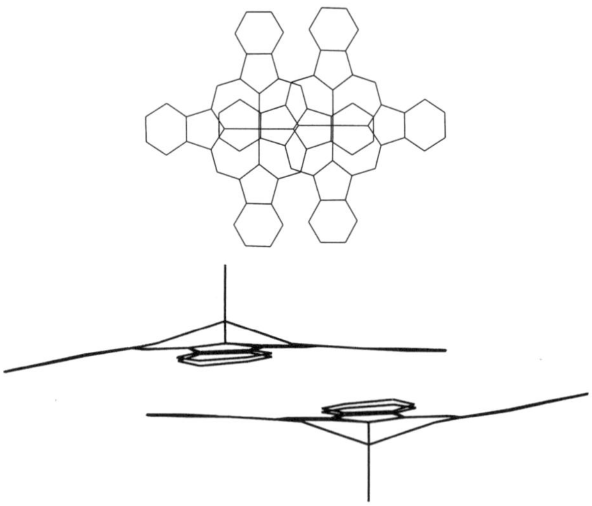

(c) II-PcTiO, BITSAY01

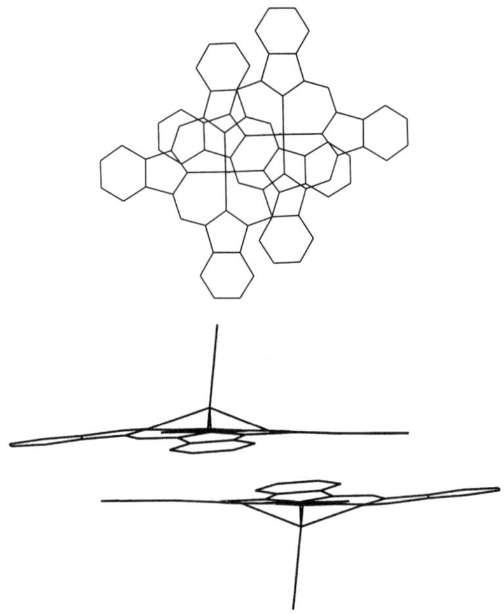

(d) PcInI, BIJQOA

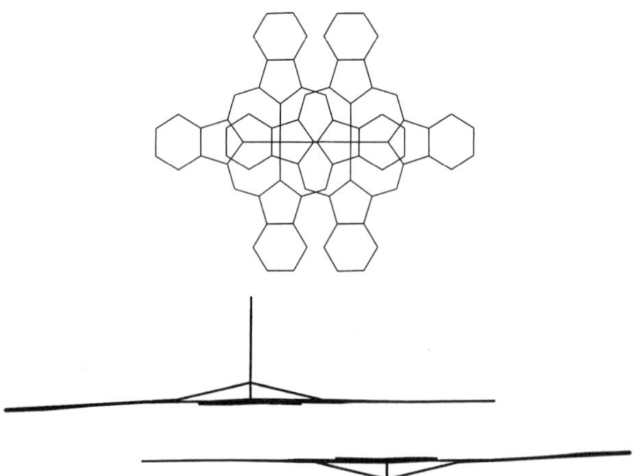

(e) PcGaCl, CUWNUD

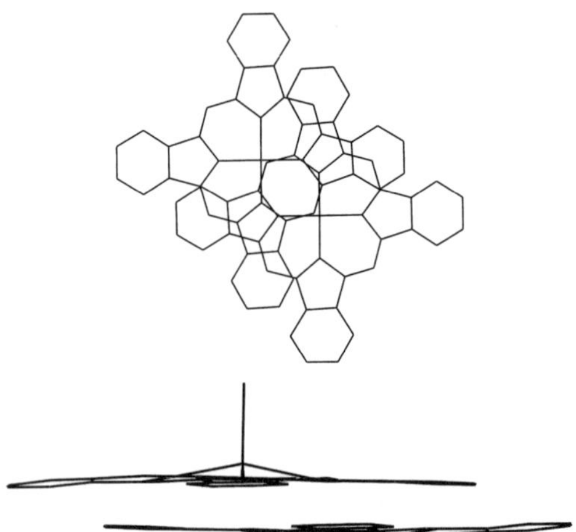

(f) PcAlCl, #1995-01

Figure 26. Perpendicular and side-on views of the concave–concave overlapping pairs of phthalocyanine complexes with out-of-plane coordinated central atoms.

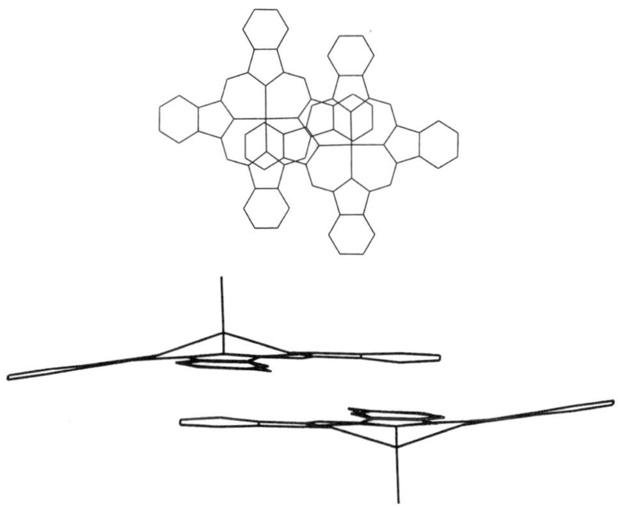

(g) I-PcTiO, BITSAY

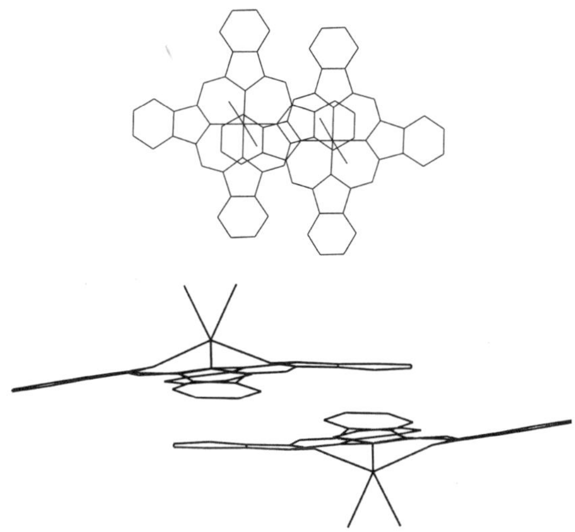

(h) $PcTiCl_2$, DAGXUE

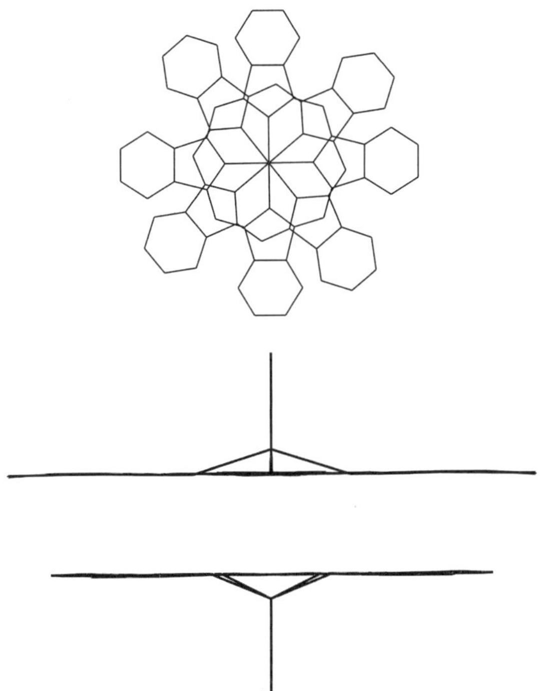

(i) PcZnCl, DAMTEQ

Figure 26. Continued.

complex, each iodine ligand of the antimony complex $PcSb(\text{-I-})_2SbPc$ (NILVUZ, App. 14:15–16) is μ-coordinated to two Sb atoms.[183] In contrast to the other *cis*-6-coordinated complexes, the coordination geometry around the central atoms is unsymmetrical. In $PcTiCl_2$ (DAGXUE, Figure 25a), the axial ligands are symmetrically coordinated with regards to the phthalocyanine macrocycle, while in NILVUZ (Figure 25b), the axial ligands are oriented towards one half of the phthalocyanine ring. It is not clear from the crystal structure if a long-range interaction between the antimony central atom and an azamethine N atom of a neighboring macrocycle (4.85 Å) could feasibly be considered as an additional ligand. If this were the case, then the ligand geometry around the central atom would resemble that of $PcZrI_2(PN)$. One year later, Kubiak and Ejsmont reported the isolation of an bismuth analogue to the NILVUZ structure, $PcSb(\text{-I-})_2SbPc$ (HOXCIG, App. 14:1–3), with a similarly strong asymmetry.[184]

In 1998, Ejsmont and Kubiak reported the structure of PcVI (SAXZEW, App. 12:59), in which the vanadium atom is located 0.322 Å above the N_4 plane.[185] The I_3 ligand in the molybdenyl phthalocyanine, $[PcMo{=}O][I_3]$ (FEWPEC, App. 12:36), is μ-coordinated between the oxygen of one phthalocyanine molecule and the Mo atom of a neighboring molecule, resulting in the formation of 1D stacks.[186] In the same year, Janczak and Kubiak reported the structure of PcInI (BIJQOA, App. 12:20), where the In atom is located 0.737 Å above the N_4 plane.[187] The same research group reported the structure of $[PcSb][I_3]$ (HOVCOK,[188] App. 12:43) where the Sb atom is located 0.966 Å above the plane of the isoindole nitrogen atoms. In contrast to the FEWPEC structure, the iodine anion coordinates unsymmetrically side-on to the macrocycle. The structure of (PcFe-I)-I-I-(I-FePc) (FIQFUG, App. 12:14–15), reported by Janczak *et al.* in 1999,[189] resembles the PcFe-Cl analog (DAPCIG, App. 12:12–13). The iodine molecule coordinates to the axial ligands of the iron center atoms. In 1999, a novel type of bismuth phthalocyanine complex has been reported by the same authors, $(PcBi)_4(Bi_4I_{16})$ (LOKSAF,[190] App. 16:1–3). In this complex, 4 PcBi

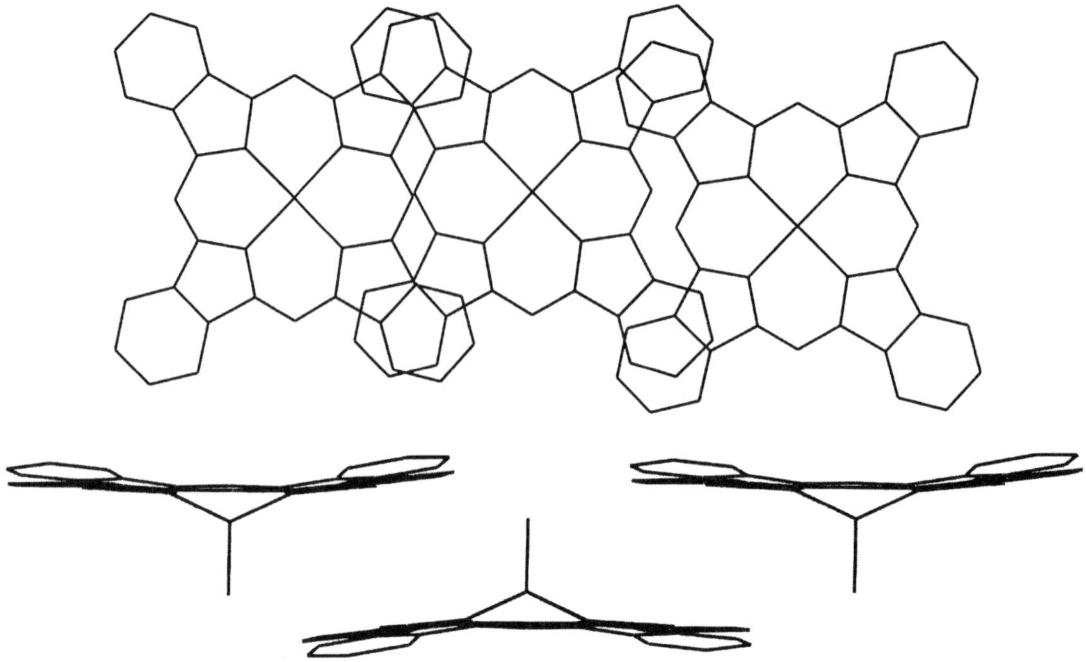

(a) II-PcTiO, BITSAY01

Figure 27. Perpendicular and side-on views on the basic arrangement patterns of convex–convex overlapping phthalocyanine complexes with an out-of-plane coordinated central atoms.

units coordinate around a Bi_4I_{16} unit. The coordination number of ligands located around the bismuth atom is seven. In the same year, Schweiger *et al.* reported two structures of thallium phthalocyanines.[191] Similar to other $PcMHal_2$ complexes, the chlorine ligands in $[PcTlCl_2][NBu_4][0.5Et_2O]$ (BOFQES, App. 14:21–22) are oriented along one of the N_{br} molecular axes. Several lithium, magnesium, and zinc phthalocyanines with triphenylphosphinoxide as axial ligands were reported by Hückstädt *et al.* in 2000,[136] $Pc(Li\text{-}OP\phi_3)_2$ (MAKJEN, App. 13:6) is a complex similar to $PcTl_2$ (HEBHOL, App. 13:9) with a metal and an axial ligand on each side, $[PcLi][Li(OP\phi_3)_4][EtOH,H_2O]$ (MAKHEL, App. 11:37–38) is a CN4 complex and $[PcMg\text{-}OP\phi_3][2py]$ (MAKHOV, App. 12:26), $[PcMg\text{-}OP\phi_3][CH_2Cl_2]$(MAKHUB, App. 12:27) $[PcMg\text{-}OP\phi_3][THF]$(MAKJAJ, App. 12:28), $PcMg\text{-}OP\phi_3$ (MAKHIP, App. 12:33), and $[PcZn\text{-}OP\phi_3][Et_2O,1\text{-}ClN]$ (MAKHAH, App. 12:62) are typical CN5 complexes. In 2001, Tutaß *et al.* reported a series of monomeric and dimeric yttrium phthalocyanines;[192] $[PcY(Cl)(py)(OP\phi_3)]$ $[0.5Et_2O]$ (BORQUU, App. 16:21), $[PcY(Cl)(dmf)_3]$ $[Et_2O]$ (BORQII, App. 17:10–11), and $[PcY(Cl)$ $(\text{-}OSMe_2)_4][Et_2O,2MeOH]$ (BORQOO, App. 17:12–13) are monomers while $PcY(\text{-}(Cl)_2\text{-},(O(Hpy)_2)_2\text{-})YPc$ (BOSQAB, App. 17:14–15) and PcY(L,L′,-Cl-,-Cl-)YPc (BOSJAU, App. 17:16–17) are dimers. In the same year, Janczak and Kubiak reported another dimeric phthalocyanine with two μ-coordinated iodine ligands, similar to NILVUZ and HOXCIG, $PcSb(\text{-}Cl\text{-})_2SbPc$ (WOVREE,[193] App. 14:17–18). In the same year, four crystal structures of PcMg (XETZIF, App. 11:29–32) were obtained at different temperatures by Mizuguchi[60,61] and Janczak and Kubiak.[59] The lower the temperature is, the further out-of-plane is the magnesium located. The "7-coordinate" tin phthalocyanine complex, $PcSn[O\text{-}C(O)\text{-}C_{10}H_{21}]_2$ (CCDC173622, App. 17:6–7), reported by Silver *et al.*,[128] is actually a CN8*cis* complex similar to BOFQAO (App. 17:8–9). While in BOFQAO, the bond lengths of the metal-axial ligands are approximately identical, those in CCDC173622 are different from each other and the bidentate ligand coordinates asymmetrically to the tin atom. In 2002, Janczak and Idemori reported another $[PcM]_4[M_4I_{16}]$ structure with antimony central atoms (M=Sb; YOHMAJ,[194] App. 16:18–19).

2. Solid-State Arrangement

As mentioned earlier, all of the *cis* complexes (CN5, CN6*cis*, CN7*cis*, CN8*cis*) have phthalocyanine macrocycles that adopt a "bent" geometry with two strongly different sides. The concave side is that which comprises the macrocycle-component alone, while the convex side

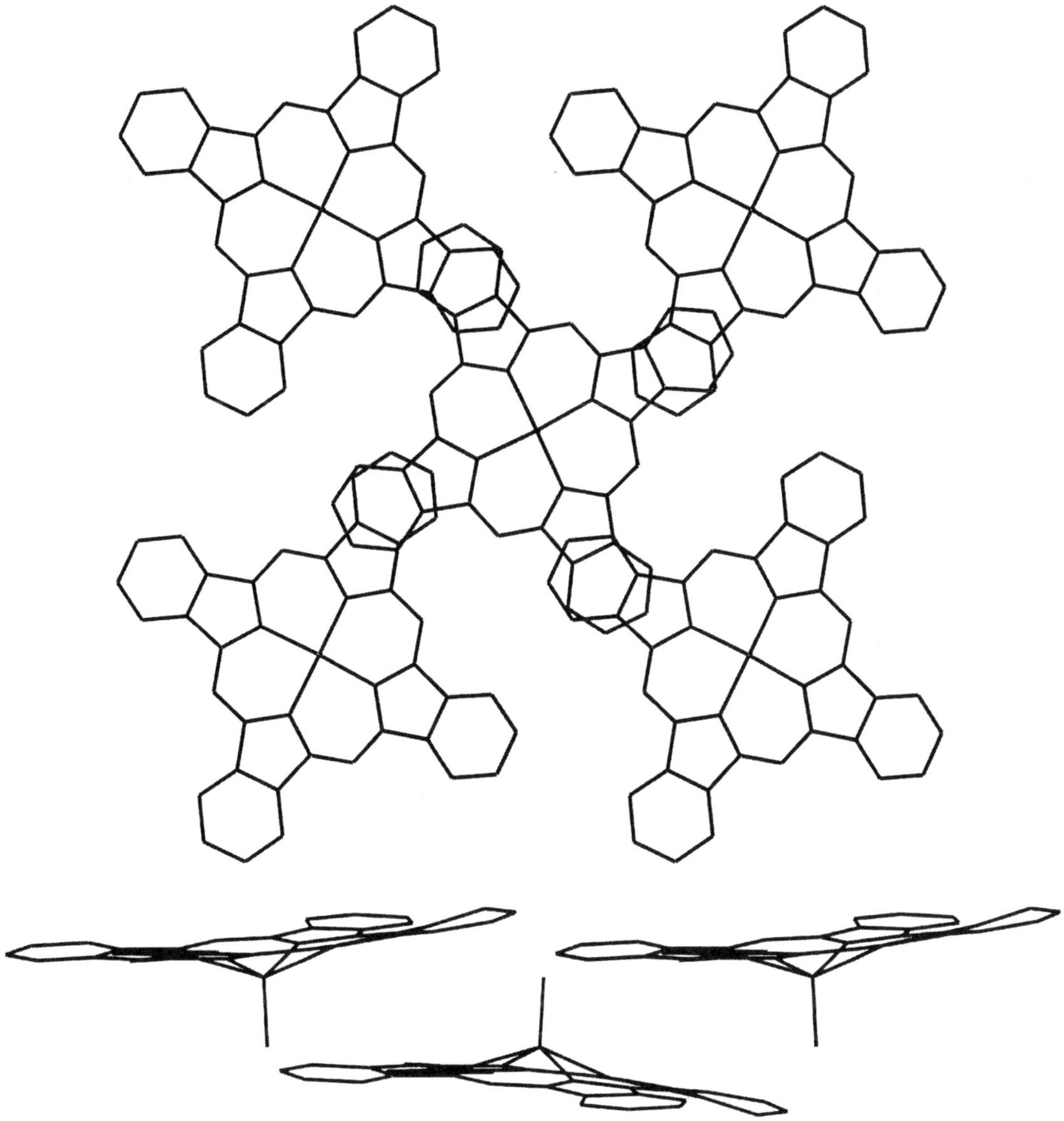

(b) I-PcTiO, BITSAY

Figure 27. Continued.

contains the macrocycle, the central atom and coordinated axial ligand(s). In these complexes, different sides have different interactions with neighbors, which therefore lead to different crystal environments. The exception is magnesium phthalocyanine, where the Mg atom is an out-of-plane coordinated, yet statistically sits on both sides of the phthalocyanine ring. The macrocycle is planar and both sides have crystallographically identical environments (Figure 15c,d).

Cis complexes generally interact with each other via similar sides, either concave–concave or convex–convex. Exceptions to this include the monoclinic polymorph of PcPb, the C polymorphs of PcTiO, and PcReN. These complexes interact via the two different sides, i.e., concave–convex. This is due to interactions between both the axial ligands and the central atom, as in PcTiO and PcReN, or between the central atom and two sets of isoindole nitrogen atoms, as in monoclinic PcPb. These complexes could also be considered as one-dimensional polymers, similar to $[PcGaF]_n$.

The concave–concave interactions in CN5 complexes with only slightly deformed macrocycles resemble the interactions observed in the CN4 complexes (Figure 26 and 11). Molecules overlap either along the N_i molecular axis, as in the SNPHCY01, BITSAY01, CUWNUD complexes, along the N_{br} molecular axis in

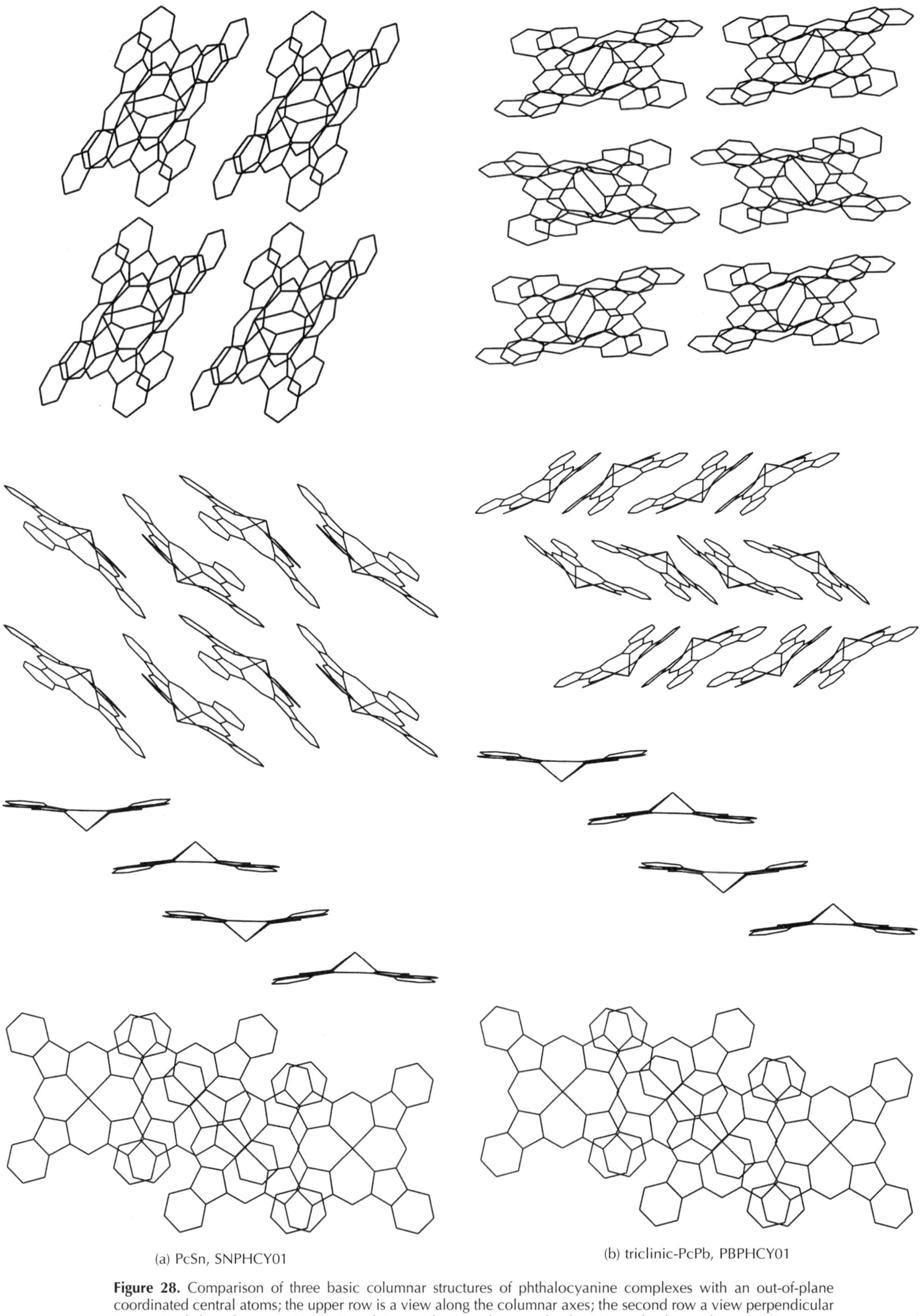

(a) PcSn, SNPHCY01

(b) triclinic-PcPb, PBPHCY01

Figure 28. Comparison of three basic columnar structures of phthalocyanine complexes with an out-of-plane coordinated central atoms; the upper row is a view along the columnar axes; the second row a view perpendicular to a vertical sheet from row 1; row 3 and 4 are side-on and perpendicular views of columns from row 1.

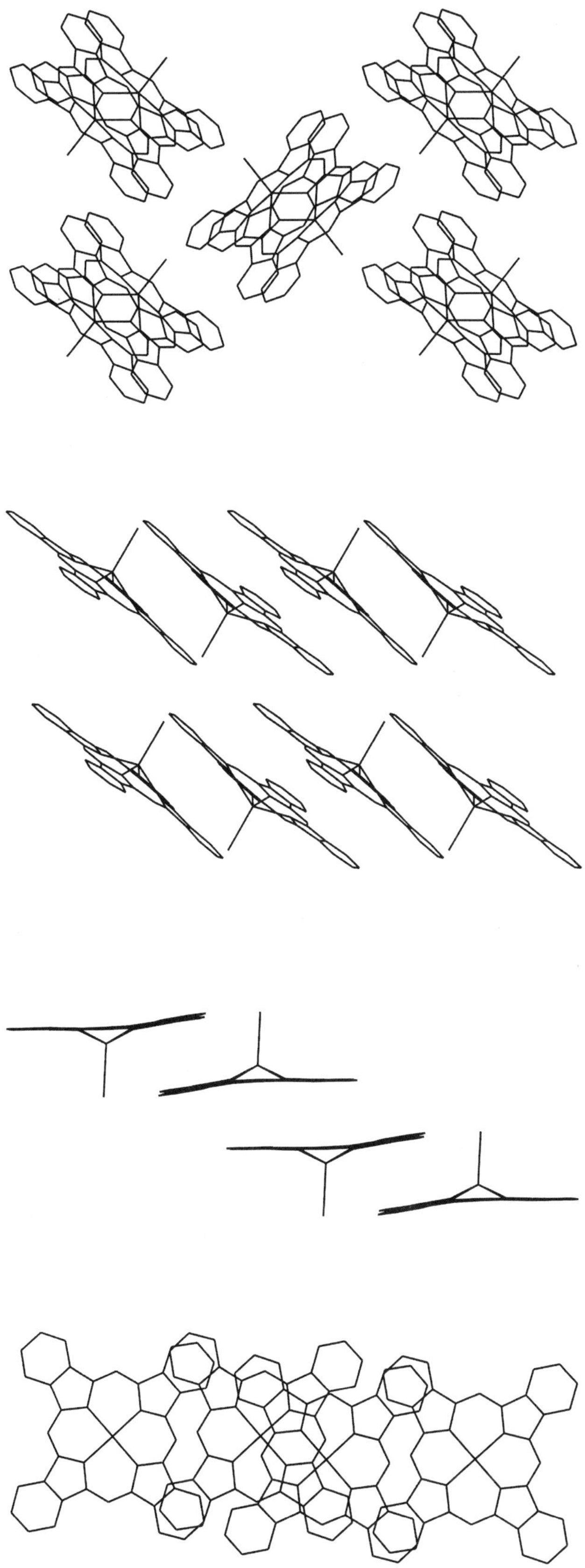

(c) PcInI, BIJQOA

Figure 28. Continued.

BIJQOA, #1995-01, or along an intermediate axis as in PBPHCY01, BITSAY, or DAGXUE. A fourth option includes the possibility of complete overlap with staggered macrocycles, as seen in DAMTEQ.

The basic interactions of the convex sides of the macrocycles are shown in Figure 27. Either one molecule coordinates to two others (Figure 27a), corresponding to an overlap along the N_{br} molecular axis, or it coordinates to four molecules (Figure 27b). While the latter leads to layer-like arrangements, the former is found in columnar-type arrangements. Three main columnar-type arrangements are known (Figure 28a–c, upper row): the column axes form either a tetragonal network with all columns identical, as in SNPHCY01 (left side), a tetragonal network with columns of different orientation along one network axis, as in PBPHCY01 (middle), or a pseudo-tetragonal network, as seen in BIJQOA (right side). The second row in Figure 28 shows a side view onto the vertical columns from the upper row; the columns are parallel in SNPHCY01 and PBPHCY01 while alternately related in BIJQOA. Molecules adopt a slipped-stacked arrangement in the columns; a parallel view of which is given in the third row, and perpendicular view in the fourth row. In SNPHCY01, the molecule pairs are typically stepwise-dislocated, where the dislocation is smaller in PBPHCY01 and still smaller in BIJQOA. This dislocation is due to the different directions of overlap in the convex and the concave pairs, ranging from between *ca.* 135° in SNPHCY01 and *ca.* 15° in BIJQOA. A comparison of the slipped-stacked column structures in CN4, CN5, CN5*, and CN6*tr* complexes are shown in Figure 15. Due to the different overlap directions mentioned above, the stacking geometries of CN5 slipped-stacked columns shows a reduced linearity compared to the CN4 or CN6*tr* complexes.

Layer-type arrangements in crystal structures of several CN5 molecules are shown in Figures 29–34. The difference between the structures of the PcAlCl (Figure 29), I-PcGaCl (Figure 30), and PcZnCl (Figure 31) complexes lies in the differently overlapping geometries of the molecules in the concave–concave layers. Two molecules of PcAlCl (#1995-01) overlap along the N_{br} molecular axis; in I-PcGaCl, the molecules overlap along the N_i molecular axis, while in PcZnCl they are staggered. While the molecules in convex–convex layers of PcAlCl, PcGaCl, and PcZnCl are not observed to overlap, those in BITSAY (Figure 32), BITSAY02 (Figure 33), and BITSAY03 (Figure 34) do. In BITSAY and BITSAY02, one molecule is found to overlap with four molecules from the neighboring layer.

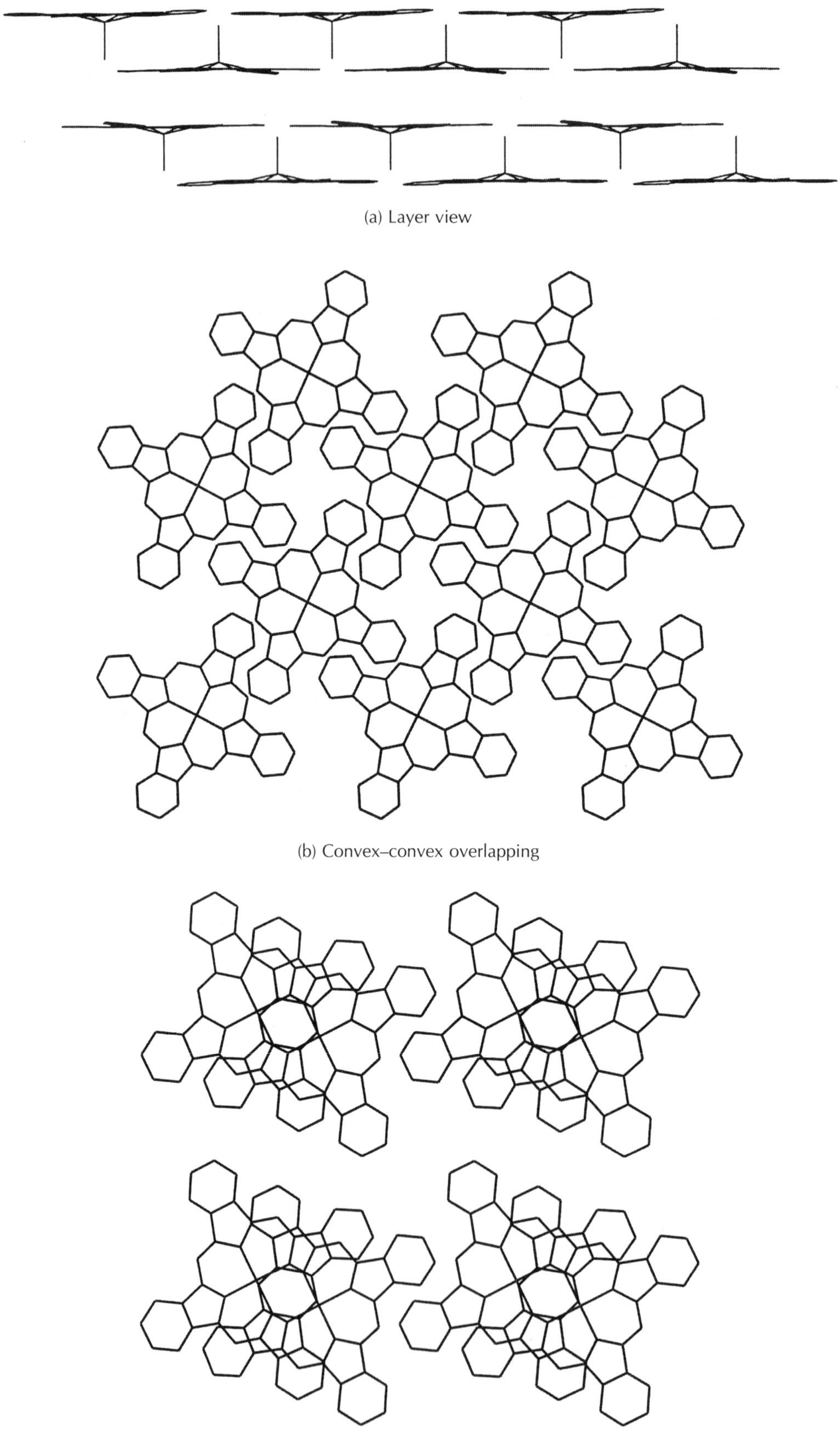

(a) Layer view

(b) Convex–convex overlapping

(c) Concave–concave overlapping

Figure 29. The crystal structure of PcAlCl, #1995-01.[155,156]

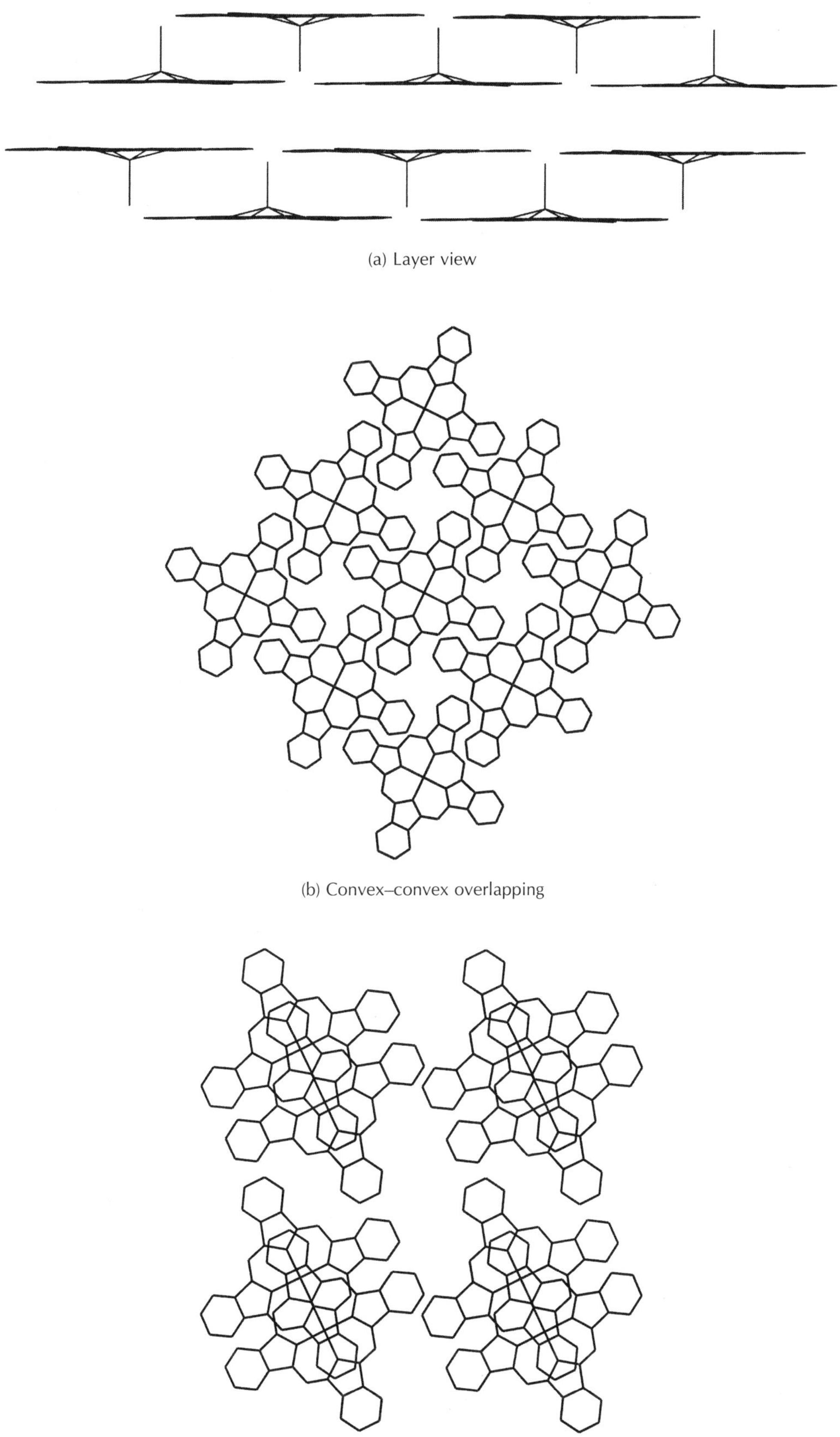

(a) Layer view

(b) Convex–convex overlapping

(c) Concave–concave overlapping

Figure 30. The crystal structure of PcGaCl, CUWNUD.[154]

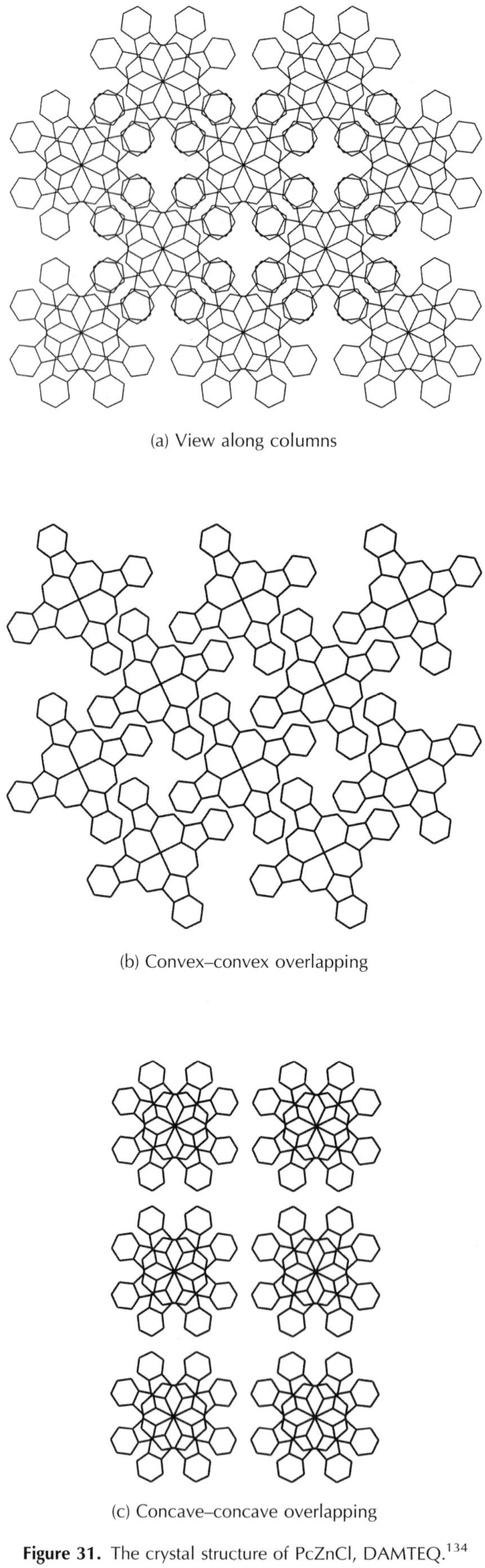

Figure 31. The crystal structure of PcZnCl, DAMTEQ.[134]

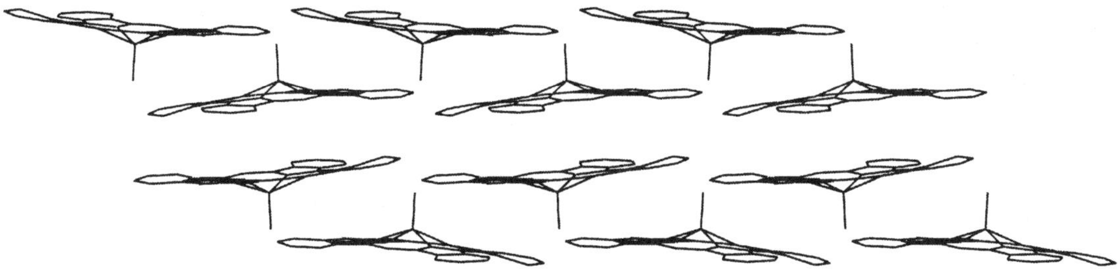

(a) Layer view

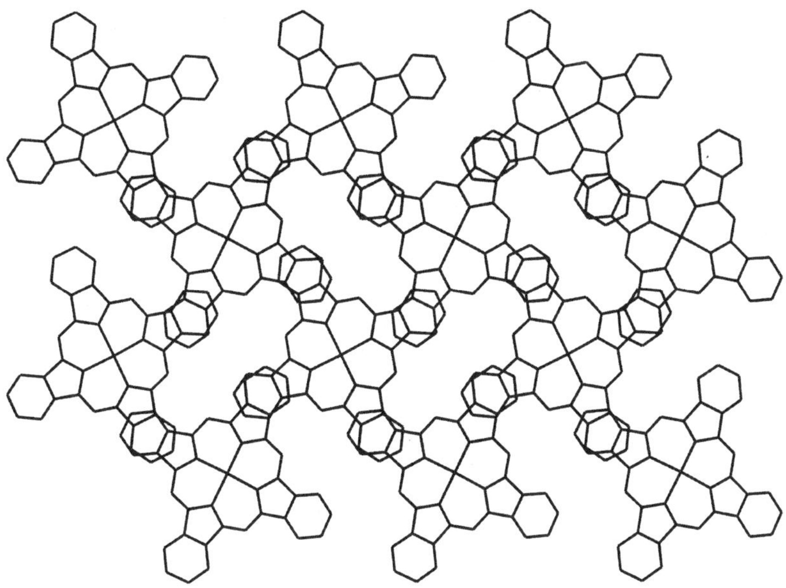

(b) Convex–convex overlapping

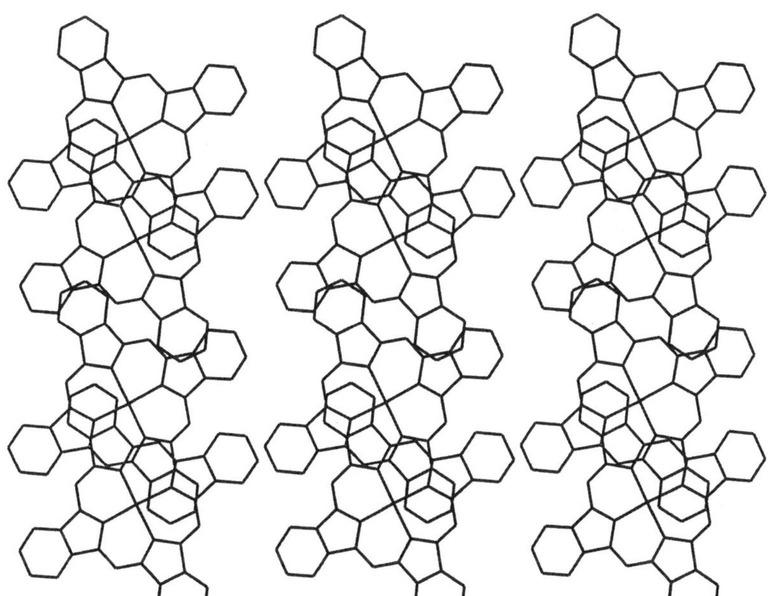

(c) Concave–concave overlapping

Figure 32. The crystal structure of I-PcTiO, BITSAY.[2]

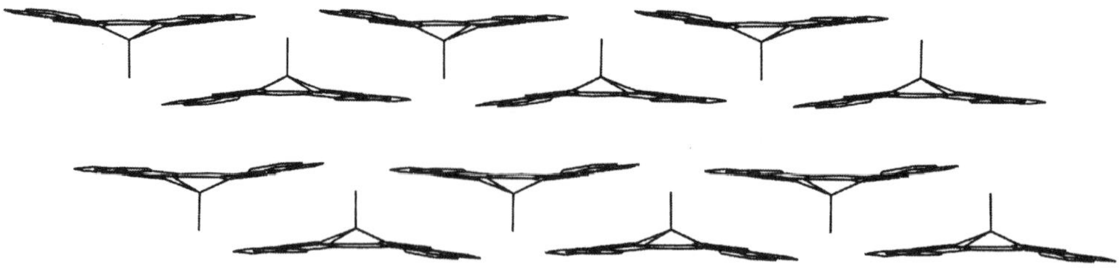

(a) Layer view

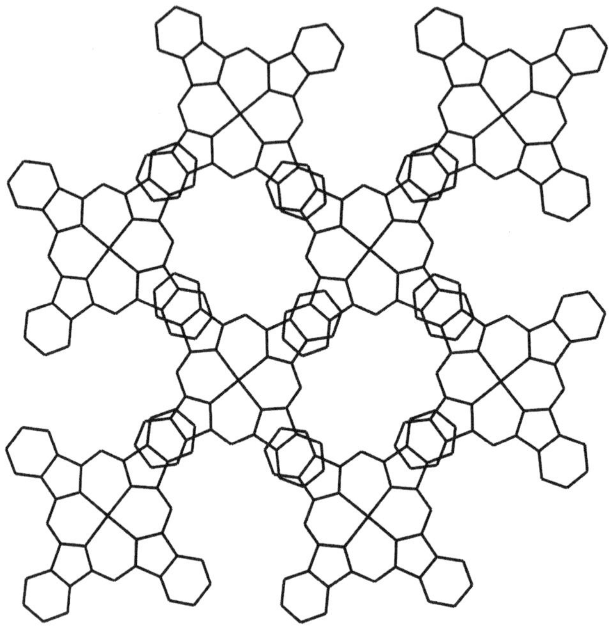

(b) Convex–convex overlapping

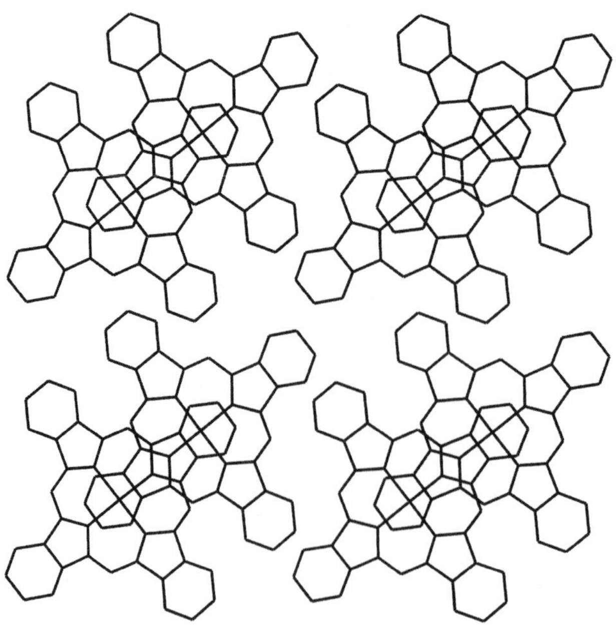

(c) Concave–concave overlapping

Figure 33. The crystal structure of Y-PcTiO, BITSAY02.[165]

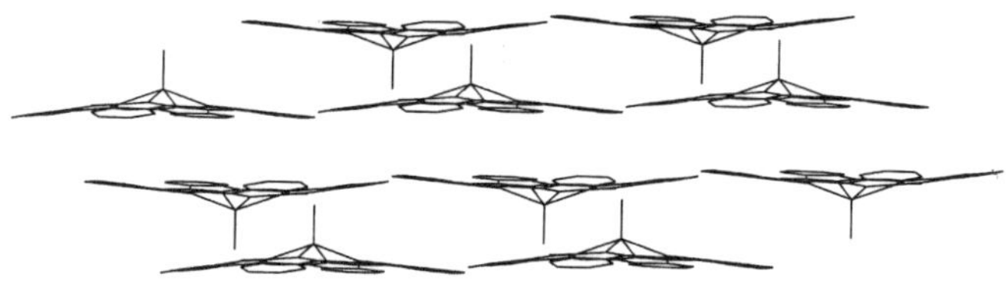

(a) Layer view

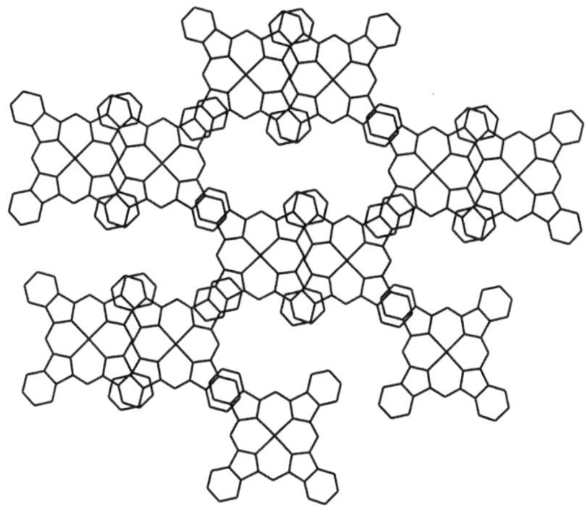

(b) Convex–convex overlapping

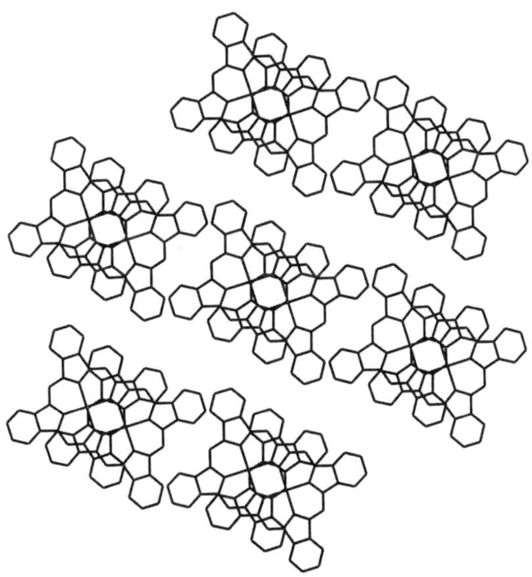

(c) Concave–concave overlapping

Figure 34. The crystal structure of IV-PcTiO, BITSAY03.[164]

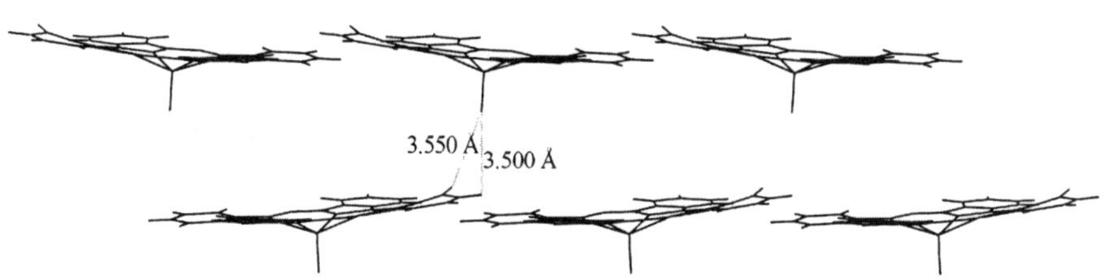

Figure 35. Interaction between aromatic hydrogen (layer 1) and oxygen atoms (layer 3) of PcTiO complexes in the I-polymorph. Layer 2 with inversed molecules is omitted for clarity.

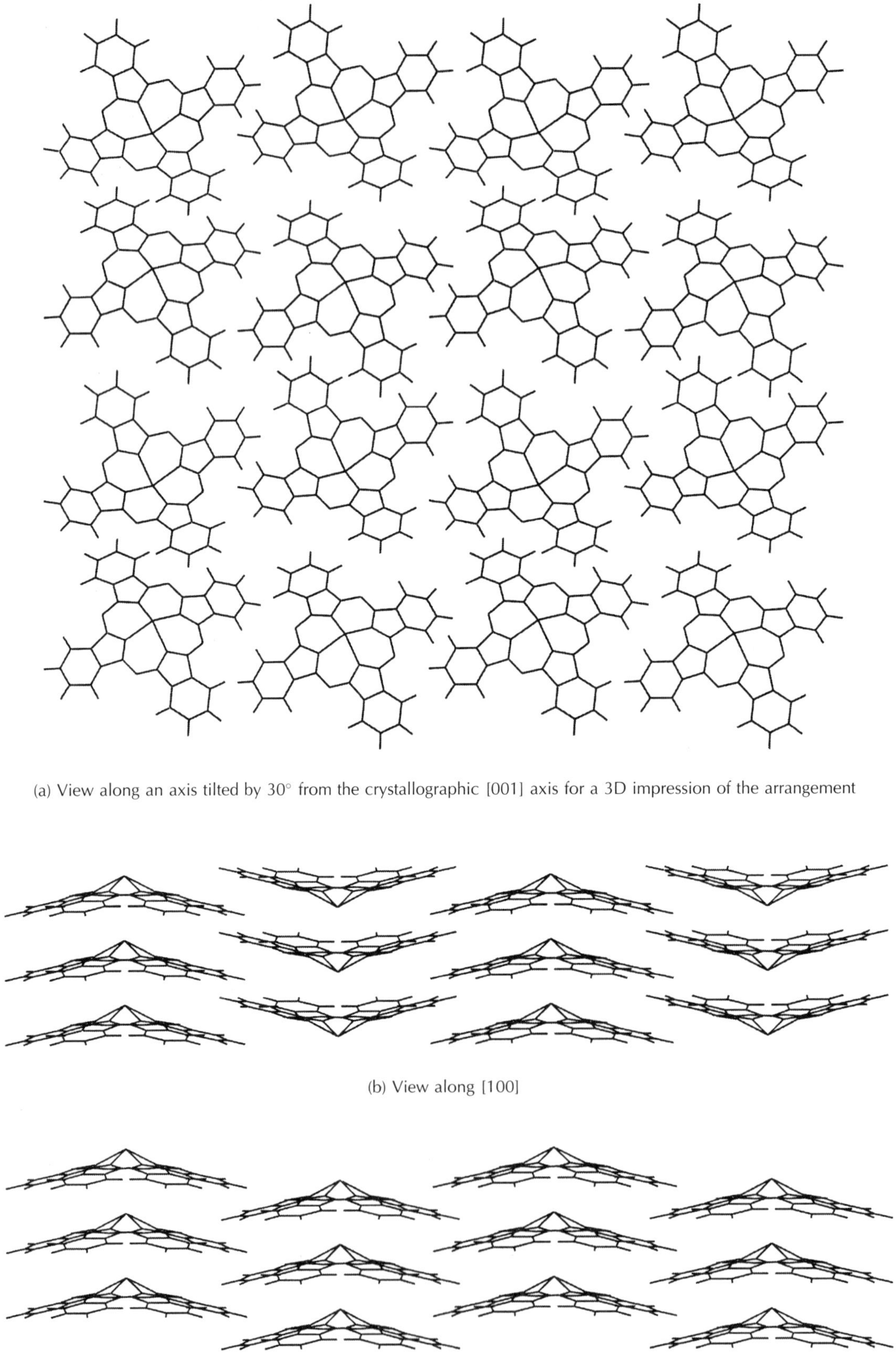

(a) View along an axis tilted by 30° from the crystallographic [001] axis for a 3D impression of the arrangement

(b) View along [100]

(c) View along [010]

Figure 36. The crystal structure of the monoclinic PcPb, PBPHCY.[133]

Table 13. Hydrogen Bonding in Convex–Convex Overlap in Columnar Structures

Compound	Intracolumnar hydrogen bonding/Å < 3.5Å	Intercolumnar hydrogen bonding/Å < 3.5Å
PcSn (SNPHCY01)	3.304, 3.346	3.237
II-PcTiO (BITSAY01)	2.650, 2.768	2.512
Tri-PcPb (PBPHCY01)	n/a	n/a

Table 14. Hydrogen Bonding in Convex–Convex Overlap in Layer Structures

Compound	Hydrogen bonding/Å < 3.0Å
PcAlCl (#1995-01)	2.788, 2.788, 2.869, 2.907
PcGaCl (CUWNUD)	2.742, 2.822, 2.852, 2.916
PcZnCl (DAMTEQ)	2.865, 2.865, 2.865, 2.865
I-PcTiO (BITSAY)	2.409, 2.613, 2.798, 2.954
Y-PcTiO (BITSAY02)	2.618, 2.677, 2.710, 2.730
IV-PcTiO (BITSAY03)	2.407, 2.795, 2.518, 2.862

Table 15. Structural Description of Complexes with an Out-of-plane Coordinated Central Atom (Monomers)

Compound	Code [Ref.]	Figure	Structural description
[PcAs][I_3]	CCDC180524[303]	12:1	Slipped-stacked columns along [100]; neighbor columns along [001] parallel, differently oriented along [010]; concave overlap along N_{br} molecular axis, convex overlap along intermediate axis
[PcCd-Cl][NEt_4][py]	#1997-01[304]	12:6–7	Isolated molecules in layers parallel (100)
[C_{2v}-1,2-NcCo-py][2py]	2001-03[135]	19:5–6	Slipped-stacked double-columns along [001], view parallel to the macrocycles along [570]; Co is hexa coordinated, py-Co-N_{br} (adjacent macrocycle)
PcCu-$OReO_3$	QEMMIE[306]	12:9	Layers parallel (100), bands along [120] or [−120]; Cu is hexa coordinated, O-Cu-N_{br} (adjacent macrocycle)
[PcFe-Me] [Na.cntnd.18c.6][THF,ϕCl]	GEDKAB[161]	12:16	Bands along [001], honeycomb structure where six bands form a channel for the counterions
PcIn-Sϕ	1998-03[308]	12:21–22	Concave overlapping double-bands along [100], concave overlap along N_i molecular axis
[PcMg-OH_2][2py]	MGPCPH[145]	12:23	Concave overlapping double-bands along [001], concave overlap along N_i molecular axis; layers parallel (010)
[PcMg(OH_2)]	#2002-06[309]	12:24	Sheets parallel (001); columns along [100]; strong hydrogen bonding present; stacks consist of fish-bone stacked concave pairs; concave overlap along N_{br} molecular axis
[PcMg(OH_2py)]$_2$[py]	#2002-07[309]	12:25	Columns along [101], columns made from alternately up and down pointing molecules; every second upward (and downward, respectively) pointing molecule is crystallographically different; in column nearly full concave overlap, shift along N_i molecular axis; convex overlap shift along N_{br} molecular axis; the hydrogen atom of water is strongly bonded to the azamethine of an adjacent convex molecule (1.590/1.579 Å); Car-H...O: 2.409/2.399 and 2.537/2.547 Å)
[PcMg-OPϕ_3][2py]	MAKHOV[136]	12:26	Double-layers parallel (010); concave OV along intermediate axis
[PcMg-OPϕ_3][CH_2Cl_2]	MAKHUB[136]	12:27	Double-layers parallel (001); strong overlap along N_{br} axis
[PcMg-OPϕ_3][THF]	MAKJAJ[136]	12:28	Double-layers parallel (001); strong overlap along N_{br} axis
PcMg-OPϕ_3	MAKHIP[136]	12:33	Double-layers parallel (001); molecules in adjacent layers different orientation; slipped stacked columns either along [110] or [−110]
[PcMg-Cl][PNP][Me_2CO]	TAYSIV[172]	12:34	Double-layers parallel (001); concave overlapping double-bands along [100], concave overlap along N_i molecular axis
[PcMo=O][I_3]	FEWPEC[186]	12:36	Sheets of stacked phthalocyanine double-bands and iodine along [010]; neighboring sheets along [100] are shifted by $\pm b$
[PcSb][I_3]	HOVCOK[188]	12:43	Double-columns along [100]; iodine between columns; orientation of molecules in adjacent columns is different; strong concave overlap along N_{br} molecular axis; convex overlap is along an intermediate molecular axis; convex and concave overlap are zigzag slipped-stacked

(Continued)

Table 15. Continued

Compound	Code [Ref.]	Figure	Structural description
[PcZn-H_2N-C_6H_{13}][0.5H_2N-C_6H_{13}]	ZNPTCY[144]	12:60	Slipped-stacked columns along [001]; identical orientation along [100]; fish-bone structure along [010]; in column overlap along intermediate molecular axis
[PcZn-NC-Fe(dppe)Cp][0.5CH_2Cl_2,THF]	WOHTES[314]	12:61	Layers parallel to [101] and [100]; concave overlap nearly along the N_i molecular axis
[PcZn-OPϕ_3][Et_2O,1-ClN]	MAKHAH[136]	12:62	Layers parallel to [101] and [100]; wave-like layers; concave overlap nearly along the N_i molecular axis
[PcInF_2][NBu_4][H_2O]	NAPJIX[178]	14:4–5	Sheets parallel to [101] and [010]; slipped-stacked bands along [010]; concave overlap nearly along the N_i molecular axis
[PcIn(-NCO)$_2$][PNP]	ZUFZOP[173]	14:6–7	Double-layers parallel (001) or slipped-stacked columns along [010]; concave overlap along N_i molecular axis; convex overlap along N_{br} molecular axis
[PcIn(-O,-O′CO)][NBu_4]	RERGOK[177]	14:8–9	Isolated concave overlapping double-bands along [010]; concave overlap along N_i molecular axis
[PcIn(O_2catCl_4)][NBu_4][CH_2Cl_2]	1998-02[308]	14:10–11	Double-layers parallel to [−110] and [001]; concave overlap nearly along N_i molecular axis
[Pc$TlCl_2$][NBu_4][0.5Et_2O]	BOFQES[191]	14:21–22	Layers made of slipped-stacked bands along [010]
[PcHf(Oϕ)$_3$][NBu_4][0.5Et_2O]	CCDC165833[400]	16:6–7	Molecular bands along [010]; neighboring bands in a layer parallel (100) are differently oriented
[PcHfF_3][NBu_4]	POQYID[179]	16:4–5	Isostructural to [Pc$TlCl_2$][NBu_4][0.5Et_2O]
[PcNb(O,O_2Naphth)][CH_2Cl_2]$_{0.5}$	1998-01[308]	16:8	Double-layers parallel (001), concave overlap nearly along N_i molecular axis
[PcY(Cl)(py)(OPϕ_3)][0.5Et_2O]	BORQUU[192]	16:21	Layers parallel (010); neighboring molecules are alternately oriented
PcZrI_2(PN)$_1$	NOFNUR[176,403]	16:22–23	Similar to [PcMg-OH_2][2py]; concave overlapping double-bands along [001], concave overlap along N_i molecular axis; layers parallel (010)
[PcZr(Oϕ)$_3$][NBu_4][Et_2O]	CCDC165838[400]	16:24–25	Isostructural to [PcHf(Oϕ)$_3$][NBu_4][Et_2O]$_{0.5}$
[Pc$ZrCl_3$][PNP]	YOKWEA[169]	16:26–27	Double-bands of concave overlapping molecules along [100]; overlap along an intermediate molecular axis
[(PcTaF_3)$_{1-x}$(PcTaF_4)$_x$][NBu_4]	NAPRAX[179]	16:28–29,17:1	Isostructural to [Pc$TlCl_2$][NBu_4][0.5Et_2O]
[PcIn(-ONO-)$_2$][NBu_4]	ZEMBEY[404]	17:2–3	Layered parallel a layer made from [010] and [101]
PcLu(OAc,H_2O,H_2O)	DICCAT[159]	17:4–5	Sheets parallel (010); sheets made from double-bands along [001]; orientation of molecules towards sheet plane alternating in neighboring sheets; zigzag slipped-stacked columns along [100]; concave overlap along intermediate molecular axis
PcSn[O-C(O)-$C_{10}H_{21}$]$_2$	CCDC173622[128]	17:6–7	Sheets parallel (010); zigzag slipped-stacked columns along [001]; concave overlap along N_i molecular axis
[PcTl(-ONO-)$_2$][NBu_4]	BOFQAO[191]	17:8–9	Isostructural to [PcIn(-ONO-)$_2$][NBu_4]
[PcY(Cl)(dmf)$_3$][Et_2O]	BORQII[192]	17:10–11	Hexagonal lattice of double-columns along [100]; no concave–concave or convex–convex overlap in double-column; in double-column macrocycles pointing in an angle of 90 ° to each other
[PcY(Cl)(-OSMe_2)$_4$][Et_2O,2MeOH]	BORQOO[192]	17:12–13	Double-layers parallel (001), consists of double-bands along [100]; overlap along an intermediate axis
[PcZr(O_2CatCl_4)$_2$][NBu_4]$_2$	CCDC165834[400]	17:20–21	Layers parallel (001); molecules in layer alternately up and down oriented

In BITSAY03, however, one molecule overlaps with a combination of one and two neighbor molecules. Furthermore, the orientation of the molecules in neighboring layers in the latter structure is identical, while the orientation of the molecules in BITSAY and BITSAY02 is not. The concave–concave overlap in all three structures is also different, the overlap reportedly occurring along the N_i molecular axis in BITSAY02, along the N_{br} axis in BITSAY03, and along an intermediate direction in BITSAY. An additional feature common to both I-PcTiO (BITSAY) and PcMoO (CUTGIH) complexes is the occurrence of non-parallel macrocycles in neighboring convex layers, which is thought to be due to the interaction between the aromatic hydrogen atoms and the axial ligand of the molecules in every first and third layer (Figure 35). Neighboring molecules in concave layers are parallel.

In contrast to the alternate head–head and tail–tail arrangements of the molecules that occur in layered or slipped-stacked columns, the arrangement in monoclinic PcPb is considerably different (Figure 36). The molecules are arranged head-to-tail in 1D columns parallel to the crystallographic *c*-axis (Figure 36a). Along the *b*-axis, the orientation of the columns changes alternately between up and down (Figure 36b). Along the

Table 16. Structural Description of Complexes with an Out-of-plane Coordinated Central Atom (Dimers and Polymers)

Compound	Code [Ref.]	Figure	Structural description
PcAl-O-AlPc	DAYYOR[122]	12:4–5	Slipped-stacked columns along [100]; nearly full overlap along N_i molecular axis
PcFe-Cl-I-I-Cl-FePc	DAPCIG[157]	12:12–13	Bands along [010] are interconnected by Iodine molecules; connection direction is [111]; overlap is along N_i molecular axis
PcFe-I-I-I-I-FePc	FIQFUG[189]	12:14–15	Isostructural to PcFe-Cl-I-I-Cl-FePc
$PcBi(\text{-}I\text{-})_2BiPc$	HOXCIG[184]	14:1–3	Slipped-stacked columns along [010]
$PcSb(\text{-}Cl\text{-})_2SbPc$	WOVREE[193]	14:17–18	Similar structure as $PcBi(\text{-}I\text{-})_2BiPc$
$PcSb(\text{-}I\text{-})_2SbPc$	NILVUZ[183]	14:15–16	Similar structure as $PcBi(\text{-}I\text{-})_2BiPc$
$[PcNb(\text{-}O\text{-},\text{-}O\text{-},\text{-}Br\text{-})NbPc][Br_3][1\text{-}ClN]$	SAMDUF[124,162]	16:15–16	Zigzag slipped-stacked columns along [100]; counterion and 1-ClN in channels between columns; overlap in column along N_i molecular axis
$[PcNb(\text{-}O\text{-},\text{-}O\text{-},\text{-}Br\text{-})NbPc][Br_3][1\text{-}ClN]$	1986-01[402]		Isostructural with $[PcNb(\text{-}O\text{-},\text{-}O\text{-},\text{-}Br\text{-})NbPc][Br_3]$ [1-ClN]
$[PcNb(\text{-}O\text{-},\text{-}O\text{-},\text{-}Br\text{-})NbPc][Br_3][PN]$	SAMFAN[124,162]	16:17	Isostructural with $[PcNb(\text{-}O\text{-},\text{-}O\text{-},\text{-}Br\text{-})NbPc][Br_3]$ [1-ClN]
$PcY(\text{-}(Cl)_2\text{-},\text{-}(O(Hpy)_2)_2\text{-})YPc$	BOSQAB[192]	17:14–15	Bands along [100]; the orientation of molecules in neighboring bands is different
PcY(L,L′,-Cl-,-Cl-)Ypc	BOSJAU[192]	17:16–17	Linear columns of slipped-stacked molecules along [100]
$[PcZr(\text{-}(OH)_2\text{-},\text{-}(O,Oala)_2\text{-})ZrPc]$ $[DMF_2,Et_2O,H_2O]$	CCDC165829[192]	17:18–19	Slipped-stacked columns along [001]
$[PcZr(\text{-}OC(O)C)_2][NEt_4]_2[CH_2Cl_2]_2$	CCDC165836[400]	17:22–23	Sheets parallel (001); double-bands along [010]; in bands concave overlap along intermediate axis
$[PcBi]_4(Bi_4I_{16})$	LOKSAF[190]	16:1-3	A layered structure is visible when looking along [110]; a view along [11,6,2] in one layer shows the arrangement of the four phthalocyanine macrocycles around the inorganic part of the crystal
$[PcSb]_4[Sb_4I_{16}]$	YOHMAJ[194]	16:18–19	Similar structure as $[PcBi]_4(Bi_4I_{16})$

a-axis, the orientation of the column is identical, but the location of molecules in neighboring columns is shifted by $\pm b/2$ (Figure 36c).

As mentioned earlier, hydrogen bonding between the axial ligands and the aromatic hydrogen atoms governs the arrangement in convex–convex structures; an overview of the hydrogen bonding in the case of CN5 complexes is given in Tables 13 and 14. The arrangement of CN6*cis*, CN7*cis*, and CN8*cis* complexes are more complicated and only a few neutral phthalocyanine complexes are known to exist: $PcTiCl_2$, the two polymorphs of $PcNbCl_2$, $PcNbCl_3$, and $PcTaCl_3$.

3. Phthalocyanines with Large Axial Ligands

Large axial ligands, cocrystallizing solvents, and the presence of other molecules, have a strong influence on the crystal structure. An overview of these structures is given in Table 15 (monomers) and Table 16 (dimers and tetramers).

B. *BIS*- AND *TRIS*-PHTHALOCYANINES

1. Introduction and Molecular Structures

A CN8*bis* phthalocyanine complex consists of a central metal out-of-plane coordinated to two phthalocyanine ligands, sometimes also described as a pc-metal-pc sandwich. Such molecules are named "*bis*-phthalocyanines" to distinguish them from the phthalocyanine dimers. *tris*-Phthalocyanines systems, Pc-M-Pc-M-Pc, are also known, e.g., Pc_3In_2 (PIFDEN,[53] App. 18:37–38), Pc_3Bi_2 (DIQPIC,[51] App. 18:5, and DIQPIC01,[52] App. 18:6), and $[(15cr5)_4Pc]_3Lu_2$ (LORKIM,[195] App. 18:53–55). In the *tris*-phthalocyanines, the inner phthalocyanine macrocycle is μ-coordinated to the two central metals. The periodic table of *bis*- and *tris*-phthalocyanines complexes (CN8*bis*) with the number of structures given next to each element is shown in Figure 37, elements in parentheses indicate that only unit-cell data but no atomic coordinates are known. A compilation of the basic crystallographic data of the structures is given in Appendix 1 and their molecular structures are shown in Appendices 18 and 23.

The two phthalocyanine macrocycles in *bis*-phthalocyanines are rotated by a staggering angle α (Figure 38), assuming undistorted macrocycles with 4-fold symmetry. If the local symmetry is lower, than the actual staggering angle should either reflect the symmetry of the molecule or should be taken as an idealized angle between the two nearest lobes of the two macrocycles in the molecule. Within the literature, no investigation of the influence of distortion on the staggering angle has been undertaken. In all known cases is the central

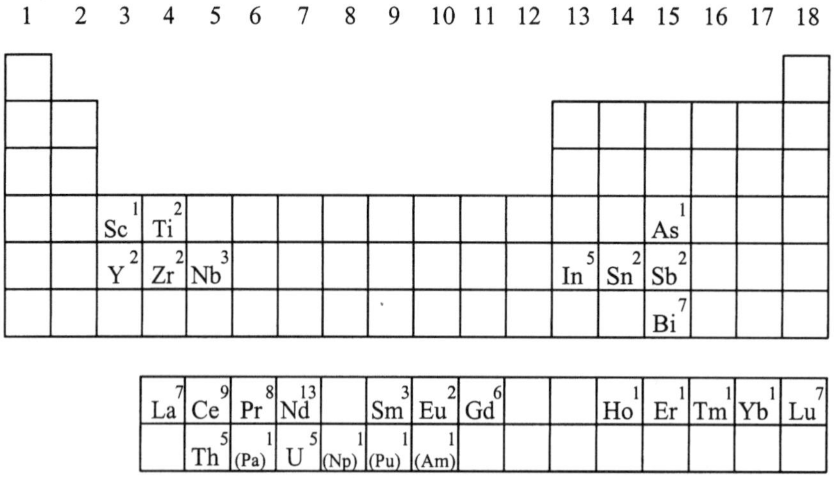

Figure 37. Periodic table of *bis*- and *tris*-phthalocyanine complexes with number of structures given next to each element. Elements in parentheses indicate that only unit-cell data but no atomic coordinates are known.

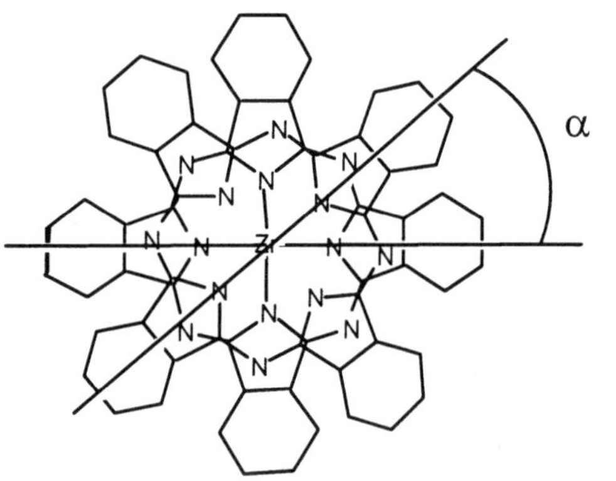

Figure 38. Staggering angle α of *bis*-phthalocyanine complexes.

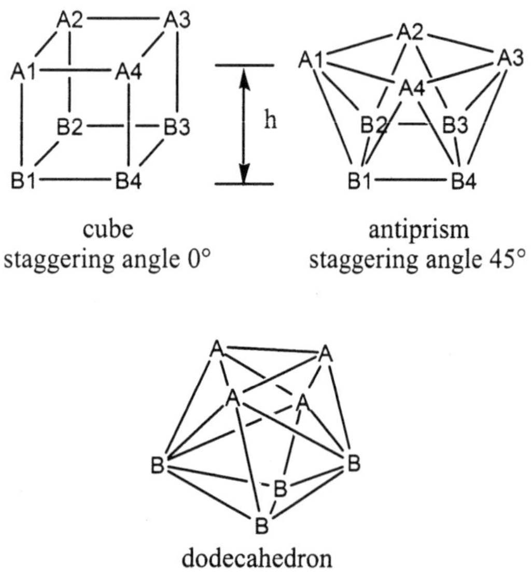

Figure 39. Ligand geometries of atoms of coordination number 8. In *bis*-phthalocyanines, atoms A1–A4 corresponds to the isoindole nitrogen atoms of macrocycle A, and atoms B1–B4 to those of macrocycle B.

metal coordinated to all of the isoindole nitrogens in each phthalocyanine macrocycle. In general, the coordination geometry around a metal with coordination number 8 can be either cubic, antiprismatic, or dodecahedral (Figure 39). All but the last geometry are allowed in the case of the *bis*-phthalocyanines. The latter is prohibited due to the rigidity of the phthalocyanine ligands. In the case of the cubic coordination geometry, all atoms of the first phthalocyanine ligand would lie above the identical atoms of the second ligand; whereas in the case of the antiprismatic arrangement, only the C_α atoms of the isoindole units in the two ligands would overlap. As the height of the coordination polyhedron (h in Figure 39) decreases, the energy of the cubic or antiprismatic arrangements become more unfavorable and, therefore, the intermediate staggering angles become more favorable. It is suggested,[196] that the staggering angle should not only depend on the height of the coordination polyhedron or the distance between the best planes of the two phthalocyanine macrocycles, but also on the crystal structure (isomorphous structures show identical staggering angles; see Table 18). As the energy differences between the different geometries for coordination number 8 are relatively small, it seems possible that the influences of the crystal structure (i.e., the interaction with neighbors) might influence the molecular geometry, e.g., the staggering angle. Despite the fact that *bis*-phthalocyanines are known since over 30 years, it is only since recently that an effort has been undertaken to explain the influence of the molecular and crystal structures on the staggering angle.[23,197,198] The suggestion that the crystal structure might have

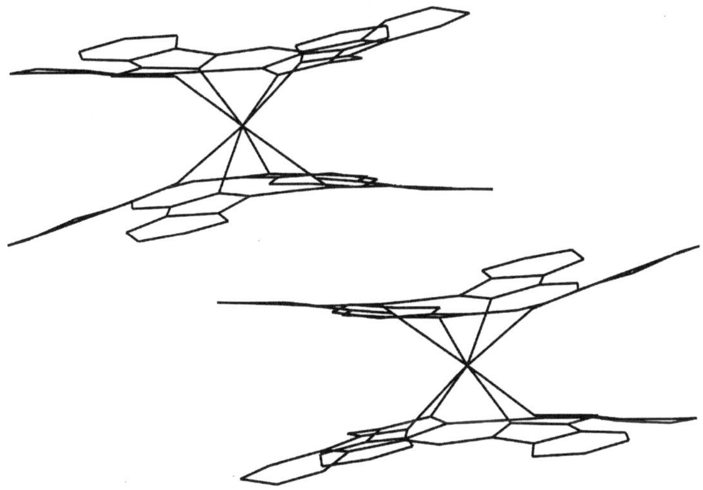

Figure 40. Flattening of the saucerform of the macrocycles in Pc_2U molecules due to interaction with a neighbor molecule.

Table 17. Overview of the Subgroups of CN8*bis* Complexes

CCDC	Molecule	Space group	Angle [°]	Reference
Neutral complexes				
*	Pc_2Th	B2/b	38	202,203
PHALCU01	Pc_2U	B2/b	38	202,203
DPCYTH03	Pc_2Th	C2/c	37	207
DPCYTH	Pc_2Th	C2/c	38	203
PHALCU02*	Pc_2U	C2/c	–	413
PHALCU*	Pc_2U	C2/c	37	200
PCNPAB*	Pc_2Pa	C2/c	–	413
JONZAN	Pc_2Ce	C2/c	38	217
SNPTCY01	Pc_2Sn	C2/c	38	223
DPCYTH01	Pc_2Th	C2/c	38	204
DPCYTH02*	Pc_2Th	C2/c	–	413
CIZGIB02	Pc_2Nd {β}	C2/c	38	196
CIZGIB06	Pc_2Nd (ox1)	C2/c	42.3	210
KOBRUO	Pc_2Pr {β1}	I2/c	42	211
SOTGAJ	Pc_2Zr	P-1	39.7(42)	215
GAWBEL	Pc_2Gd	$P2_12_12_1$	*ca* 41	212
ZAKZOA	Pc_2In	$P2_12_12_1$	41.2	224
DULZAL	Pc_2Lu {γ}	$P2_12_12_1$	41	208
GAWBAH	Pc_2LuH	$P2_12_12_1$	41	212
CIZGIB03*	Pc_2Nd {γ(1)}	$P2_12_12_1$	–	196
SNPTCY	Pc_2Sn	$P2_12_12_1$	42	201
YUHSUP	Pc_2Er {α}	P4/nnc	41.4	225
CIZGIB08	Pc_2Nd	P4/nnc	41.3	222
KOBRUO02	Pc_2Pr	P4/nnc	41.7	222
CIZGIB*	Pc_2Nd {α}	P-42m,P422	–	196
SAFRAS*	$[Pc_2Nd][Cl_xO_{1-x}]$	P4nc	–	213
SAFPUK	$[Pc_2Nd][O]$	P4nc	38.1	213
CIZGIB07	Pc_2Nd {α}	P4nc	38	209
ZUWDAW	$[Pc_2Bi][CH_2Cl_2]$	Pnma	45	228
RIYREW*	$[Pc_2In][CH_2Cl_2]$	Pnma	–	229
ZEHTUB	$[Pc_2La][CH_2Cl_2]$	Pnma	45	226
DICBUM	$[Pc_2Lu][CH_2Cl_2]$	Pnma	45	159
PTCNND	$[Pc_2Nd][CH_2Cl_2]$	$P2_12_12_1$	45	205
JUVJIT	$[Pc_2Y][CH_2Cl_2]$	Pnma	45	216
PIFDEN	Pc_3In_2	P-1	37.5	53
DIQPIC	Pc_3Bi_2	P-1	36.4	51

(Continued)

Table 17. Continued

CCDC	Molecule	Space group	Angle [°]	Reference
DIQPIC01	Pc_3Bi_2	P-1	38.3	52
VODDIB	$[(H_{13}C_6S)_8Pc]_2Lu$	C2/c	?	247
LORKIM	$[(15cr5)_4Pc]_3Lu_2][18DMSO,2CHCl_3]$	P2/n	43.7	195
Cationic complexes				
SEWRER	$[Pc_2Ti][1\text{-}ClN]$	C2/c	45	214
NOCYEJ	$[Pc_2Nb][I_3][0.5I_2,3.5ClN]$	P-1	45	231
BECTOS	$[Pc_2As][I_2]$	P4/mcc	40	233
KOJQUV	$[Pc_2Bi][I_{1.5}]$	P4/mcc	39.4	232
NOFMAW	$[Pc_2In][I_2]$	P4/mcc	40	409
PIBJIT	$[Pc_2Ti][I_3]_{0.66}$	P4/mcc	41.1	219
HOTWOC	$[Pc_2U][I_{5/3}]$	P4/mcc	40	234
BECSUX	$[Pc_2Yb][I_2]$	P4/mcc	40	233
NOFMEA	$[Pc_2Zr][I_2]$	P4/mcc	40.6	409
RABDEZ	$[Pc_2Ce][BF_4]_{0.33}$	P4/nnc	41	230
WAYSUK*	$[Pc_2Pr][Br_{1.5}]$ {α}	P4/nnc	41	218
WAYSIH	$[Pc_2Pr][Br_{1.5}]$ {α}	P4/nnc	–	218
ZOMWUT	$[Pc_2Sm][Br_{1.45}]$ {α}	P4/nnc	41	227
ZOMXAA	$[Pc_2Sm][ClO_4]_{0.63}$ {α}	P4/nnc	41	227
MEGNOB	$[Pc_2U][I_3]_{2/3}$	P4/nnc	40.7	235
Anionic complexes				
QOCJOH	$[Pc_2La][NBu_4]$	P4/ncc	4.1	198
QOCJIB	$[Pc_2Ce][NBu_4]$	P4/ncc	4.7	198
QOCJEX	$[Pc_2Bi][NBu_4]$	P4/ncc	6	198
NAKXUS	$[Pc_2Nd][NBu_4]$	P4/ncc	6.2	197
NAKYAZ	$[Pc_2Gd][NBu_4]$	$P42_12$	34.4	197
QOCLUP	$[Pc_2In][I][PNP]_2[2Et_2O]$	P2/c	35.8	198
QOCLOJ	$[Pc_2Bi][Br][PNP]_2[2Et_2O]$	P2/c	37.7	198
QOCLID	$[Pc_2Sb][Br][PNP]_2[2Et_2O]$	C2/c	37.8	198
QOCKAU	$[Pc_2Bi][NBu_4]$	$Pna2_1$	41.8	198
QOCKUO	$[Pc_2Sc][NBu_4]$	$Pna2_1$	42.1	198
GAVZUY	$[Pc_2Lu][NBu_4][2DMF]$	P-1	43	212
QOCKEY	$[Pc_2Pr][NBu_4][0.75MeOH]$	$Pna2_1$	43.1	198
NAKYED	$[Pc_2Ho][NBu_4][H_2O]$	$P42_12$	43.2	197
OBUJIE	$[Pc_2Gd][C_4(CN)_6][2ClB]$	P-1	43.3	236
QOCMAW	$[Pc_2La][NBu_4][EtOH]$	P42/n	43.3	198
QOCKIC	$[Pc_2Sm][NBu_4][MeOH]$	$Pna2_1$	43.3	198
YEGTUZ	$[Pc_2Tm][PNP][0.5H_2O]$	$P2_1/c$	43.3	220
YEGTOT	$[Pc_2Gd][PNP][0.33H_2O]$	$P2_1/c$	43.6	220
QOCLEZ	$[Pc_2Sb][NBu_4][2thf]$	P-1	44.6	198
QOCLAV	$[Pc_2Pr][NBu_4][2py]$	P-1	44.8	198
QOCKOI	$[Pc_2Y][NBu_4]$	$Pna2_1$	44.8	198
YEGTIN	$[Pc_2La][PNP]$	$P2_1/c$	44.9	220
QOCJUN	$[Pc_2La][NPe_4]$	$P2_1/m$	45	198
NAKYIH	$[Pc_2Lu][NBu_4][H_2O]$	P4/nmm	45	197
#2002-10	$[Pc_2Ce][PNP]$	$P2_1/c$		221

*No atomic coordinates available.

a strong influence on the molecular structure has already been discussed in the case of the deformation of $PcSnCl_2$, which it is thought to be due to hydrogen bonding. Other molecular deformations are always accompanied by hydrogen bonding of the deformed parts to neighboring molecules; the interaction between overnext layers in I-PcTiO (Figure 35), which was found because of a strongly out-of-plane located aromatic carbon atom, or a strongly twisted benzo ring in $PcNbCl_2$ where the deformation points to the neighboring chlorine atoms, has already been mentioned.

Since the early work of Linstead and coworkers,[32] and the first single-crystal structures reported in the late 60s and early 70s,[199–201] many new structures have been published. The first publicly known structure of a *bis*-phthalocyanine complex, Pc_2U (PHALCU), was published by Gieren and Hoppe in 1971.[200] The atom coordinates of this structure have never been reported, and the original data can no longer be retrieved. Fortunately, a second structure of Pc_2U was later obtained by Kirin *et al.* in 1973 (PHALCU01,[202,203] App. 18:117–118). Unit-cell and other reported structural data suggests that both Pc_2U structures are isomorphous. The main features of the molecular structure are a uranium atom located equidistantly between the two phthalocyanine ligands. The phthalocyanine rings are rotated *ca.* 37° towards each other and their deformation is similar to the *cis* complexes, giving the molecule the appearance of a double saucer-shape. Similar to the flattening of the phthalocyanine macrocycle in phthalocyanine complexes with an out-of-plane coordinated central atom, the saucer deformation of a phthalocyanine ligand in Pc_2U appears flattened when it interacts with an adjacent molecule (Figure 40 and 42).

In 1973, the crystal structure of tin *bis*-phthalocyanine, Pc_2Sn (SNPTCY, App. 18:103–104), was published by Bennett *et al.*[199,201] The two phthalocyanine moieties are rotated by 41°, and separated by 3.42 Å. Compared to Pc_2U, a similar but significantly smaller deformation from a uniform double saucer-shape is found. In addition to the already mentioned structure of uranium *bis*-phthalocyanine, Kirin *et al.* published in 1974 the isomorphous structure of thorium *bis*-phthalocyanine, Pc_2Th (DPCYTH,[203] App. 18:107–108). The same deformation of the phthalocyanine rings is also apparent, where the staggering angle is 38°. A nearly identical structure of thorium *bis*-phthalocyanine (with slightly different unit-cell parameters) was later reported in 1978 by Kobayashi (DPCYTH01,[204] App. 18:109–110).

The first structure of a lanthanide *bis*-phthalocyanine, $[Pc_2Nd][CH_2Cl_2]$ (PTCNND, App. 18:67–68) was published in 1980 by Kasuga *et al.*[205,206] By comparison with structures of other $[Pc_2M][CH_2Cl_2]$ *bis*-phthalocyanines, published after 1980, it can be concluded that the reported space group, $P2_12_12_1$, is actually Pnma. The atomic coordinates, which are not reported in the original communication, are reprinted in Appendix 38. In 1984, Darovskikh *et al.* published a paper about the polymorphism of Pc_2Nd.[196] They report the unit-cell data for α-, β-, and γ-modifications of Pc_2Nd, as well as the full structure of β-Pc_2Nd (CIZGIB02, App. 18:69–70). In this last structure, the phthalocyanine moieties are rotated by 38° with respect to each other. From their findings, and from a comparison with earlier structures, Darovskikh *et al.* assumed that the staggering angle associated with *bis*-phthalocyanines is not determined by the distance of the macrocycles from each other, but rather by the character of the packing of the molecules. DeCian *et al.* published the structure of $[Pc_2Lu][CH_2Cl_2]$ (DICBUM, App. 18:51–52) in 1985,[159] (probably) isomorphous with $[Pc_2Nd][CH_2Cl_2]$ (PTCNND, App. 18:67–68). A rerefined structure of Pc_2Th was later published in 1985 by Darovskikh *et al.* (DPCYTH03,[207] App. 18:111–112). The structure was obtained from the same isometric crystal as used by Kirin *et al.*[203] The structure of γ-Pc_2Lu (DULZAL, App. 18:59–60), which is isomorphous to that of Pc_2Sn reported by Bennett *et al.*, was first reported by Darovskihk *et al.* in 1986.[208] They found the lutetium atom to be surrounded by the eight isoindole nitrogen atoms in the form of a distorted antiprism, where the Lu atom is slightly displaced from the center of the polyhedron towards one of the ligands.

The structure of the tetragonal α-phase of Pc_2Nd (CIZGIB07, App. 18:75–76) was published by Darovskikh *et al.* in 1986.[209] The author suggested that the Nd atom is distributed over two crystallographic positions and that the molecule Pc_2Nd looses its individual character in the α-phase. A second polymorphic structure of Pc_2Nd was published by Jullien *et al.* in 1986, (CIZGIB06,[210] App. 18:73–74). The transformation of the unit-cell from C2/c to I2/c reveils that there is a similarity with the β1-Pc_2Pr structure (KOBRUO[211]). This can also be seen in the arrangement pattern analysis below (see Section IV.A.2). The atomic coordinates of $NdPc_2(ox1)$ have never been published, but were obtained from one of the authors and are contained in Appendix 39. Each phthalocyanine ring within the molecule adopts an S-shaped deformation, as shown in App. 18:73. The S-shaped deformation is related to the interaction of the molecule with its neighbors. Moussavi *et al.* reported the crystal structures of Pc_2Gd, "Pc_2LuH," and $[Pc_2Lu][NBu_4][2DMF]$ in 1988,[212] in which they found Pc_2Gd (GAWBEL, App. 18:31–32) to be isomorphous with both DULZAL and SNPTCY. The structure of Pc_2LuH (GAWBAH, App. 18:61–62) is isomorphous to the γ-Pc_2M structures.

$[Pc_2Lu][NBu_4][2DMF]$ (GAVZUY, App. 18:63–64) is the first structure of an anionic *bis*-phthalocyanine. Due to influences of the packing, the molecule is unsymmetrically deformed. Mossoyan-Deneux *et al.* reported the crystal structures of $[Pc_2Nd][O]$ (SAFPUK, App. 18:79–80) and $[Pc_2Nd][Cl_xO_{1-x}]$ (SAFRAS) in 1988.[213] The atomic coordinates of these structures have never been published. Those of SAFPUK were obtained from the authors and are reprinted in Appendix 40. The oxygen or the chlorine atoms are

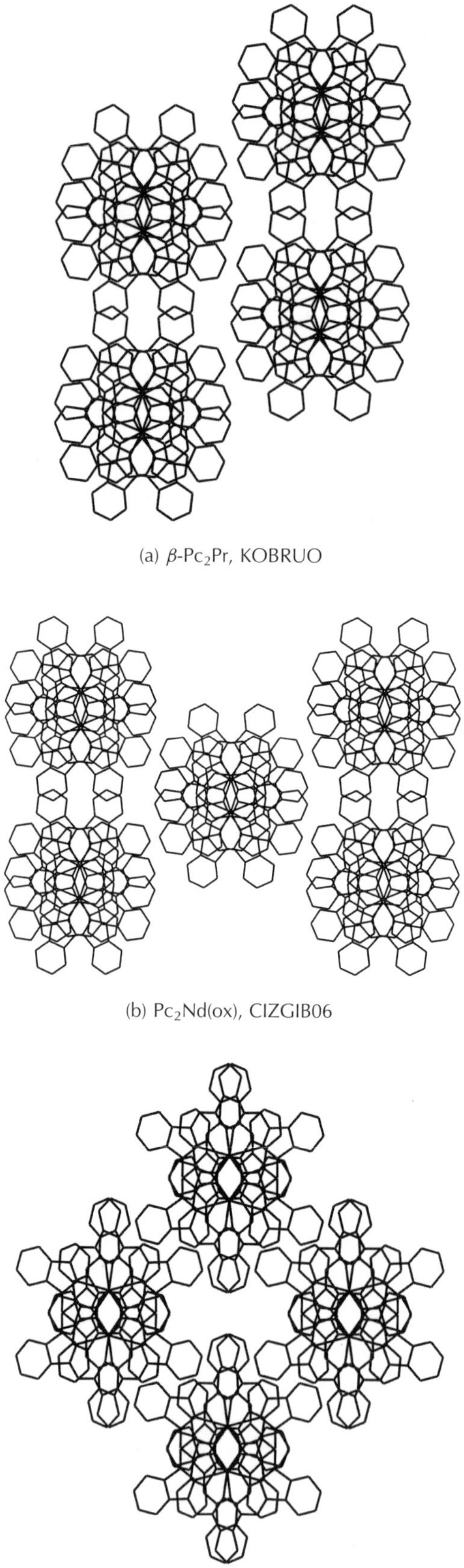

(a) β-Pc_2Pr, KOBRUO

(b) $Pc_2Nd(ox)$, CIZGIB06

(c) Pc_2Ce, JONZAN

Figure 41. View along the column axes of selected *bis*-phthalocyanine complexes.

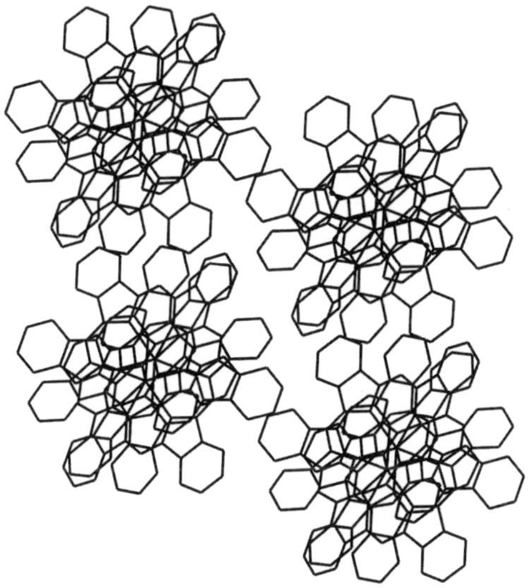

(d) Pc_2Zr, SOTGAJ along [0–11]

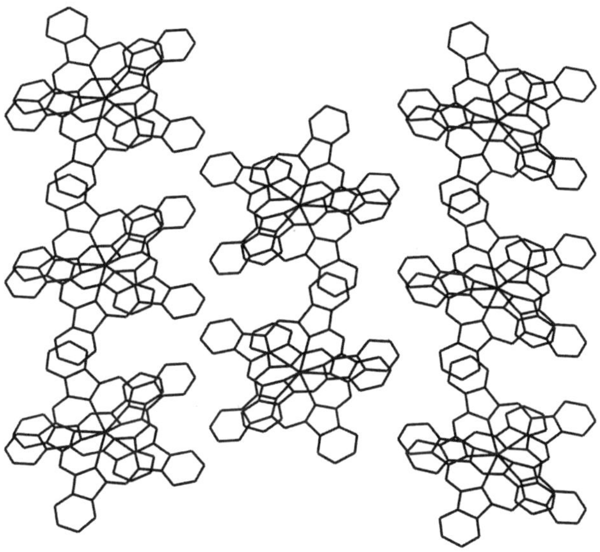

(e) Pc_2Sn, SNPTCY

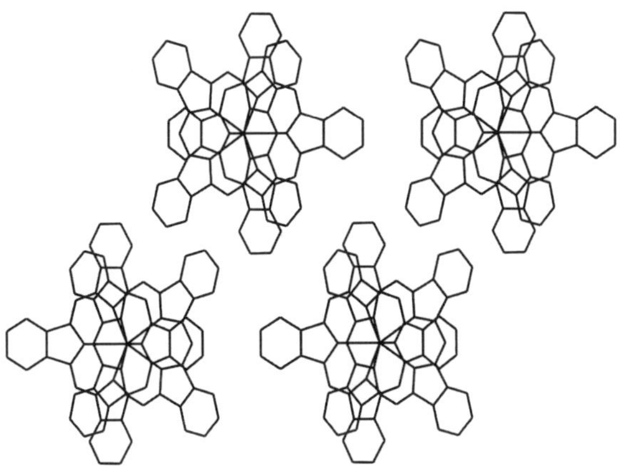

(f) $[Pc_2Lu][CH_2Cl_2]$, DICBUM

Figure 41. Continued.

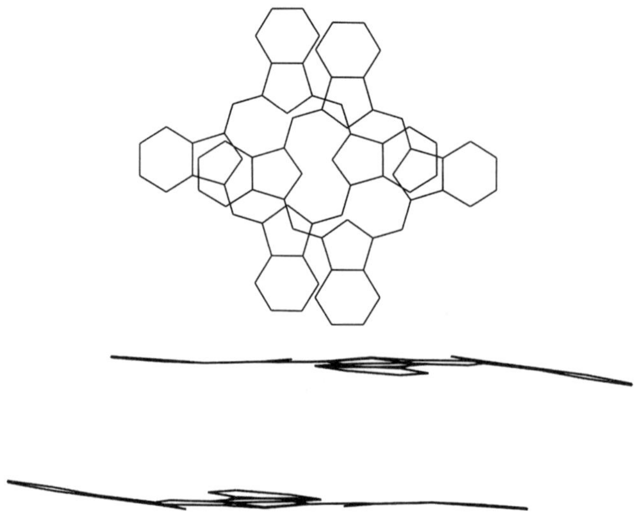

(a) β-Pc_2Pr, KOBRUO

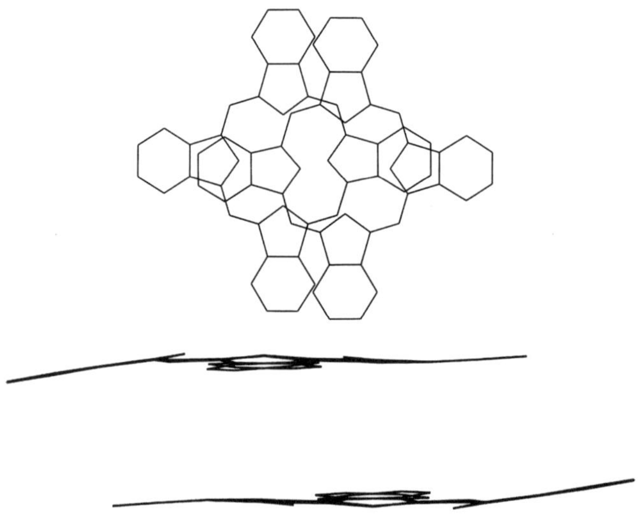

(b) $Pc_2Nd(ox)$, CIZGIB06

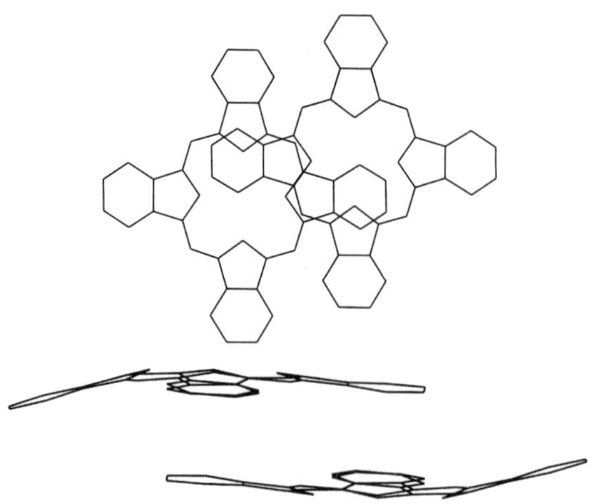

(c) Pc_2Ce, JONZAN

Figure 42. Perpendicular and side-on views of the overlapping geometry of *bis*-phthalocyanine complexes.

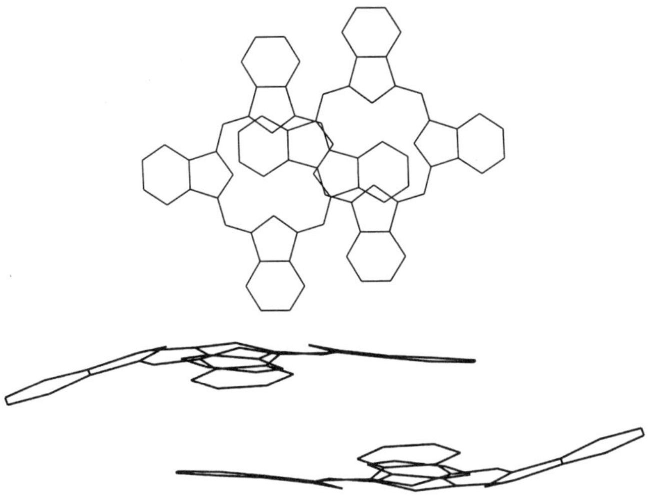

(d) Pc_2Zr, SOTGAJ

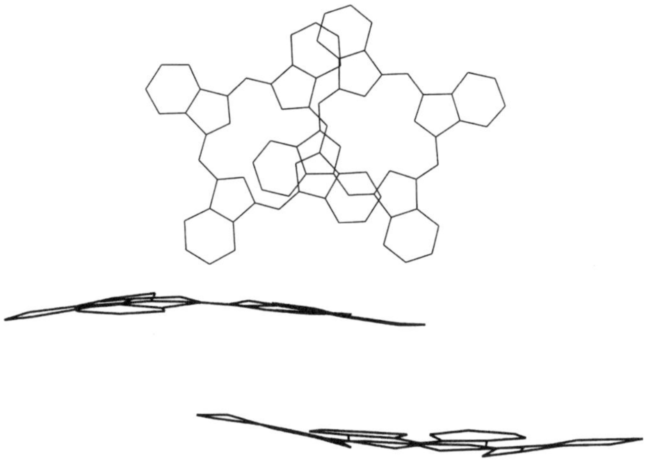

(e) Pc_2Sn, SNPTCY

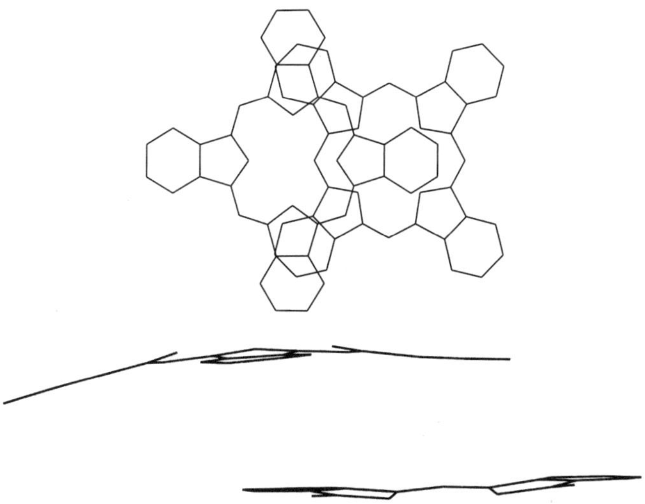

(f) $[Pc_2Lu][CH_2Cl_2]$, DICBUM

Figure 42. Continued.

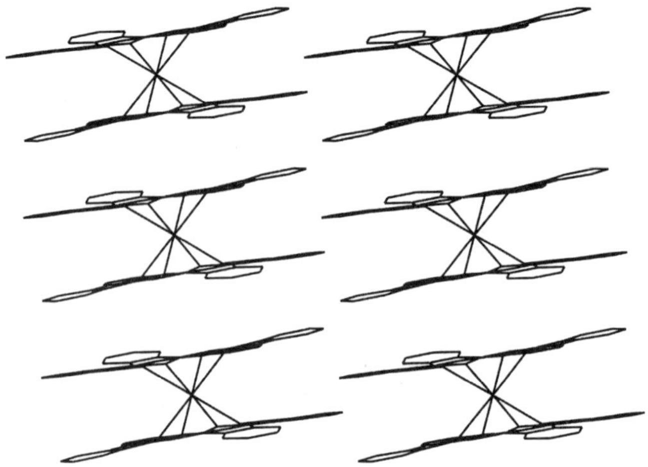

(a) β-Pc_2Pr, KOBRUO

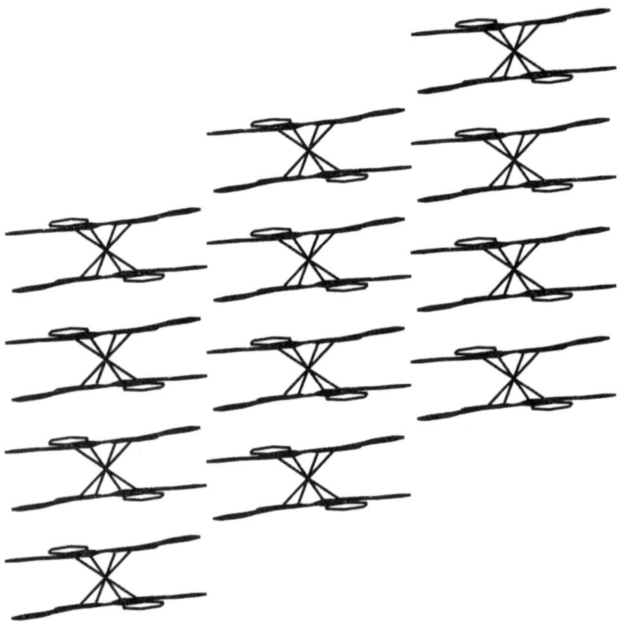

(b) $Pc_2Nd(ox)$, CIZGIB06

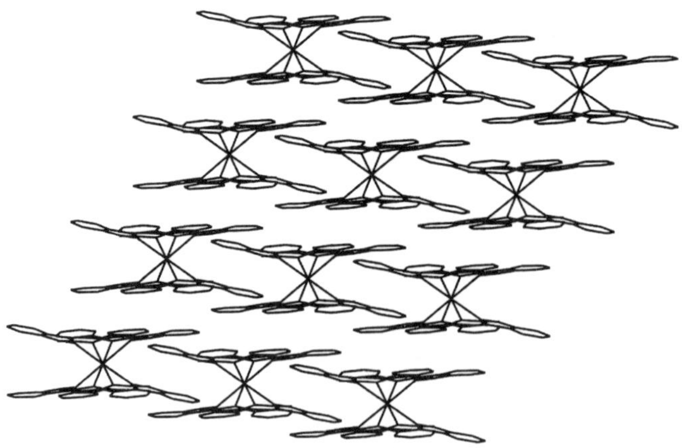

(c) Pc_2Ce, JONZAN

Figure 43. Perpendicular view onto one (DICBUM: two) sheet(s) of *bis*-phthalocyanine molecules.

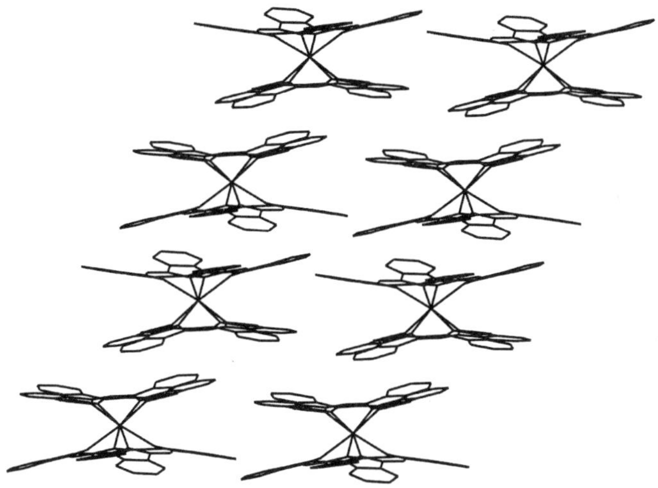

(d) Pc_2Zr, SOTGAJ

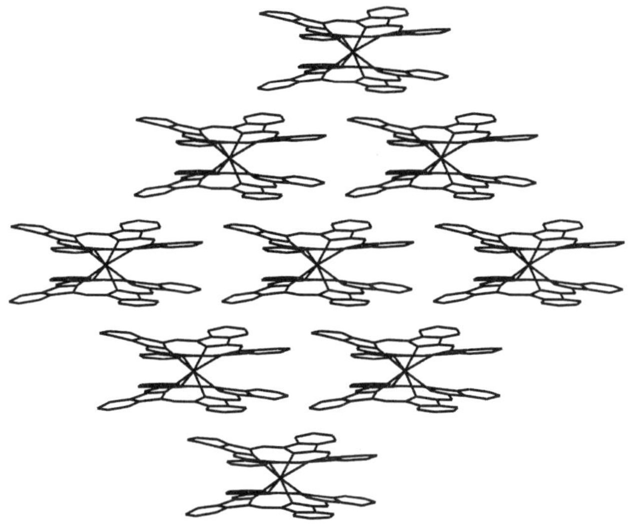

(e) Pc_2Sn, SNPTCY

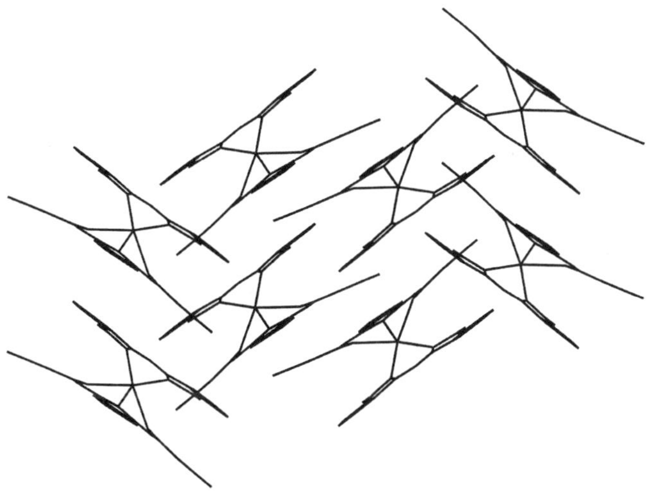

(f) $[Pc_2Lu][CH_2Cl_2]$, DICBUM

Figure 43. Continued.

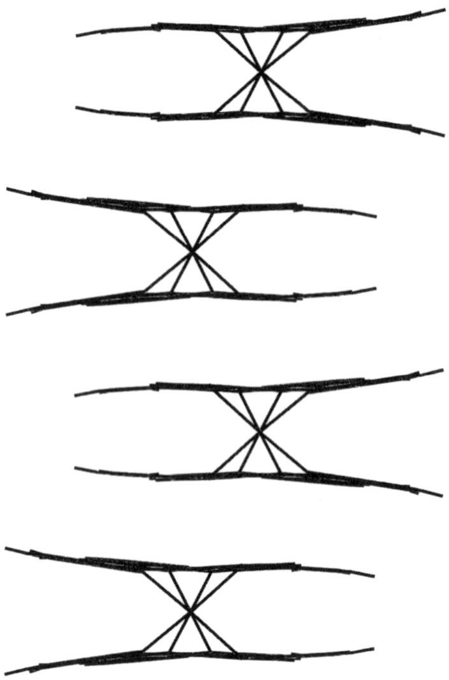

(a) β-Pc_2Pr, KOBRUO

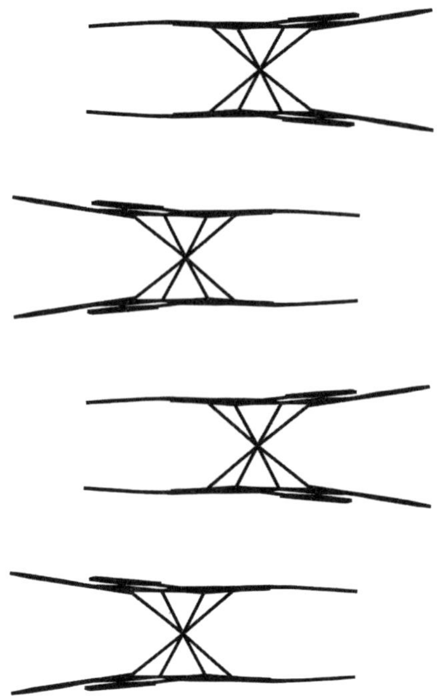

(b) $Pc_2Nd(ox)$, CIZGIB06

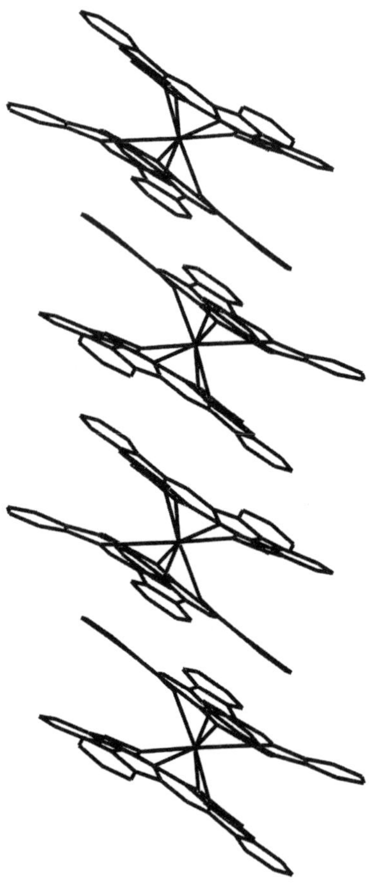

(c) Pc_2Ce, JONZAN

(d) Pc_2Zr, SOTGAJ

Figure 44. Arrangement pattern of slipped-stacked molecules in columns of *bis*-phthalocyanine molecules.

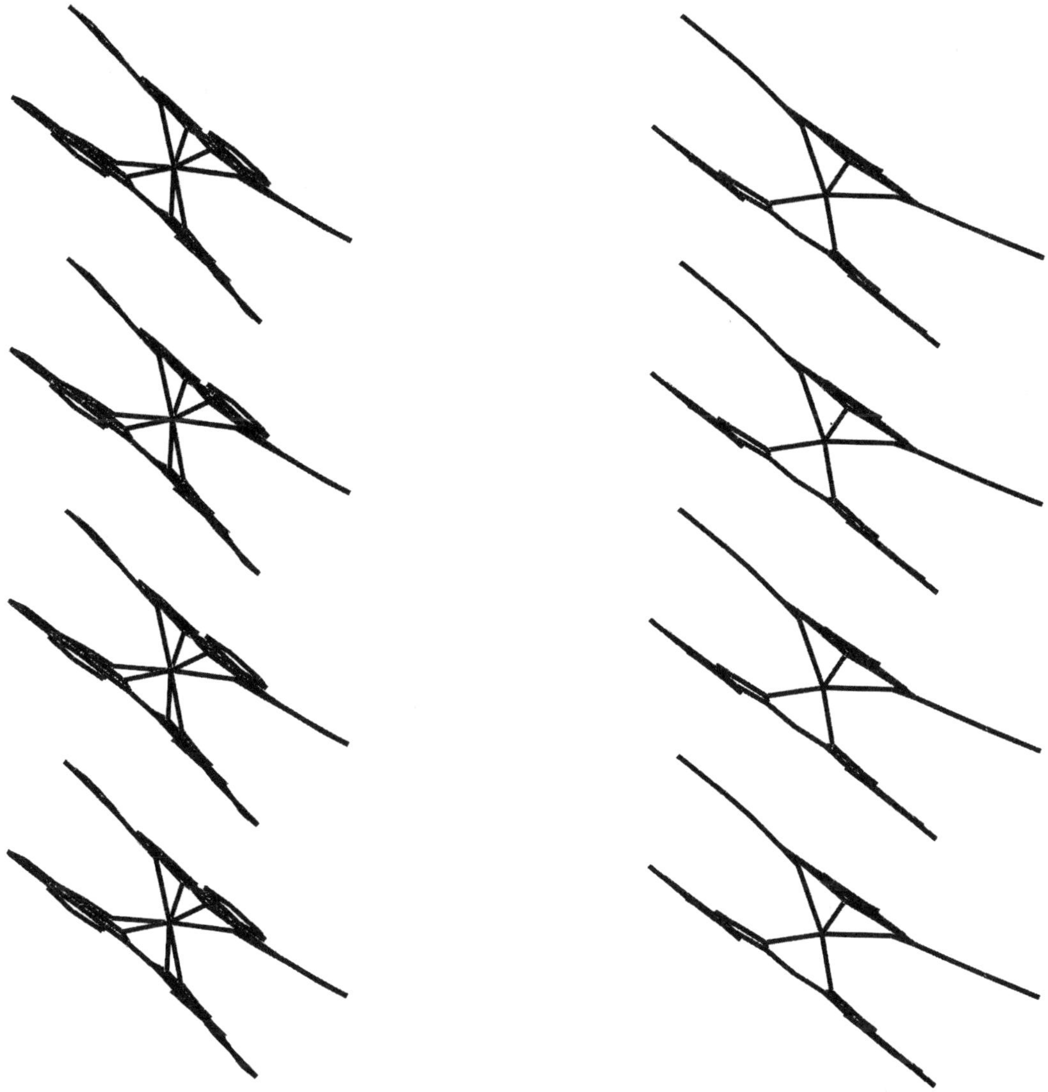

(e) Pc_2Sn, SNPTCY

(f) $[Pc_2Lu][CH_2Cl_2]$, DICBUM

Figure 44. Continued.

located between two neighboring molecules. Due to strong interaction with the adjacent molecules, the phthalocyanine moieties are nearly planar as in most other cationic phthalocyanines too. Ercolani *et al.* reported in 1990 that inter-ligand bonding between the two phthalocyanine macrocycles occurs in a *bis*-phthalocyanine complex if the size of the central atom is very small and its Lewis acidity is strong.[214] The inter-ligand bonds in $[Pc_2Ti][1\text{-}ClN]$ (SEWRER, App. 21:15–16) break up the aromaticity of the phthalocyanines and its absorption spectra show that it is actually not a phthalocyanine system anymore. Some similarities consist with the Pcc macrocycles as shown in Figure 49 and App. 21:7–12. In 1991, Silver *et al.* published a crystal structure of Pc_2Zr (SOTGAJ,[215] App. 18:129–130). The deformation of the molecular structure due to interaction with its neighbors is visible in App. 18:129. The staggering angle is 42°. Others recalculated the angle to be 39.7°.[23] Also in 1991, another $[Pc_2M][CH_2Cl_2]$ structure where M represents Y was reported by Paillaud *et al.*, (JUVJIT,[216] App. 18:123–124). The crystal structure of the β1-phase of Pc_2Pr (KOBRUO, App. 18:83–84) has been published by Darovsky *et al.* in 1991.[211] The staggering angle is 42°, markedly different from the 37–38° of other C2/c structures but is very similar to the Pc_2Nd structure of Jullien *et al.*, CIZGIB06.[210] The crystal structure of Pc_2Ce (JONZAN, App. 18:15–16) has been reported by Haghighi *et al.* in 1992.[217] The staggering angle is 38° and the structure is isomorphous with Pc_2Th and Pc_2U.

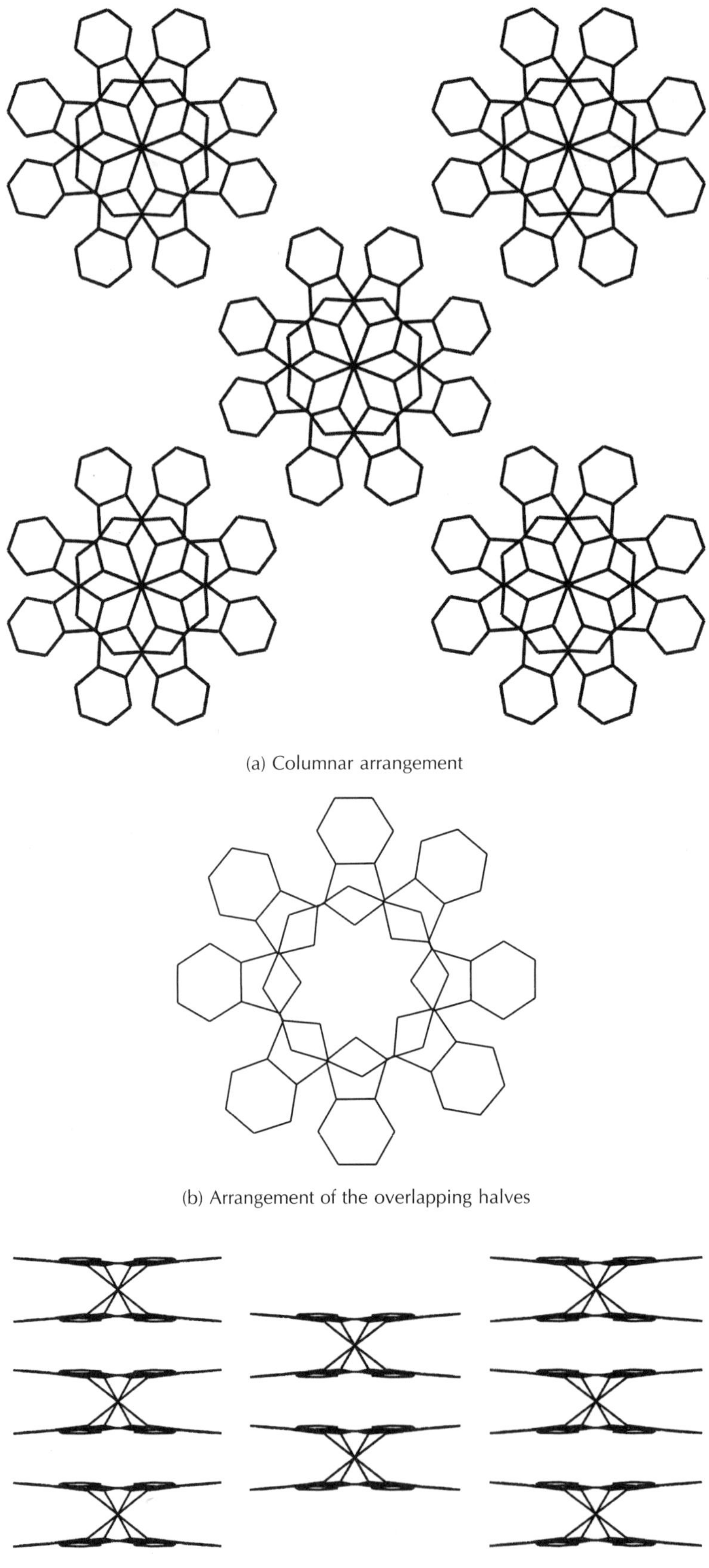

(a) Columnar arrangement

(b) Arrangement of the overlapping halves

(c) Side-on view

Figure 45. Structure of erbium *bis*-phthalocyanine YUHSUP.

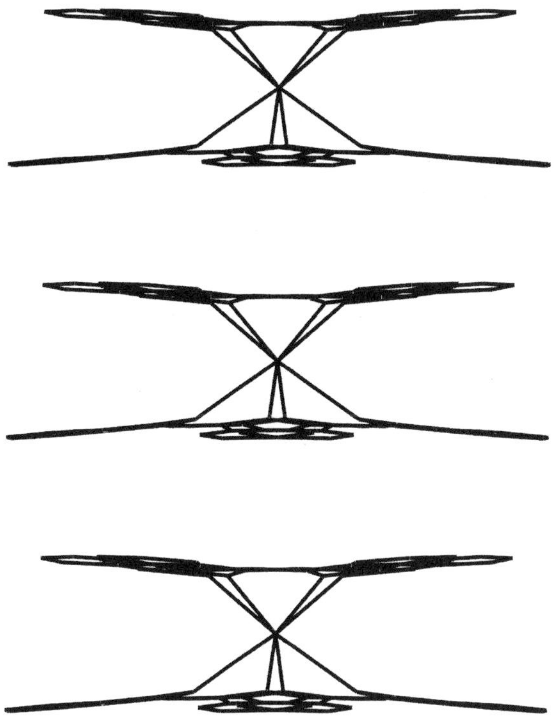

(d) 1 column

Figure 45. Continued.

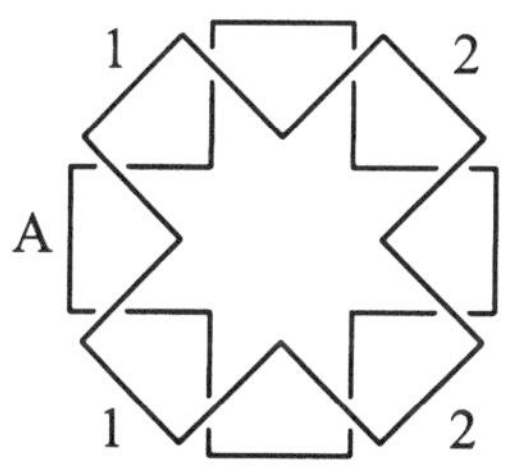

Figure 46. Nomenclature of the overlapping geometries of *bis*-phthalocyanines; e.g., A1 indicates that the overlapping with the molecule below occurs along lobe A while the overlapping with the molecule above occurs along lobe 1.

The structure of a polyhalogenide of Pc_2Pr in the tetragonal α-phase, $[Pc_2Pr][Br_{1.5}]$ (WAYSIH, App. 18:89–90), has been published by Haghighi *et al.* in 1993.[218] The phthalocyanine moieties are planar and the staggering angle is 41°. The first structure of a dimetal triphthalocyaninate or *tris*-phthalocyanine, Pc_3In_2 (PIFDEN, App. 18:37–38) has been reported by Janczak and Kubiak in 1993.[53] While each In atom coordinates symmetrically to the isoindole nitrogen atoms of the outer phthalocyanine moiety with a bond lengths of *ca.* 2.21 Å, it coordinates with slightly longer bonds (2.35 Å) to only two isoindole nitrogen atoms of the inner moiety. Similar to the deformation in the case of phthalocyanines with an out-of-plane coordinated central atom, the connected isoindole units bent away from the In atom (App. 18:37–38). The structure of the "repaired Pc_2Ti, SEWRER" is reported by Capobianchi *et al.* in 1993.[219] While the neutral Pc_2Ti shows extreme deformation due to inter-ligand C–C bonds, the oxidized $[Pc_2Ti]^{0.66+}$ in $[Pc_2Ti][I_3]_{0.66}$ has nearly planar phthalocyanine macrocycles (PIBJIT, App. 18:113–114). The molecular structures of $[Pc_2La][PNP]$ (YEGTIN, App. 18:49–50), $[Pc_2Gd][PNP][0.33H_2O]$ (YEGTOT, App. 18:27–28), $[Pc_2Tm][PNP][0.5H_2O]$ (YEGTUZ, App. 18:115–116) reported in 1994 by Haghighi *et al.*[220] and $[Pc_2Ce][PNP]$ (#2002-10,[221] App. 18:19–20) show the same kind of deformation, where three quarter of the phthalocyanine macrocycles of the molecule are nearly planar while one quarter retains the shuttlecock deformation. All four crystal structures are very similar, those of the waterless La and Ce complexes are isomorphous and those of the water-containing complexes of Gd and Tm too.

bis-Phthalocyanines in the α-form (Pc_2Nd) have been reported by Darovskikh *et al.* in 1986[209] and by Mossoyan-Deneux *et al.* in 1988.[213] Two new examples, α-Pc_2Nd (CIZGIB08, App. 18:71–72) and α-Pc_2Pr (KOBRUO02, App. 18:81–82), were published by Darovsky *et al.* in 1994.[222] While the staggering angle of the examples reported by Darovskikh and Mossoyan-Deneux is 38°, those of the Darovsky examples is 41°, indicating different structures despite similar unit-cell parameters. A polymorphic structure of Pc_2Sn (SNPTCY01, App. 18:105–106) was found by Janczak and Kubiak in 1994.[223] The structure is isomorphic with the structures of Pc_2Ce, Pc_2U, and Pc_2Th. Janczak *et al.* reported the crystal structure of Pc_2In (ZAKZOA,[224] App. 18:35–36), which is isomorphous with the other $P2_12_12_1$ structures.

The only ordered structure of an α-Pc_2M complex has been reported by Ostendorp *et al.* in 1995.[225] The structure of Pc_2Er (YUHSUP, App. 18:23–24) is very similar to the CIZGIB08 and KOBRUO02 structures reported by Darovsky and coworkers. All three structures have staggering angles of *ca.* 41° but the phthalocyanine macrocycles in Pc_2Er are less planar and the Er atom is not disordered over two sites compared to those of Nd and Pr. Ostendorp and Homborg reported the structure of $[Pc_2La][CH_2Cl_2]$ (ZEHTUB,[226] App. 18:43–44), which is isomorphous with the other Pnma structures of identical composition. The first *bis*-phthalocyanine complexes with staggering angles different from 38–41° and 45° have been obtained by Koike *et al.*;[197] $[Pc_2Gd][NBu_4]$ (NAKYAZ, App. 18:25–26)

Table 18. Structural Description of Complexes with an Out-of-plane Coordinated Central Atom (*bis*-Phthalocyanines)

Compound	Code [Ref.]	Figure	Structural description
$[Pc_2As][I_2]$	BECTOS[233]	18:1–2	Linear stacks along [001]; tetragonal lattice; counterion in channels between four stacks
$[Pc_2Bi][Br][PNP]_2[2Et_2O]$	QOCLOJ[198]	18:7–8	Layers parallel (001); in layer tetragonal lattice
$[Pc_2Bi][NBu_4]$	QOCJEX[198]	18:9–10	Linear stacks along [001]; tetragonal lattice; in stack alternately *bis*-phthalocyanine and counterion
$[Pc_2Bi][NBu_4]$	QOCKAU[198]	18:11–12	Stacks along [001]; hexagonal lattice; in stack alternately *bis*-phthalocyanine and counterion
$[Pc_2Bi][I_{1.5}]$	KOJQUV[232]	18:13–14	Linear stacks along [001]; tetragonal lattice; counterion in channels between four stacks
$[Pc_2Ce][NBu_4]$	QOCJIB[198]	18:17–18	Isostructural to $[Pc_2Bi][NBu_4]$, QOCJEX
$[Pc_2Ce][PNP]$	#2002-10[221]	18:19–20	Bands along [100], honeycomb structure where six bands form a channel for the counterions
$[Pc_2Ce][BF_4]_{0.33}$	RABDEZ[230]	18:21–22	Linear stacks along [001]; tetragonal lattice; counterion in channels between four stacks
$[Pc_2Gd][NBu_4]$	NAKYAZ[197]	18:25–26	Linear stacks along [001]; tetragonal lattice; in stack alternately *bis*-phthalocyanine and counterion
$[Pc_2Gd][PNP][0.33H_2O]$	YEGTOT[220]	18:27–28	Structure similar to $[Pc_2Ce][PNP]$
$[Pc_2Gd][C_4(CN)_6][2ClB]$	OBUJIE[236]	18:29–30	Slipped-stacked along [100]; in stack every second molecule identical; "hexagonal" lattice; counterions and solvent molecules between stacks
$[Pc_2Ho][NBu_4][H_2O]$	NAKYED[197]	18:33–34	Similar to $[Pc_2Gd][NBu_4]$
$[Pc_2In][I][PNP]_2[2Et_2O]$	QOCLUB[198]	18:39–40	Isostructural to $[Pc_2Bi][Br][PNP]_2[2Et_2O]$
$[Pc_2In][I_2]$	NOFMAW[409]	18:41–42	Similar to $[Pc_2As][I_2]$ but In atom symmetrically located between two rings
$[Pc_2La][NBu_4]$	QOCJOH[198]	18:45–46	Isostructural to $[Pc_2Bi][NBu_4]$, QOCJEX
$[Pc_2La][NPe_4]$	QOCJUN[198]	18:47–48	Resembling $[Pc_2Bi][NBu_4]$, QOCKAU
$[Pc_2La][PNP]$	YEGTIN[220]	18:49–50	Isostructural with $[Pc_2Ce][PNP]$
$[Pc_2Lu][NBu_4][_2DMF]$	GAVZUY[212]	18:63–64	Linear stacks along [001]; hexagonal lattice; in stack alternately *bis*-phthalocyanine and counterion
$[Pc_2Lu][NBu_4][H_2O]$	NAKYIH[197]	18:65–66	Similar to $[Pc_2Gd][NBu_4]$
$[Pc_2Nb][I_3][0.5I_2,3.5ClN]$	NOCYEJ[231]	21:13–14	Tetragonal lattice of double-columns along [001]
Pc_2Nd	CIZGIB08[222]	18:71–72	Linear stacks along [001], tetragonal lattice
Pc_2Nd {α}	CIZGIB07[209]	18:75–76	Linear stacks along [001], tetragonal lattice
$[Pc_2Nd][NBu_4]$	NAKXUS[197]	18:77–78	Similar to $[Pc_2Gd][NBu_4]$
$[Pc_2Nd][O]$	SAFPUK[213]	18:79–80	Linear stacks along [001], tetragonal lattice
Pc_2Pr	KOBRUO02[222]	18:81–82	Linear stacks along [001], tetragonal lattice
$[Pc_2Pr][NBu_4][0.75MeOH]$	QOCKEY[198]	18:85–86	Similar to $[Pc_2Bi][NBu_4]$, QOCKAU
$[Pc_2Pr][NBu_4][2py]$	QOCLAV[198]	18:87–88	Similar to $[Pc_2Lu][NBu_4][_2DMF]$
$[Pc_2Pr][Br_{1.5}]$ {α}	WAYSIH[218]	18:89–90	
$[Pc_2Sb][NBu_4][2THF]$	QOCLEZ[198]	18:91–92	Similar to $[Pc_2Lu][NBu_4][_2DMF]$
$[Pc_2Sb][Br][PNP]_2[2Et_2O]$	QOCLID[198]	18:93–94	Isostructural to $[Pc_2Bi][Br][PNP]_2[2Et_2O]$
$[Pc_2Sc][NBu_4]$	QOCKUO[198]	18:95–96	Isostructural with $[Pc_2Ce][PNP]$
$[Pc_2Sm][NBu_4][MeOH]$	QOCKIC[198]	18:97–98	Similar to $[Pc_2Bi][NBu_4]$, QOCKAU
$[Pc_2Sm][Br_{1.45}]$ {α}	ZOMWUT[227]	18:99–100	Linear stacks along [001]; tetragonal lattice; counterion in channels between four stacks
$[Pc_2Sm][ClO_4]_{0.63}$ {α}	ZOMXAA[227]	18:101–102	Linear stacks along [001]; tetragonal lattice; counterion in channels between four stacks
$[Pc_2Ti][1\text{-}ClN]$	SEWRER[214]	21:15–16	Tightly intercepted bands along [010]
$[Pc_2Ti][I_3]_{0.66}$	PIBJIT[219]	18:113–114	Linear stacks along [001]; tetragonal lattice; counterion in channels between four stacks
$[Pc_2Tm][PNP][0.5H_2O]$	YEGTUZ[220]	18:115–116	Isostructural with $[Pc_2Ce][PNP]$????
$[Pc_2U][I_3]_{0.66}$	MEGNOB[235]	18:119–120	Linear stacks along [001]; tetragonal lattice; counterion in channels between four stacks
$[Pc_2U][I_{5/3}]$	HOTWOC[234]	18:121–122	Linear stacks along [001]; tetragonal lattice; counterion in channels between four stacks
$[Pc_2Y][NBu_4]$	QOCKOI[198]	18:125–126	Similar to $[Pc_2Bi][NBu_4]$, QOCKAU
$[Pc_2Yb][I_2]$	BECSUX[233]	18:127–128	Linear stacks along [001]; tetragonal lattice; counterion in channels between four stacks
$[Pc_2Zr][I_2]$	NOFMEA[409]	18:131–132	Linear stacks along [001]; tetragonal lattice; counterion in channels between four stacks

34.4°, $[Pc_2Ho][NBu_4][H_2O]$ (NAKYED, App. 18:33–34) 43.2°, $[Pc_2Lu][NBu_4][H_2O]$ (NAKYIH, App. 18:65–66) 45.0°, and $[Pc_2Nd][NBu_4]$ (NAKXUS, App. 18:77–78) 6.2°. From these data, they concluded that the staggering angle depends on the π–π interaction strength between the two phthalocyanine rings. They calculated best planes for the whole macrocycles and best planes for the isoindole nitrogen atoms and found a relationship between the distances between those planes and the staggering angles of these macrocycles. Engel rejected these conclusions as premature as the authors did not include many important structures in their data-set for calculation and because their theory cannot explain the occurrence of polymorphic structures where *bis*-phthalocyanine complexes with the same central atoms show different staggering angles.[23] Koike *et al.* synthesized $[Pc_2Lu][NBu_4][H_2O]$ which has a staggering angle of 45° but they did not try to explain the differences with $[Pc_2Lu][NBu_4][2DMF]$ (GAVZUY) with a staggering angle of 43°. They mentioned Pc_2Sn (SNPTCY) with a staggering angle of 42° but excluded the polymorphic structure SNPTCY01 with a staggering angle of 38°. A similar rebuttal comes from Hückstädt *et al.*[198] From the data in Table 17, it can be concluded that the assumption of Darovskikh that the staggering angle depends on the arrangement of the molecules in the solid-state is correct[196] and any discussion, which treats the staggering angle as a molecular property only, leads to wrong conclusions.

Ostendorp *et al.*[227] found two tetragonal structures of Pc_2Sm, α-$[Pc_2Sm][Br_{1.45}]$ (ZOMWUT, App. 18:99–100), and $[Pc_2Sm][ClO_4]_{0.63}$ (ZOMXAA, App. 18:101–102). The phthalocyanine macrocycles are nearly planar and the staggering angle is *ca.* 41°. The authors discuss the tetragonal structures of oxidized monophthalocyanines and *bis*-phthalocyanines in detail. Ostendorp and Homborg[228] reported the structure of $[Pc_2Bi][CH_2Cl_2]$ (ZUWDAW, App. 18:3–4), which is isomorphous with the other Pnma structures of identical composition. The unit-cell data of an isomorphous Indium complex (RIYREW) are reported by Ostendorp and Homborg in 1996.[229] Ostendorp *et al.* reported in 1996 that $[Pc_2Ce][BF_4]_{0.33}$ (RABDEZ, App. 18:21–22), also crystallizes in a tetragonal structure.[230] Donzello *et al.* published the structure of $[Pc_2Nb][I_3][0.5I_2,3.5(1\text{-}ClN)]$ (NOCYEJ,[231] App. 21:13–14), with highly distorted molecules, similar to the molecules in $[Pc_2Ti][1\text{-}ClN]$, SEWRER.[214] The iodine-doped, disordered structure of bismuth(III) diphthalocyanine, $[Pc_2Bi][I_{1.5}]$ (KOJQUV, App. 18:13–14), with planar phthalocyanine macrocycles has been reported by Janczak *et al.* in 1998.[232] $[Pc_2M][I_2]$ complexes of ytterbium (BECSUX, App. 18:127–128) and As (BECTOS, App. 18:1–2), were published by Janczak *et al.* in 1999.[233] The authors suggested that $[Pc_2As]$ should better be described as [(PcAs)(Pc)] with alternating PcAs and Pc units.

Janczak and coworkers published two papers on iodine-doped uranium *bis*-phthalocyanines; $[Pc_2U][I_{5/3}]$ (HOTWOC,[234] App. 18:121–122) in 1999 and $[Pc_2U][I_3]_{2/3}$ (MEGNOB,[235] App. 18:119–120) in 2000. In HOTWOC, the ligands are planar while in MEGNOB, they are slightly deformed. The crystal structure of a lutetium *tris*-phthalocyanine with 15-crown-5 substituents, $[(15cr5)_4Pc]_3[Lu_2]$ $[18DMSO, 2CHCl_3]$ (LORKIM, App. 18:53–55) has been published in 1999 by Troyanov *et al.*[195] Two identical structures of the *tris*-phthalocyanine, Pc_3Bi_2 (DIQPIC and DIQPIC01, App. 18:5–6) have been reported by Janczak *et al.*[51] and Benihya *et al.*[52] The Bi atoms coordinate symmetrically to the outer and inner macrocycle, which is in contrast to the unsymmetrical bonding of the In atoms to the central phthalocyanine in Pc_3In_2 (PIFDEN, App. 18:37–38).

A series of 15 anionic *bis*-phthalocyanine complexes has been synthesized and crystal structures obtained by Hückstädt *et al.*[198] in 2001; $[Pc_2Bi][Br][PNP]_2[2Et_2O]$ (QOCLOJ, App. 18:7–8), two polymorphic structures of $[Pc_2Bi][NBu_4]$ (QOCJEX, App. 18:9–10, and QOCKAU, App. 18:11–12), $[Pc_2Ce][NBu_4]$ (QOCJIB, App. 18:17–18), $[Pc_2In][I][PNP]_2[2Et_2O]$ (QOCLUP, App. 18:39–40), $[Pc_2La][NBu_4]$ (QOCJOH, App. 18:45–46), $[Pc_2La][NBu_4][EtOH]$ (QOCMAW, only unit-cell data), $[Pc_2La][NPe_4]$ (QOCJUN, App. 18:47–48), $[Pc_2Pr][NBu_4][0.75MeOH]$ (QOCKEY, App. 18:85–86), $[Pc_2Pr][NBu_4][2py]$ (QOCLAV, App. 18:87–88), $[Pc_2Sb][Br][PNP]_2[2Et_2O]$ (QOCLID, App. 18:93–94), $[Pc_2Sb][NBu_4][2thf]$ (QOCLEZ, App. 18:91–92), $[Pc_2Sc][NBu_4]$ (QOCKUO, App. 18:95–96), $[Pc_2Sm][NBu_4][MeOH]$ (QOCKIC, App. 18:97–98), and $[Pc_2Y][NBu_4]$ (QOCKOI, App. 18:125–126). In every complex salt, the phthalocyanine ligand is severely distorted from planarity and adopts either domed, saddled, waved or mixed non-planar conformations. Comparing the data of this paper and the data from Koike *et al.*, it can be seen that the *bis*-phthalocyanines can be grouped into two broad groups according to the staggering angle, one group has staggering angles around 4–6°, and the second group comprises complexes with angles between 34 and 45°. No *bis*-phthalocyanines with a staggering angle between 7 and 34° have been found yet. As will be shown later, the group of molecules with the angles between 34 and 45° can be divided into several subgroups. While this group derives from a situation where the central atom is surrounded by the isoindole nitrogen atoms forming a distorted antiprism, the group with the smaller staggering angles derives from a distorted cubic bonding of the nitrogen atoms toward the central atom. Rittenberg *et al.*[236] reported in 2001 the structure of a cationic gadolinium *bis*-phthalocyanine, $[Pc_2Gd][C_4(CN)_6][2ClB]$ (OBUJIE, App. 18: 29–30), with a staggering angle of 43.3°.

Sandwich-type mixed phthalocyaninato-porphyrinato metal complexes are recently discussed in detail in two reviews by Jiang *et al.*[237] and Buchler and Ng.[54] A compilation of the basic crystallographic data of these structures is given in Appendices 1 and 6a and their molecular structures are shown in Appendices 18 and 23.

2. Solid-State Arrangement

CN8*bis* complexes are principally complexes with two identical sides. Crystallographically however, they are different due to different spatial orientations of the two macrocycles in the molecule. Therefore, the arrangements in both directions perpendicular to the molecule are different. A comparison of the unit-cell data from Appendix 1 and the data of Table 17, it can be seen that the neutral *bis*-phthalocyanines can be divided into seven main subgroups, illustrated by the structures of Pc_2Pr, Pc_2Nd, Pc_2Ce, Pc_2Zr, Pc_2Sn, $[Pc_2Lu][CH_2Cl_2]$ and Pc_2Er (in Table 17 in bold captions). In all these structures, with the exception of Pc_2Er, the molecules are arranged in columns of slipped-stacked molecules. The view along the column axes is shown in Figure 41. The relationship between some of these structures can be seen when only the overlapping macrocycles are investigated (Figure 42). KOBRUO and CIZGIB06, as well as JONZAN and SOTGAJ have similar structures, in addition to which, SNPTCY bears a close resemblance to DICBUM. The structural similarity of the members of the different groups can also be seen when looking perpendicularly onto the columns, and parallel to the macrocycles (Figure 43). The stacking angle of the macrocycle in relation to the column axis is around 80° for KOBRUO and CIZGIB06, and exactly 90° in YUHSUP (Figure 45). This angle is considerably smaller in the structures of JONZAN and SOTGAJ, where the molecules are much more inclined towards the stacking axis than in the former structures. The structure of SNPTCY consists of sheets of molecules, which are rotated by nearly 90° towards each other. A view perpendicular to one of the sheets is shown in Figure 43e. Each sheet consists of parallel columns of slipped-stacked molecules. DICBUM also consists of sheets, but the columns are not parallel but are rather arranged in a fish-bone structure. Neighboring sheets are related by inversion symmetry. Figures 44 and 45 show the presence of several types of columnar arrangements: 1) KOBRUO/CIZGIB06 (Figure 44a,b), 2) JONZAN/SOTGAJ (Figure 44c,d), 3) SNPTCY/DICBUM (Figure 44e,f) and 4) YUHSUP (Figure 45d). In the first two groups, every second molecule of a column is identical, while in groups 3 and 4, every molecule is identical. The differences between group 1 and 2 lie in the directions of overlap with the neighboring molecules below and above (Figure 46). In group 1, the overlap directions is referred to as A1, i.e., the overlap direction with the adjacent molecule below is along the lobe A while the overlap with the molecule above is along lobe 1. Group 2 overlap is referred to as A2. The molecules in YUHSUP are fully overlapped, as well as being staggered.

3. Ionic Complexes

The arrangement of ionic *bis*-phthalocyanines is detailed in Table 18.

IV. Phthalocyanine-Related Complexes

A. SUBSTITUTED PHTHALOCYANINES

It has been a long myth that peripherally substituted phthalocyanine complexes would never crystallize as single crystals. However, since the first structures were discovered in the early 1990s,[41,42] many new structures have been reported.[45,166,171,195,238–248] An overview is given in Figure 47. A compilation of the basic crystallographic data of the structures is given in Appendix 1 and their molecular structures are shown in Appendices 10–18.

B. NAPHTHALOCYANINES

Only few structures of naphthalocyanines are known which are compiled in Figure 48.[104,249–252] A compilation of the basic crystallographic data of the structures is given in Appendix 2 and their molecular structures are shown in Appendix 19.

C. PHTHALOCYANINE-RELATED ISOINDOLE COMPLEXES

The phthalocyanine is a member of a series of monocyclic circularly linked isoindole complexes with azamethine nitrogen bridges (Figure 49). In the case of

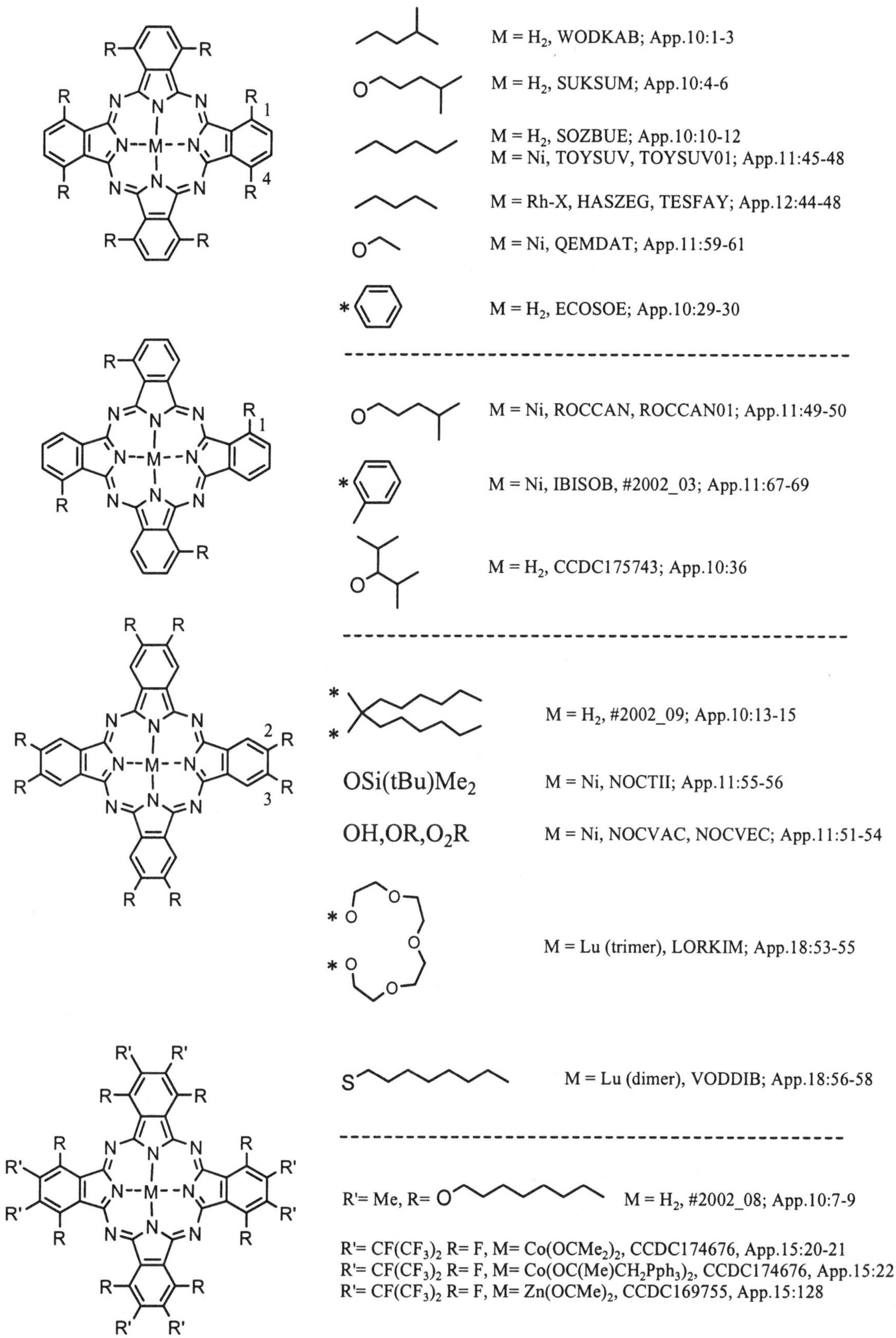

Figure 47. Overview of substituted phthalocyanines where crystal structures have been reported.

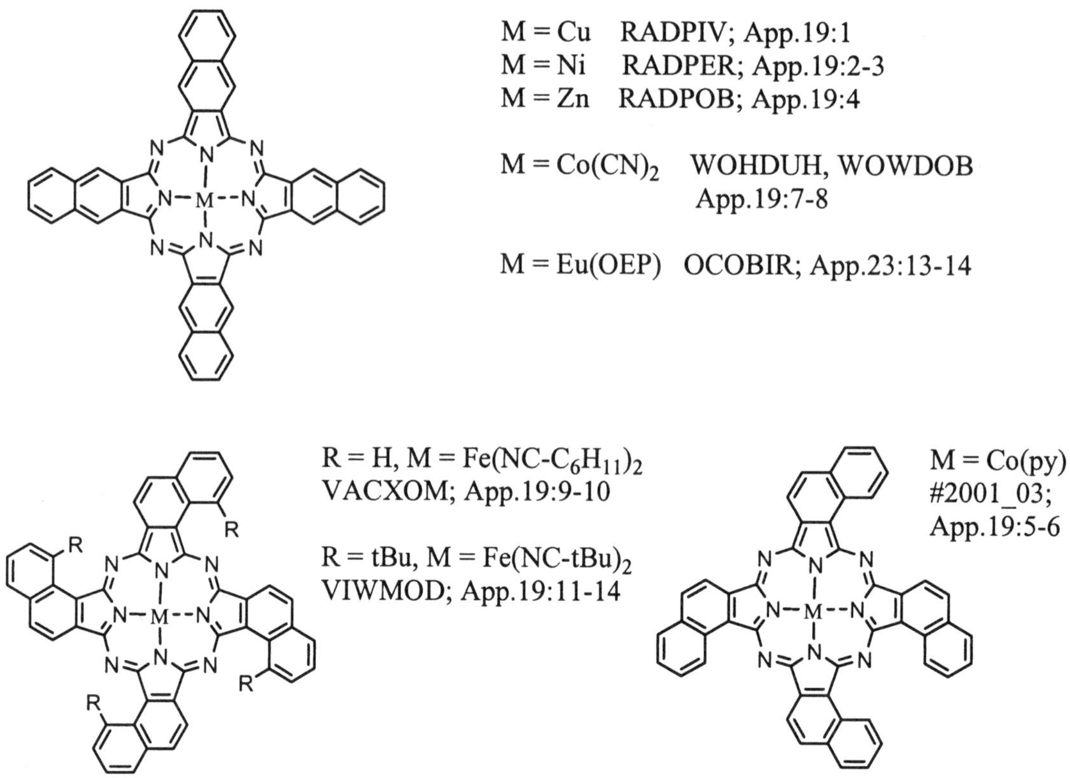

Figure 48. Overview of naphthalocyanines where crystal structures have been reported.

the phthalocyanines, four isoindole units are linked. Other members of this series are the subphthalocyanines with three units and the superphthalocyanines with five units. While the phthalocyanine macrocycle can coordinate to a variety of central atoms, subphthalocyanine complexes are known only with boron and superphthalocyanine complexes only with an uranium central atom. Larger macrocycles, so-called extended phthalocyanines, with six or more isoindole units are also reported in literature but have not been characterized by single-crystal structure analysis. Extended macrocycles may also coordinate with several central atoms.

During the synthesis of phthalocyanine complexes, sometimes bi- and tri-cyclic macrocycles with different kind of arrangement of isoindole units around the central atom were found. They are often referred to as bicyclic phthalocyanines despite the fact that they are no phthalocyanines, as they do not contain a conjugated aromatic system. While they were known already since long time,[253] the first crystal structure has been discovered by Gingl and Strähle in 1990.[254] Recently, several new members have been obtained.[255–258] In these complexes the half of a phthalocyanine coordinates perpendicularly to two α-carbon atoms of a phthalocyanine macrocycle as shown in Figure 49d–f and Appendix M. Three different configurations are known, referred to as 1,4-coordination in Figure 49d, as 1,5-coordination in Figure 49e and as 1,6-coordination in Figure 49f. 1,4-Coordination has been found in PccNbCl (KEYPAF,[254] App. 21:7), 1,5-coordination in PccIn (LIFJOZ,[255] App. 21:10), PccGd (ZENGUU,[256] App. 21:8), and PccTl (ZOLHUD,[257] App. 21:9) while the 1,6-coordination has been found in PccBi (QOWHIT,[258] App. 21:11–12). The latter macrocycle could also be described as a complex where one isoindole unit coordinates perpendicularly to two α-carbon atoms of a superphthalocyanine macrocycle as shown in Figure 49g. Other coordination pattern, e.g., a 1,3-coordination, might also exist but have not been found yet. At the moment also, only even-numbered bicyclic complexes are known, 4+2 isoindole units in the case of bicyclic phthalocyanines. Using the nomenclature above, 4+4 complexes then are "tricyclic phthalocyanines." Usually, however, they are referred to as *bis*-phthalocyanines with inter-ligand bonds. Like the bicyclic phthalocyanines, they are actually no phthalocyanines as the aromaticity of the ringsystem is disrupted. Such complexes are known for small central atoms like titanium (SEWRER,[214] App. 21:15–16) or niobium (NOCYEJ,[231] App. 21:13–14). A compilation of the basic crystallographic data of the structures is given in Appendix 4 and their molecular structures are shown in Appendix 21.

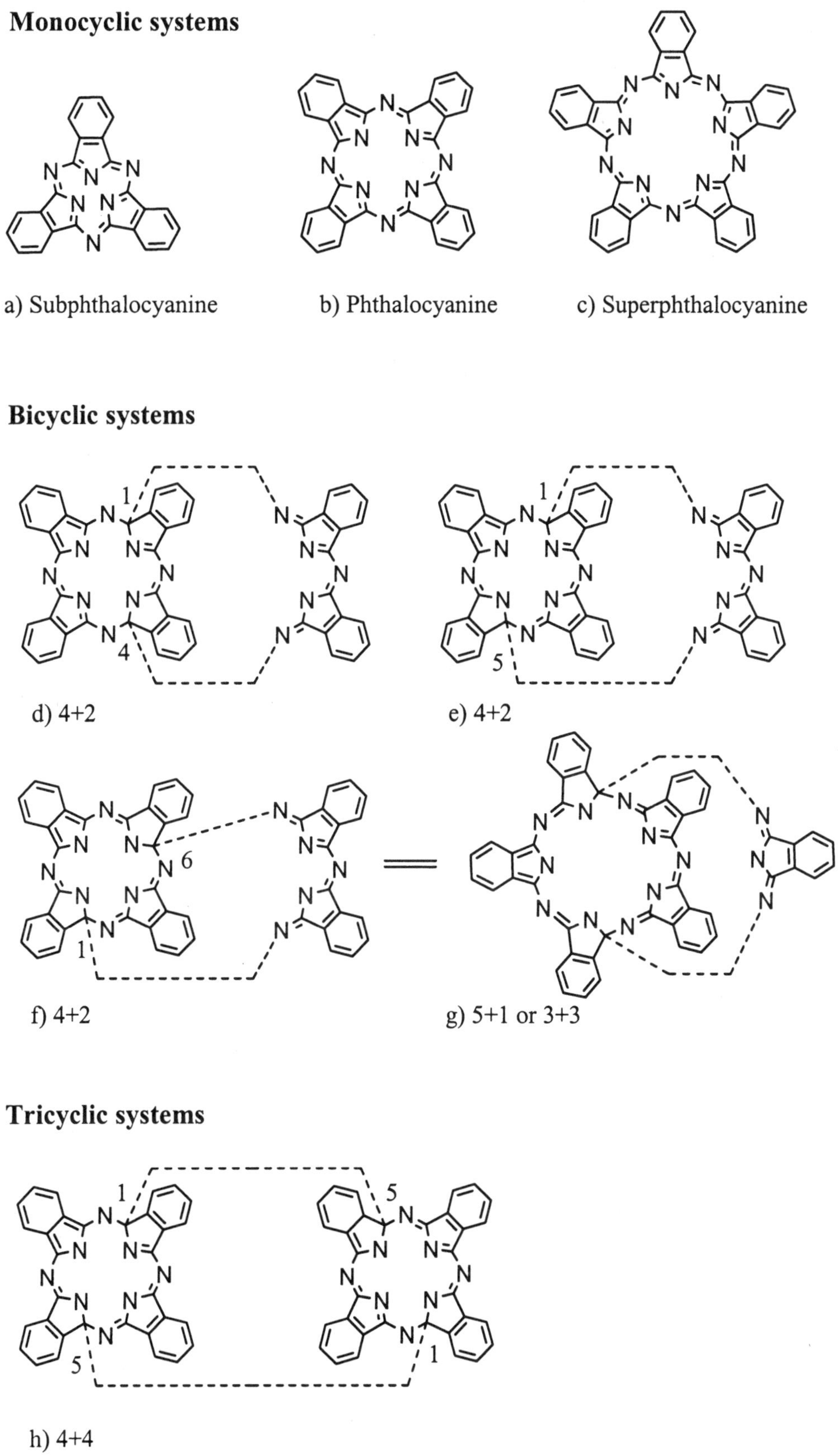

Figure 49. Overview of cyclic isoindole complexes where crystal structures have been reported.

D. SUB-[259] AND SUPERPHTHALOCYANINES[260]

A subphthalocyanine complex is composed of three isoindole subunits, in contrast to the four subunits of a phthalocyanine complex. At the moment, only boron coordinated complexes are known but subphthalocyanine complexes of other central atoms as P, As cannot be ruled out. The unknown "metal-free subphthalocyanine" ($SubpcH_2$) would certainly be very unstable as the central cavity is much smaller than the cavity in phthalocyanines and cannot accommodate two hydrogen atoms. However, a [SubpcH] anion could theoretically exist. The first single-crystal structure of a subphthalocyanine complex, SubpcB-Cl (PHTLBC, App. 20:1) has been obtained by Kietaibl already in 1974.[138] However, it was only after the ring-extension reaction of subphthalocyanine into phthalocyanine became known that several groups restarted studying these materials and obtained new crystal structures. Single-crystal structures are known for subphthalocyanine complexes axially substituted with chloro,[138,261] phenyl,[137,139,141] hydroxy,[141] alkoxy,[137,140,141,262,263] acetoxy,[141] and siloxy.[137] Additionally, four structures with a substituted subphthalocyanine macrocycle have been reported, monobenzosubphthalocyanine[264] dodecafluorosubphthalocyanine[265] and, very recently, the *cis*- and *trans*-isomer of a hexadecafluoro substituted binuclear subphthalocyanine has been reported.[261]

Similar to the deformation of the phthalocyanine macrocycle in a phthalocyanine complex with an out-of-plane coordinated central atoms, the subphthalocyanine macrocycle adopts a non-planar, saucer-type structure. From the data available, it is not possible to see an influence of the axial ligand onto the molecular structure of the macrocycle. The geometry of the subphthalocyanine macrocycle varies much less than

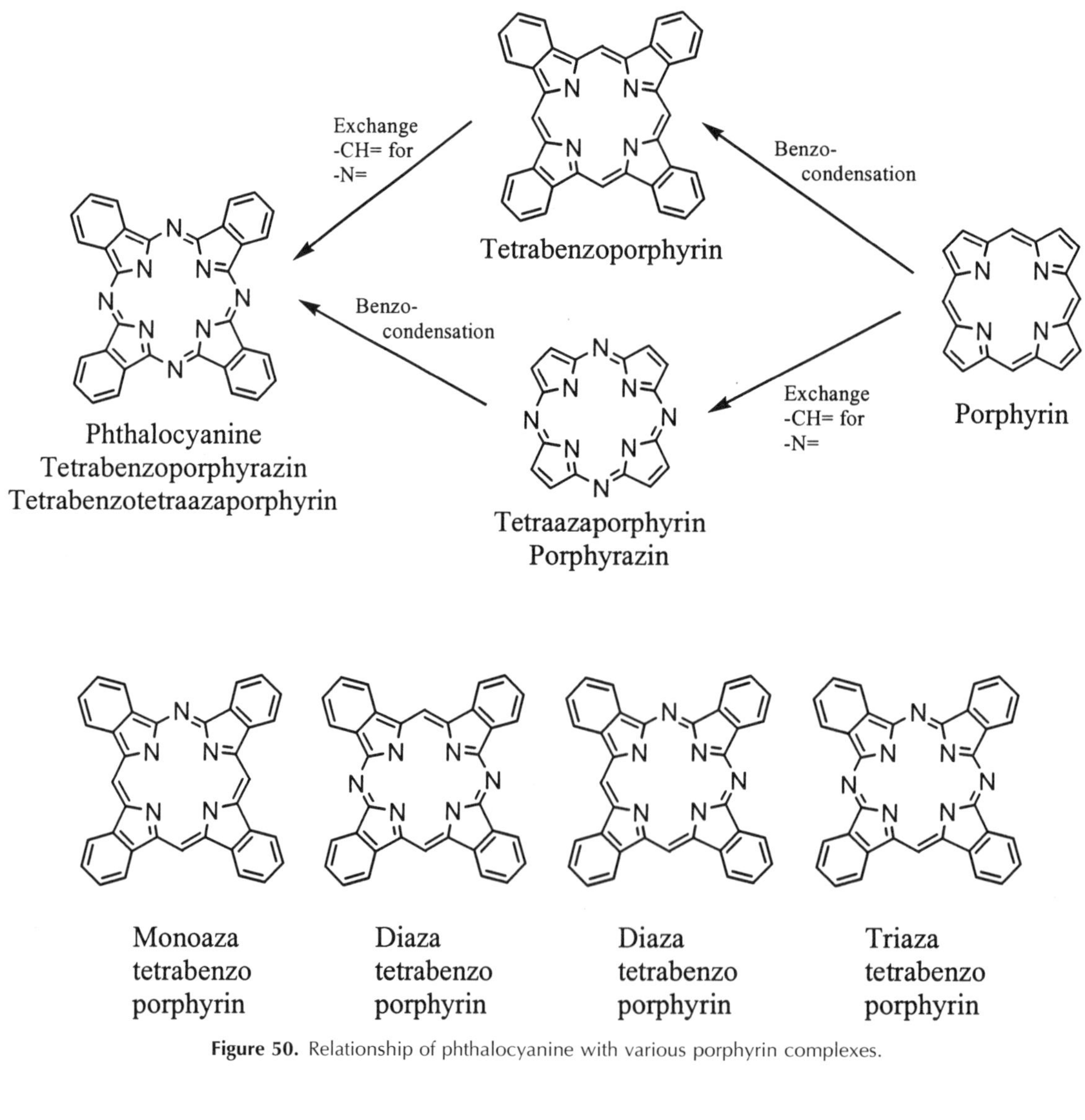

Figure 50. Relationship of phthalocyanine with various porphyrin complexes.

those of the phthalocyanines.[23,137] The overcrowded inner cavity of a subphthalocyanine complex can best be seen by the increased angle of C_{α}-N_i-C_{α} which is due to the nitrogen atoms pushed back by the central atom. Subphthalocyanine in analogy to phthalocyanine complexes with an out-of-plane coordinated central atom has two different sides, the macrocycle-only side (concave side) and the side with the boron and the axial ligand (convex side). Concave and convex sides show different arrangement pattern. Overlapping of concave sides produces two different patterns, either the head-to-tail or the head-to-head overlap as shown in Engel *et al.*[137] The overlap of convex sides is governed by hydrogen bonding similar to the case of phthalocyanines with an out-of-plane coordinated central atom. The fluorine-substituted complex, #2002-04, stacks similar to monoclinic PcPb in linear 1D columns.[265] The only known superphthalocyanine complex which is structurally characterized is SuperpcUO_2[260] where five isoindole units, as shown in Figure 49c, are coordinated around a linear uranyl unit O=U=O. It can be seen in App. 20:19–20 that the macrocycle is heavily deformed. A compilation of the basic crystallographic data of the sub- and superphthalocyanine structures is given in Appendix 3 and their molecular structures are shown in Appendix 20.

E. PORPHYRAZINES[266–268]

Phthalocyanines are structurally related to the porphyrins, as shown in Figure 50. Two formal structural routes lead from the porphyrin to the phthalocyanine macrocycle, either by a) first exchanging the methine groups of a porphyrin for azamethine groups and then extending the macrocycle by benzo-condensation or b) by the reverse way. The intermediate of the first route is called tetraazaporphyrin or porphyrazine and shown in Figure 50. While many groups either work in the field of porphyrins or phthalocyanines, only few groups are interested in porphyrazine chemistry. Most of the known porphyrazine single-crystal structures have been reported by the groups of Barrett and Hoffman. A compilation of the basic crystallographic data of the porphyrazine structures is given in Appendix 5 and their molecular structures are shown in Appendix 22.

F. OTHERS

Phthalocyanine complexes are also structurally related to complexes were either one or every second isoindole unit is exchanged for another unit (Figure 51). While some of these complexes can be seen as aromatic benzo-condensated porphyrazine complexes, complexes where the pyrrole-type unit is exchanged for pyridine

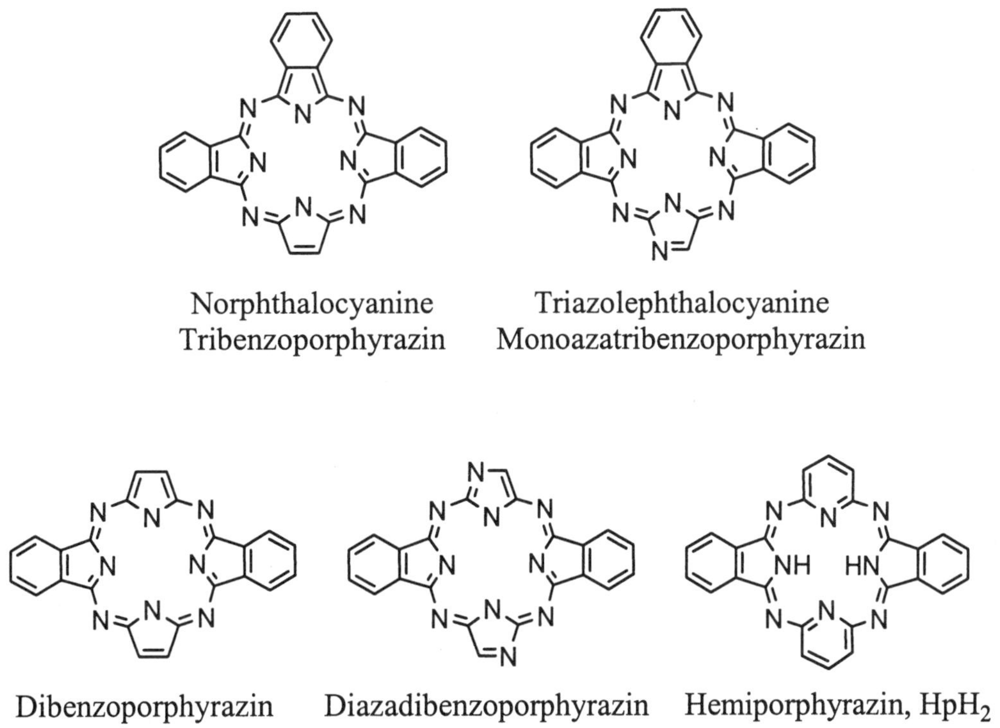

Figure 51. Relationship of benzoporphyrazin complexes with norphthalocyanine, triazophthalocyanine, and hemiporphyrazin complexes.

or similar macrocycles, are not aromatic anymore. Such complexes are referred to as hemiporphyrazine derivatives and will not be treated in this review despite their close structural relationship with phthalocyanine complexes.

V. Conclusions

Despite the vast amount of reported structures, very few is known about the phthalocyanine structures in general. Especially limited is the knowledge of the arrangement of molecules in the solid state. The only group which is sufficiently described, is the group of CN4 structures but, still, there are some remaining "white areas" which need to be explored. With the exception of cationic phthalocyanines with fractional charges and the radical PcLi, phthalocyanine macrocycles are arranged in slipped-stacked columns. In these columns, every molecule is identical if both sides of the macrocycle are identical as in, e.g., PcCu, $PcSiCl_2$ or very similar as in, e.g., PcMoOCl or Pc_2Zr. The *bis*-phthalocyanines with their chemically identical but crystallographically different sides display a variety of slipped-stacked zigzag and linear columnar arrangements. If both sides are different as in, e.g., PcTiO, many different structures can be found due to the variety of ways, in which the axial ligand can interact with the neighboring molecules.

Additional cocrystallizing molecules as e.g., solvent molecules, or counterions if the phthalocyanine complex is an ion, or large axial ligands have a strong influence on the arrangement of the phthalocyanine macrocycles. A good example for such an influence is the series of complexes of similar composition $[PcCo(CN)_2][2LM]$, where LM represents H_2O, $CHCl_3$, $CHBr_3$, or DMSO, which display strongly different arrangements. Large groups coordinated to the central atom reduce the possibility of interaction between phthalocyanine macrocycles of neighbor molecules and the arrangement is mainly determined by interactions between atoms of the axial ligands. Molecules, where the axial ligands are coordinated on one side and the phthalocyanine macrocycle on the other side of the central atom tend to crystallize in pairs with a strong overlap of the macrocycle sides.

In this review, it could be shown that complexes with rather different molecular structures (e.g., PcCu, $PcSiCl_2$, II-PcTiO, Pc_2Zr) crystallize in similar arrangements based on slipped-stacked columns. Interaction between the hydrogen atoms of the aromatic benzo rings of the phthalocyanine macrocycle and the axial ligands can strongly orient neighboring molecules. Hydrogen bonding as a Coulomb force can extend over a long distance and it can be considered as one of the major interactions which determine the arrangement of phthalocyanine macrocycles. The example of the orientation of PcTiO molecules through an interaction between the axial ligand and aromatic hydrogen atoms through an intermediate layer of neighbor molecules is certainly a convincing example for the capability of hydrogen bonds.

There are many examples that the crystal arrangement can also influence and deform the molecular structure, a fact which has been overlooked within literature. There are only three other studies, which take this account into consideration.[23,196,198] In most cases of molecular deformations, they are accompanied by hydrogen bonding of the deformed parts towards the neighbor molecules. Further investigation is necessary, e.g., a mapping of the deformation of the molecular structure onto the solid-state arrangement. Such an investigation eventually needs the development of new methods of structural description.

ACKNOWLEDGMENTS

This work is dedicated to Gerhard Ebel and Manfred Schüler, my former chemistry teachers at the Nordpfalzgymnasium in Kirchheimbolanden, Germany, as a recognition of their capability to make chemistry classes an interesting and joyful event.

The first work on this collection started in 1993, when I worked as Visiting Researcher in the field of photosensitive phthalocyanine complexes. I soon found out that everybody knew that the photosensitivity depends on the arrangement of the molecules in the solid state but nobody actually had any information about how phthalocyanine molecules are arranged. During the collection of these structural data, I had the pleasure to get into contact with many researchers who contributed as early as 1961 and/or are still contributing to the phthalocyanine structural chemistry.

I greatly appreciate that many of them not only send me additional information on their work but also entrusted me with unpublished materials to use it for this review. I especially want to thank Prof. Heiner Homborg, who was very open about his unpublished results and who regularly gave me updates on newly found structures. But without the friendly and helpful collaboration of Ms. Gill Heale from the Cambridge

Crystallographic Data Centre in UK while checking and rechecking crystal structures this review would have never been finished.

I also want to thank the colleagues who gave their permission to reproduce some of their work: Prof. Everly B. Fleischer,[34] Dr. Edward C. Bissell,[40] Dr. David R. Swift,[269] Dr. Sax A. Mason,[31] Dr. Anthony D'Addario,[270] Prof. Kuninobu Kasuga,[205] Dr. Mireille Mossoyan-Deneux,[210,213] Prof. Joachim Strähle,[120,124] Dr. Thomas Krämer,[120] Dr. Franz Gingl,[124] Prof. Joel S. Miller,[271] Prof. Naoki Sato,[156] and Dr. Masahiro Hasegawa.[155,156]

Additionally, I am thankful for the wealth of information which I received from: Dr. Michael Chen, Prof. Michael Cook, Dr. Dominique Dempf, Prof. William S. Durfee, Prof. Claudio Ercolani, Prof. Jean Fischer, Dr. Jeffrey P. Fitzgerald, Dr. Madeleine Helliwell, Prof. Bernard Hoskins, Dr. Heiner Hückstädt, Prof. Tamotsu Inabe, Prof. James A. Ibers, Prof. Seiji Isoda, Dr. Jan Janczak, Dr. Atsushi Kakuta, Prof. Malcom E. Kenney, Prof. Nagao Kobayashi, Prof. Takashi Kobayashi, Prof. Rolf Köhler, Prof. Ryszard Kubiak, Prof. Gerhard Lincke, Dr. Neil McKeown, Prof. Jin Mizuguchi, Dr. Kenji Okada, Prof. David Phillips, Tom DiPietro, Ms. Jean Reid, Prof. W. Robert Scheidt, Dr. David Watson, Dr. David Williams, Dr. Kenneth J. Wynne, Prof. Kyuya Yakushi, and Prof. Peter Zugenmaier.

I appreciated very much the helpful comments of Dr. Jonathan Wright during manuscript preparation and I am very grateful to Dr. Pierre Bassoul (ESPCI, France) for introducing me to the interesting world of crystal structures and for helping me so many times.

And finally, I would like to thank Dr. Chyongjin Pac who gave me the unique opportunity and his support to write the 1997 review.[23]

Note Added in Proof

In 1967, Raidt published his doctoral thesis on the crystallography of phthalocyanines.[461] Raidt discusses α-, β-, and γ-polymorphs of several CN4 complexes, PcCu, PcNi, and PcPt, as well as mixed crystals made from them. In addition, he investigated crystals of H_2SO_4-containing copper phthalocyanines.

His main focus, similar to the paper of Honigmann *et al.*,[74] was the search of the α-PcCu structure. He was able to find out many characteristics of the structure but could not determine it unequivocally. While trying to stabilize the metastable α-PcCu structure by doping with the stable α-PcPt, he obtained the unknown β-PcPt polymorph (PcPt doped with PcCu). It crystallizes in $P2_1/a$ with $a = 20.3,3$ Å, 4.75,? Å, $c = 15.0,5$ Å and $\beta = 122.0,?^\circ$. Raidt became aware of Brown's suggestion that the assignment of $P2_1/a$ by Robertson should be replaced by C2/c, as referenced in Suito and Uyeda,[462] and reinvestigated the PcPt structure. He could confirm Robertson's assignment of $P2_1/a$. A comparison with the work of Honigmann *et al.*, led Raidt to conclude that several polymorphs should exist for PcPt, a fact which was later proven by Brown.[77] Raidt found that the reported structures differ from each other in the *b* lattice constant, 3.76 Å (Honigmann structure), 3.81 Å (Robertson), and 3.96 Å (Raidt). The structures later published by Brown have similar constants, 3.818 Å (PTPOCY10), and 3.969 Å (PTPOCY11). However, the structure similar to the Robertson structure (PTPOCY11) has the larger b constant and *vice versa*. It seems as if the b constant is not linked to the full crystal structure of the polymorph. Raidt suggested that the b constant is linked to the amount of chlorinated by-product in the sample. He then investigated mixed crystals of PcCu–PcNi, PcCu–PcH_2, and PcCu–PcPt, but did not found any new structures with the exception of the β-PcPt polymorph (PcPt doped with PcCu). Raidt also obtained single-crystals of PcCu–H_2SO_4 but could not determine their structure due to their instability under air. Instead, he could crystallographically observe the step-wise transformation of these crystals into α-PcCu.

In 2000, Simon and Bassoul worked out a new concept of the description of crystal structures, which is based upon topology and isohedral tilings. They used this concept for the description of the structure of the β-polymorphs of phthalocyanines.[463]

Appendices

Appendix 1. Basic Crystallographic Data of Crystal Structures of Phthalocyanines[a–e]

CamRef	CN#	Crystal description	Figure	Space group	Axis a / Å	Axis b / Å	Axis c / Å	α / deg	β / deg	γ / deg	Z	V	Reference
WOKDAB	1	$(i\text{-}H_{11}C_5)_8PcH_2$	10:1–3	C2/c	20.433,5	33.873,4	10.025,4	90	110.12,2	90	4	6515.1,31	45
SUKSUM	1	$(i\text{-}H_{11}C_5O)_8PcH_2$	10:4–6	P-1	10.665,4	11.108,4	15.597,6	73.08,3	87.16,3	87.93,3	1	1765.3,12	41
#2002-08	1	$(Me)_8(H_{17}C_8O)_8PcH_2$	10:7–9	P-1	17.840,5	21.110,8	14.039,7	91.07,4	95.99,3	93.90,3	2	5244,3	238
SOZBUE	1	$(n\text{-}H_{13}C_6)_8PcH_2$	10:10–12	P-1	9.2668,9	9.7323,8	19.991,2	100.964,7	91.166,7	95.899,7	1	1759.2,8	42
ECOSOE	1	$(\phi)_8PcH_2$	10:29–30	P-1	15.5258,5	15.8918,7	16.4589,2	87.287,2	67.873,2	61.084,3	2	3247.1,2	239
#2002-09	1	$\{(H_{13}C_6)_2C\}_4PcH_2$	10:13–15	P-1	15.7306,16	20.621,3	21.527,4	62.462,10	82.358,17	86.470,13	3	6136.7,15	238
*nc	1	PcH_2 {> 7kBar}	–	?	23.72,?	3.91,?	16.12,?	90	130.3,20	90			79
PHTHCY04	1	PcH_2 {α}	10:16	C2/n	26.124,3	3.801,1	23.889,3	90	94.18,2	90	4	2365.8,6	97
PHTHCY08	1	PcH_2 {α}	10:17	C2/c	26.121,4	3.7970,7	23.875,3	90	94.16,2	90	4	2361.7,8	98
#1971-03 (App. 26)	1	PcH_2 {β, N}	10:18	$P2_1/a$	19.86,?	4.73,4	14.80,?	90	122.0,?	90	2	1180or73	31
PHTHCY*nc	1	PcH_2 {β, N}	–	$P2_1/a$	19.86,?	4.73,4	14.80,?	90	122.0,?	90	2	1180or73	38
PHTHCY01	1	PcH_2 {β}	10:19	$P2_1/a$	19.85,?	4.72,?	14.8,?	90	122.25,?	90	2	1173,?	1,57,58
#1970-02 (App. 25)	1	PcH_2 {β}	10:20	$P2_1/a$	19.847,4	4.727,1	14,792,3	90	121.97,3	90			40
PHTHCY03	1	PcH_2 {β}	10:21	$P2_1/c$	14.803,3	4.729,2	19.870,4	90	122.02,2	90	2	1179.3,2	96
PHTHCY07	1	PcH_2 {β}	10:22	$P2_1/a$	19.870,7	4.731,7	14.813,7	90	121.98,4	90	2	1181,1	29,30
PHTHCY02	1	PcH_2 {β}	10:23	$P2_1/n$	14.796,6	4.7325,6	17.357,7	90	104.32,2	90	4	1177.6,7	102
*nc	1	PcH_2 {β}	–	$P2_1/a$	19.847,4	4.727,1	14,792,3	90	121.97,3	90			80
*nc	1	PcH_2 {τ}	–	monocl./ orthorhomb	23.26,?	4.85,?	21.11,?				4		100
PHTHCY05	1	PcH_2 {X, R}	10:24	$P2_1/a$	21.4,?	23.7,?	4.9,?	90	30.24,?	90	2		99
PHTHCY06	1	PcH_2 {X, R}	10:25	$P2_1/a$	10.63,?	23.15,?	4.89,?	90	95.98,?	90	2	1194.2,?	101
#1997-03	1	PcH_2 {X, R}	10:26	$Pna2_1$	21.0,1	4.91,5	23.1,1	90	90	90	4	2382,?	102
*nc	1	PcH_2 {X}	–	–	23.86,?	4.93,?	9.94,?	90	98.4,?	90			92
GAKWUK	1	PcH_2/Cl_4PcH_2 {α}	10:27	C2/n	26.460,1	3.760,2	24.275,1	90	93.53,1	90	4	2410.2,2	93
DIJYEA*nc	1	$[PcH_2][I_3]_{0.33}$	–	P4/mcc	13.979,6	13.979,6	6.502,3	90	90	90	2		272
LATLIB	1	$[PcH_2][I_3]_{0.33}$	10:28	P4/mcc	13.931,2	13.931,2	6.411,1	90	90	90	2	1244.2,6	273
CCDC184514	1	$[PcH_2][IBr_2]$	10:34	$P2_1/a$	8.0279,9	21.258,2	18.1439,2	90	113.975,2	90	4	2828.8,5	114
CCDC184515	1	$[PcH_2][IBr_2][Br][1\text{-}BrN]$	10:35	P-1	8.4724,6	13.5794,10	13.8403,10	90.854,1	103.417,1	97.049,1	1	1535.61,19	114
*nc	1	$[PcH_2]_2[AsF_6]_3[1\text{-}ClN]$	–	triclinic	16.355,7	16.502,8	9.781,3	109.01,4	97.34,4	64.45,3			274
CCDC168923	1	$[PcH_2]_3[AsF_6]_2[1\text{-}ClN]$	10:31–32	P-1	9.6571,3	15.9677,6	16.4953,6	111.933,2	98.638,2	105.583,2	1	2181.60,13	275
CCDC168921	1	$[PcH_2]_3[(SbF_6)_2][1\text{-}ClN]$	10:33	P-1	9.6765,9	16.0495,14	16.7247,15	112.6061,16	99.3131,19	105.3728,16	1	2208.2,3	275
CCDC175743	1	$[((i\text{-}Pr)_2CHO)_4PcH_2][(HOOC)_2]$	10:36	$P2_1/c$	15.8962,15	8.8141,9	21.695,2	90	99.911,2	90	2	2994.4,5	175,248
BZAZPN	1	$MATBPH_2$	10:37	$P2_1/a$	17.63,2	6.58,1	12.50,2	90	123.4,1	90	2		68
COBGOP*nc	1	PcLiH	–	$P2_1/a$	19.081,9	4.996,3	14.437,7	90	119.04,4	90			276
COPTCY07	4	PcCo {α, R}	11:1–2	P-1	12.090,8	3.754,2	12.800,9	88.96,6	90.97,6	95.09,7	1		107
COPTCY03*nc	4	PcCo {α}	–	orthorhombic	25.4,1	3.77,2	24.6,1	90	90	90	4	2358	277
COPTCY06	4	PcCo {β, N115K}	11:3	$P2_1/c$	14.489,9	4.763,4	19.156,16	90	120.76,2	90	2	1136,3	95
COPTCY01	4	PcCo {β, N4.3K}	11:4	$P2_1/c$	14.495,5	4.742,4	19.107,5	90	120.76,2	90	2	1128.56,?	27,28
COPTCY08	4	PcCo {β, R}	11:5	$P2_1/a$	14.5982,9	4.7937,3	19.4348,11	90	120.782,3	90	2		107
COPTCY05	4	PcCo {β, X115K}	11:6	$P2_1/c$	14.542,4	4.790,1	19.215,5	90	120.74,2	90		1150.4,5	94
COPTCY10	4	PcCo {β, X295K}	11:7	$P2_1/c$	14.542,2	4.7731,5	19.352,2	90	120.824,7	90	2	1153.49,?	27,83,84
COPTCY02*nc	4	PcCo {β}	–	$P2_1/a$	20.2,?	4.77,?	15.0,?	90	121.3,?	90	2	1235,?	58

*nc	4	PcCo {β}	–	P2$_1$/a	19.39,?	4.79,?	14.57,?	90	120.66,?	90		1164,?	73
VIYFEO	4	[PcCo][AsF$_6$]$_{0.5}$	11:8	P4/mcc	14.234,2	14.234,2	6.296,2	90	90	90	2		278
DIDTEP	4	[PcCo][I$_3$]$_{0.33}$	11:9	P4/mcc	13.9266,12	13.9266,12	6.2466,11	90	90	90	2	1212,?	112
NOYYIJ	4	[PcCo][NBu$_4$][Me$_2$CO]	11:10	P2$_1$/c	12.370,1	23.370,3	15.952,8	90	93.55,2	90	4	4603,2	279
CCDC156503	4	[PcCo$_{0.25}$Ni$_{0.75}$][AsF$_6$]$_{0.5}$	11:11	Pnc2	14.033,4	28.54,1	6.435,2	90	90	90	4	2577,2	280
CCDC156502	4	[PcCo$_{0.45}$Ni$_{0.55}$][AsF$_6$]$_{0.5}$	11:12	P4/mcc	14.226,2	14.226,2	6.414,1	90	90	90	2	1298.1,3	280
*nc	4	PcCr {α}	–	–	25.3,1	3.76,2	25.0,1	90	90	90	4	2374	277
*nc	4	PcCr {β}	–	–	19.7,1	4.85,2	14.8,1	90	120	90	2	1225	277
#1999-01	4	PcCr {β}	11:13–14	P2$_1$/n	14.5752,3	4.8211,1	17.1911,9	90	105.772,3	90			63
#1982-01	4	Cl$_{16}$PcCu	11:15	Cmm	19.62,?	26.04,?	3.76,?	90	116.5,?	90			88,89,281
*nc	4	Cl$_1$PcCu {α1}	–	P1 or P-1	12.2,?	3.74,?	13.2,?	90	90.7,?	95.6,?			74
*nc	4	Cl$_4$PcCu {α}	–	P2$_1$/a	26.3,?	3.67,?	13.9,?	90	94,?	90			74
RADPIV	4	NcCu	19:1	P2$_1$/c	6.873,2	7.906,1	30.385,9	90	91.02,2	90	2	1650.9,2	104
CUPOCY01*nc	4	PcCu {α}	–	–	25.9,1	3.80,1	24.7,1	90.00,2	90.09,2	95.60,2			282
*nc	4	PcCu {α}	–	P4/m	17.367,?	17.367,?	12.790,?	90	90	90			283
*nc	4	PcCu {β}	–	P2$_1$/a	19.406,5	4.7904,1	14.627,4	90	120.565,8	90	2		284,285
*nc	4	PcCu {β}	–	P2$_1$/a	19.6,?	4.79,?	14.6,?	90	120.6,?	90	2	1180,?	13,57,58
CUPOCY10	4	PcCu {β}	11:16	P2$_1$/a	19.407,?	4.790,?	14.628,?	90	120.93,?	90	2	1166,?	76,286
*nc	4	PcCu {β}	–	–	19.61,?	4.80,?	14.74,?		121.51,?			1180,?	73
*nc	4	PcCu {β}	–	P2$_1$/a	19.6,?	4.79,?	14.6,?	90	120.6,?	90			74
CYNCUB*nc	4	tspcCu	–	–	26.9,15	3.8,3	25.6,15	89.9,2	90.0,2	95.5,2			282
CCDC168920	4	[PcCu]$_3$[AsF$_6$]$_2$[1-ClN]	11:18–19	P-1	9.6671,8	16.0579,14	16.3332,14	111.5453,14	98.3260,16	105.8881,16	1	2181.2,3	275
CCDC168919	4	[PcCu]$_3$[(SbF$_6$)$_2$][1-ClN]	11:20	P-1	9.6660,10	16.0559,17	16.7070,17	112.5314,16	99.4653,18	105.3764,18	1	2203.8,4	275
FEGJOQ	4	[PcCu][I$_3$]$_{0.33}$	11:17	P4/mcc	13.888,12	13.888,12	6.390,8	90	90	90	2	1232,?	113
JECGON	4	[tatbpCu]$_3$[ReO$_4$]$_2$[1-ClN]	–	P-1	9.167,6	15.849,10	16.065,10	65.87,2	72.11,3	84.01,2	1		287
*nc	4	PcFe {β}	–	P2$_1$/a	17.302,1	4.7934,4	14.610,1	90	105.605,8	90	2		284,285
FEPHCY01*nc	4	PcFe {β}	–	P2$_1$/a	20.2,?	4.77,?	15.0,?	90	121.6,?	90	2	1231,?	58
FEPHCY	4	PcFe {β}	11:21	P2$_1$/a	19.392,5	4.786,2	14.604,4	90	120.85,1	90	2	1163.7,?	81
*nc	4	PcFe {β}	–	P2$_1$/a	19.39,?	4.78,?	14.55,?	90	120.74,?	90		1159,?	73
FEDXOB	4	[PcFe][K.cntnd.222]	11:24–25	P2$_1$/n	22.748,8	17.312,8	12.037,5	90	92.48,1	90	4	4736,?	288
FEDXUH	4	[PcFe][Li.cntnd.18cr6] [Li(THF).cntnd.18cr6]	11:23	P-1	12.841,3	18.107,5	12.325,3	97.24,1	91.19,1	94.53,1	2	2833,?	288
DAPCEC	4	[PcFe][I$_3$]$_{0.33}$	11:22	P4/mcc	13.84,2	13.84,2	6.77,1	90	90	90	2	1320,?	157
DIXNIH02	4	PcLi	11:26	P4/nnc	19.575,3	19.575,3	6.491,3	90	90	90	4	2487.9,27	91
DIXNIH01*nc	4	PcLi {m}	–	–	19.357,15	4.935,6	14.555,15	90	120.36,9	90			90
DIXNIH*nc	4	PcLi {t}	–	–	13.851,1	13.851,1	6.488,2	90	90	90		1244.7,?	90
RESYET	4	[PcLi][F][NBu$_4$]$_2$[0.5H$_2$O]	11:27–28	P-1	8.6480,1	12.620,2	14.866,5	82.44,2	87.01,2	75.02,1	1	1553.4,6	115
RESYAP	4	[PcLi][NBu$_4$][THF]	11:29–30	P2$_1$/c	12.455,5	23.396,5	16.120,5	90	94.986,5	90	4	4680,3	115
ZEPKIO	4	[PcLi][PNP][CH$_2$Cl$_2$]	11:31–32	P-1	14.499,1	14.899,3	15.904,2	67.19,1	88.00,1	67.78,1	2	2906.2,7	289
ZEPKOU	4	[PcLi][PNP][H$_2$O]	11:33–35	P2$_1$/n	14.469,6	26.629,4	29.864,8	90	101.66,3	90	4	11269,6	289
DIXNAZ*nc	4	[PcLi][Br$_3$]$_{0.33}$	–	–	13.894,1	13.894,1	6.443,2	90	90	90		1243.8,?	90
DIXNED*nc	4	[PcLi][Cl$_3$]$_{0.33}$	–	–	13.912,1	13.912,1	6.476,2	90	90	90		1253.4,?	90
DIXMUS*nc	4	[PcLi][I$_3$]$_{0.33}$	–	–	14.036,1	14.036,1	6.378,1	90	90	90		1256.5,?	90
DIXMUS10	4	[PcLi][I$_3$]$_{0.33}$	11:36	P4/mcc	14.057,3	14.057,3	6.397,1	90	90	90	2	1264.0,4	290
MAKHEL	4	[PcLi][Li(OPϕ_3)$_4$] [EtOH,H$_2$O]	11:37–38	P2$_1$2$_1$2$_1$	19.177,4	19.279,4	24.116,5	90	90	90	4	8916,4	136
MNPHCY21	4	PcMn {β, 116K, 1.Refine.}	11:39	P2$_1$/c	14.590,3	4.741,1	19.301,5	90	120.79,1	90			85
#1980-01	4	PcMn {β, 116K, 2.Refine.}	11:40	P2$_1$/c	14.590,3	4.741,1	19.301,5	90	120.79,1	90			85

(*Continued*)

Appendix 1. Continued

CamRef	CN#	Crystal description	Figure	Space group	Axis a / Å	Axis b / Å	Axis c / Å	α / deg	β / deg	γ / deg	Z	V	Reference
MNPHCY11	4	PcMn {β, 116K}	11:41	$P2_1/c$	14.590,3	4.741,1	19.301,5	90	120.79,1	90	2	1146.9,?	83–85
MNPHCY02	4	PcMn {β, 295K}	11:42	$P2_1/c$	14.576,5	4.755,1	19.362,7	90	120.70,2	90	2	1153.9,?	84
MNPHCY03	4	PcMn {β, N, 5.8K}	11:43	$P2_1/c$	14.601,5	4.69,1	19.327,5	90	121.10,1	90	2	1133.3,?	86,291
MNPHCY	4	PcMn {β, RT}	11:44	$P2_1/a$	19.400,4	4.761,2	14.613,3	90	120.74,1	90	2	1160.0,?	81
MNPHCY04*nc	4	PcMn {β}	–	$P2_1/a$	20.2,?	4.75,?	15.1,?	90	121.7,?	90	2	1233,?	58
*nc	4	PcMn {β}	–	$P2_1/a$	19.17,?	4.77,?	14.43,?	90	119.4,?	90		1148,?	73
TOYSUV	4	$(H_{13}C_6)_8$PcNi {103K}	11:45–46	$P2_1/n$	22.347,?	37.246,?	18.834,?	90	115.185,?	90	8	14186,?	240
TOYSUV01	4	$(H_{13}C_6)_8$PcNi {295K}	11:47–48	C2/c	22.55,2	37.14,1	18.994,7	90	114.60,5	90	8	14463,29	240
ROCCAN, ROCCAN01	4	(i-$H_{11}C_5O)_4$PcNi	11:49–50	$P2_1/n$	9.364,3	13.833,4	18.063,2	90	98.17,2	90	2	2315.9,8	241,242
NOCTII	4	$(Me_3SiO)_8$PcNi	11:55–56	$P2_1/c$	14.4045,2	23.1604,3	15.4997,2	90	100.799,1	90	2	5079.34,12	243
IBISOB	4	(o-tol)$_4$PcNi	–	P-1	14.2704,2	23.6456,4	7.8483,1	90.572,1	93.3680,6	80.487,1	2	2607.34,7	245
#2002-03	4	(o-tol)$_4$PcNi	11:67–69	P1	7.7271,6	14.2403,9	23.622,2	80.586,8	90.192,8	87.654,9	2	2562.0,3	244
NOCVEC	4	$(RO_2)_2(RO)_2(HO)_2$PcNi	11:51–52	P-1	18.099,3	18.415,4	18.466,3	108.238,9	116.276,8	95.711,8	1	5032.3,16	243
NOCVAC	4	$(RO_2)_2(RO)_2(HO)_2$PcNi	11:53–54	P-1	18.664,5	17.566,5	18.708,7	110.78,3	116.59,3	96.35,2	1	4854,3	243
RADPER	4	NcNi	19:2–3	$P2_1/c$	6.8800,9	7.8730,8	30.48,2	90	90.50,9	90	2	1650.8,2	104
*nc	4	PcNi { < 7kBar}	–	$B2_1/a$	19.8,?	4.71,?	24.7,?	90	101,?	90			78
*nc	4	PcNi { > 7kBar}	–	?	23.1,?	3.76,?	25.0,?	90	96,?	90			78
*nc	4	PcNi { > 7kBar}	–	?	23.1,?	3.76,?	16.1,?	90	129.5,1	90			79
NIPHTC	4	PcNi {β}	11:57	$P2_1/a$	19.9,?	4.71,?	14.9,?	90	121.9,?	90	2	1186,?	13,14,57,58
*nc	4	PcNi {β}	–	$P2_1/a$	19.48,?	4.70,?	14.82,?	90	122.33,?	90		1147,?	73
QEMTAD	4	[(EtO)$_8$CpPcNi][PF_6]	11:59–61	P-1	11.143,2	19.801,4	34.248,7	104.92,3	92.97,3	99.56,3	2	7165,2	246
FIYSEL*nc	4	[PcNi][AsF_6]$_{0.5}$	–	Pcc2	14.061,2	14.214,2	6.467,7	90	90	90	1	1292.6,14	274,292–294
KARDAI	4	[PcNi][AsF_6]$_{0.5}$	11:58	Pnc2	14.015,1	28.485,2	6.466,3	90	90	90			293,294
*nc	4	[PcNi][$(BF_4)_{0.33}$]	–	?	13.97,2	13.97,2	6.48,1	90	90	90			296,297
*nc	4	[PcNi][$(BF_4)_{0.33}$]	–	P4/mcc	13.97,2	13.97,2	6.48,1	90	90	90			296,298
FUFJOF*nc	4	[PcNi][$(ClO_4)_{0.42}$]	–	P4/mcc	13.957,3	13.957,3	6.4672,9	90	90	90	2		297,299
FUFJOF01*nc	4	[PcNi][$(ClO_4)_{0.42}$] {153K}	–	P4/mnc	19.543,8	19.543,8	6,416,4	90	90	90			299
FANZUP	4	[PcNi][SbF_6]$_{0.5}$	11:62	Pnc2	14.113,1	28.710,2	6.441,2	90	90	90	2	2609.7,7	292–295
DOXCOI	4	[PcNi][Br_3]$_{0.33}$	11:63	P4/mcc	13.793,7	13.793,7	6.431,4	90	90	90	2	1224,2	300
NIPHTI	4	[PcNi][I_3]$_{0.33}$	11:64	P4/mcc	13.936,6	13.936,6	6.488,3	90	90	90	2	1260.1,?	111
CCDC156502	4	[PcNi$_{0.55}$Co$_{0.45}$][AsF_6]$_{0.5}$	11:12	P4/mcc	14.226,2	14.226,2	6.414,1	90	90	90	2	1298.1,3	280
CCDC156503	4	[PcNi$_{0.75}$Co$_{0.25}$][AsF_6]$_{0.5}$	11:11	Pnc2	14.033,4	28.54,1	6.435,2	90	90	90	4	2577,2	280
CCDC168918	4	[PcNi]$_3$[$(ReO_4)_2$][1-ClN]	11:65	P-1	9.7359,11	15.849,2	15.985,3	109.902,6	107.082,10	97.156,7	1	2146.5,5	275
CCDC168922	4	[PcNi]$_3$[$(SbF_6)_2$][1-ClN]	11:66	P-1	9.665,5	16.017,7	16.628,8	112.427,9	99.741,10	105.382,10	1	2186.0,18	275
PTPOCY02	4	PcPt	11:70	$P2_1/a$	23.9,?	3.81,?	16.9,?	90	129.6,?	90	2	1186,?	13,15,57,58
PTPOCY10	4	PcPt {α}	11:71	C2/n	26.29,?	3.818,?	23.92,?	90	94.6,?	90	4	2393,?	77
*nc	4	PcPt {α}	–	A2/a or Aa	24.1,?	3.76,?	26.1,?	90	94.6,?	90			74
*nc	4	PcPt {α}	–	$P2_1/a$	23.9,?	3.81,?	16.9,?	90	129.6,?	90			74
*nc	4	PcPt {α}	–	C2/c	26.29,?	3.818,?	34.09,?	90	135.6,?	90			77
*nc	4	PcPt {α}	–	I2/a	23.92,?	3.818,?	34.09,?	90	129.8,?	90			77
PTPOCY11	4	PcPt {γ}	11:72	$B2_1/a$	25.78,?	3.969,?	23.16,?	90	94.6,?	90	4	2362,?	77
*nc	4	PcPt {γ}	–	$P2_1/a$	23.16,?	3.969,?	16.62,?	90	129.4,?	90	2		77
JECGAZ	4	[PcPt][$(ClO_4)_{0.5}$]	11:73	P4/mcc	14.062,1	14.062,1	6.510,1	90	90	90	2		294,301
PHPRSI	4	PcSi$_{0.17}$	11:74–75	B2/b	26.078,9	23.949,9	3.887,9	90	90	93.46,3	4		87
RADPOB	4	NcZn	19:4	$P2_1/c$	6.850,2	7.997,4	30.273,4	90	91.88,2	90	2	1657.5,8	104

PTHCZN	4	PcZn {β}	11:76	$P2_1/a$	19.274,5	4.8538,15	14.553,4	90	120.48,2	90	2		82
*nc	4	PcZn {β}	–	$P2_1/a$	19.13,?	4.86,?	14.50,?	90	120.1,?	90		1166,?	73
#2002-01	4,5,6	$[(O_3ReO–CuPc)_2PcCu]$	12:10–11	P-1	12.6202,6	13.7596,7	14.0294,7	64.6376,9	64.2570,9	66.0549,9	1	1909.22,16	302
ZZZNHK*nc	4?	PcBe	–	$P2_1/a$	21.2,?	4.84,?	14.7,?	90	121.0,?	90	2	1293,?	58
CUWPAL	5	PcAl–Cl	12:2	P-1	13.776,6	13.775,4	14.059,5	98.36,3	108.60,3	90.16,3	4	2498.2,?	154
#1995-01 (App. 28)	5	PcAl–Cl	12:3	P-1	13.746,2	14.059,2	13.746,2	98.37,?	90.01,1	108.620,9	4	2483.0,6	155,156
DAYYOR	5	PcAl-O-AlPc	12:4–5	P-1	7.694,2	12.520,3	12.705,3	91.03,2	94.54,2	90.37,2	1	1219.8,?	122
CCDC180524	5	$[PcAs][I_3]$	12:1	$P2_1/n$	8.652,2	17.558,4	19.668,4	90	95.42,3	90	4	2974.4,11	303
#1997-01	5	$[PcCd-Cl][NEt_4][py]$	12:6–7	Pbca	19.033,5	19.285,5	21.992,5	90	90	90	8	8072.2,35	304
#2001-03	5	$[C_{2v}$-1,2-NcCo-py][2py]	19:5–6	Pccn	29.649,1	31.385,1	10.080,1	90	90	90	8	9379.5,7	135
YEJJIG	5	PcCr-I	12:8	$P2_1/n$	13.173,3	10.424,2	19.435,4	90	99.76,3	90	4	2630.1,10	305
QEMMIE	5	$PcCu–OReO_3$	12:9	$P2_1/c$	20.455,4	11.947,2	27.327,6	90	126.86,3	90	8	5343.4,19	306
DAPCIG	5	PcFe-Cl-I-I-Cl-FePc	12:12–13	P-1	12.270,7	13.202,7	8.335,6	92.80,3	97.17,3	101.97,2	1	1306,?	157
FIQFUG	5	PcFe-I-I-I-I-FePc	12:14–15	P-1	8.452,2	13.055,3	13.211,3	102.01,2	91.71,2	96.88,2	1	1413.4,6	189
CAVLOA*nc	5	PcFe-O-FePc	–	Ccc2	19.171,9	44.907,21	25.668,5	90	90	90			307
GEDKAB	5	[PcFe-Me][Na.cntnd.18cr6] [THF,ϕCl]	12:16	$P2_1/n$	13.139,3	21.501,6	18.631,5	90	104.91,2	90	4	5086,?	161
CUWNUD	5	PcGa–Cl	12:17	P-1	9.301,3	11.272,1	13.143,3	105.46,2	105.61,2	96.80,2	2	1252.8,?	154
CUWNUD01	5	PcGa–Cl {II, R}	12:18	P-1	12.93,?	12.93,?	7.73,?	107.05,?	102.48,?	91.72,?	2		174
NALSOI*nc	5	PcGa–OH {V, R}	–	P-1	11.52,?	12.75,?	8.89,?	95.87,?	96.00,?	69.45,?	2		167,168
NALSOI02	5	PcGa–OH {V, R}	12:19	P-1	11.52,?	12.75,?	8.89,?	95.87,?	96.00,?	69.45,?	2		167
NALSOI01	5	PcGa–OH {XI, R}	–	$P2_1/c$	16.63,?	26.32,?	7.16,?	90	129.93,?	90	4		174
BIJQOA	5	PcIn-I	12:20	$P2_1/n$	12.797,3	10.776,2	19.475,4	90	98.05,3	90	4	2659.1,10	187
#1998-03	5	PcIn–Sϕ	12:21–22	P-1	9.450,3	12.760,5	13.280,5	111.430,2	90.970,2	93.610,1			308
MGPCPH	5	$[PcMg-OH_2][2py]$	12:23	$P2_1/n$	17.098,3	16.951,3	12,449,3	90	105.880,3	90		3470,?	145
#2002-06	5	$[PcMg-OH_2]$	12:24	$P2_1/a$	13.526,2	13.927,2	13.559,2	90	104.678,7	90	4	2470.9,6	309
#2002-07	5	$[PcMg-OH_2]_2[3py]$	12:25	P-1	13.586,8	15.640,9	17.667,10	102.96,5	112.59,4	104.01,5	4	3144,3	309
MAKHOV	5	$[PcMg-OP\phi_3][2py]$	12:26	$P2_1/n$	12.873,3	20.832,2	17.970,1	90	98.14,3	90	4	4771,3	136
MAKHUB	5	$[PcMg-OP\phi_3][CH_2Cl_2]$	12:27	P-1	12.384,2	12.938,2	14.671,2	73.68,1	67.97,1	87.66,1	2	2085.8,6	136
MAKJAJ	5	$[PcMg-OP\phi_3][THF]$	12:28	P-1	12.358,2	13.413,2	14.532,2	73.57,1	68.09,2	88.09,1	2	2136.4,5	136
MAKHIP	5	$PcMg-OP\phi_3$	12:33	C2/c	19.882,5	17.523,5	23.282,3	90	108.89,2	90	8	7676,5	136
TAYSIV	5	$[PcMg–Cl][PNP][Me_2CO]$	12:34	P-1	12.268,2	12.464,3	19.267,3	100.86,2	92.775,14	91.97,2	2	2887.1,9	172
CUTGIH	5	PcMo=O	12:35	$P2_1/c$	13.441,3	13.292,2	14.041,3	90	103.81,1	90	4		153,310
FEWPEC	5	$[PcMo=O][I_3]$	12:36	Cmcm	14.441,3	15.194,3	14.085.3	90	90	90	4(or2)	3090.5,11	186
PBPHCY02*nc	5	PcPb	–	$P2_1/b$	25.56,?	29.32,?	7.41,?	90	97.3,?	90			151
*nc	5	PcPb	–	$P2_1$-C2	14.716,4	13.710,2	12.806,4	90	97.4,2	90	4	2562,?	62
PBPHCY	5	PcPb {monoclinic}	12:37–38	$P2_1/b$	25.48,8	25.48,8	3.73,1	90	90	90	4		133
PBPHCY01	5	PcPb {triclinic}	12:39–40	P-1	13.123,7	16.131,6	12.889,5	94.22,4	96.20,4	114.19,2	4		132,148
NEKRAW	5	PcRe–RePc	12:41–42	P-1	7.799,3	12.563,7	12.689,10	89.97,5	94.14,5	106.39,4	1	1189,1	181
TESFAY	5	$(H_{11}C_5)_8$PcRh-H-RePc $(C_5H_{11})_8$	12:44–45	C2/c	36.579,2	19.312,1	18.772,2	90	101.08,4	90	8	13014,2	171
HASZEG	5	$(H_{11}C_5)_8$PcRh–Me	12:46–48	$P2_1/n$	14.704,3	25.872,5	17.264,3	90	108.11,0	90	4	6242,28	166
HOVCOK	5	$[PcSb][I_3]$	12:43	$P2_1/n$	8.645,2	17.661,4	19.771,4	90	95.19,3	90	4	3006.2,11	188
#1971-02 (App. 27)	5	PcSn	12:49	P-1	12.060,1	12.618,1	8.675,1	95.89,1	95.08,1	68.17,1	2	1217,?	31
SNPHCY*nc	5	PcSn	–	P-1	12.060,1	12.618,1	8.675,1	95.89,1	95.08,1	68.17,1	2	1217,?	142
SNPHCY01	5	PcSn	12:50–51	P-1	12.048,3	12.630,3	8.671,3	95.85,5	95.10,5	68.25,5	2	1217.2,3	143
BITSAY04	5	PcTi=O {C,R}	12:52	Cc	25.19,?	3.85,?	25.46,?	90	90.3,?	90	4 or 8		3
*nc	5	PcTi=O {I, MD}	–	$(P2_1/c)$	13.56,?	13.09,?	14.08,?	89.98,?	105.48,?	90.03,?			312

(*Continued*)

Appendix 1. Continued

CamRef	CN#	Crystal description	Figure	Space group	Axis *a* / Å	Axis *b* / Å	Axis *c* / Å	α / deg	β / deg	γ / deg	Z	V	Reference
*nc	5	PcTi=O {I, R}	–	$P2_1/c$	13.43,?	13.25,?	13.94,?	90	103.70,?	90	4		165
BITSAY	5	PcTi=O {I}	12:53	$P2_1/c$	13.411,6	13.230,3	13.810,4	90	103.72,3	90	4	2381.2,?	2
*nc	5	PcTi=O {I}	–	$P2_1/c$	13.431,7	13.258,3	13.905,4	90	103.78,3	90	4	2404.7,?	2
BITSAY01	5	PcTi=O {II}	12:54	P-1	12.166,4	12.584,5	8.641,3	96.28,3	95.03,4	67.86,4	2	1216.4,?	2
BITSAY03	5	PcTi=O {IV, R}	12:55	P-1	10.83,?	13.12,?	9.96,?	72.28,?	77.25,?	104.48,?	2		164
*nc	5	PcTi=O {Y, MD}	–	$(P2_1/c)$	14.14,?	14.14,?	15.00,?	90.01,?	117.98,?	89.97,?			312
BITSAY02	5	PcTi=O {Y, R}	12:56	$P2_1/c$	13.85,?	13.92,?	15.14,?	90	120.22,?	90	4		165
NELCIQ	5	[PcTl-I][CH_2Cl_2]	12:57	Pnma	13.778,3	14.649,2	14.907,1	90	90	90	4	3008.7,8	180
*nc	5	PcV=O	–	P-1	14.0,?	14.0,?	10.6,?	109.2,?	133.9,?	90	2		313
VPHTHC	5	PcV=O {II}	12:58	P-1	12.027,6	12.571,8	8.690,5	96.04,5	94.80,5	68.20,5	2	1212,?	150
*nc	5	PcV=O {II}	–	P-1	12.032,?	12.579,?	8.712,?	96.15,?	94.88,?	68.16,?	2		149
SAXZEW	5	PcV-I	12:59	$P2_1/n$	13.135,3	10.309,2	19.363,4	90	99.99,3	90	4	2582.17,97	185
ZNPTCY	5	[PcZn-H_2N-C_6H_{13}] [0.5H_2N-C_6H_{13}]	12:60	$P2_1/c$	12.40,2	15.76,2	20.05,5	90	93.2,5	90	4	3902.6,?	144
WOHTES	5	[PcZn-NC-Fe(dppe)Cp] [0.5CH_2Cl_2,THF]	12:61	$P2_1/n$	21.620,4	12.172,2	22.861,5	90	94.75,3	90	4		314
MAKHAH	5	[PcZn–OPϕ_3][Et_2O,1-ClN]	12:62	P-1	12.754,2	17.102,3	21.805,3	72.95,1	87.00,1	89.26,1	2	4541,1	136
DAMTEQ	5	PcZn-Cl	12:63	P4/nnc	13.85,1	13.85,1	13.09,1	90	90	90	4	2511,?	134,315
ZUHCIO	5	[PcZn–ONO][PNP][H_2O]	12:64	P-1	12.6439,15	13.677,2	17.107,3	67.357,10	85.069,12	86.429,10	2	2718.9,?	170
BANCOI	5*	[Pc(K(18cr6))$_2$][1.5C_6H_6]	13:1–3	P-1	11.878,5	20.915,8	13.056,4	90.93,3	97.36,3	90.64,3	2		316
BANCUO	5*	Pc[K(diglyme)]$_2$	13:4	C2/c	24.734,8	7.579,5	22.165,5	90	112.15,2	90	4		316
BARMAI	5*	Pc[K(DMF)$_2$]$_2$	13:5	P-1	7.342,1	12.388,1	12.768,1	97.17,2	99.76,1	102.02,1	1	1104,2	317
MAKJEN	5*	Pc(Li-OPϕ_3)$_2$	13:6	P-1	10.533,2	11.870,2	12.661,3	78.85,2	68.33,2	65.79,2	1	1336.3,5	136
XETZIF01	5*	PcMg {120K}	12:29	$P2_1/c$	14.368,3	4.898,1	18.904,4	90	119.86,3	90	2	1153.7,4	59
XETZIF02	5*	PcMg {200K}	12:30	$P2_1/c$	14.395,3	4.909,1	18.953,4	90	119.86,3	90	2	1161.5,4	59
XETZIF03	5*	PcMg {260K}	12:31	$P2_1/c$	14.417,3	4.913,1	19.002,4	90	119.85,3	90	2	1167.4,4	59
XETZIF	5*	PcMg {273K}	12:32	$P2_1/a$	18.971,2	4.916,2	14.401,2	90	119.890,9	90	2	1164.3,6	60,61
*nc	4,5*[d)]	PcMg {α}	–	monocl.	26.26,9	3.82,4	23.96,9	90	94.4,4	90	4	2400,?	62
*nc	4,5*[d)]	PcMg {β}	–	monocl.	20.07,5	4.70,5	15.24,5	90	122.3,2	90	2	1217,?	62
QASKAW	5*	Pc(Na–OPϕ_3)$_2$	13:7	P-1	10.3026,10	12.119,1	13.000,2	80.038,12	67.167,10	65.332,10	1	1359.3,3	50
CALWIV	5*	Pc(SmL_2)$_2$	13:8	P-1	12.941,6	14.680,4	21.205,4	88.22,2	86.54,3	71.32,3	2	3808.9,?	318,319
HEBHOL	5*	PcTl_2	13:9	Cmca	25.173,4	7.353,2	13.765,3	90	90	90	4	2547.9,7	49
*nc	5?[e)]	PcCo-CN {EM}	–	–	6.60,?		7.69,?		126,?				320
FIYSUB*nc	5?[e)]	PcCo-CN	–	P-1	7.65,3	9.85,4	9.82,3	90.8,3	107.3,3	107.2,3	1	670.35,?	321–323
*nc	5?[e)]	PcMo	–	$P2_1/c$	12.079,3	14.444,2	14.716,3	90	109.1,1	90			310
*nc	5?[e)]	PcW	–	$P2_1/c$	12.069,?	14.401,?	14.712,?	90	109,?	90			310
HOXCIG	6*cis*	PcBi(-I-)$_2$BiPc	14:1–3	P-1	10.496,2	11.069,2	12.078,2	85.72,3	81.32,3	74.53,3	2	1336.0,4	184
NAPJIX	6*cis*	[PcInF_2][NBu_4][H_2O]	14:4–5	$P2_1/n$	13.081,3	13.936,2	23.972,2	90	97.79,1	90	4	4330,1	178
ZUFZOP	6*cis*	[PcIn(-NCO)$_2$][PNP]	14:6–7	P-1	12.944,2	14.897,3	18.563,2	72.03,1	74.78,1	74.41,1	2	3216,1	173
RERGOK	6*cis*	[PcIn(-O,-O'CO)][NBu_4]	14:8–9	$P2_1/c$	13.219,2	14.002,3	23.792,4	90	97.7,1	90	4	4388.1,14	177
#1998-02	6*cis*	[PcIn(O_2catCl_4)][NBu_4] [CH_2Cl_2]	14:10–11	P-1	15.400,3	18.818,3	19.872,5	79.008,2	73.783,2	69.666,1			308
BEJTIT	6*cis*	Pc$NbCl_2$	14:12–13	$P2_1/c$	12.557,2	10.315,2	20.571,7	90	90.59,2	90	4	2664.3,?	152
BEJTIT01	6*cis*	Pc$NbCl_2$	14:14	P-1	8.707,2	12.288,3	12.801,3	80.72,2	85.43,2	76.62,2	2	1313.7,6	163
WOVREE	6*cis*	PcSb(-Cl-)$_2$SbPc	14:17–18	P-1	9.560,2	10.806,2	12.462,2	91.96,3	96.21,3	92.37,3	2		193
NILVUZ	6*cis*	PcSb(-I-)$_2$SbPc	14:15–16	P-1	10.458,2	10.925,2	12.097,2	86.17,3	81.83,3	76.31,3	2	1328.4,4	183

DAGXUE	*6cis*	$PcTiCl_2$	14:19–20	P-1	8.744,9	12.244,4	12.835,4	79.73,3	95.32,6	103.27,6	2	1314,1	158
BOFQES	*6cis*	$[PcTlCl_2][NBu_4][0.5Et_2O]$	14:21–22	$P2_1/n$	13.146,1	14.204,5	24.900,3	90	93.88,1	90	4	4639,2	191
ZUVYOE	6tr	$[PcAl(\text{-}ONO)_2][NBu_4]$	15:1	C2/c	23.151,5	23.095,5	8.596,1	90	109.207,14	90	4	4340.2,14	324
WOHDUH	6tr	$[2,3\text{-}NcCo(CN)_2][2DCB]$	19:7	P-1	10.684,1	15.005,2	8.068,1	92.37,1	95.72,1	70.08,1	1	1209.9,3	251
WOWDOB	6tr	$[2,3\text{-}NcCo(CN)_2][CH_3CN]$	19:8	P-1	9.864,2	12.475,4	8.827,2	107.83,2	93.02,2	102.30,2	1	1002.0,6	251
*nc	6tr	$[PcCo(4\text{-}mepy)_2][2mepy]$	–	Pbca	10.395,2	25.069,4	17.992,3	90	90	90	4		325
MPCOPC10	6tr	$[PcCo(4\text{-}mepy)_2][2mepy]$	15:2	Pbca	10.395,1	25.069,3	17.992,2	90	90	90	4		326
YAVPEQ	6tr	$[PcCo(CN)_2][2CHBr_3]$	15:3	$P2_1/c$	7.938,3	18.066,3	13.020,2	90	97.23,2	90	2	1852.3,7	327,328
ZODWAQ	6tr	$[PcCo(CN)_2][2CHCl_3]$	15:4	P-1	10.771,5	11.008,5	8.431,4	96.89,4	108.87,3	107.95,3	1	872.7,8	329
DOVYUI	6tr	$[PcCo(CN)_2][2DMSO]$	15:5	P-1	10.948,2	11.032,2	8.187,2	97.42,2	107.76,2	108.96,1	1	861.4,3	330
SILPEI	6tr	$[PcCo(CN)_2][2H_2O]$	15:6	C2/m	14.173,4	14.250,2	7.701,2	90	115.19,2	90	2	1407.3,6	331,332
DOWBAS*nc	6tr	$[PcCo(CN)_2][CH_3CN]$	–	Cc	8.01,4	19.39,2	19.02,4	90	95.3,3	90	4	2939,14	330
FODLOZ	6tr	$[PcCo(py,\text{-}SCN)][CHCl_3]$	15:7–8	$P2_1/n$	19.381,14	18.316,9	10.115,6	90	90.79,5	90	4	3590.3,?	333
DOVZAP	6tr	$PcCo(CN)_2$	15:9	$P2_1/a$	13.583,7	14.067,4	14.920,4	90	107.75,3	90	4	2715,1	330
KEHXEA	6tr	$PcCoCl_2$	15:10	$P2_1/n$	9.060,5	14.640,2	10.268,5	90	97.73,1	90	2	1349.8,7	123
NOFGAQ	6tr	[NC-(Pc)Co-CN-Co(Pc)-CN] [B[c]PT]	15:11	$P2_1/a$	17.680,4	20.692,4	17.599,3	90	96.31,2	90	4	6399,1	334
VENDAT10	6tr	$[K][PcCo(CN)_2][5CH_3CN]$	15:12	P-1	15.577,3	16.626,3	10.645,2	94.89,2	99.22,2	138.37,1	1	1717.3,7	331,332,335
VENDAT	6tr	$[K][PcCo(CN)_2][5CH_3CN]$	–	P-1	15.577,3	16.626,3	10.645,2	94.89,2	99.22,2	138.37,1	1		
NIRPAF	6tr	$[PcCo(CN)_2]$ [PTMA][MeCN] {113K}	15:13	Pnnm	13.668,1	7.310,1	16.371,1	90	90	90	2	1635.7,1	336
NIRDOH	6tr	$[PcCo(CN)_2]$ [PTMA][MeCN] {295K}	15:14	Pnnm	13.871,1	7.350,1	16.296,1	90	90	90	2	1661.3,1	336
BAPZOH	6tr	$[PcCo(CN)_2][PXX]_2$	15:15	$P2_1/a$	17.743,2	29.344,3	7.678,3	90	102.36,2	90	4	3904,1	130,337
SOPQIX	6tr	$[PcCo(CN)_2][p\text{-}BPYNN]$	15:17–18	$P2_1/c$	15.830,3	11.201,4	25.709,3	90	90.29,1	90	4	4558.4,17	338
SOPQET	6tr	$[PcCo(CN)_2][p\text{-}PPYNN]$	15:16	$P2_1/c$	15.585,6	11.015,4	26.111,3	90	89.95,2	90	4	4482,3	338
–	6tr	$[PcCo(CN)_2][TMTTF]$ $[2CH_3CN]$	–	P-1	11.343,5	12.153,4	8.714,3	108.98,3	104.68,3	82.02,3	1	1097,1	339
–	6tr	$[PcCo(CN)_2][TTF]$	–	P42/n	21.440,5	21.440,5	7.488,2	90	90	90	4	3442,1	339
JAMRAQ	6tr	$[PcCo(CN)_2][P\phi_4]_{0.5}$	15:19	P42/n	21.676,8	21.676,8	7.474,4	90	90	90	2	3511,3	339,340
CCDC174676	6tr	$[F_8((F_3C)_2FC)_8PcCo$ $(OCMe_2)_2]$	15:20–21	$P2_1/c$	15.2500,9	20.690,1	12.0890,7	90	102.121,1	90	2	3729.3,4	341
CCDC174677	6tr	$[F_8((F_3C)_2FC)_8PcCo(OC(Me)$ $(CH_2P\phi_3)_2)$	15:22	$P2_1/n$	13.477,3	13.281,?	28.758,7	90	91.126,5	90	2	5146,2	341
#2001-08	6tr	[py-(Pc)Cr-O-Cr(Pc)-py][2py]	15:23	$P2_12_12_1$	12.845,2	22.611,2	23.584,7	90	90	90			342
KEHWUB	6tr	$PcCrCl_2$	15:24	$P2_1/n$	9.137,3	14.734,2	10.364,2	90	97.36,1	90	2	1383.8,7	123
TOPVID	6tr	$[PcCr(\text{-}ONO)_2][PNP]$	15:25	P-1	10.225,3	12.101,2	13.248,3	85.60,2	67.49,2	67.80,2	2	1397.6,5	343
QEMMOK	6tr	$PcCu(ReO_4)_2$	15:26	$P2_1/c$	8.9825,18	9.3147,19	17.712,4	90	95.96,3	90	2	1474.0,5	306
QILWAJ	6tr	$PcCu(SbF_6)_2$	15:27	P-1	8.596,4	9.150,5	10.724,5	92.528,9	101.917,9	99.127,9	1	812.3,7	344
PTCYFE*nc	6tr	$[PcFe(4\text{-}mepy)_2][2mepy]$	–	Pbca	25.19,2	17.86,2	10.30,2	90	90	90	4		345
PTCYFE01	6tr	$[PcFe(4\text{-}mepy)_2][2mepy]$	15:28	Pbca	10.315,1	25.006,3	17.876,2	90	90	90	4		326
PHTHFE	6tr	[PcFe(DMF,CO)][DMF]	15:29	$P2_1/n$	9.86,1	17.35,3	19.79,4	90	87.9,2	90	4	3383,?	346
WOHVAQ	6tr	$[PcFe(NC\text{-}Ru(P\phi_3)_2Cp)_2]$ $[1.5CH_2Cl_2]$	15:30	C2/c	25.941,5	25.999,5	17.352,3	90	119.50,3	90	4	10186,3	314
SPCNFE	6tr	$[PcFe(\text{-}OSMe_2)_2][2DMSO]$	15:31	$P2_1/c$	7.993,2	16.175,7	16.086,7	90	104.53,3	90	2	2013,?	347
WOHTIW	6tr	$[PcFeCN,NC\text{-}Ru(P\phi_3)_2Cp]$ $[2CHCl_3]$	15:32	P-1	13.570,4	15.394,4	18.178,6	93.98,2	103.32,3	106.34,2	2	3466,2	314

(Continued)

Appendix 1. Continued

CamRef	CN#	Crystal description	Figure	Space group	Axis a / Å	Axis b / Å	Axis c / Å	α / deg	β / deg	γ / deg	Z	V	Reference
FUNGEA	6tr	[R-(Pc)Fe-C-Fe(Pc)-R] [3Me$_2$CO] {+M}	15:33	C2/c	35.309,9	14.503,15	23.161,12	90	103.81,13	90	4	11517,?	348
KITPIM	6tr	[R-(Pc)Fe-O-Fe(Pc)-R] [3Me$_2$CO] {+M}	15:34	C2/c	35.236,5	14.173,2	23.220,3	90	103.56,10	90	4	11272.8,?	349
VAFMIY	6tr	Br-(Pc)Fe-N-Fe(Pc)-Br	15:35	P4/nnc	13.84,4	13.84,4	13.05,4	90	90	90	2	2500,22	350
VACXOM	6tr	D_{4h}-1,2-NcFe(NC-C_6H_{11})$_2$	19:9–10	P2$_1$/c	12.224,5	10.277,4	19.326,8	90	98.451,3??	90	2	2401.5,?	249
XENLAD	6tr	PcFe(4mepip)$_2$	15:36–37	P2$_1$/a	11.792,2	17.213,3	12.645,3	90	102.58,3	90	4	2505.0,9	351
KEHXAW	6tr	PcFeCl$_2$	15:38	P2$_1$/n	9.023,2	14.588,2	10.216,5	90	97.58,1	90	2	1333.2,7	123
RERGEA	6tr	py-(Pc)Fe-C-Fe(Pc)-py	15:39	P2$_1$2$_1$2$_1$	12.692,3	22.186,2	23.653,2	90	90	90	4	6660,2	352
VIWMOD	6tr	tBu$_4$-D_{4h}-1,2-NcFe(NC-tBu)$_2$	19:11–14	C2/c	26.856,?	11.004,?	24.617,?	90	102.43,?	90			250
PAZREN	6tr	THF-(Pc)Fe-C-Fe(Pc)-THF	15:40	P2$_1$2$_1$2$_1$	12.642,1	22,361,7	23,629,3	90	90	90	4	6680,2	353
PAZRAJ	6tr	THF-(Pc)Fe-C-Fe(tpp)-THF	23:1	P22$_1$2$_1$	14.487,3	20.753,5	25.803,7	90	90	90	4	7758,3	353
#2001-09	6tr	[PcFe(im)$_2$][PNP]	15:41–42	P-1	12.848,5	13.162,3	30.234,14	79.833,3	84.217,3	81.727,2			354
NAPHIV	6tr	[F-(Pc)Fe-C-Fe(Pc)-F] [NBu$_4$]$_2$[3H$_2$O]	15:43	P2$_1$/n	15.814,1	22.690,5	25.127,3	90	98.27,1	90	4	8922,2	355
NEKLUK	6tr	[N$_3$-(Pc)Fe-N-Fe(Pc)-N$_3$] [PNP]$_2$[I$_2$][2Et$_2$O]	15:44	C2/c	34.567,9	20.237,9	21.251,5	90	119.79,2	90	4	12902,7	356
DEFNUX	6tr	[PcFe(CN)$_2$][PNP]	15:45	P-1	12.940,4	12.320,2	10.322,3	65.22,8	68.24,15	76.34,9	1	1381.2,?	129,357
DODXUP	6tr	[PcFe(CN)$_2$][PNP]$_2$[CH$_2$Cl$_2$]	15:46	P-1	14.306,4	13.205,4	12.081,4	98.77,9	92.15,8	92.60,8	1	2251,?	358
WOPZOQ	6tr	[PcFe(CN)$_2$][Pϕ_4]$_{0.5}$	15:47	P42/n	21.722,2	21.722,2	7.448,2	90	90	90	2	3514.4,5	359
	6tr	[PcFe(CN)$_2$][PTMA]	–	P2$_1$/a	8.703,1	21.870,4	18.401,3	90	98.62,1	90	4	3462.7,9	336
NIRDUN	6tr	[PcFe(CN)$_2$][PTMA] [MeCN] {123K}	15:48	Pnnm	13.850,1	7.249,1	16.155,1	90	90	90	2	1622.1,2	336
NIRDUN01	6tr	[PcFe(CN)$_2$][PTMA] [MeCN] {295K}	15:49	Pnnm	13.900,1	7.332,1	16.317,1	90	90	90	2	16.62.9,1	336
NELCOW	6tr	[PcFe(OMe)$_2$][PNP]$_2$[1Me$_2$CO, 1.5MeOH]	15:50	P-1	13.160,5	15.480,5	17.140,5	97.54,5	91.79,5	95.44,5	2	3442,2	360
YUGKOA	6tr	[H$_2$O-(Pc)Fe-N-Fe(tpp)-OH$_2$] [I$_5$][2THF]	23:2	Cc	27.13,1	12.032,5	25.596,41	90	90.69,6	90	4	8298,20	361
WOHVEU	6tr	[PcFe(CN-Ru(Pϕ_3)$_2$Cp) (NC-Ru(Pϕ_3)$_2$Cp)][SbF$_6$]	15:51	P2$_1$/n	26.902,8	14.788,4	29.957,11	90	108.08,2	90	4	11329,6	314
WOHTOC	6tr	[PcFe(NC-Fe(dppe)Cp)$_2$] [SbF$_6$]	15:52	P2$_1$/n	18.935,4	9.017,1	25.538,7	90	97.53,2	90	2	9322,2	314
WOHTUI	6tr	[PcFe(NC-Ru(Pϕ_3)$_2$Cp)$_2$] [SbF$_6$]	15:53	P-1	14.348,3	16.021,3	25.072,5	81.22,3	81.26,3	79.73,3	2	5558,2	314
#2001-01	6tr	[THF-(Pc)Fe-N-Mn (TPP)-THF][I$_5$][2THF]	23:3	Cc	27.322,?	12.153,?	25.824,?	90	96.47,?	90			362
BAMSEN, BAMSEN10	6tr	[-Ga(Pc)F-]$_n$	15:54	P-1	3.8711,3	12.601,1	12.793,1	90.271,1	96.42,1	91.28,1	1	620	121,122
BUDHUD*nc	6tr	[-Ge(Pc)O-]$_n$	–	P4/m	13.27,5	13.27,5	3.53,2	90	90	90	4		363
*nc	6tr	[-Ge(Pc)O-]$_n$	–	I4/m	18.76,5	18.76,5	3.57,2	90	90	90	2		363
#1985-01 (App. 29)	6tr	PcGe(OH)$_2$ {α}	15:55	P2$_1$/c	6.721,3	11.734,5	16.530,9	90	105.97,5	90	2	1253,?	120
#1985-02 (App. 30)	6tr	PcGe(OH)$_2$ {β}	15:56	P2$_1$/n	6.812,1	14.541,2	12.842,2	90	94.38,1	90	2	1268.3,?	120
*nc	6tr	PcGe(OH)$_2$ {γ}	–	triclinic	7.43,?	13.03,?	6.86,?	101.0,?	103.4,?	97.5,?			125
RESFIE	6tr	PcGeI$_2$	15:57	P-1	7.522,2	8.663,2	11.349,2	72.61,3	78.96,3	87.31,3	1	692.6,3	119

*nc	6tr	HO-Ge(Pc)-O-Si(Pc)-OSi(C_6H_{13})$_3$	–	Pn	14.28,?	14.28,?	23.40,?	90	90		2	4772,?	364
BUDPOF*nc	6tr	[-Ge(Pc)O-]$_n$(I_3)$_{0.35}$	–	P4/mcc	13.96,5	13.96,5	6.96,4	90	90	90	2		365
GODLEQ	6tr	[PcGeI$_2$][I$_2$]	15:58	P-1	7.716,2	8.597,2	12.540,3	80.19,3	74.83,3	81.74,3	1	786.8,3	366
GUDZUA	6tr	[PcIr(Cl,-CC(O)C)][H_2O]	15:59	P1n1	12.452,3	12.339,3	15.352,4	90	100.75,2	90	4	2317.3,10	367
RENJID	6tr	[PcIr(py,I)][py]	15:60	P2$_1$/n	10.258,1	17.589,3	19.723,2	90	92.95,1	90	4	3553.9,8	368
TOCLIG	6tr	py-(Pc)Ir-Ir(Pc)-py	15:61	P2$_1$2$_1$2$_1$	12.385,4	23.266,4	23.707,4	90	90	90	4	6831.2,28	369
JAHSOA	6tr	[PcIr-(acido,N$_3$)][NBu$_4$][0.25Et$_2$O,0.5CH$_2$Cl$_2$]	15:62	P-1	16.267,8	17.938,3	18.335,4	74.77,2	73.73,3	84.25,3	4	4954,3	370
RILVOX	6tr	[PcIr(-NO$_2$)$_2$][PNP]	15:63	P-1	10.649,1	12.357,1	12.841,1	75.47,1	67.88,1	65.31,1	1	1413.5,2	371
LIQVAI	6tr	[PcIr(OH,OMe)][NBu$_4$]	15:64	P2$_1$/n	12.510,1	19.561,3	18.489,2	90	90.42,1	90	4		372
QOPCON	6tr	PcMg(HOCH$_2$CH$_2$OMe)$_2$	15:65	P2$_1$/n	12.37,1	8.078,6	17.974,9	90	109.70,7	90	2	1690,2	373
QOPCUT	6tr	[PcMg(OH$_2$)$_2$][2NMP]	15:66	P2$_1$/n	6.876,2	21.427,4	12.994,6	90	102.20,2	90	2	1871,1	373
CCDC166120	6tr	[PcMg(py)$_2$]	15:67	P2$_1$/c	9.918,2	19.964,4	9.141,2	90	111.89,3	90	2	1679.4,6	374
ROPTAR	6tr	(dppe)(Cp)Fe-(Pc)Mn-O-Mn(Pc)-NC-Fe(dppe)Cp	15:68	C2/c	51.762,2	13.696,2	32.491,2	90	128.01,2	90	4	18149.5,30	375
PILBER	6tr	[-(Pc)Mn-NC-C(CN)-CN-]$_n$	15:69	C2/c	16.265,2	12.539,2	17.450,3	90	123.93,1	90	4	2952.9,?	376
PILBAN	6tr	[-(Pc)Mn-TCNE-]$_n$ {β}	15:70	P-1	11.769,2	12.468,3	18.561,3	71.62,2	76.59,2	89.13,2	1	2509.2,?	376
XEYREY	6tr	[PcMn(NO)(OPϕ_3)]	15:71	P2$_1$2$_1$2$_1$	12.470,1	14.175,3	22.366,5	90	90	90	4	3954,1	377
PHCPMP10	6tr	[py-(Pc)Mn-O-Mn(Pc)-py][2py]	15:72–73	P2$_1$2$_1$2$_1$	22.635,5	23.850,5	12.808,3	90	90	90	4	6717.1,?	378,379
XEYQOH	6tr	[PcMn(NO)(-NCO)][PNP]	15:74	P-1	10.378,1	12.372,1	13.117,2	76.95,1	67.60,1	65.75,1	1	1414.9,3	377
XEYQIB	6tr	[PcMn(NO)(-ONO)][PNP]	15:75	P-1	10.293,2	12.265,2	13.067,2	77.22,3	67.63,2	65.92,2	1	1388.5,4	377
XEYQUN	6tr	[PcMn(NO)$_2$][PNP]	15:76	P-1	10.424,2	12.337,3	12.993,2	85.39,2	67.92,1	65.77,1	1	1406.6,5	377
#1996-02 (App. 31)	6tr	[PcMn(NCMe)$_2$][C$_4$(CN)$_6$]	15:77	P-1	8.302,20	8.787,2	14.446,5	82.52,2	73.21,3	74.31,2	1	969.8,?	271
#1990-03 (App. 32)	6tr	PcMo(=O,-Cl)	15:78	P2$_1$/n	10.504,3	13.974,2	8.989,3	90	95.25,2	90	2	1313.9,?	124
RIKSIN	6tr	[PcMoBr$_2$][PNP]	15:79	P-1	10.506,1	12.436,2	12.918,2	76.186,1	67.890,1	68.689,1	1	1446.5,4	380
#1997-04	6tr	[PcOsCl$_2$][PNP][0.33PNPOH]	15:82	I23	22.25,5	22.250,6	22.250,5	90	90	90	12	11015,20	131
#1997-05	6tr	[PcOsCl$_2$][PNP]	15:83	P-1	10.534,2	12.287,2	12.905,1	85.636,8	68.077,11	66.060,13	2	1410.6,4	131
PCPCOS	6tr	PcOs(CO,py)	15:80	P2$_1$/n	11.966,9	15.705,3	17.749,5	90	107.92,4	90	4	3125,4	381
#1996-01	6tr	[PcOs(NO,C$_6$H$_{12}$O$_2$)][BF$_4$]	15:81	P2$_1$/n	12.864,8	16.246,4	16.827,3	90.04,2	92.22,3	90.19,3	4	3514.1,24	382
XEYRIC	6tr	PcRe(NO)(OPϕ_3)	15:84	P2$_1$2$_1$2$_1$	12.432,1	14.142,3	22.487,3	90	90	90	4	3953,1	377
NAPPAV	6tr	PcRe(Pϕ_3)$_2$	15:85	P-1	11.512,2	12.795,2	12.858,2	64.42,2	79.45,2	72.74,1	1	1628.1,5	383
DOSWIR	6tr	[PcRe(CN)$_2$][NBu$_4$][0.5Me$_2$CO]	15:86	P2$_1$/n	12.117,4	16.741,6	23.471,6	90	92.28,2	90	4	4757,3	384
DOSWOX	6tr	[PcRe(CN)$_2$][PNP]	15:87	P-1	10.356,2	12.364,2	12.982,2	76.06,1	67.70,1	65.40,1	1	1391.4,3	384
GEQCOU	6tr	[PcReF$_2$][0.33(PNP)F][2H$_2$O]	15:88	I23	21.836,2	21.836,2	21.836,2	90	90	90	6	10412,2	385
XEYRAU	6tr	[PcRe(NO)$_2$][PNP]	15:89	P-1	10.393,3	12.251,3	12.896,4	85.17,2	67.79,2	65.97,2	1	1383.5,7	377
XEYROI	6tr	[PcRe(NO)(OPϕ_3)][I$_3$][CH$_2$Cl$_2$]	15:90	P2$_1$/n	13.994,5	20.291,4	17.459,4	90	98.39,2	90	4	4904,2	377
#1997-02	6tr	[PcRh(OϕCl)$_2$][NBu$_4$]	15:92	P-1	11.231,8	13.486,6	18.929,7	89.499,3	83.57,5	75.706,5			386
NEXSIS	6tr	[py-(Pc)Rh-Rh(Pc)-py][2C$_6$H$_6$]	15:91	Ia-3	35.960,7	35.960,7	35.960,7	90	90	90	24	46500,16	387
RERGIE	6tr	py-(Pc)Ru-C-Ru(Pc)-py	15:93	P2$_1$2$_1$2$_1$	13.002,3	22.635,6	23.901,6	90	90	90	4	7034,4,?	352
RUJQES	6tr	[PcRu(-NO$_2$)$_2$][NBu$_4$]$_2$	15:94	P2$_1$/n	15.114,4	22.34,3	18.206,11	90	90.88,5	90	4	6146,9	388
GEQDIP	6tr	[PcRu(py,CN)][NBu$_4$]	15:95	Pca2$_1$	28.319,5	29.850,3	24.566,7	90	90	90	16	20766,7	389
#1970-01 (App. 34)	6tr	(Me$_3$SiO)$_2$MeSiO-((Pc)Si-O)$_3$-SiMe(OSiMe$_3$)$_2$	15:97	P-1	14.642,6	15.831,7	14.318,9	90.986,29	94.866,29	128.049,30			269
QUDQIP	6tr	[-(Pc)Si-(OSiMe$_2${CH$_2$}$_7$SiMe$_2$O-)]$_n$	15:98	Pbcm	9.704,2	20.042,2	21.491,4	90	90	90	8	4179.5,12	390,391
BUDHOX*nc	6tr	[-(Pc)Si-O-]$_n$	–	Ibam	13.80,5	27.59,5	6.66,4	90	90	90	4		363

(*Continued*)

Appendix 1. Continued

CamRef	CN#	Crystal description	Figure	Space group	Axis *a* / Å	Axis *b* / Å	Axis *c* / Å	α / deg	β / deg	γ / deg	Z	V	Reference
*nc	6tr	$[\text{-(Pc)Si-O-}]_n$	–	Pbam	13.80,5	????	3.33,?	90	90	90			363
NEYCEZ	6tr	$[\text{PcSi(-O-C(O)-Fc)}_2][2\text{CHCl}_3]$	15:96	$P2_1/n$	10.417,6	21.046,9	23.130,6	90	94.22,4	90	4	5057,4	392
VIKZAQ	6tr	$[\text{PcSi(OMe)}_2][0.48\text{H}_2\text{O}]$	15:99	P-1	11.606,4	7.583,3	8.304,3	96.46,3	105.23,3	74.87,3	1	680.0,4	127
SIOPCY	6tr	$[\text{PcSiOSiEt}_3\text{,Me}][0.5\text{C}_6\text{H}_{12}]$	–	P-1	12.050,6	14.892,6	12.003,5	117.17,2	95.12,2	94.33,2	2		449
#1971-01 (App. 33)	6tr	$[\text{PcSiOSiEt}_3\text{,Me}][0.5\text{C}_6\text{H}_{12}]$	15:100	P-1	12.050,6	14.892,6	12.003,5	117.17,2	95.12,2	94.33,2	2		270
DOHYAA	6tr	$\text{Me}_2\text{(tBu)SiO-(Pc)Si-O-Si(Pc)-OSi(tBu)Me}_2$	15:101	Pbcn	21.670,8	13.724,5	23.031,9	90	90	90	4	6849.4,?	393
CCDC173623	6tr	$\text{PcSi(O-C(O)-C}_{10}\text{H}_{21})_2$	15:102–103	P-1	8.9976,14	15.522,2	19.031,2	105.797,10	100.621,11	93.939,12	2	2493,12	128
*nc	6tr	PcSi(OH)_2	–	triclinic	7.27,?	13.07,?	6.88,?	102.5,?	104.2,?	97.4,?			394
QEKBEN01	6tr	PcSi(OH)_2 {I}	15:104	P-1	12.992,1	7.2830,8	6.861,1	104.413,7	101.757,8	96.973,6			126
QEKBEN	6tr	PcSi(OH)_2 {II}	15:105	$P2_1/n$	12.7494,5	14.5778,6	6.7727,3	90	94.353,2	90			126
SIMPLC10	6tr	$\text{PcSi(OSiMe}_3)_2$	15:106	$P2_1/c$	18.7610,9	9.8385,4	19.531,1	90	89.6427,8	90	4		395
VILGOM	6tr	PcSi(thff)_2	15:107	P-1	10.234,6	10.266,7	9.761,6	98.93,2	117.94,2	68.83,2	1	845,2	127
#2001-02	6tr	$\text{PcSi[OCH}_2\text{-C}_6\text{H}_3\text{((OCH}_2\text{CH}_2\text{O)}_3\text{Me)}_2]_2$	15:111–112	P-1	13.064,4	16.318,3	8.208,2	92.02,2	96.05,2	79.55,2	1	1710.8,7	396
NOLXOB	6tr	$\text{PcSi[OCH}_2\text{-C}_6\text{H}_3\text{(OCH}_2\phi)_2]_2$	15:113–114	P-1	14.079,3	16.291,5	12.710,3	103.62,2	109.30,2	71.83,2	1	2585,1	397
CAVZAA	6tr	$\text{PcSi[Oepiglobulyl]}_2$	15:115	P1	9.608,1	10.742,1	14.468,1	107.121,1	102.757,1	104.349,1	1	1311.1,4	398
QUDQOV	6tr	$\text{PcSi[OSi(C}_6\text{H}_{13})_3]_2$	15:118–119	P1	10.062,7	12.419,4	13.788,6	92.86,3	107.29,4	94.75,4	2	1634.4,15	390,391
QUDQUB	6tr	$\text{PcSi[OSiMe}_2\text{(C}_8\text{H}_{17})]_2$	15:120–121	P1	9.758,1	9.974,2	13.562,1	78.85,1	73.69,1	78.88,1	2	1229.4,2	390,391
ACUMOA	6tr	$\text{PcSi[O(CO)(CH}_2)_2\phi\text{(3,4-(OMe)}_2)_2$	15:110	P1	9.730,1	10.118,1	13.580,1	69.94,1	88.48,1	61.66,1	1	1090.18,17	399
ACUMIU	6tr	$\text{PcSi[O(CO)-(3-thienyl)]}_2$	15:109	$P2_1/c$	10.078,1	22.591,2	8.243,1	90	101.59,1	90	2	1838.4,3	399
ACUMEQ	6tr	$\text{PcSi[O(CO)}\phi\text{-4tBu]}_2$	15:108	R-3	36.327,8	36.327,8	8.841,2	90	90	120	9	10104,4	399
VIKVEQ	6tr	PcSiCl_2	15:116	$P2_1/n$	10.261,1	14.618,2	9.072,1	90	98.51,1	90	2	1345.8,3	127
CCDC173621	6tr	PcSiCl_2	15:117	$P2_1/n$	8.997,2	14.582,4	10.129,4	90	97.74,3	90	2	1316.9,7	128
FAGMAB*nc	6tr	$[\text{-(Pc)Si-O-}]_n\text{(BF}_4)_{0.36}$	–	P4/mcc	13.70,7	13.70,7	6.58,4	90	90	90	2		298
BUDPIZ*nc	6tr	$[\text{-(Pc)Si-O-}]_n\text{(Br}_3)_{0.37}$	–	P4/mcc	13.97,5	13.97,5	6.60,4	90	90	90	2		365
BUDPEV*nc	6tr	$[\text{-(Pc)Si-O-}]_n\text{(I}_3)_{0.37}$	–	P4/mcc	13.97,5	13.97,5	6.60,4	90	90	90	2		365
FAGMEF*nc	6tr	$[\text{-(Pc)Si-O-}]_n\text{(PF}_6)_{0.36}$	–	P4/mcc	13.98,6	13.98,6	6.58,4	90	90	90	2		298
FAGMIJ*nc	6tr	$[\text{-(Pc)Si-O-}]_n\text{(SbF}_6)_{0.36}$	–	P4/mcc	14.31,4	14.31,4	6.58,4	90	90	90	2		298
BUDJAL*nc	6tr	$[\text{-(Pc)Sn-O-}]_n$	–	P4/m	12.81,5	12.81,5	3.82,2	90	90	90	1		363
CPHCYT*nc	6tr	PcSnCl_2	–	A2/a	21.104,2	11.060,1	11.392,1	90	96.04,1	90	4		116
#1972-01 (App. 35)	6tr	PcSnCl_2	15:122–123	A2/a	21.104,2	11.060,1	11.392,1	90	96.04,1	90	4		117
RESFOK	6tr	PcSnI_2	15:124	P-1	7.618,2	8.551,2	11.602,2	73.34,3	77.20,3	87.05,3	1	706.0,3	119
RESFOK01	6tr	PcSnI_2	15:125	C2/c	11.581,2	11.141,2	21.962,4	90	97.94,3	90	4	2806.5,9	118
#2001-05	6tr	$[\text{PcWCl}_2][\text{PNP}]$	15:126	P-1	10.499,1	12.331,1	12.855,2	85.499,1	68.085,1	65.67,1			342
#1990-02 (App. 36)	6tr	PcW(=O,-Cl)	15:127	$P2_1/n$	10.504,4	13.910,2	8.944,3	90	95.32,3	90	2	1301.2,?	124
LOKSAF	7*cis*	$\text{(PcBi)}_4\text{(Bi}_4\text{I}_{16})$	16:1–3	P-1	15.651,3	15.989,3	17.193,3	71.89,3	61.49,3	68.62,3	1	3471.0,11	190
CCDC165833	7*cis*	$[\text{PcHf(O}\phi)_3][\text{NBu}_4][0.5\text{Et}_2\text{O}]$	16:6–7	P-1	11.719,3	12.11,5	22.65,4	89.15,3	81.983,3	73.498,1			400
POQYID	7*cis*	$[\text{PcHfF}_3][\text{NBu}_4]$	16:4–5	$P2_1/n$	13.517,1	13.856,1	23.384,2	92.67,1	90	90	4	4375,1	182
#1998-01	7*cis*	$[\text{PcNb(O,O}_2\text{Naphth)}][\text{CH}_2\text{Cl}_2]_{0.5}$	16:8	P2/c	19.447,4	10.771,1	26.781,7	90	107.443,2	90			308
NEXVAN	7*cis*	PcNbCl_3	16:9–10	P-1	12.086,2	13.710,2	8.345,1	93.49,2	100.95,2	79.79,2	2	1335.6,4	163
#1990-01 (App. 37)	7*cis*	PcNbCl_3	16:11–12	P-1	8.360,1	12.090,2	13.655,1	79.85,1	93.48,1	101.06,?	2	1332.8,?	124
CAPQOZ	7*cis*	$[\text{PcNbF}_2\text{O}][\text{NBu}_4]$	16:13–14	$P2_1/n$	13.460,5	13.820,5	23.360,5	90	92.640,5	90	4	4341,2	401

SAMDUF	7*cis*	[PcNb(-O-,-O-,-Br-)NbPc][Br$_3$][1-ClN]	16:15–16	P2$_1$/c	15.563,2	23.948,5	18.415,2	90	109.06,1	90	4	6487.1,?	124,162
#1986-01	7*cis*	[PcNb(-O-,-O-,-Br-)NbPc][Br$_3$][1-ClN]	–	P2$_1$/c	15.563,2	23.948,5	18.415,2	90	109.06,1	90	4	6484,?	402
SAMFAN	7*cis*	[PcNb(-O-,-O-,-Br-)NbPc][Br$_3$][PN]	16:17	P2$_1$/c	15.573,8	23.305,9	18.511,6	90	108.18,4	90	4	6382.4	124,162
YOHMAJ	7*cis*	[PcSb]$_4$[Sb$_4$I$_{16}$]	16:18–19	P-1	15.963,3	16.051,3	16.552,3	88.59,3	62.54,3	69.16,3	1	3466.9,11	194
GIFZUQ	7*cis*	PcTaCl$_3$	16:20	P-1	8.360,2	12.077,4	13.688,4	79.86,3	93.34,2	100.97,2	2	1335.1,?	160
	7*cis*	PcTaCl$_3$	–	P-1	8.364,2	12.083,4	13.691,4	79.85,3	93.34,2	100.96,2	2	1335.1,?	124
BORQUU	7*cis*	[PcY(Cl)(py)(OPϕ_3)][0.5Et$_2$O]	16:21	P-1	12.613,2	14.165,3	15.639,5	103.64,2	95.41,3	114.66,3	2	2408,1	192
NOFNUR	7*cis*	PcZrI$_2$(PN)	16:22–23	P2$_1$/n	9.965,2	21.188,4	16.792,3	90	93.10,3	90	4	3540.2,1	176,403
CCDC165838	7*cis*	[PcZr(Oϕ)$_3$][NBu$_4$][Et$_2$O]	16:24–25	P-1	11.735,3	12.132,4	22.612,4	89.379,2	82.119,2	73.765,2			400
YOKWEA	7*cis*	[PcZrCl$_3$][PNP]	16:26–27	P2$_1$/n	15.219,4	20.262,10	20.719,4	90	93.46,2	90	4		169
NAPRAX	7–8*cis*	[(PcTaF$_3$)$_{1-x}$(PcTaF$_4$)$_x$][NBu$_4$]	16:28-29,17:1	P2$_1$/n	13.368,2	13.787,2	23.069,3	90	93.35,1	90	4	4244,1	179
ZEMBEY	8*cis*	[PcIn(-ONO-)$_2$][NBu$_4$]	17:2–3	P2$_1$/n	14.608,3	17.326,3	18.520,3	90	100.00,1	90	4	4616,1	404
DICCAT	8*cis*	PcLu(OAc,H$_2$O,H$_2$O)	17:4–5	P-1	12.033,3	20.807,6	14.011,3	90	95.34,2	90	4	3492,?	159
CCDC173622	8*cis*	PcSn[O-C(O)-C$_{10}$H$_{21}$]$_2$	17:6–7	P-1	13.206,5	15.539,4	12.762,4	98.15,3	103.26,3	100.93,3	2	2455.3,13	128
BOFQAO	8*cis*	[PcTl(-ONO-)$_2$][NBu$_4$]	17:8–9	P2$_1$/n	14.496,2	17.293,5	18.293,2	90	98.76,1	90	4	4532,1	191
BORQII	8*cis*	[PcY(Cl)(DMF)$_3$][Et$_2$O]	17:10–11	P2$_1$2$_1$2$_1$	11.547,2	13.710,3	27.518,6	90	90	90	4	4356,1	192
BORQOO	8*cis*	[PcY(Cl)(-OSMe$_2$)$_4$][Et$_2$O,2MeOH]	17:12–13	P2$_1$/c	13.127,5	13.043,2	31.120,9	90	98.69,3	90	4	5267,3	192
BOSQAB	8*cis*	PcY(-(Cl)$_2$-,-(O(Hpy)$_2$)$_2$-)YPc	17:14–15	P2$_1$/c	12.087,2	16.470,3	18.320,3	90	90.62,2	90	4	3647,1	192
BOSJAU	8*cis*	PcY(L,L′,-Cl-,-Cl-)YPc	17:16–17	P-1	11.146,1	12.209,1	13.994,2	110.46,1	96.30,1	98.96,1	1	1734.5,4	192
CCDC165829	8*cis*	[PcZr(-(OH)$_2$-,-(O,Oala)$_2$-)ZrPc][2DMF,Et$_2$O,H$_2$O]	17:18–19	P-1	14.5665,4	15.9154,3	21.7057,8	70.5393,2	80.2422,3	63.8517,2			400
CCDC165834	8*cis*	[PcZr(O$_2$CatCl$_4$)$_2$][NBu$_4$]$_2$	17:20–21	P2$_1$/c	18.9845,3	18.3872,2	23.7193,5	90	112.9661,1	90			400
CCDC165836	8*cis*	[PcZr(-OC(O)C)$_2$][NEt$_4$]$_2$[CH$_2$Cl$_2$]$_2$	17:22–23	P-1	13.827,3	14.987,3	15.046,5	86.827,3	62.986,6	77.15,5			400
BECTOS	8bis	[Pc$_2$As][I$_2$]	18:1–2	P4/mcc	13.926,2	13.926,2	6.433,1	90	90	90	1	1247.6,3	233
ZUWDAW	8bis	[Pc$_2$Bi][CH$_2$Cl$_2$]	18:3–4	Pnma	28.176,5	22.913,3	7.925,1	90	90	90	4	5116,1	228
DIQPIC	8bis	Pc$_3$Bi$_2$	18:5	P-1	11.217,2	13.616,3	14.931,3	107.67,3	100.14,3	112.82,3	1	1887.4,7	51
DIQPIC01	8bis	Pc$_3$Bi$_2$	18:6	P-1	11.2179,5	13.6180,6	14.8834,8	107.756,5	99.725,5	112.932,4	1	1883.8,?	52
QOCLOJ	8bis	[Pc$_2$Bi][Br][PNP]$_2$[2Et$_2$O]	18:7–8	P2/c	13.709,1	13.554,1	32.061,6	90	101.53,1	90	2	5837,1	198
QOCJEX	8bis	[Pc$_2$Bi][NBu$_4$]	18:9–10	P4/ncc	16.937,2	16.937,2	22.416,4	90	90	90	4	6430,1	198
QOCKAU	8bis	[Pc$_2$Bi][NBu$_4$]	18:11–12	Pna2$_1$	24.962,4	25.527,4	11.525,5	90	90	90	4	7344,4	198
KOJQUV	8bis	[Pc$_2$Bi][I$_{1.5}$]	18:13–14	P4/mcc	13.997,2	13.997,2	6.549,1	90	90	90	1	1283.05,12	232
SACNOZ	8bis	[(Pc)Ce(T(pMeOP)P)][2CHCl$_3$,C$_5$H$_{12}$]	23:4	P2$_1$/c	16.190,5	32.356,9	16.267,5	90	118.85,2	90	4	7463.7,?	405
JUDPAZ	8bis	[(Pc)Ce(T(pMeOP)P)Ce(Pc)][2tol]	23:5	P-1	14.569,4	15.863,4	11.962,3	111.77,2	96.23,2	100.51,2	1	2467.8,?	406
JUDNUR	8bis	[(TPP)Ce(Pc(OMe)$_8$)Ce(TPP)][6CH$_2$Cl$_2$]	23:6	P2$_1$/n	20.373,6	16.009.5	18.729,6	90	95.37,2	90	2	6081.7,?	406
JONZAN	8bis	Pc$_2$Ce	18:15–16	C2/c	18.783,12	18.739,16	15.618,10	90	114.30,7	90	4	5010,2	217
TANMUQ	8bis	[(TPP)Ce(Pc)Gd(OEP)][2ClB]	23:7	P-1	17.212,5	19.777,6	15.050,5	91.10,2	104.46,2	89.63,2	2	4959.8,?	407
QOCJIB	8bis	[Pc$_2$Ce][NBu$_4$]	18:17–18	P4/ncc	16.864,1	16.864,1	22.554,3	90	90	90	4	6414,1	198
#2002-10	8bis	[Pc$_2$Ce][PNP]	18:19–20	P2$_1$/c	14.8734,3	20.835,3	26.2743,2	90	93.2792,1	90			221
RABDEZ	8bis	[Pc$_2$Ce][BF$_4$]$_{0.33}$	18:21–22	P4/nnc	19.643,3	19.643,3	6.525,3	90	90	90	2	2517,2	230

(Continued)

Appendix 1. Continued

CamRef	CN#	Crystal description	Figure	Space group	Axis a / Å	Axis b / Å	Axis c / Å	α / deg	β / deg	γ / deg	Z	V	Reference
SACNUF	8bis	[PcCeTPP][$SbCl_6$]	23:8	P-1	15.272,4	16.249,4	13.566,4	92.05,2	94.40,2	73.64,2	2	3220.3,?	405
YUHSUP	8bis	Pc_2Er {α}	18:23–24	P4/nnc	19.907,5	19.907,5	6.430,5	90	90	90	2	2548.1,22	225
OCOBIR	8bis	[2,3-NcEuOEP][4$CHCl_3$]	23:13–14	Pnma	27.9422,8	26.124,1	10.9643,7	90	90	90	4	8003,1	252
GAWBEL	8bis	Pc_2Gd	18:31–32	$P2_12_12_1$	10.556,3	50.724,9	8.875,3	90	90	90		4752.1	212
TANMUQ	8bis	[(OEP)Gd(Pc)Ce(TPP)][2ClB]	23:7	P-1	17.212,5	19.777,6	15.050,5	91.10,2	104.46,2	89.63,2	2	4959.8,?	407
NAKYAZ	8bis	[Pc_2Gd][NBu_4]	18:25–26	$P42_12$	17.399,8	17.399,8	11.377,6	90	90	90	2	3444.1,23	197
YEGTOT	8bis	[Pc_2Gd][PNP][0.33H_2O]	18:27–28	$P2_1/c$	15.172,8	20.826,2	25.876,3	90	95.19,3	90	4	8143,4	220
ZAZQEW	8bis	[(TPP)Gd(Pc)][$SbCl_6$] [2CH_2Cl_2,H_2O]	23:9	P-1	18.190,5	16.100,5	14.130,4	104.87,2	104.17,2	107.72,2	2	3567.7,?	408
OBUJIE	8bis	[Pc_2Gd][$C_4(CN)_6$][2ClB]	18:29–30	P-1	13.2324,2	15.6369,3	18.2546,3	110.770,8	97.788,1	97.772,9	2	???	236
NAKYED	8bis	[Pc_2Ho][NBu_4][H_2O]	18:33–34	$P42_12$	18.096,5	18.096,5	11.079,6	90	90	90	2	3628.0,22	197
RIYREW*nc	8bis	[Pc_2In][CH_2Cl_2]	–	Pnma	27.973,17	22.872,7	7.901,7	90	90	90	4		229
ZAKZOA	8bis	Pc_2In	18:35–36	$P2_12_12_1$	10.569,3	50.868,11	8.899,2	90	90	90	4	4784.1,?	224
PIFDEN	8bis	Pc_3In_2	18:37–38	P-1	14.492,3	13.179,3	11.493,2	113.96,2	97.15,2	66.13,2	1	1832.2,6	53
QOCLUP	8bis	[Pc_2In][I][PNP]$_2$[2Et_2O]	18:39–40	P2/c	13.749,2	13.614,3	31.898,5	90	101.13,2	90	2	5858,2	198
NOFMAW	8bis	[Pc_2In][I_2]	18:41–42	P4/mcc	13.984,2	13.984,2	6.393,1	90	90	90	1	1250.2,3	409
ZEHTUB	8bis	[Pc_2La][CH_2Cl_2]	18:43–44	Pnma	28.365,5	22.935,3	8.026,3	90	90	90	4	5221,2	226
TUPWEG	8bis	[PcLaTPyP][OH] [$CHCl_3$,MeOH]	23:10	P-1	12.894,3	14.143,4	18.352,5	95.83,3	105.15,3	98.95,4	2	3155.4,?	410
ZAZQAS	8bis	[TPPLaPc][2CH_2Cl_2]	23:11	$P2_1/c$	12.938,5	18.674,6	26.114,9	90	90.90,2	90	4	6308.5,?	408
QOCJOH	8bis	[Pc_2La][NBu_4]	18:45–46	P4/ncc	16.839,1	16.839,1	22.666,3	90	90	90	4	6427,1	198
QOCMAW	8bis	[Pc_2La][NBu_4][EtOH]	–	P42/n	28.15,?	28.15,?	18.40,?	90	90	90	8	14581,?	198
QOCJUN	8bis	[Pc_2La][NPe_4]	18:47–48	$P2_1/m$	11.717,2	22.762,5	14.197,2	90	108.84,1	90	2	3583,1	198
YEGTIN	8bis	[Pc_2La][PNP]	18:49–50	$P2_1/c$	14.84,2	20.881,7	26.35,1	90	92.97,7	90	4	8156,13	220
LORKIM	8bis	[(15cr5)$_4$Pc]$_3$$Lu_2$] [18DMSO,2$CHCl_3$]	18:53–55	P2/n	18.435,4	16.289,2	43.875,9	90	94.00,2	90	2	13143,4	195
VODDIB	8bis	[($H_{13}C_6S$)$_8$Pc]$_2$Lu	18:56–58	C2/c	31.5575,15	32.7547,15	20.4841,10	90	127.1190,10	90	4	16887.5,14	247
DICBUM	8bis	[Pc_2Lu][CH_2Cl_2]	18:51–52	Pnma	28.242,9	22.877,8	8.050,4	90	90	90	4	5201,?	159
DULZAL	8bis	Pc_2Lu {γ}	18:59–60	$P2_12_12_1$	10.553,2	50.795,1	8.959,2	90	90	90	4		208
GAWBAH	8bis	Pc_2LuH	18:61–62	$P2_12_12_1$	10.567,3	50.683,9	8.849,3	90	90	90	4	4739,?	212
GAVZUY	8bis	[Pc_2Lu][NBu_4][2DMF]	18:63–64	P-1	22.390,8	15.371,6	11.590,4	109.35,2	95.33,2	92.54,2	2	3735,?	212
NAKYIH	8bis	[Pc_2Lu][NBu_4][H_2O]	18:65–66	P4/nmm	18.171,3	18.171,3	10.987,4	90	90	90	2	1458.48,?	197
NOCYEJ	8bis	[Pc'_2Nb][I_3][0.5I_2,3.5ClN]	21:13–14	P-1	17.702,3	18.708,3	13.504,2	106.00,2	93.39,2	83.26,1	2	4267.5,13	231
#2001-06	8bis	[Pc_2Nb][IBr_2]	?	?	?	?	?	?	?	?	?	?	411
#2001-07	8bis	[Pc_2Nb][IBr_2][I_2]	?	?	?	?	?	?	?	?	?	?	411
DULBIV	8bis	[(Pc)Nd(Pc)Nd(T(4-OMe)PP)] [1,1,2-trichloroethane]	23:12	P4/n	20.248,6	20.248,6	13.098,6	90	90	90	2		412
PTCNND (App. 38)	8bis	[Pc_2Nd][CH_2Cl_2]	18:67–68	$P2_12_12_1$ (Pnma)	8.030,4	22.925,7	28.315,7	90	90	90	4	5212.43,3	205,206
CIZGIB08	8bis	Pc_2Nd	18:71–72	P4/nnc	19.544,2	19.544,2	6.514,1	90	90	90	2	2488,?	222
CIZGIB06 (App. 39)	8bis	Pc_2Nd (ox1)	18:73–74	C2/c	28.29,1	14.187,5	13.325,5	90	115.35,4	90	4	4833,?	210
CIZGIB*nc	8bis	Pc_2Nd {α}	–	P-42m,P422	27.601,?	27.601,?	6.581,?	90	90	90			196
CIZGIB07	8bis	Pc_2Nd {α}	18:75–76	P4nc	19.52,1	19.52,1	6.581,5	90	90	90	2		209
CIZGIB02	8bis	Pc_2Nd {β}	18:69–70	C2/c	19.010,?	19.066,?	15.538,?	90	116.1,?	90	4		196
CIZGIB03*nc	8bis	Pc_2Nd {γ(1)}	–	$P2_12_12_1$	10.672,?	51.572,?	8.470,?	90	90	90	4		196

CIZGIB04*nc	8bis	Pc_2Nd {γ(2)}	–		36.444,?	12.811,?	9.233,?	90	95.8,?	90	4		196
CIZGIB05*nc	8bis	Pc_2Nd {γ(3)}	–		19.780,?	18.699,?	9.233,?	90	93.7,?	90			196
NAKXUS	8bis	$[Pc_2Nd][NBu_4]$	18:77–78	P4/ncc	17.0545,6	17.0545,6	22.807,3	90	90	90	4	6633.6,10	197
SAFRAS*nc	8bis	$[Pc_2Nd][Cl_xO_{1-x}]$	–	P4nc	19.700,5	19.700,5	6.510,3	90	90	90	2	2529,3	213
SAFPUK (App. 40)	8bis	$[Pc_2Nd][O]$	18:79–80	P4nc	19.514,5	19.514,5	6.511,3	90	90	90	2	2480,3	213
PCNPAB*nc	8bis	Pc_2Pa	–	C2/c	18.84,1	18.79,1	15.68,1	90	113.65,5	90	4		413
KOBRUO02	8bis	Pc_2Pr	18:81–82	P4/nnc	20.066,7	20.066,7	6.463,2	90	90	90	2	2602,?	222
KOBRUO01*nc	8bis	Pc_2Pr {α}	–		19.52,1	19.52,1	6.59,1	90	90	90			218
KOBRUO	8bis	Pc_2Pr {β1}	18:83–84	I2/c	13.383,2	14.154,3	25.644,7	90	83.32,?	90	4	4824.6,2	211
QOCKEY	8bis	$[Pc_2Pr][NBu_4][0.75MeOH]$	18:85–86	$Pna2_1$	25.003,5	25.476,3	11.561,3	90	90	90	4	7364,3	198
QOCLAV	8bis	$[Pc_2Pr][NBu_4][2py]$	18:87–88	P-1	11.895,1	15.335,1	22.473,2	90.99,1	101.26,1	109.06,1	2	3785,1	198
WAYSUK*nc	8bis	$[Pc_2Pr][Br_2]$ {α}	–		19.63,1	19.63,1	6.48,1	90	90	90	1		218
WAYSIY	8bis	$[Pc_2Pr][Br_{1.5}]$ {α}	18:89–90	P4/nnc	19.634,5	19.634,5	6.485,2	90	90	90	2		218
WAYSOE*nc	8bis	$[Pc_2Pr][I_2]$ {α}	–		19.66,1	19.66,1	6.42,1	90	90	90			218
QOCLEZ	8bis	$[Pc_2Sb][NBu_4][2THF]$	18:91–92	P-1	11.659,3	15.364,4	22.504,5	91.25,2	100.94,2	109.23,2	2	3721,1	198
QOCLID	8bis	$[Pc_2Sb][Br][PNP]_2[2Et_2O]$	18:93–94	C2/c	19.209,2	19.322,6	31.251,5	90	92.62,1	90	4	11587,4	198
QOCKUO	8bis	$[Pc_2Sc][NBu_4]$	18:95–96	$Pna2_1$	25.100,19	25.353,3	11.380,10	90	90	90	4	7242,6	198
QOCKIC	8bis	$[Pc_2Sm][NBu_4][MeOH]$	18:97–98	$Pna2_1$	25.243,5	25.508,4	11.501,3	90	90	90	4	7405,3	198
ZOMWUT	8bis	$[Pc_2Sm][Br_{1.45}]$ {α}	18:99–100	P4/nnc	19.638,3	19.638,3	6.460,1	90	90	90	2	2491,1	227
ZOMXAA	8bis	$[Pc_2Sm][ClO_4]_{0.63}$ {α}	18:101–102	P4/nnc	19.757,3	19.757,3	6.474,1	90	90	90	2	2527,1	227
SNPTCY	8bis	Pc_2Sn	18:103–104	$P2_12_12_1$	10.547,1	50.743,4	8.9046,9	90	90	90	4		199,201
SNPTCY01	8bis	Pc_2Sn	18:105–106	C2/c	18.754,4	18.759,4	15.357,3	90	115.06,4	90	4	4894.1,2	223
DPCYTH	8bis	Pc_2Th	18:107–108	C2/c	18.74,4	18.73,3	15.79,3	90	113.3,5	90			203
DPCYTH01	8bis	Pc_2Th	18:109–110	C2/c	18.92,1	18.81,1	15.92,1	90	116.?,?	90			204
DPCYTH02*nc	8bis	Pc_2Th	–	C2/c	18.93,1	18.81,1	15.92,1	90	116.00,6	90	4		413
DPCYTH03	8bis	Pc_2Th	18:111–112	C2/c	18.768,?	18.767,?	15.815,?	90	113.2,?	90	4		207
DPCYTH*nc	8bis	Pc_2Th	–	B2/b	18.74,4	15.79,3	18.73,3	90	113.3,5	90	4		202,203
SEWRER	8bis	$[Pc'_2Ti][1\text{-}ClN]$	21:15–16	C2/c	16.327,3	18.568,4	19.022,4	90	94.50,1	90	4	5748.9,20	214
PIBJIT	8bis	$[Pc_2Ti][I_3]_{0.66}$	18:113–114	P4/mcc	13.841,3	13.841,3	6.370,3	90	90	90	1	1220.3,7	219
YEGTUZ	8bis	$[Pc_2Tm][PNP][0.5H_2O]$	18:115–116	$P2_1/c$	15.301,11	20.817,4	25.833,6	90	95.71,4	90	4	8187,6	220
PHALCU*nc	8bis	Pc_2U	–	C2/c	18.74,1	18.73,3	15.61,3	90	113.6,9	90	4		200
PHALCU01	8bis	Pc_2U	18:117–118	B2/b	18.73,4	15.60,3	18.70,4	90	113.7,5	90	4		202,203
PHALCU02*nc	8bis	Pc_2U	–	C2/c	18.77,2	18.76,2	15.66,1	90	113.57,6	90	4		413
MEGNOB	8bis	$[Pc_2U][I_3]_{0.66}$	18:119–120	P4/nnc	19.731,2	19.731,2	6.507,1	90	90	90	2	2533.3,5	235
HOTWOC	8bis	$[Pc_2U][I_{5/3}]$	18:121–122	P4/mcc	13.947,2	13.947,2	6.501,1	90	90	90	1	1264.6,3	234
JUVJIT	8bis	$[Pc_2Y][CH_2Cl_2]$	18:123–124	Pnma	28.083,9	22.910,8	7.921,4	90	90	90	4	5096,?	216
QOCKOI	8bis	$[Pc_2Y][NBu_4]$	18:125–126	$Pna2_1$	25.153,4	25.497,3	11.483,1	90	90	90	4	7364,2	198
BECSUX	8bis	$[Pc_2Yb][I_2]$	18:127–128	P4/mcc	13.927,2	13.927,2	6.409,1	90	90	90	1	1243.1,3	233
SOTGAJ	8bis	Pc_2Zr	18:129–130	P-1	13.410,2	13.400,7	16.340,11	68.68,1	65.92,1	74.74,1	2	2474.4,?	215
NOFMEA	8bis	$[Pc_2Zr][I_2]$	18:131–132	P4/mcc	13.941,2	13.941,2	6.487,1	90	90	90	1	1260.8,3	409

[a]Abbreviations are compiled in Table 1.
[b]*nc indicates that no coordinates are available or known.
[c]? indicates that error of the data is not known.
[d]The crystallographic data would indicate a coordination number of 4, recently found structures indicate a coordination number of 5.
[e]No full crystal structure exists and the coordination number is not known.

Appendix 2. Basic Crystallographic Data of Crystal Structures of Naphthalocyanine Complexes[a–c]

CamRef	CN#	Crystal description	Figure	Space group	Axis a / Å	Axis b / Å	Axis c / Å	α / deg	β / deg	γ / deg	Z	V	Reference
RADPIV	4	NcCu	19:1	$P2_1/c$	6.873,2	7.906,1	30.385,9	90	91.02,2	90	2	1650.9,2	104
RADPER	4	NcNi	19:2–3	$P2_1/c$	6.8800,9	7.8730,8	30.48,2	90	90.50,9	90	2	1650.8,2	104
RADPOB	4	NcZn	19:4	$P2_1/c$	6.850,2	7.997,4	30.273,4	90	91.88,2	90	2	1657.5,8	104
#2001-03	5	[C_{2v}-1,2-NcCo-py][2py]	19:5–6	Pccn	29.649,1	31.385,1	10.080,1	90	90	90	8	9379.5,7	135
WOHDUH	6tr	[2,3-NcCo$(CN)_2$][2DCB]	19:7	P-1	10.684,1	15.005,2	8.068,1	92.37,1	95.72,1	70.08,1	1	1209.9,3	251
WOWDOB	6tr	[2,3-NcCo$(CN)_2$][CH_3CN]	19:8	P-1	9.864,2	12.475,4	8.827,2	107.83,2	93.02,2	102.30,2	1	1002.0,6	251
VACXOM	6tr	D_{4h}-1,2-NcFe$(NC\text{-}C_6H_{11})_2$	19:9–10	$P2_1/c$	12.224,5	10.277,4	19.326,8	90	98.451,3??	90	2	2401.5,?	249
VIWMOD	6tr	tBu_4-D_{4h}-1,2-NcFe$(NC\text{-}tBu)_2$	19:11–14	C2/c	26.856,?	11.004,?	24.617,?	90	102.43,?	90			250
OCOBIR	8bis	[2,3-NcEuOEP][4$CHCl_3$]	23:13–14	Pnma	27.9422,8	26.124,1	10.9643,7	90	90	90	4	8003,1	252

[a]Abbreviations are compiled in Table 1.
[b]*nc indicates that no coordinates are available or known.
[c]? indicates that error of the data is not known.

Appendix 3. Basic Crystallographic Data of Crystal Structures of Sub- and Superphthalocyanine Complexes[a–d]

CamRef	CN#	Crystal description	Figure	Space group	Axis a / Å	Axis b / Å	Axis c / Å	α / deg	β / deg	γ / deg	Z	V	Reference
PHTLBC	4t	SubpcB-Cl	20:1	Pnma	12.123,?	14.824,?	10.378,?	90	90	90			138
KAJGIL	4t	SubpcB-OH	20:2	C2/c	26.276,5	11.817,2	11.987,4	90	92.42,2	90	8	3718.7,16	141
KAJGOR	4t	[SubpcB-OH][2H_2O]	20:3	P-1	8.714,5	10.745,2	12.873,6	67.83,3	72.39,4	82.43,5	2	1063.7,8	141
KAJQAN	4t	[SubpcB-OAc] [0.5H_2O,EtOH]	20:4	C2/c	28.290,3	11.020,2	15.790,2	90	90.650,12	90	4	4922.3,12	141
KAJQER	4t	[SubpcB-OAc][H_2O,py]	20:8	C2/c	28.271,2	10.894,1	15.759,3	90	90.97,2	90	4	4852.8,11	141
RUGJOS	4t	SubpcB-OEt	20:9	$P2_1/n$	16.577,?	14.483,?	9.051,?	90	102.39,?	90			140
#1998-04	4t	SubpcB–OMe	20:13–4	$P2_1/n$	11.558,3	15.834,2	12.099,2	90	110.15,1	90	4	2078.6,7	137
#1998-05	4t	SubpcB–OMe	20:15–6	Pnma	12.317,3	15.279,2	10.5637,8	90	90	90	4	1988.1,6	137
#1992-01 (App. 41)	4t	SubpcB–OMe	20:17	Pnma	12.339,2	15.290,4	10.567,3	90	90	90	4	1983.5,7	262,263
KAJPEQ	4t	SubpcB–OtBu	20:7	$P2_1/c$	14.382,3	8.286,5	19.838,10	90	102.40,4	90	4	2308.7,18	141
#1998-06	4t	SubpcB–OSi(tBu)Me_2	20:12	$P2_1/a$	12.639,4	12.302,6	18.048,3	90	100.27,2	90	4	2761,2	137
#2002-04	4t	F_{12}SubpcB–Oϕ	20:18	$P2_1/c$	11.6416,12	10.8634,11	21.851,2	90	98.050,2	90	4	2736.2,5	265
#1998-07	4t	SubpcB–ϕ	20:11	Pbca	17.475,9	18.32,1	14.366,9	90	90	90	8	4598,4	137
YOKMUG	4t	SubpcB–ϕ	–	Pbca	14.322,?	17.455,?	18.280,?	90	90	90			139
YOKMUG01	4t	SubpcB–ϕ	20:10	Pbca	14.263,2	17.359,3	18.295,2	90	90	90	8	4529.7,11	141
XAQGEB	4t	MonobenzosubpcB-Cl	20:5–6	$P2_1/n$	20.9556,6	7.8004,2	26.5274,8	90	112.716,1	90			264
#2002-11	4t	[Cis-binucl-$F_{16}H_2$ $(SubpcBCl)_2$][2.5CH_2Cl_2]	20:21	$P2_1/c$	14.4237,13	31.6302,28	10.2000,10	90	101.354,3	90			261
#2002-12	4t	[Trans-binucl-$F_{16}H_2$ $(SubpcBCl)_2$][2CH_2Cl_2]	20:22	P-1	10.8638,11	13.4945,15	17.2990,18	107.608,4	91.300,5	101.347,4			261
CIMINU10	7tr	SuperpcUO_2	20:19–20	$P2_1/c$	8.210,3	21.667,7	18.462,5	90	103.16,2	90	4		260

[a]Abbreviations are compiled in Table 1.
[b]*nc indicates that no coordinates are available or known.
[c]? indicates that error of the data is not known.
[d]4t = tetrahedral.

Appendix 4. Basic Crystallographic Data of Crystal Structures of Phthalocyanine-Related Isoindole Complexes[a–c]

CamRef	CN#	Crystal description	Figure	Space group	Axis *a* / Å	Axis *b* / Å	Axis *c* / Å	*α* / deg	*β* / deg	*γ* / deg	Z	V	Reference
MESJOJ	4	$(MeO)_2$(rpc)Ni	21:1–3	P-1	7.160,2	12.853,3	16.936,3	93.33,2	99.98,2	105.28,4	2	1471.9,6	414
MESJUP	4	$(EtO)_2$(rpc)Ni	21:4–6	P-1	7.950,2	13.080,4	15.395,5	109.12,3	97.99,3	94.14,3	2	1486.0,8	414
QOWHIT	6	PccBi	21:11–12	P-1	12.544,2	13.654,1	14.868,1	73.83,1	78.75,1	68.20,1	2	2258.8,2	258
ZENGUU	6	PccGd	21:8	P-1	11.683,2	12.388,3	15.978,3	75.62,3	86.00,3	65.12,3	2		256
LIFJOZ	6	PccIn	21:10	P-1	11.693,?	12.408,?	15.998,?	75.56,?	85.63,?	64.92,?			255
KEYPAF	7	PccNbCl	21:7	C2/c	22.899,3	10.436,2	18.539,2	90	123.09,3	90			254
ZOLHUD	6	PccTl	21:9	P-1	11.677,2	12.391,3	16.047,3	75.65,3	86.44,3	65.33,3	2		257
NOCYEJ	8	$[Pc_2Nb][I_3][0.5I_2,3.5ClN]$	21:13–14	P-1	17.702,3	18.708,3	13.504,2	106.00,2	93.39,2	83.26,1	2	4267.5,13	231
SEWRER	8	$[Pc_2Ti][1\text{-}ClN]$	21:15–16	C2/c	16.327,3	18.568,4	19.022,4	90	94.50,1	90	4	5748.9,20	214

[a]Abbreviations are compiled in Table 1.
[b]*nc indicates that no coordinates are available or known.
[c]? indicates that error of the data is not known.

Appendix 5. Basic Crystallographic Data of Crystal Structures of Porphyrazin Complexes[a–c]

CamRef	CN#	Crystal description	Figure	Space group	Axis *a* / Å	Axis *b* / Å	Axis *c* / Å	*α* / deg	*β* / deg	*γ* / deg	Z	V	Reference
JEWRAE	4	$[(tBu_2SnS_2)_4PazNi][C_7H_8][1.25CH_2Cl_2]$	22:1–2	C2/c	21.911,3	21.265,2	14.822,3	90	94.44,2	90	4	6885,3	415
JEWRAE10	4	$[(tBu_2SnS_2)_4PazNi][C_7H_8][1.25CH_2Cl_2]$	22:3	C2/c	21.911,3	21.265,2	14.822,3	90	94.44,2	90	4	6885,1	416
VUMZOS	5	$[(MeS)_8PazMg(OH_2)]$	22:4	Pnma	14.279,3	26.403,6	8.257,2	90	90	90	4	3113,1	416
PAGYAX	5	[PazFeCl]	22:5	$P2_1/n$	13.324,4	19.425,6	13.381,4	90	90.09,2	90	4		417
HATDEL	4	$[(dppeNiS_2)_4PazNi][4DMF]$	22:6	$Cmc2_1$	24.630,5	20.607,4	30.675,6	90	90	90	4	15569,5	418
JUXNOF	5	$[(Et)_8N_1PorFeCl]$	22:7	P-1	10.051,2	13.746,3	14.712,3	66.50,3	80.72,3	75.93,3	2	1803.4,6	419
WETGAD	5	$[(cr_{18}S_2)_4PazMg(OH_2)]$	22:8	P1/P-1	9.584,3	17.672,2	19.620,4	84.904,14	85.21,2	89.29,2	2	3298.3,13	420,421
WETJUA	1	$[(dppf)PdS_2NorpcH_2][CH_2Cl_2]$	22:46–47	C2/c	37.916,7	10.807,1	29.588,4	90	109.851,11	90	8	11404,3	43
PITROZ	5	$Et_8PazRhMe$	22:9	$P2_1/c$	9.889,2	22.660,4	14.781,3	90	106.85,2	90	4		422
POGKEB	1	$[(Me_2N)_8\text{-}2secoPazH_2\text{-}2,3\text{-}dione]$	22:10	P1	6.270,4	11.210,7	14.108,9	103.90,2	102.77,2	101.46,2	1	905,?	423
POGKEB01	1	$[(Me_2N)_8\text{-}2secoPazH_2\text{-}2,3\text{-}dione]$	22:10	P1	6.270,4	11.210,7	14.108,9	103.90,2	102.77,2	101.46,2	1	905,?	424
POFNON	1	$[(Bn_2N)_8PazMg(O\text{-}C(OEt)CH_3)$	22:11–12	$P2_1/n$	18.939,4	27.813,6	22.582,5	90	99.56,3	90	4	11730,4	425
POFNUT	4	$[\{(allyl)(Bn)N\}_8PazNi][C_6H_5Cl]$	22:13–14	$P2_1/n$	17.453,2	10.634,2	25.256,3	90	109.11,2	90	2	4429,?	425
WETGAD10	5	$[(cr_{18}S_2)_4PazMg(OH_2)]$	22:8	P-1	9.584,3	17.672,2	19.620,4	84.904,14	85.21,2	89.29,2	2	3298.4,14	421
ZAPNEJ	1	$[(Me_2N)_8PazH_2]_2[C60]$	22:15	$P422_12$	22.349,3	22.349,3	20.928,4	90	90	90	4	10.556,3	426
ZAPNAF	1	$[(Me_2N)_8PazH_2][TCNQ]$	22:16	$P2_1/n$	13.466,2	7.124,2	23.405,5	90	104.48,2	90	2	2174.0,6	426
SURDIS	4	$[(Ag\text{-}contng\text{-}crownS_2)_4Ag_4PazNi][8BF_4]$	22:17	P-1	15.377,8	17.685,6	19.002,7	104.94,3	95.57,4	99.71,4	2	4867,8	427
	4	$[(Ag\text{-}contng\text{-}crownS_2)_4Ag_4PazNi][8BF_4]$	–	cubic	30.200,5	30.200,5	30.200,5	90	90	90			427
TISDOO	5	$[(EtS)_8PazMnCl]$	22:18	$P2_1/c$	15.941,2	18.055,2	15.797,3	90	117.71,1	90	4	4025.4,8	428
TISDUU	5	$[(EtS)_8PazFeCl]$	22:19	$P2_1/c$	15.832,2	18.039,1	15.791,2	90	117.20,1	90	4	4011.0,8	428
TOGPEK	4	$[(EtS)_8PazMn]$	22:20	$P2_1/n$	10.305,3	22.709,6	16.708,3	90	91.4,2	90	4	3908,2	429
WETJUA10	1	$[(dppf)PdS_2NorpcH_2]$	22:46–47	C2/c	37.916,7	10.807,1	29.588,4	90	109.851,1	90	8	11404,3	44

(Continued)

Appendix 5. Continued

CamRef	CN#	Crystal description	Figure	Space group	Axis a / Å	Axis b / Å	Axis c / Å	α / deg	β / deg	γ / deg	Z	V	Reference
TOZCIU	4	[(teeda)PtS_2NorpcNi][0.5$CHCl_3$]	22:48	P1	13.160,12	16.526,10	20.374,11	113.80,4	92.93,6	98.34,6	4	3982,5	44
TOZCOA	4	[tr($(Et_3P)_2PtS_2)_2$PazNi][$C_6H_4Cl_2]_2$	22:40	$P2_1/c$	13.006,4	18.645,8	14.469,6	90	105.86,3	90	2	3375,2	430
ZUQDOE	5	[$(Et)_8$PazFe($OClO_3$)]	22:21	P-1	10.398,4	14.307,9	14.303,6	78.73,4	69.52,4	69.37,4	2	1859,2	431
POWLUI	4	[$(Me_2N)_8$-2secoPazZn-2,3-dione]	22:22	P-1	11.516,1	12.446,1	15.125,1	113.47,1	98.28,1	96.10,1	1	1935.6,2	424
ROPLUD	1	[$(RO_2)_4PazH_2$]	22:23	$P2_12_12$	16.895,5	23.715,4	14.396,2	90	90	90	4	5768,2	432
NULSUI	8bis	$(OETAP)_2Zr$	22:50	$P2_1/n$	15.405,3	15.250,3	25.600,5	90	92.63,2	90	4	6007.6,8	433
NULSOC	8bis	(OETAP)Zr(OEP)	22:51	$P2_1/n$	15.425,3	15.265,3	25.783,5	90	92.43,3	90	4	6065.5,8	433
HOGGAL	4	[$(Pr)_6$((mnt)PdNMe$)_2$PazNi]	22:24	$P2_1/n$	17.435,4	17.982,2	30.577,2	90	104.27,2	90	4	9290.7,1	434
NUGSUD	5	[$(Me_2N)_8$PazMn(Et_2dtc)]	22:25–26	C2/c	22.480,3	20.128,2	19.423,3	90	102.97,1	90	8	8564,2	435
NUGTAK	5	[$(Me_2N)_8$PazMnCl]	22:27	P-1	10.216,6	13.426,8	15.683,14	114.59,5	96.82,6	101.87,5	2	1863,2	435
VACQEV	1	[((4-tBu-$\phi)_2)_3$(Cp_2MoS_2)$PazH_2$] [2.5MeOH]	22:28	P-1	17.732,2	23.126,3	23.633,3	107.61,1	91.41,1	110.09,1	4	8586,2	436
PUNTEX	5	[$(EtS)_8$PazMn(SH)]	22:29	$P2_1/c$	15.876,4	17.932,4	15.642,3	90	117.61,2	90	4	3946.0,1	437
HEXJAV	4	[($cr_{18}N_2Bn_2)_4$PazMg][0.75EtOAc]	22:30	$P2_1/n$	24.419,9	17.824,8	25.512,10	90	99.43,3	90	4	10954,7	438
POWNEU	1	[$(Pr_2)_2${(2-PrO$)_2$Bz$\}_2PazH_2$	22:31	$P2_12_12_1$	9.599,3	20.902,3	22.349,3	90	90	90	4	4484,1	439
JUTQIY	4	(4-tBu-ϕ-$)_6$Seco'PazNi	22:32	P-1	14.923,1	16.310,1	17.626,1	82.070,1	71.520,1	80.380,1	2	3995.0,4	440
HOSYUJ	4	[$(EtS)_8$PazCo]	22:33	$P2_1/n$	10.260,5	22.650,5	16.720,5	90	91.680,5	90	4	3884,2	441
FOPYEO01	4	[$(Me_2N)_8$PazCu$]_2$[C_{60}]173K	22:34	P-1	13.892,2	14.229,2	15.840,2	67.006,8	83.601,8	65.078,8	1	2608.6,6	442
FOPYEO	4	[$(Me_2N)_8$PazCu$]_2$[C_{60}]293K	22:35	P-1	13.957,2	14.3197,13	15.9550,12	67.374,5	83.752,7	65.031,7	1	2662.7,4	442
FOPZEP01	4	[$(Me_2N)_8$PazNi][C_{60}]123K	22:36	$P2_12_12_1$	15.413,1	18.337,1	21.855,1	90	90	90	4	6177.0,6	442
FOPZEP	4	[$(Me_2N)_8$PazNi][C_{60}]293K	22:37	$P2_12_12_1$	15.480,2	18.6037,7	22.0248,6	90	90	90	4	6343.0,8	442
QOTPIY	1	[(($Me_2N)_2)_3$(benzo)$PazH_2$]	22:39	R-3	28.242,4	28.242,4	22.215,3	90	90	120	18	15345,4	443
QOTPOE	4	[cis(($Me_2N)_2)_2$(benzo$)_2$PazMg]	22:41	P-1	10.0138,13	12.074,2	14.3671,9	94.816,10	98.341,9	98.634,13	2	1689.2,4	443
QOTPUK	4	[$(Me_2N)_2$(benzo$)_3$PazMg]	22:49	P-1	10.5857,13	12.827,2	13.1407,10	107.292,14	99.978,11	104.792,11	2	1586.1,4	443
#2001-04	1	[((4-tBu-$\phi)_2)_2$($Cp_2VS_2)_2PazH_2$]	22:38	$P2_1/c$	13.4103,?	12.8078,?	24.4431,?	90	97.4744,?	90	2	4162.58789,4	444
ECOSUK	6tr	[Ac_2pazCo(OH_2)(OC_6H_{13})][acetone]	22:42–43	C2/c	31.460,9	20.602,4	28.041,3	90	115.608,1	90	8	16389,5	239
#2002-05	4	[$(pr_2)_2$($\phi_2$benzo$)_2$PazCo][tol]	22:44	P-1	12.890,2	14.5300,8	16.270,4	69.290,8	69.750,4	84.370,4	2	2673.1,8	445
UDIHEU	1	[((PrO(CO)($CH_2)_3)_2)_2$((iPrO$)_2$ benzo$)_2PazH_2$]	22:45	P-1	16.5650,10	19.3140,10	32.720,4	88.067,7	75.552,8	85.939,6	6	10110,2	446
ICISIW	8bis	[($pr_8Paz)_2$Ce]	22:52–53	Cc	32.094,2	9.617,1	27.840,1	90	112.66,1	90	4	7930.1,8	447
#2002-02	5	[TBCSiOSi$(Et)_3$]	22:55–56	$P2_1/n$	8.9139,5	24.062,3	14.934,2	90	102.499,1	90	4	3127.2,6	445
BUDQOC	6tr	[(4-tBu-$\phi)_8$CzP($OMe)_2$] [2CH_2Cl_2,0.75H_2O]	22:54	$P2_1/c$	15.852,4	23.245,5	25.777,6	90	104.835,4	90	4	9181.5,33	448

[a]Abbreviations are compiled in Table 1.
[b]*nc indicates that no coordinates are available or known.
[c]? indicates that error of the data is not known.

Appendix 6a. Basic Crystallographic Data of Crystal Structures of Phthalocyanine–Porphyrin Complexes[a–c]

CamRef	CN#	Crystal description	Figure	Space group	Axis a / Å	Axis b / Å	Axis c / Å	α / deg	β / deg	γ / deg	Z	V	Reference
PAZRAJ	6tr	THF-(Pc)Fe-C-Fe(tpp)-THF	23:1	$P22_12_1$	14.487,3	20.753,5	25.803,7	90	90	90	4	7758,3	353
YUGKOA	6tr	[H_2O-(Pc)Fe-N-Fe(tpp)-OH_2][I_5][2THF]	23:2	Cc	27.13,1	12.032,5	25.596,41	90	90.69,6	90	4	8298,20	361
#2001-01	6tr	[THF-(Pc)Fe-N-Mn(TPP)-THF][I_5][2THF]	23:3	Cc	27.322,?	12.153,?	25.824,?	90	96.47,?	90			362
SACNOZ	8bis	[(Pc)Ce(T(pMeOP)P)][2$CHCl_3$,pentane]	23:4	$P2_1/c$	16.190,5	32.356,9	16.267,5	90	118.85,2	90	4	7463.7,?	405
JUDPAZ	8bis	[(Pc)Ce(T(pMeOP)P)Ce(Pc)][2tol]	23:5	P-1	14.569,4	15.863,4	11.962,3	111.77,2	96.23,2	100.51,2	1	2467.8,?	406
JUDNUR	8bis	[(TPP)Ce(Pc($OMe)_8$)Ce(TPP)][6CH_2Cl_2]	23:6	$P2_1/n$	20.373,6	16.009.5	18.729,6	90	95.37,2	90	2	6081.7,?	406
TANMUQ	8bis	[(TPP)Ce(Pc)Gd(OEP)][2ClB]	23:7	P-1	17.212,5	19.777,6	15.050,5	91.10,2	104.46,2	89.63,2	2	4959.8,?	407
SACNUF	8bis	[PcCeTPP][$SbCl_6$]	23:8	P-1	15.272,4	16.249,4	13.566,4	92.05,2	94.40,2	73.64,2	2	3220.3,?	405
ZAZQEW	8bis	[(TPP)Gd(Pc)][$SbCl_6$][2CH_2Cl_2,H_2O]	23:9	P-1	18.190,5	16.100,5	14.130,4	104.87,2	104.17,2	107.72,2	2	3567.7,?	408
TUPWEG	8bis	[PcLaTPyP][OH][$CHCl_3$,MeOH]	23:10	P-1	12.894,3	14.143,4	18.352,5	95.83,3	105.15,3	98.95,4	2	3155.4,?	410
ZAZQAS	8bis	[TPPLaPc][2CH_2Cl_2]	23:11	$P2_1/c$	12.938,5	18.674,6	26.114,9	90	90.90,2	90	4	6308.5,?	408
DULBIV	8bis	[(Pc)Nd(Pc)Nd(T(4-OMe)PP)] [1,1,2-trichloroethane]	23:12	P4/n	20.248,6	20.248,6	13.098,6	90	90	90	2		412
OCOBIR	8bis	[2,3-NcEuOEP][4$CHCl_3$]	23:13–14	Pnma	27.9422,8	26.124,1	10.9643,7	90	90	90	4	8003,1	252

[a]Abbreviations are compiled in Table 1.
[b]*nc indicates that no coordinates are available or known.
[c]? indicates that error of the data is not known.

Appendix 6b. Basic Crystallographic Data of Crystal Structures (Added in Proof)[a–d]

CamRef	CN#	Crystal description	Figure	Space group	Axis a / Å	Axis b / Å	Axis c / Å	α / deg	β / deg	γ / deg	Z	V	Reference
	1	$PcH_2\{\beta\}$	8	$P2_1/a$	19.85,?	4.72,?	14.8,?	90	122.25,?	90			34
	4	[TBPNi]$_3$[AsF_6]$_2$[1-ClN]	—	P–1	16.019,2	16.470,3	9.853,5	108.30,2	96.10,3	66.06,1	1	2255.0,?	294,459
	4	[tatbpCu][I_3]$_{0.33}$	—	P4/mcc	13.998,5	13.998,5	6.424,3	90	90	90	2	1259,?	455
	4	[tbpH_2][I_3]$_{0.33}$	—	P4/mcc	14.173,10	14.173,10	6.463,4	90	90	90	2	1298,?	454
	6tr	[PcCrI_2][I_2]	—	P-1	7.851,2	8.402,2	12.668,3	80.32,3	74.06,3	82.33,3	1	788.7,3	466
	6tr	PcSi$(OH)_2$	—	triclinic	7.27,?	13.07,?	6.88,?	102.5,?	104.2,?	97.4,?			125
	6tr	PcSi$(OtBu)_2$	—	$Pn2_1/a$ or Pnma	19.011,?	20.194,?	9.144,?	90	90	90	4		464
	6tr	PcSi$((OCH_2CH_2)_3OMe)_2$	—	P-1	9.01	10.25	13.23	93.9	94.7	105.7	1	1044	465
CCDC169755	6tr	[$F_8((F_3C)_2FC)_8$PcZn$(OCMe_2)_2$	—	$P2_1/c$	15.3731,4	20.8988,5	12.2247,7	90	102.271,1	90	2	3837.8,2	341
	4t	F_{12}SubpcB-Cl	—	$P2_1/c$	11.2997,11	10.6022,11	19.1563,19	90	95.507,7	90	4	2284.4,4	261

[a]Abbreviations are compiled in Table1.
[b]*nc indicates that no coordinates are available or known.
[c]? indicates that error of the data is not known.
[d]4t = tetrahedral.

Appendix 7. Cross-Reference Table 1: Crystal Structures by Element

CamRef	Crystal description	CN#	Figure
Hydrogen			
PHTHCY04[97]	PcH_2 {α}	1	10:16
PHTHCY08[98]	PcH_2 {α}	1	10:17
PHTHCY01[1,57,58]	PcH_2 {β}	1	10:19
#1970-02[40]	PcH_2 {β}	1	10:20
#1971-03[31]	PcH_2 {β, N}	1	10:18
PHTHCY02[102]	PcH_2 {β}	1	10:23
PHTHCY03[96]	PcH_2 {β}	1	10:21
PHTHCY07[29,30]	PcH_2 {β}	1	10:22
PHTHCY05[99]	PcH_2 {X, R}	1	10:24
PHTHCY06[101]	PcH_2 {X, R}	1	10:25
#1997-03[102]	PcH_2 {X, R}	1	10:26
GAKWUK[93]	PcH_2/Cl_4PcH_2 {α}	1	10:27
SOZBUE[42]	$(n\text{-}H_{13}C_6)_8PcH_2$	1	10:10–12
WOKDAB[45]	$(i\text{-}H_{11}C_5)_8PcH_2$	1	10:1–3
#2002-09[238]	$\{(n\text{-}H_{13}C_6)_2C\}_4PcH_2$	1	10:13–15
#2002-08[238]	$(Me)_8(H_{17}C_8O)_8PcH_2$	1	10:7–9
SUKSUM[41]	$(i\text{-}H_{11}C_5O)_8PcH_2$	1	10:4–6
ECOSOE[239]	$(\phi)_8PcH_2$	1	10:29–30
LATLIB[273]	$[PcH_2][I_3]_{1/3}$	1	10:28
CCDC168921[275]	$[PcH_2]_3[(SbF_6)_2][1\text{-}ClN]$	1	10:33
CCDC168923[275]	$[PcH_2]_3[(AsF_6)_2][1\text{-}ClN]$	1	10:31–32
CCDC184514[114]	$[PcH_2][IBr_2]$	1	10:34
CCDC184515[114]	$[PcH_2][IBr_2][Br]$ [1-BrN]	1	10:35
BZAZPN[68]	$MATBPH_2$	1	10:37
WETJUA[43]	$[(dppf)PdS_2NorpcH_2][CH_2Cl_2]$	1	22:46–47
WETJUA10[44]	$[(dppf)PdS_2NorpcH_2][CH_2Cl_2]$	1	22:46–47
#2001-04[444]	$[((4\text{-}tBu\text{-}\phi)_2)_2(Cp_2VS_2)_2PazH_2]$	1	22:38
POGKEB[423]	$[(Me_2N)_8\text{-}2secoPazH_2\text{-}2,3\text{-}dione]$	1	22:10
POGKEB01[424]	$[(Me_2N)_8\text{-}2secoPazH_2\text{-}2,3\text{-}dione]$	1	22:10
POWNEU[439]	$[(Pr_2)_2\{(2\text{-}PrO)_2Bz\}_2PazH_2$	1	22:31
QOTPIY[443]	$[((Me_2N)_2)_3(benzo)PazH_2]$	1	22:39
ROPLUD[432]	$[(RO_2)_4PazH_2]$	1	22:23
UDIHEU[446]	$[((PrO(CO)(CH_2)_3)_2)_2((iPrO)_2benzo)_2PazH_2]$	1	22:45
ZAPNAF[426]	$[(Me_2N)_8PazH_2][TCNQ]$	1	22:16
ZAPNEJ[426]	$[(Me_2N)_8PazH_2]_2[C_{60}]$	1	22:15
VACQEV[436]	$[((4\text{-}tBu\text{-}\phi)_2)_3(Cp_2MoS_2)PazH_2][2.5MeOH]$	1	22:28
Lithium			
DIXNIH02[91]	PcLi	4	11:26
DIXMUS10[290]	$[PcLi][I_3]_{1/3}$	4	11:36
MAKHEL[136]	$[PcLi][Li(OP\phi_3)_4][EtOH,H_2O]$	4	11:37–38
RESYAP[115]	$[PcLi][NBu_4][THF]$	4	11:29–30
RESYET[115]	$[PcLi][F][NBu_4]_2[0.5H_2O]$	4	11:27–28
ZEPKIO[289]	$[PcLi][PNP][CH_2Cl_2]$	4	11:31–32
ZEPKOU[289]	$[PcLi][PNP][H_2O]$	4	11:33–34
MAKJEN[136]	$Pc(Li\text{-}OP\phi_3)_2$	5*	13:6
Sodium			
QASKAW[50]	$Pc(Na\text{-}OP\phi_3)_2$	5*	13:7
Potassium			
BANCOI[316]	$[Pc(K(18cr6))_2][1.5C_6H_6]$	5*	13:1–3
BANCUO[316]	$Pc[K(diglyme)]_2$	5*	13:4–5
BARMAI[317]	$Pc[K(DMF)_2]_2$	5*	13:5
Rubidium, Caesium			
–			
Berrylium			
–			
Magnesium			
XETZIF[60,61]	PcMg {273K}	5*	12:32

(*Continued*)

Appendix 7. Continued

CamRef	Crystal description	CN#	Figure
XETZIF01[59]	PcMg {120K}	5*	12:29
XETZIF02[59]	PcMg {200K}	5*	12:30
XETZIF03[59]	PcMg {260K}	5*	12:31
TAYSIV[172]	[PcMg-Cl][PNP][Me_2CO]	5	12:34
#2002-06[309]	PcMg(OH_2)	5	12:24
#2002-07[309]	[PcMg(OH_2py)]$_2$[py]	5	12:25
MGPCPH[145]	[PcMg-OH_2][2py]	5	12:23
MAKHIP[136]	PcMg-OPϕ_3	5	12:33
MAKHOV[136]	[PcMg-OPϕ_3][2py]	5	12:26
MAKHUB[136]	[PcMg-OPϕ_3][CH_2Cl_2]	5	12:27
MAKJAJ[136]	[PcMg-OPϕ_3][THF]	5	12:28
CCDC166120[374]	[PcMg(py)$_2$]	6tr	12:24
QOPCON[373]	PcMg($HOCH_2CH_2OMe$)$_2$	6tr	15:65
QOPCUT[373]	[PcMg(OH_2)$_2$][NMP]$_2$	6tr	15:66
WETGAD[420,421]	[(cr18S_2)$_4$PazMg(OH_2)]	5	22:8
HEXJAV[438]	[(cr18N_2Bn_2)$_4$PazMg][0.75EtOAc]	4	22:30
QOTPOE[443]	[cis((Me_2N)$_2$)$_2$(benzo)$_2$PazMg]	4	22:41
QOTPUK[443]	[(Me_2N)$_2$(benzo)$_3$PazMg]	4	22:49
WETGAD10[421]	[(cr18S_2)$_4$PazMg(OH_2)]	5	22:8
VUMZOS[416]	[(MeS)$_8$PazMg(OH_2)]	5	22:42–43
POFNON[425]	[(Bn_2N)$_8$PazMg(O-C(OEt)CH_3)]	5	22:11–12
Calcium, Strontium, Barium			
–			
Scandium			
QOCKUO[198]	[Pc_2Sc][NBu_4]	8	18:95–96
Yttrium			
BORQII[192]	[PcY(Cl)(DMF)$_3$][Et_2O]	8*cis*	17:10–11
BORQOO[192]	[PcY(Cl)(-$OSMe_2$)$_4$][Et_2O,2MeOH]	8*cis*	17:12–13
BORQUU[192]	[PcY(Cl)(py)(OPϕ_3)][0.5Et_2O]	8*cis*	16:21
BOSJAU[192]	PcY(L,L′,-Cl-,-Cl-)Ypc	8*cis*	17:16–17
BOSQAB[192]	PcY(-(Cl)$_2$-,-(O(Hpy)$_2$)$_2$-)YPc	8*cis*	17:14–15
JUVJIT[216]	[Pc_2Y][CH_2Cl_2]	8bis	18:123–124
QOCKOI[198]	[Pc_2Y][NBu_4]	8bis	18:125–126
Titanium			
BITSAY[2]	PcTi=O {I}	5	12:53
BITSAY01[2]	PcTi=O {II}	5	12:54
BITSAY02[165]	PcTi=O {Y, R}	5	12:56
BITSAY03[164]	PcTi=O {IV, R}	5	12:55
BITSAY04[3]	PcTi=O {C, R}	5	12:52
DAGXUE[158]	Pc$TiCl_2$	6*cis*	14:19–20
SEWRER[214]	[Pc_2Ti][1-ClN]	8bis	21:15–16
PIBJIT[219]	[Pc_2Ti][I_3]$_{0.66}$	8bis	18:113–114
Zirconium			
NOFNUR[176,403]	PcZrI_2(PN)	7*cis*	16:22–23
CCDC165829[400]	[PcZr(-(OH)$_2$-,-(O,Oala)$_2$-)ZrPc][DMF_2,Et_2O,H_2O]	8*cis*	17:18–19
CCDC165834[400]	[PcZr(O_2CatCl_4)$_2$][NBu_4]$_2$	8*cis*	17:20–21
CCDC165838[400]	[PcZr(Oϕ)$_3$][NBu_4][Et_2O]	7*cis*	16:24–25
CCDC165836[400]	[PcZr(-OC(O)C)$_2$][NEt_4]$_2$[CH_2Cl_2]$_2$	8*cis*	17:22–23
YOKWEA[169]	[Pc$ZrCl_3$][PNP][2CH_2Cl_2]	7*cis*	16:26–27
SOTGAJ[215]	Pc_2Zr	8bis	18:129–130
NOFMEA[409]	[Pc_2Zr][I_2]	8bis	18:131–132
NULSUI[433]	(OETAP)$_2$Zr	8bis	22:50
NULSOC[433]	(OETAP)(OEP)Zr	8bis	22:51
Hafnium			
POQYID[182]	[PcHfF_3][NBu_4]	7*cis*	16:4–5
CCDC165833[400]	[PcHf(Oϕ)$_3$][NBu_4][0.5Et_2O]	7*cis*	16:6–7
Vanadium			
SAXZEW[185]	PcV-I	5	12:59
VPHTHC[150]	PcV=O {II}	5	12:58

(*Continued*)

Appendix 7. Continued

CamRef	Crystal description	CN#	Figure
Niobium			
BEJTIT[152]	$PcNbCl_2$	6*cis*	14:12–13
BEJTIT01[163]	$PcNbCl_2$	6*cis*	14:14
#1990-01[124]	$PcNbCl_3$	7*cis*	16:11–12
NEXVAN[163]	$PcNbCl_3$	7*cis*	16:9–10
#1998-01[308]	$[PcNb(O,O_2Naphth)][CH_2Cl_2]_{0.5}$	7*cis*	16:8
#1986-01[402]	$[PcNb(\text{-O-,-O-,-Br-})NbPc][Br_3][1\text{-}ClN]$	7*cis*	–
SAMDUF[124,162]	$[PcNb(\text{-O-,-O-,-Br-})NbPc][Br_3][1\text{-}ClN]$	7*cis*	16:15–16
SAMFAN[124,162]	$[PcNb(\text{-O-,-O-,-Br-})NbPc][Br_3][PN]$	7*cis*	16:17
CAPQOZ[401]	$[PcNbF_2O][NBu_4]$	7*cis*	16:13–14
#2001-06[411]	$[Pc_2Nb][IBr_2]$	?	–
#2001-07[411]	$[Pc_2Nb][IBr_2][I_2]$	?	–
KEYPAF[254]	PccNbCl	7	21:7
NOCYEJ[231]	$[Pc_2Nb][I_3][0.5I_2,3.5ClN]$	8*bis*	21:13–14
Tantalum			
GIFZUQ[160]	$PcTaCl_3$	7*cis*	16:20
NAPRAX[179]	$[(PcTaF_3)_{1-x}(PcTaF_4)_x][NBu_4]$	7*cis*/8*cis*	16:28–29,17:1
Chromium			
#1999-01[63]	PcCr $\{\beta\}$	4	11:13–14
YEJJIG[305]	PcCr-I	5	12:8
KEHWUB[123]	$PcCrCl_2$	6tr	15:24
#2001-08[342]	[py-(Pc)Cr-O-Cr(Pc)-py][2py]	6tr	15:23
TOPVID[343]	$[PcCr(\text{-ONO})_2][PNP]$	6tr	15:25
Molybdenum			
CUTGIH[153,310]	PcMo=O	5	12:35
FEWPEC[186]	$[PcMo{=}O][I_3]$	5	12:36
RIKSIN[380]	$[PcMoBr_2][PNP]$	6tr	15:79
#1990-03[124]	PcMo(=O,-Cl)	6tr	15:78
Tungsten			
#1990-02[124]	PcW(=O,-Cl)	6tr	15:127
#2001-05[342]	$[PcWCl_2][PNP]$	6tr	15:126
Manganese			
MNPHCY[81]	PcMn $\{\beta$, RT$\}$	4	11:44
MNPHCY02[84]	PcMn $\{\beta$, 295K$\}$	4	11:42
MNPHCY03[86,291]	PcMn $\{\beta$, N, 5.8K$\}$	4	11:43
MNPHCY11[83–85]	PcMn $\{\beta$, 116K$\}$	4	11:41
MNPHCY21[85]	PcMn $\{\beta$, 116K, 1.Refinement$\}$	4	11:39
#1980-01[85]	PcMn $\{\beta$, 116K, 2.Refinement$\}$	4	11:40
XEYQIB[377]	[PcMn(NO)(-ONO)][PNP]	6tr	15:75
XEYQOH[377]	[PcMn(NO)(-NCO)][PNP]	6tr	15:74
XEYQUN[377]	$[PcMn(NO)_2][PNP]$	6tr	15:76
XEYREY[377]	$[PcMn(NO)(OP\phi_3)]$	6tr	15:71
#1996-02[271]	$[PcMn(NCMe)_2][C_4(CN)_6]$	6tr	15:77
PHCPMP10[378,379]	[py-(Pc)Mn-O-Mn(Pc)-py][2py]	6tr	15:72–73
ROPTAR[375]	(dppe)(Cp)Fe-(Pc)Mn-O-Mn(Pc)-NC-Fe(dppe)Cp	6tr	15:68
PILBAN[376]	$[\text{-(Pc)Mn-TCNE-}]_n$ $\{\beta\}$	6tr	15:70
PILBER[376]	$[\text{-(Pc)Mn-NC-C(CN)-CN-}]_n$	6tr	15:69
#2001-01[362]	$[THF\text{-}(Pc)Fe\text{-}N\text{-}Mn(TPP)\text{-}THF][I_5][2THF]$	6tr	23:3
TOGPEK[429]	$[(EtS)_8PazMn]$	4	22:20
TISDOO[428]	$[(EtS)_8PazMnCl]$	5	22:18
PUNTEX[437]	$[(EtS)_8PazMn(SH)]$	5	22:29
NUGSUD[435]	$[(Me_2N)_8PazMn(Et_2dtc)]$	5	22:25–26
NUGTAK[435]	$[(Me_2N)_8PazMnCl]$	5	22:27
Rhenium			
NEKRAW[311]	PcRe-RePc	5	12:41–42
NAPPAV[383]	$PcRe(P\phi_3)_2$	6tr	15:85
DOSWIR[384]	$[PcRe(CN)_2][NBu_4][0.5Me_2CO]$	6tr	15:86
DOSWOX[384]	$[PcRe(CN)_2][PNP]$	6tr	15:87
GEQCOU[385]	$[PcReF_2][0.33(PNP)F][2H_2O]$	6tr	15:88

(Continued)

Appendix 7. Continued

CamRef	Crystal description	CN#	Figure
XEYRAU[377]	[PcRe(NO)$_2$][PNP]	6tr	15:89
XEYRIC[377]	PcRe(NO)(OPϕ_3)	6tr	15:84
XEYROI[377]	[PcRe(NO)(OPϕ_3)][I$_3$][CH$_2$Cl$_2$]	6tr	15:90
Iron			
FEPHCY[81]	PcFe {β}	4	11:21
DAPCEC[157]	[PcFe][I$_3$]$_{1/3}$	4	11:22
FEDXOB[288]	[PcFe][K.cntnd.222]	4	11:24–25
FEDXUH[288]	[PcFe][Li.cntnd.18cr6][Li(THF).cntnd.18c6]	4	11:23
DAPCIG[157]	PcFe-Cl-I-I-Cl-FePc	5	12:12–13
FIQFUG[189]	PcFe-I-I-I-I-FePc	5	12:14–15
GEDKAB[161]	[PcFe-Me][Na.cntnd.18cr6][THF,ϕCl]	5	12:16
KEHXAW[123]	PcFeCl$_2$	6tr	15:38
PTCYFE01[326]	[PcFe(4-mepy)$_2$][2mepy]	6tr	15:28
PHTHFE[346]	[PcFe(DMF,CO)][DMF]	6tr	15:29
SPCNFE[347]	[PcFe(-OSMe$_2$)$_2$][2DMSO]	6tr	15:31
XENLAD[351]	PcFe(4mepip)$_2$	6tr	15:36–37
#2001-09[354]	[PcFe(im)$_2$][PNP]	6tr	15:41–42
DEFNUX[129,357]	[PcFe(CN)$_2$][PNP]	6tr	15:45
DODXUP[358]	[PcFe(CN)$_2$][PNP]$_2$[CH$_2$Cl$_2$]	6tr	15:46
NIRDUN[336]	[PcFe(CN)$_2$][PTMA][MeCN] {123K}	6tr	15:48
NIRDUN01[336]	[PcFe(CN)$_2$][PTMA][MeCN] {295K}	6tr	15:49
WOPZOQ[359]	[PcFe(CN)$_2$][Pϕ_4]$_{0.5}$	6tr	15:47
NELCOW[360]	[PcFe(OMe)$_2$][PNP]$_2$[1Me$_2$CO, 1.5MeOH]	6tr	15:50
WOHTIW[314]	[PcFeCN,NC-Ru(Pϕ_3)$_2$Cp][2CHCl$_3$]	6tr	15:32
WOHTOC[314]	[PcFe(NC-Fe(dppe)Cp)$_2$][SbF$_6$]	6tr	15:52
WOHTUI[314]	[PcFe(NC-Ru(Pϕ_3)$_2$Cp)$_2$][SbF$_6$]	6tr	15:53
WOHVAQ[314]	[PcFe(NC-Ru(Pϕ_3)$_2$Cp)$_2$][1.5CH$_2$Cl$_2$]	6tr	15:30
WOHVEU[314]	[PcFe(CN-Ru(Pϕ_3)$_2$Cp)(NC-Ru(Pϕ_3)$_2$Cp)][SbF$_6$]	6tr	15:51
FUNGEA[348]	[R-(Pc)Fe-C-Fe(Pc)-R][3Me$_2$CO] {+M}	6tr	15:33
RERGEA[352]	py-(Pc)Fe-C-Fe(Pc)-py	6tr	15:39
NAPHIV[355]	[F-(Pc)Fe-C-Fe(Pc)-F][NBu$_4$]$_2$[3H$_2$O]	6tr	15:43
PAZRAJ[353]	THF-(Pc)Fe-C-Fe(tpp)-THF	6tr	23:1
PAZREN[353]	THF-(Pc)Fe-C-Fe(Pc)-THF	6tr	15:40
VAFMIY[350]	Br-(Pc)Fe-N-Fe(Pc)-Br	6tr	15:35
YUGKOA[361]	[H$_2$O-(Pc)Fe-N-Fe(tpp)-OH$_2$][I$_5$][2THF]	6tr	23:2
NEKLUK[356]	[N$_3$-(Pc)Fe-N-Fe(Pc)-N$_3$][PNP]$_2$[I$_2$][2Et$_2$O]	6tr	15:44
KITPIM[349]	[R-(Pc)Fe-O-Fe(Pc)-R][3Me$_2$CO] {+M}	6tr	15:34
VACXOM[249]	1,2-NcFe(NC-C$_6$H$_{11}$)$_2$	6tr	19:9–10
VIWMOD[250]	(tBu)$_4$-1,2-NcFe(NC-tBu)$_2$	6tr	19:11–14
JUXNOF[419]	[(Et)$_8$N$_1$PorFeCl]	5	22:7
PAGYAX[417]	[PazFeCl]	5	22:5
TISDUU[428]	[(EtS)$_8$PazFeCl]	5	22:19
ZUQDOE[431]	[(Et)$_8$PazFe(OClO$_3$)]	5	22:21
Ruthenium			
GEQDIP[389]	[PcRu(py,CN)][NBu$_4$]	6tr	15:95
RERGIE[352]	py-(Pc)Ru-C-Ru(Pc)-py	6tr	15:93
RUJQES[388]	[PcRu(-NO$_2$)$_2$][NBu$_4$]$_2$	6tr	15:94
Osmium			
PCPCOS[381]	PcOs(CO,py)	6tr	15:80
#1996-01[382]	[PcOs(NO,C$_6$H$_{12}$O$_2$)][BF$_4$]	6tr	15:81
#1997-04[131]	[PcOsCl$_2$][PNP][0.33PNPOH]	6tr	15:82
#1997-05[131]	[PcOsCl$_2$][PNP]	6tr	15:83
Cobalt			
COPTCY01[27,28]	PcCo {β, N4.3K}	4	11:4
COPTCY05[94]	PcCo {β, X115K}	4	11:6
COPTCY06[95]	PcCo {β, N115K}	4	11:3
COPTCY07[107]	PcCo {α, R}	4	11:1–2
COPTCY08[107]	PcCo {β, R}	4	11:5
COPTCY10[27,83,84]	PcCo {β, X295K}	4	11:7
DIDTEP[112]	[PcCo][I$_3$]$_{1/3}$	4	11:9

(*Continued*)

Appendix 7. Continued

CamRef	Crystal description	CN#	Figure
VIYFEO[278]	$[PcCo][(AsF_6)_{0.5}]$	4	11:8
CCDC156502[280]	$[PcCo_{0.45}Ni_{0.55}][(AsF_6)_{0.5}]$	4	11:12
CCDC156503[280]	$[PcNi_{0.75}Co_{0.25}][(AsF_6)_{0.5}]$	4	11:11
NOYYIJ[279]	$[PcCo][NBu_4][Me_2CO]$	4	11:10
KEHXEA[123]	$PcCoCl_2$	6tr	15:10
MPCOPC10[326]	$[PcCo(4\text{-mepy})_2][2mepy]$	6tr	15:2
FODLOZ[333]	$[PcCo(py,\text{-}SCN)][CHCl_3]$	6tr	15:7–8
DOVZAP[330]	$PcCo(CN)_2$	6tr	15:9
SILPEI[331,332]	$[PcCo(CN)_2][2H_2O]$	6tr	15:6
ZODWAQ[329]	$[PcCo(CN)_2][2CHCl_3]$	6tr	15:4
YAVPEQ[327,328]	$[PcCo(CN)_2][2CHBr_3]$	6tr	15:3
DOVYUI[330]	$[PcCo(CN)_2][2DMSO]$	6tr	15:5
JAMRAQ[339,340]	$[PcCo(CN)_2][P\phi_4]_{0.5}$	6tr	15:19
NIRDOH[336]	$[PcCo(CN)_2]$[PTMA][MeCN] {295K}	6tr	15:14
NIRPAF[336]	$[PcCo(CN)_2]$[PTMA][MeCN] {113K]	6tr	15:13
SOPQET[338]	$[PcCo(CN)_2]$[p-PPYNN]	6tr	15:16
SOPQIX[338]	$[PcCo(CN)_2]$[p-BPYNN]	6tr	15:17–18
BAPZOH[130,337]	$[PcCo(CN)_2][PXX]_2$	6tr	15:15
NOFGAQ[334]	[NC-(Pc)Co-CN-Co(Pc)-CN][B[c]PT]	6tr	15:11
VENDAT	$[K][PcCo(CN)_2][5CH_3CN]$	6tr	
VENDAT10[331,332,335]	$[K][PcCo(CN)_2][5CH_3CN]$	6tr	15:12
CCDC174676[341]	$[F_8((F_3C)_2FC)_8PcCo(OCMe_2)_2]$	6tr	15:20–21
CCDC174677[341]	$[F_8((F_3C)_2FC)_8PcCo(OC(Me)(CH_2P\phi_3)_2]$	6tr	15:22
#2001-03[135]	[NcCo-py][2py]	6tr	19:5–6
WOHDUH[251]	$[2,3\text{-}NcCo(CN)_2][2DCB]$	6tr	19:7
WOWDOB[251]	$[2,3\text{-}NcCo(CN)_2][CH_3CN]$	6tr	19:8
HOSYUJ[441]	$[(EtS)_8PazCo]$	4	22:33
#2002-05[445]	$[(pr_2)_2(\phi_2benzo)_2PazCo][tol]$	6tr	22:44
ECOSUK[239]	$[Ac_2pazCo(OH_2)(OC_6H_{13})][Me_2CO]$	6tr	22:42–43
Rhodium			
HASZEG[166]	$(H_{11}C_5)_8PcRh\text{-}Me$	5	12:46–48
TESFAY[171]	$(H_{11}C_5)_8PcRh\text{-}H\text{-}RhPc(C_5H_{11})_8$	5	12:44–45
NEXSIS[387]	$[py\text{-}(Pc)Rh\text{-}Rh(Pc)\text{-}py][2C_6H_6]$	6tr	15:91
#1997-02[386]	$[PcRh(O\phi Cl)_2][NBu_4]$	6tr	15:92
PITROZ[422]	$Et_8PazRhMe$	5	22:9
Iridium			
GUDZUA[367]	$[PcIr(Cl,\text{-}CC(O)C)][H_2O]$	6tr	15:59
JAHSOA[370]	$[PcIr\text{-}(acido,N_3)][NBu_4][0.25Et_2O,0.5CH_2Cl_2]$	6tr	15:62
LIQVAI[372]	$[PcIr(OH,OMe)][NBu_4]$	6tr	15:64
RENJID[368]	[PcIr(py, I)][py]	6tr	15:60
RILVOX[371]	$[PcIr(\text{-}NO_2)_2][PNP]$	6tr	15:63
TOCLIG[369]	py-(Pc)Ir-Ir(Pc)-py	6tr	15:61
Nickel			
NIPHTC[13,14,57,58]	PcNi {β}	4	11:57
IBISOB[245]	$(o\text{-}tol)_4PcNi$	4	
#2002-03[244]	$(o\text{-}tol)_4PcNi$	4	11:67–69
ROCCAN,ROCCAN01[241,242]	$(i\text{-}H_{11}C_5O)_4PcNi$	4	11:49–50
TOYSUV[240]	$(H_{13}C_6)_8PcNi$ {103K}	4	11:45–46
TOYSUV01[240]	$(H_{13}C_6)_8PcNi$ {295K}	4	11:47–48
NOCTII[243]	$(Me_3SiO)_8PcNi$	4	11:55–56
NOCVAC[243]	$(RO_2)_2(RO)_2(HO)_2PcNi$	4	11:53–54
NOCVEC[243]	$(RO_2)_2(RO)_2(HO)_2PcNi$	4	11:51–52
QEMTAD[246]	$[(EtO)_8CpPcNi][PF_6]$	4	11:59–61
NIPHTI[111]	$[PcNi][I_3]_{1/3}$	4	11:64
DOXCOI[300]	$[PcNi][Br_3]_{1/3}$	4	11:63
KARDAI[293,295]	$[PcNi][(AsF_6)_{0.5}]$	4	11:58
FANZUP, FANZUP10[292–295]	$[PcNi][(SbF_6)_{0.5}]$	4	11:62
CCDC168918[275]	$[PcNi]_3[(ReO_4)_2][1\text{-}ClN]$	4	11:65
CCDC168922[275]	$[PcNi]_3[(SbF_6)_2][1\text{-}ClN]$	4	11:66
RADPER[104]	2,3-NcNi	4	19:2–3

(*Continued*)

Appendix 7. Continued

CamRef	Crystal description	CN#	Figure
FOPZEP[442]	$[(Me_2N)_8PazNi][C_{60}]293K$	4	22:37
FOPZEP01[442]	$[(Me_2N)_8PazNi][C_{60}]123K$	4	22:36
HATDEL[418]	$[(dppeNiS_2)_4PazNi][4DMF]$	4	22:6
HOGGAL[434]	$[(Pr)_6((mnt)PdNMe)_2PazNi]$	4	22:24
JEWRAE[415]	$[(tBu_2SnS_2)_4PazNi][C_7H_8][1.25CH_2Cl_2]$	4	22:1–2
JEWRAE10[416]	$[(tBu_2SnS_2)_4PazNi][C_7H_8][1.25CH_2Cl_2]$	4	22:3
JUTQIY[440]	$(4\text{-}tBu\text{-}\phi\text{-})_6$Seco'PazNi	4	22:32
POFNUT[425]	$[\{(allyl)(Bn)N\}_8PazNi][\phi Cl]$	4	22:13–14
SURDIS[427]	$[(Ag\text{-}contng\text{-}crownS_2)_4Ag_4PazNi][8BF_4]$	4	11:49
TOZCIU[44]	$[(teeda)PtS_2NorpcNi][0.5CHCl_3]$	4	22:48
TOZCOA[430]	$[tr((Et_3P)_2PtS_2)_2PazNi][C_6H_4Cl_2]_2$	4	22:40
MESJOJ[414]	$(MeO)_2(rpc)Ni$	4	21:1–3
MESJUP[414]	$(EtO)_2(rpc)Ni$	4	21:4–6
Palladium			
–			
Platinum			
PTPOCY02[13,15,57,58]	PcPt	4	11:70
PTPOCY10[77]	PcPt $\{\alpha\}$	4	11:71
PTPOCY11[77]	PcPt $\{\gamma\}$	4	11:72
JECGAZ[294,301]	$[PcPt][(ClO_4)_{0.5}]$	4	11:73
Copper			
CUPOCY10[76,286]	PcCu $\{\beta\}$	4	11:16
#1982-01[88,89,281]	$Cl_{16}PcCu$	4	11:15–6
FEGJOQ[113]	$[PcCu][I_3]_{1/3}$	4	11:17
QEMMIE[306]	$PcCu\text{-}OReO_3$	5	12:9
QEMMOK[306]	$PcCu(ReO_4)_2$	6tr	15:26
QILWAJ[344]	$PcCu(SbF_6)_2$	6tr	15:27
#2002-01[302]	$[(O_3ReO\text{-}CuPc)_2PcCu]$	6tr	12:10–11
CCDC168919[275]	$[PcCu]_3[(SbF_6)_2][1\text{-}ClN]$	4	11:20
CCDC168920[275]	$[PcCu]_3[(AsF_6)_2][1\text{-}ClN]$	4	11:18–19
JECGON[287]	$[tatbpCu]_3[ReO_4]_2[1\text{-}ClN]$	4	
RADPIV[104]	2,3-NcCu	4	19:1
FOPYEO[442]	$[(Me_2N)_8PazCu]_2[C_{60}]293K$	4	22:35
FOPYEO01[442]	$[(Me_2N)_8PazCu]_2[C_{60}]173K$	4	22:34
Silver, Gold			
–			
Zinc			
PTHCZN[82]	PcZn $\{\beta\}$	4	11:76
DAMTEQ[134,315]	PcZn-Cl	5	12:63
MAKHAH[136]	$[PcZn\text{-}OP\phi_3][Et_2O,1\text{-}ClN]$	5	12:62
WOHTES[314]	$[PcZn\text{-}NC\text{-}Fe(dppe)Cp][0.5CH_2Cl_2,THF]$	5	12:61
ZNPTCY[144]	$[PcZn\text{-}H_2N\text{-}C_6H_{13}][0.5H_2N\text{-}C_6H_{13}]$	5	12:60
ZUHCIO[170]	$[PcZn\text{-}ONO][PNP][H_2O]$	5	12:64
CCDC169755[374]	$[F_8((F_3C)_2FC)_8PcZn(OCMe_2)_2]$	6tr	
RADPOB[104]	2,3-NcZn	4	19:4
POWLUI[424]	$[(Me_2N)_8\text{-}2secoPazZn\text{-}2,3\text{-}dione]$	4	22:22
Cadmium			
#1997-01[304]	$[PcCd\text{-}Cl][NEt_4][py]$	5	12:6–7
Mercury			
–			
Boron			
PHTLBC[138]	SubpcB-Cl	4t	20:1
#1992-01[262,263]	SubpcB-OMe	4t	20:17
#1998-04[137]	SubpcB-OMe	4t	20:13–4
#1998-05[137]	SubpcB-OMe	4t	20:15–6
RUGJOS[140]	SubpcB-OEt	4t	20:9
#1998-06[137]	$SubpcB\text{-}OSi(tBu)Me_2$	4t	20:12
KAJGIL[141]	SubpcB-OH	4t	20:2

(*Continued*)

Appendix 7. Continued

CamRef	Crystal description	CN#	Figure
KAJGOR[141]	[SubpcB-OH][$2H_2O$]	4t	20:3
KAJPEQ[141]	SubpcB-OtBu	4t	20:7
KAJQAN[141]	[SubpcB-OAc][$0.5H_2O$,EtOH]	4t	20:4
KAJQER[141]	[SubpcB-OAc][H_2O,py]	4t	20:8
#1998-07[137]	SubpcB-ϕ	4t	20:11
YOKMUG[139]	SubpcB-ϕ	4t	
YOKMUG01[141]	SubpcB-ϕ	4t	20:10
#2002-04[265]	F_{12}-SubpcB-Oϕ	4t	20:18
XAQGEB[264]	SubbpcB-Cl	4t	20:5–6
#2002-11[261]	[Cis-binuclear-$F_{16}H_2$(SubpcBCl)$_2$][$2.5CH_2Cl_2$]	4t	20:21
#2002-12[261]	[Trans-binuclear-$F_{16}H_2$(SubpcBCl)$_2$][$2CH_2Cl_2$]	4t	20:22
Aluminium			
CUWPAL[154]	PcAl-Cl	5	12:2
#1995-01[155,156]	PcAl-Cl	5	12:3
DAYYOR[122]	PcAl-O-AlPc	5	12:4–5
ZUVYOE[324]	[PcAl(-ONO)$_2$][NBu_4]	6tr	15:1
Gallium			
CUWNUD[154]	PcGa-Cl	5	12:17
CUWNUD01[174]	PcGa-Cl {II, R}	5	12:18
NALSOI01[174]	PcGa-OH {XI, R}	5	
NALSOI02[167]	PcGa-OH {V, R}	5	12:19
BAMSEN[121]	[-Ga(Pc)F-]$_n$	6tr	15:54
BAMSEN10[122]	[-Ga(Pc)F-]$_n$	6tr	15:54
Indium			
BIJQOA[187]	PcIn-I	5	12:20
#1998-03[308]	PcIn-Sϕ	5	12:21–22
#1998-02[308]	[PcIn(O_2catCl$_4$)][NBu_4][CH_2Cl_2]	6*cis*	14:10–11
NAPJIX[178]	[PcInF_2][NBu_4][H_2O]	6*cis*	14:4–5
RERGOK[177]	[PcIn(-O,-O'CO)][NBu_4]	6*cis*	14:8–9
ZUFZOP[173]	[PcIn(-NCO)$_2$][PNP]	6*cis*	14:6–7
ZEMBEY[404]	[PcIn(-ONO-)$_2$][NBu_4]	8*cis*	17:2–3
ZAKZOA[224]	Pc_2In	8bis	18:35–36
NOFMAW[409]	[Pc_2In][I_2]	8bis	18:41–42
QOCLUP[198]	[Pc_2In][I][PNP]$_2$[$2Et_2O$]	8bis	18:39–40
PIFDEN[53]	Pc_3In_2	8bis	18:37–38
LIFJOZ[255]	PccIn	6	21:10
Thallium			
NELCIQ[180]	[PcTl-I][CH_2Cl_2]	5	12:57
HEBHOL[49]	$PcTl_2$	5*	13:9
BOFQES[191]	[PcTlCl_2][NBu_4][$0.5Et_2O$]	6*cis*	14:21–22
BOFQAO[191]	[PcTl(-ONO-)$_2$][NBu_4]	8*cis*	17:8–9
ZOLHUD[257]	PccTl	6	21:9
Silicon			
PHPRSI[87]	$PcSi_{0.17}$	4?	11:74–75
VIKVEQ[127]	$PcSiCl_2$	6tr	15:116
CCDC173621[128]	$PcSiCl_2$	6tr	15:117
QEKBEN[126]	PcSi(OH)$_2$ {II}	6tr	15:105
QEKBEN01[126]	PcSi(OH)$_2$ {I}	6tr	15:104
SIMPLC10[395]	PcSi(OSiMe$_3$)$_2$	6tr	15:106
VIKZAQ[127]	[PcSi(OMe)$_2$][$0.48H_2O$]	6tr	15:99
VILGOM[127]	PcSi(thff)$_2$	6tr	15:107
SIOPCY[449]	[PcSiOSiEt$_3$,Me][$0.5C_6H_{12}$]	6tr	
#1971-01[270]	[PcSiOSiEt$_3$,Me][$0.5C_6H_{12}$]	6tr	15:100
CCDC173623[128]	PcSi(O-C(O)-$C_{10}H_{21}$)$_2$	6tr	15:102–103
ACUMEQ[399]	PcSi[O(CO)ϕ-4tBu]$_2$	6tr	15:108
ACUMIU[399]	PcSi[O(CO)-(3-thienyl)]$_2$	6tr	15:109
NEYCEZ[392]	[PcSi(-O-C(O)-Fc)$_2$][$2CHCl_3$]	6tr	15:96
ACUMOA[399]	PcSi[O(CO)(CH$_2$)$_2\phi$(3,4-(OMe)$_2$]$_2$	6tr	15:110
NOLXOB[397]	PcSi[OCH$_2$–C_6H_3(OCH$_2\phi$)$_2$]$_2$	6tr	15:113–114

(*Continued*)

Appendix 7. Continued

CamRef	Crystal description	CN#	Figure
#2001-02[396]	$PcSi[OCH_2–C_6H_3((OCH_2CH_2O)_3Me)_2]_2$	6tr	15:111–112
QUDQOV[390,391]	$PcSi[OSi(C_6H_{13})_3]_2$	6tr	15:118–119
QUDQUB[390,391]	$PcSi[OSiMe_2(C_8H_{17})]_2$	6tr	15:120–121
CAVZAA[398]	$PcSi[Oepiglobulyl]_2$	6tr	15:115
DOHYAA[393]	$Me_2(tBu)SiO$-(Pc)Si-O-Si(Pc)-OSi(tBu)Me_2	6tr	15:101
#1970-01[269]	$(Me_3SiO)_2MeSiO$–((Pc)Si-O)$_3$–SiMe$(OSiMe_3)_2$	6tr	15:97
QUDQIP[390,391]	[-(Pc)Si-(OSi$Me_2\{CH_2)_7SiMe_2O$-)]n	6tr	15:98
#2002-02[450]	[TBCSiOSi(Et)$_3$]	5	22:55–56
Germanium			
#1985-01[120]	$PcGe(OH)_2$ {α}	6tr	15:55
#1985-02[120]	$PcGe(OH)_2$ {β}	6tr	15:56
RESFIE[119]	$PcGeI_2$	6tr	15:57
GODLEQ[366]	$[PcGeI_2][I_2]$	6tr	15:58
Tin			
#1971-02[31]	PcSn	5	12:49
SNPHCY01[143]	PcSn	5	12:50–51
#1972-01[117]	$PcSnCl_2$	6tr	15:122–123
RESFOK[119]	$PcSnI_2$	6tr	15:124
RESFOK01[118]	$PcSnI_2$	6tr	15:125
CCDC173622[128]	PcSn[O-C(O)-$C_{10}H_{21}]_2$	8*cis*	17:6–7
SNPTCY[201]	Pc_2Sn	8bis	18:103–104
SNPTCY01[223]	Pc_2Sn	8bis	18:105–106
Lead			
PBPHCY[133]	PcPb {monoclinic}	5	12:37–38
PBPHCY01[132,148]	PcPb {triclinic}	5	12:39–40
Phosphorus			
BUDQOC[448]	[(4-tBu-$\phi)_8CzP(OMe)_2][2CH_2Cl_2$,0.75$H_2O$]	6tr	22:54
Arsenic			
CCDC180524[303]	$[PcAs][I_3]$	5	12:1
BECTOS[233]	$[Pc_2As][I_2]$	8bis	18:1–2
Antimony			
HOVCOK[188]	$[PcSb][I_3]$	5	12:43
NILVUZ[183]	PcSb(-I-)$_2$SbPc	6*cis*	14:15–16
WOVREE[193]	PcSb(-Cl-)$_2$SbPc	6*cis*	14:17–18
YOHMAJ[194]	$[PcSb]_4[Sb_4I_{16}]$	7*cis*	16:18–19
QOCLEZ[198]	$[Pc_2Sb][NBu_4]$[2THF]	8bis	18:91–92
QOCLID[198]	$[Pc_2Sb][Br][PNP]_2[2Et_2O]$	8bis	18:93–94
Bismuth			
HOXCIG[184]	PcBi(-I-)$_2$BiPc	6*cis*	14:1–3
LOKSAF[190]	$(PcBi)_4(Bi_4I_{16})$	7*cis*	16:1–3
ZUWDAW[228]	$[Pc_2Bi][CH_2Cl_2]$	8bis	18:3–4
KOJQUV[232]	$[Pc_2Bi][I_{1.5}]$	8bis	18:13–14
QOCLOJ[198]	$[Pc_2Bi][Br][PNP]_2[2Et_2O]$	8bis	18:7–8
QOCJEX[198]	$[Pc_2Bi][NBu_4]$	8bis	18:9–10
QOCKAU[198]	$[Pc_2Bi][NBu_4]$	8bis	18:11–12
DIQPIC[51]	Pc_3Bi_2	8bis	18:5
DIQPIC01[52]	Pc_3Bi_2	8bis	18:6
QOWHIT[258]	PccBi	6	21:11–12
Selenium, Tellurium			
–			
Lanthanum			
YEGTIN[220]	$[Pc_2La]$[PNP]	8bis	18:49–50
QOCJOH[198]	$[Pc_2La][NBu_4]$	8bis	18:45–46
ZEHTUB[226]	$[Pc_2La][CH_2Cl_2]$	8bis	18:43–44
QOCJUN[198]	$[Pc_2La][NPe_4]$	8bis	18:47–48
QOCMAW[198]	$[Pc_2La][NBu_4]$[EtOH]	8bis	
TUPWEG[410]	[PcLaTPyP][OH][$CHCl_3$,MeOH]	8bis	23:10
ZAZQAS[408]	[TPPLaPc][$2CH_2Cl_2$]	8bis	23:11

(Continued)

Appendix 7. Continued

CamRef	Crystal description	CN#	Figure
Cerium			
JONZAN[217]	Pc_2Ce	8bis	18:15–16
RABDEZ[230]	$[Pc_2Ce][BF_4]_{0.33}$	8bis	18:21–22
#2002-10[221]	$[Pc_2Ce]$[PNP]	8bis	18:19–20
QOCJIB[198]	$[Pc_2Ce][NBu_4]$	8bis	18:17–18
JUDNUR[406]	$[(TPP)Ce(Pc(OMe)_8)Ce(TPP)][6CH_2Cl_2]$	8bis	23:6
JUDPAZ[406]	[(Pc)Ce(T(pMeOP)P)Ce(Pc)][2tol]	8bis	23:5
SACNOZ[405]	$[(Pc)Ce(T(pMeOP)P)][2CHCl_3$,pentane]	8bis	23:4
SACNUF[405]	$[PcCeTPP][SbCl_6]$	8bis	23:8
TANMUQ[407]	[(TPP)Ce(Pc)Gd(OEP)][2ClB]	8bis	23:7
ICISIW[447]	$[(pr_8Paz)_2Ce]$	8bis	22:52–53
Praseodymium			
KOBRUO[211]	Pc_2Pr {β1}	8bis	18:83–84
KOBRUO02[222]	Pc_2Pr	8bis	18:81–82
WAYSIY[218]	$[Pc_2Pr][Br_{1.5}]$ {α}	8bis	18:89–90
QOCKEY[198]	$[Pc_2Pr][NBu_4]$[0.75MeOH]	8bis	18:85–86
QOCLAV[198]	$[Pc_2Pr][NBu_4]$[2py]	8bis	18:87–88
Neodymium			
CIZGIB02[196]	Pc_2Nd {β}	8bis	18:69–70
CIZGIB06[210]	Pc_2Nd (ox1)	8bis	18:73–74
CIZGIB07[209]	Pc_2Nd {α}	8bis	18:75–76
CIZGIB08[222]	Pc_2Nd	8bis	18:71–72
PTCNND[205,206]	$[Pc_2Nd][CH_2Cl_2]$	8bis	18:67–68
SAFPUK[213]	$[Pc_2Nd]$[O]	8bis	18:79–80
NAKXUS[197]	$[Pc_2Nd][NBu_4]$	8bis	18:77–78
DULBIV[412]	[(Pc)Nd(Pc)Nd(T(4-OMe)PP)][1,1,2-trichloroethane]	8bis	23:12
Samarium			
CALWIV[318,319]	$Pc(SmL_2)_2$	5*	13:8
ZOMWUT[227]	$[Pc_2Sm][Br_{1.45}]$ {α}	8bis	18:99–100
ZOMXAA[227]	$[Pc_2Sm][ClO_4]_{0.63}$ {α}	8bis	18:101–102
QOCKIC[198]	$[Pc_2Sm][NBu_4]$[MeOH]	8bis	18:97–98
Europium			
OCOBIR[252]	$[2,3\text{-}NcEuOEP][4CHCl_3]$	8bis	23:13–14
Gadolinium			
GAWBEL[212]	Pc_2Gd	8bis	18:31–32
NAKYAZ[197]	$[Pc_2Gd][NBu_4]$	8bis	18:25–26
OBUJIE[236]	$[Pc_2Gd][C_4(CN)_6]$[2ClB]	8bis	18:29–30
YEGTOT[220]	$[Pc_2Gd]$[PNP]$[0.33H_2O]$	8bis	18:27–28
TANMUQ[407]	[(TPP)Ce(Pc)Gd(OEP)][2ClB]	8bis	23:7
ZAZQEW[408]	$[(TPP)Gd(Pc)][SbCl_6][2CH_2Cl_2,H_2O]$	8bis	23:9
ZENGUU[256]	PccGd	6	21:8
Terbium, Dysprosium			
–			
Holmium			
NAKYED[197]	$[Pc_2Ho][NBu_4][H_2O]$	8bis	18:33–34
Erbium			
YUHSUP[225]	Pc_2Er {α}	8bis	18:23–24
Thulium			
YEGTUZ[220]	$[Pc_2Tm]$[PNP]$[0.5H_2O]$	8bis	18:115–116
Ytterbium			
BECSUX[233]	$[Pc_2Yb][I_2]$	8bis	18:127–128
Lutetium			
DICCAT[159]	$PcLu(OAc,H_2O,H_2O)$	8*cis*	17:4–5
DULZAL[208]	Pc_2Lu {γ}	8bis	18:59–60
GAWBAH[212]	Pc_2LuH	8bis	18:61–62
DICBUM[159]	$[Pc_2Lu][CH_2Cl_2]$	8bis	18:51–52
GAVZUY[212]	$[Pc_2Lu][NBu_4]$[2DMF]	8bis	18:63–64

(*Continued*)

Appendix 7. Continued

CamRef	Crystal description	CN#	Figure
NAKYIH[197]	$[Pc_2Lu][NBu_4][H_2O]$	8bis	18:65–66
VODDIB[247]	$[(H_{13}C_6S)_8Pc]_2Lu$	8bis	18:56–58
LORKIM[195]	$[(15cr5)_4Pc]_3Lu_2][18DMSO,2CHCl_3]$	8bis	18:53–55
Thorium			
DPCYTH[203]	Pc_2Th	8bis	18:107–108
DPCYTH01[204]	Pc_2Th	8bis	18:109–110
DPCYTH03[207]	Pc_2Th	8bis	18:111–112
Uranium			
PHALCU01[202,203]	Pc_2U	8bis	18:117–118
HOTWOC[234]	$[Pc_2U][I_{5/3}]$	8bis	18:121–122
MEGNOB[235]	$[Pc_2U][I_3]_{2/3}$	8bis	18:119–120

R: Rietveld analysis
N: Neutron diffraction

Appendix 8. Cross-Reference Table 2: #YYYY-xx and CCDCxxxxxx Structures

CamRef	Crystal description	Figure
#1970-01[269]	$(Me_3SiO)_2MeSiO\text{-}((Pc)Si\text{-}O)_3\text{-}SiMe(OSiMe_3)_2$	15:97
#1970-02[40]	PcH_2 {β}	10:20
#1971-01[270]	$[PcSiOSiEt_3,Me][0.5C_6H_{12}]$	15:100
#1971-02[31]	PcSn	12:49
#1971-03[31]	PcH_2 {β, N}	10:18
#1972-01[117]	$PcSnCl_2$	15:122–123
#1980-01[85]	PcMn {β, 116K, 2.Refinement}	11:40
#1982-01[88,89,281]	$Cl_{16}PcCu$	11:15–6
#1985-01[120]	$PcGe(OH)_2$ {α}	15:55
#1985-02[120]	$PcGe(OH)_2$ {β}	15:56
#1986-01[402]	$[PcNb(\text{-O-,-O-,-Br-})NbPc][Br_3][1\text{-}ClN]$	–
#1990-01[124]	$PcNbCl_3$	16:11–12
#1990-02[124]	PcW(=O,-Cl)	15:127
#1990-03[124]	PcMo(=O,-Cl)	15:78
#1992-01[262,263]	SubpcB-OMe	20:17
#1995-01[155,156]	PcAl-Cl	12:3
#1996-01[382]	$[PcOs(NO,C_6H_{12}O_2)][BF_4]$	15:81
#1996-02[271]	$[PcMn(NCMe)_2][C_4(CN)_6]$	15:77
#1997-01[304]	$[PcCd\text{-}Cl][NEt_4][py]$	12:6–7
#1997-02[386]	$[PcRh(O\phi Cl)_2][NBu_4]$	15:92
#1997-03[102]	PcH_2 {X, R}	10:26
#1997-04[131]	$[PcOsCl_2][PNP][0.33PNPOH]$	15:82
#1997-05[131]	$[PcOsCl_2][PNP]$	15:83
#1998-01[308]	$[PcNb(O,O_2Naphth)][CH_2Cl_2]_{0.5}$	16:8
#1998-02[308]	$[PcIn(O_2catCl_4)][NBu_4][CH_2Cl_2]$	14:10–11
#1998-03[308]	PcIn-Sϕ	12:21–22
#1998-04[137]	SubpcB-OMe	20:13–4
#1998-05[137]	SubpcB-OMe	20:15–6
#1998-06[137]	$SubpcB\text{-}OSi(tBu)Me_2$	20:12
#1998-07[137]	SubpcB-ϕ	20:11
#1999-01[63]	PcCr {β}	11:13–14
#2001-01[362]	$[THF\text{-}(Pc)Fe\text{-}N\text{-}Mn(TPP)\text{-}THF][I_5][2THF]$	23:3
#2001-02[396]	$PcSi[OCH_2\text{-}C_6H_3((OCH_2CH_2O)_3Me)_2]_2$	15:111–112
#2001-03[135]	[NcCo-py][2py]	19:5–6
#2001-04[444]	$[((4\text{-}tBu\text{-}\phi)_2)_2(Cp_2VS_2)_2PazH_2]$	22:38
#2001-05[342]	$[PcWCl_2][PNP]$	15:126
#2001-06[411]	$[Pc_2Nb][IBr_2]$	–
#2001-07[411]	$[Pc_2Nb][IBr_2][I_2]$	–
#2001-08[342]	$[py\text{-}(Pc)Cr\text{-}O\text{-}Cr(Pc)\text{-}py][py]_2$	15:23
#2001-09[354]	$[PcFe(im)_2][PNP]$	15:41–42
#2002-01[302]	$[(O_3ReO\text{-}CuPc)_2PcCu]$	12:10–11

Appendix 8. Continued

CamRef	Crystal description	Figure
		(Continued)
#2002-02[450]	[TBCSiOSi(Et)$_3$]	22:55–56
#2002-03[244]	(o-tol)$_4$PcNi	11:67–69
#2002-04[265]	F$_{12}$-SubpcB-Oϕ	20:18
#2002-05[445]	[(pr$_2$)$_2$($\phi_2$benzo)$_2$PazCo][tol]	22:44
#2002-06[309]	PcMg(OH$_2$)	12:24
#2002-07[309]	[PcMg(OH$_2$py)]$_2$[py]	12:25
#2002-08[238]	(Me)$_8$(H$_{17}$C$_8$O)$_8$PcH$_2$	10:7–9
#2002-09[238]	{(n-H$_{13}$C$_6$)$_2$C}$_4$PcH$_2$	10:13–15
#2002-10[221]	[Pc$_2$Ce][PNP]	18:19–20
#2002-11[261]	[Cis-binuclear-F$_{16}$H$_2$(SubpcBCl)$_2$][2.5CH$_2$Cl$_2$]	20:21
#2002-12[261]	[Trans-binuclear-F$_{16}$H$_2$(SubpcBCl)$_2$][2CH$_2$Cl$_2$]	20:22
CCDC156502[280]	[PcCo$_{0.45}$Ni$_{0.55}$][(AsF$_6$)$_{0.5}$]	11:12
CCDC156503[280]	[PcNi$_{0.75}$Co$_{0.25}$][(AsF$_6$)$_{0.5}$]	11:11
CCDC165829[400]	[PcZr(-(OH)$_2$-,-(O,Oala)$_2$-)ZrPc][DMF2,Et$_2$O,H$_2$O]	17:18–19
CCDC165833[400]	[PcHf(Oϕ)$_3$][NBu$_4$][0.5Et$_2$O]	16:6–7
CCDC165834[400]	[PcZr(O$_2$CatCl$_4$)$_2$][NBu$_4$]$_2$	17:20–21
CCDC165836[400]	[PcZr(-OC(O)C)$_2$][NEt$_4$]$_2$[CH$_2$Cl$_2$]$_2$	17:22–23
CCDC165838[400]	[PcZr(Oϕ)$_3$][NBu$_4$][Et$_2$O]	16:24–25
CCDC166120[374]	[PcMg(py)$_2$]	12:24
CCDC168918[275]	[PcNi]$_3$[(ReO$_4$)$_2$][1-ClN]	11:65
CCDC168919[275]	[PcCu]$_3$[(SbF$_6$)$_2$][1-ClN]	11:20
CCDC168920[275]	[PcCu]$_3$[(AsF$_6$)$_2$][1-ClN]	11:18–19
CCDC168921[275]	[PcH$_2$]$_3$[(SbF$_6$)$_2$][1-ClN]	10:33
CCDC168922[275]	[PcNi]$_3$[(SbF$_6$)$_2$][1-ClN]	11:66
CCDC168923[275]	[PcH$_2$]$_3$[(AsF$_6$)$_2$][1-ClN]	10:31–32
CCDC169755[451]	[F$_2$((F$_3$C)$_2$FC)$_2$PcZn(OCMe$_2$)$_2$]	15:128
CCDC173621[128]	PcSiCl$_2$	15:117
CCDC173622[128]	PcSn[O-C(O)-C$_{10}$H$_{21}$]$_2$	17:6–7
CCDC173623[128]	PcSi(O-C(O)-C$_{10}$H$_{21}$)$_2$	15:102–103
CCDC174676[341]	[F$_2$((F$_3$C)$_2$FC)$_2$PcCo(OCMe$_2$)$_2$]	15:20–21
CCDC174677[341]	[F$_2$((F$_3$C)$_2$FC)$_2$PcCo(OC(Me)(CH$_2$Pϕ_3)$_2$]	15:22
CCDC180524[303]	[PcAs][I$_3$]	12:1
CCDC184514[114]	[PcH$_2$][IBr$_2$]	10:34
CCDC184515[114]	[PcH$_2$][IBr$_2$][Br] [1-BrN]	10:35

R: Rietveld analysis
N: Neutron diffraction

Appendix 9. Cross-Reference Table 3: Crystal Structures by Cambridge Code

CamRef	Crystal description	Figure
ACUMEQ[399]	PcSi[O(CO)ϕ-4tBu]$_2$	15:108
ACUMIU[399]	PcSi[O(CO)-(3-thienyl)]$_2$	15:109
ACUMOA[399]	PcSi[O(CO)(CH$_2$)$_2\phi$(3,4-(OMe)$_2$]$_2$	15:110
BAMSEN[121]	[-Ga(Pc)F-]$_n$	15:54
BAMSEN10[122]	[-Ga(Pc)F-]$_n$	15:54
BANCOI[316]	[Pc(K(18cr6))$_2$][1.5C$_6$H$_6$]	13:1–3
BANCUO[316]	Pc[K(diglyme)]$_2$	13:4–5
BAPZOH[130,337]	[PcCo(CN)$_2$][PXX]$_2$	15:15
BARMAI[317]	Pc[K(DMF)$_2$]$_2$	13:5
BECSUX[233]	[Pc$_2$Yb][I$_2$]	18:127–128
BECTOS[233]	[Pc$_2$As][I$_2$]	18:1–2
BEJTIT[152]	PcNbCl$_2$	14:12–13
BEJTIT01[163]	PcNbCl$_2$	14:14
BIJQOA[187]	PcIn-I	12:20
BITSAY[2]	PcTi=O {I}	12:53
BITSAY01[2]	PcTi=O {II}	12:54
BITSAY02[165]	PcTi=O {Y, R}	12:56
BITSAY03[164]	PcTi=O {IV, R}	12:55

Appendix 9. Continued

CamRef	Crystal description	Figure
		(Continued)
BITSAY04[3]	PcTi=O {C, R}	12:52
BOFQAO[191]	[PcTl(-ONO-)$_2$][NBu_4]	17:8–9
BOFQES[191]	[$PcTlCl_2$][NBu_4][0.5Et_2O]	14:21–22
BORQII[192]	[PcY(Cl)($DMF)_3$][Et_2O]	17:10–11
BORQOO[192]	[PcY(Cl)(-$OSMe_2)_4$][Et_2O,2MeOH]	17:12–13
BORQUU[192]	[PcY(Cl)(py)($OP\phi_3$)][0.5Et_2O]	16:21
BOSJAU[192]	PcY(L,L′,-Cl-,-Cl-)Ypc	17:16–17
BOSQAB[192]	PcY(-$(Cl)_2$-,-$(O(Hpy)_2)_2$-)YPc	17:14–15
BUDQOC[448]	[(4-tBu-$\phi)_8$CzP$(OMe)_2$][2CH_2Cl_2,0.75H_2O]	22:54
BZAZPN[68]	$MATBPH_2$	10:37
CALWIV[318,319]	Pc$(SmL_2)_2$	13:8
CAPQOZ[401]	[$PcNbF_2O$][NBu_4]	16:13–14
CAVZAA[398]	PcSi[Oepiglobulyl]$_2$	15:115
CIZGIB02[196]	Pc_2Nd {β}	18:69–70
CIZGIB06[210]	Pc_2Nd (ox1)	18:73–74
CIZGIB07[209]	Pc_2Nd {α}	18:75–76
CIZGIB08[222]	Pc_2Nd	18:71–72
COPTCY01[27,28]	PcCo {β, N4.3K}	11:4
COPTCY05[94]	PcCo {β, X115K}	11:6
COPTCY06[95]	PcCo {β, N115K}	11:3
COPTCY07[107]	PcCo {α, R}	11:1–2
COPTCY08[107]	PcCo {β, R}	11:5
COPTCY10[27,83,84]	PcCo {β, X295K}	11:7
CUPOCY10[76,286]	PcCu {β}	11:16
CUTGIH[153,310]	PcMo=O	12:35
CUWNUD[154]	PcGa-Cl	12:17
CUWNUD01[174]	PcGa-Cl {II, R}	12:18
CUWPAL[154]	PcAl-Cl	12:2
DAGXUE[158]	$PcTiCl_2$	14:19–20
DAMTEQ[134,315]	PcZn-Cl	12:63
DAPCEC[157]	[PcFe][I_3]$_{1/3}$	11:22
DAPCIG[157]	PcFe-Cl-I-I-Cl-FePc	12:12–13
DAYYOR[122]	PcAl-O-AlPc	12:4–5
DEFNUX[129,357]	[PcFe$(CN)_2$][PNP]	15:45
DICBUM[159]	[Pc_2Lu][CH_2Cl_2]	18:51–52
DICCAT[159]	PcLu(OAc,H_2O,H_2O)	17:4–5
DIDTEP[112]	[PcCo][I_3]$_{1/3}$	11:9
DIQPIC[51]	Pc_3Bi_2	18:5
DIQPIC01[52]	Pc_3Bi_2	18:6
DIXMUS10[290]	[PcLi][I_3]$_{1/3}$	11:36
DIXNIH02[91]	PcLi	11:26
DODXUP[358]	[PcFe$(CN)_2$][PNP]$_2$[CH_2Cl_2]	15:46
DOHYAA[393]	Me_2(tBu)SiO-(Pc)Si-O-Si(Pc)-OSi(tBu)Me_2	15:101
DOSWIR[384]	[PcRe$(CN)_2$][NBu_4][0.5Me_2CO]	15:86
DOSWOX[384]	[PcRe$(CN)_2$][PNP]	15:87
DOVYUI[330]	[PcCo$(CN)_2$][2DMSO]	15:5
DOVZAP[330]	PcCo$(CN)_2$	15:9
DOXCOI[300]	[PcNi][Br_3]$_{1/3}$	11:63
DPCYTH[203]	Pc_2Th	18:107–108
DPCYTH01[204]	Pc_2Th	18:109–110
DPCYTH03[207]	Pc_2Th	18:111–112
DULBIV[412]	[(Pc)Nd(Pc)Nd(T(4-OMe)PP)][1,1,2-trichloroethane]	23:12
DULZAL[208]	Pc_2Lu {γ}	18:59–60
ECOSOE[239]	$(\phi)_8PcH_2$	10:29–30
ECOSUK[239]	[Ac_2pazCo(OH_2)(OC_6H_{13})][Me_2CO]	22:42–43
FANZUP,FANZUP10[292–295]	[PcNi][$(SbF_6)_{0.5}$]	11:62
FEDXOB[288]	[PcFe][K.cntnd.222]	11:24–25
FEDXUH[288]	[PcFe][Li.cntnd.18cr6][Li(THF).cntnd.18c6]	11:23
FEGJOQ[113]	[PcCu][I_3]$_{1/3}$	11:17
FEPHCY[81]	PcFe {β}	11:21
FEWPEC[186]	[PcMo=O][I_3]	12:36

Appendix 9. Continued

CamRef	Crystal description	Figure
		(Continued)
FIQFUG[189]	PcFe-I-I-I-I-FePc	12:14–15
FODLOZ[333]	[PcCo(py,-SCN)][$CHCl_3$]	15:7–8
FOPYEO[442]	[$(Me_2N)_8$PazCu]$_2$[C_{60}]293K	22:35
FOPYEO01[442]	[$(Me_2N)_8$PazCu]$_2$[C_{60}]173K	22:34
FOPZEP[442]	[$(Me_2N)_8$PazNi][C_{60}]293K	22:37
FOPZEP01[442]	[$(Me_2N)_8$PazNi][C_{60}]123K	22:36
FUNGEA[348]	[R-(Pc)Fe-C-Fe(Pc)-R][3Me_2CO] {+M}	15:33
GAKWUK[93]	PcH_2/Cl_4PcH_2 {α}	10:27
GAVZUY[212]	[Pc_2Lu][NBu_4][2DMF]	18:63–64
GAWBAH[212]	Pc_2LuH	18:61–62
GAWBEL[212]	Pc_2Gd	18:31–32
GEDKAB[161]	[PcFe-Me][Na.cntnd.18cr6][THF,ϕCl]	12:16
GEQCOU[385]	[PcRe(F)$_2$][0.33(PNP)F][2H_2O]	15:88
GEQDIP[389]	[PcRu(py,CN)][NBu_4]	15:95
GIFZUQ[160]	$PcTaCl_3$	16:20
GODLEQ[366]	[$PcGeI_2$][I_2]	15:58
GUDZUA[367]	[PcIr(Cl,-CC(O)C)][H_2O]	15:59
HASZEG[166]	$(H_{11}C_5)_8$PcRh-Me	12:46–48
HATDEL[418]	[$(dppeNiS_2)_4$PazNi][4DMF]	22:6
HEBHOL[49]	$PcTl_2$	13:9
HEXJAV[438]	[$(cr18N_2Bn_2)_4$PazMg][0.75EtOAc]	22:30
HOGGAL[434]	[$(Pr)_6$((mnt)PdNMe)$_2$PazNi]	22:24
HOSYUJ[441]	[$(EtS)_8$PazCo]	22:33
HOTWOC[234]	[Pc_2U][$I_{5/3}$]	18:121–122
HOVCOK[188]	[PcSb][I_3]	12:43
HOXCIG[184]	PcBi(-I-)$_2$BiPc	14:1–3
IBISOB[245]	$(o\text{-}tol)_4$PcNi	
ICISIW[447]	[$(pr_8Paz)_2$Ce]	22:52–53
JAHSOA[370]	[PcIr-(acido,N_3)][NBu_4][0.25Et_2O,0.5CH_2Cl_2]	15:62
JAMRAQ[339,340]	[$PcCo(CN)_2$][$P\phi_4$]$_{0.5}$	15:19
JECGAZ[294,301]	[PcPt][$(ClO_4)_{0.5}$]	11:73
JECGON[287]	[tatbpCu]$_3$[ReO_4]$_2$[1-ClN]	
JEWRAE[415]	[$(tBu_2SnS_2)_4$PazNi][C_7H_8][1.25CH_2Cl_2]	22:1–2
JEWRAE10[416]	[$(tBu_2SnS_2)_4$PazNi][C_7H_8][1.25CH_2Cl_2]	22:3
JONZAN[217]	Pc_2Ce	18:15–16
JUDNUR[406]	[(TPP)Ce(Pc(OMe)$_8$)Ce(TPP)][6CH_2Cl_2]	23:6
JUDPAZ[406]	[(Pc)Ce(T(pMeOP)P)Ce(Pc)][2tol]	23:5
JUTQIY[440]	(4-tBu-ϕ-)$_6$Seco'PazNi	22:32
JUVJIT[216]	[Pc_2Y][CH_2Cl_2]	18:123–124
JUXNOF[419]	[$(Et)_8N_1$PorFeCl]	22:7
KAJGIL[141]	SubpcB-OH	20:2
KAJGOR[141]	[SubpcB-OH][2H_2O]	20:3
KAJPEQ[141]	SubpcB-OtBu	20:7
KAJQAN[141]	[SubpcB-OAc][0.5H_2O,EtOH]	20:4
KAJQER[141]	[SubpcB-OAc][H_2O,py]	20:8
KARDAI[293,295]	[PcNi][$(AsF_6)_{0.5}$]	11:58
KEHWUB[123]	$PcCrCl_2$	15:24
KEHXAW[123]	$PcFeCl_2$	15:38
KEHXEA[123]	$PcCoCl_2$	15:10
KEYPAF[254]	PccNbCl	21:7
KITPIM[349]	[R-(Pc)Fe-O-Fe(Pc)-R][3Me_2CO] {+M}	15:34
KOBRUO[211]	Pc_2Pr {β1}	18:83–84
KOBRUO02[222]	Pc_2Pr	18:81–82
KOJQUV[232]	[Pc_2Bi][$I_{1.5}$]	18:13–14
LATLIB[273]	[PcH_2][I_3]$_{1/3}$	10:28
LIFJOZ[255]	PccIn	21:10
LIQVAI[372]	[PcIr(OH,OMe)][NBu_4]	15:64
LOKSAF[190]	$(PcBi)_4(Bi_4I_{16})$	16:1–3
LORKIM[195]	[$(15cr5)_4$Pc]$_3$Lu$_2$][18DMSO,2$CHCl_3$]	18:53–55
MAKHAH[136]	[PcZn-OPϕ_3][Et_2O,1-ClN]	12:62
MAKHEL[136]	[PcLi][Li(OPϕ_3)$_4$][EtOH,H_2O]	11:37–38

Appendix 9. Continued

CamRef	Crystal description	Figure
		(Continued)
MAKHIP[136]	PcMg-OPϕ_3	12:33
MAKHOV[136]	[PcMg-OPϕ_3][2py]	12:26
MAKHUB[136]	[PcMg-OPϕ_3][CH_2Cl_2]	12:27
MAKJAJ[136]	[PcMg-OPϕ_3][THF]	12:28
MAKJEN[136]	Pc(Li-OPϕ_3)$_2$	13:6
MEGNOB[235]	[Pc_2U][I_3]$_{2/3}$	18:119–120
MESJOJ[414]	(MeO)$_2$(rpc)Ni	21:1–3
MESJUP[414]	(EtO)$_2$(rpc)Ni	21:4–6
MGPCPH[145]	[PcMg-OH_2][2py]	12:23
MNPHCY[81]	PcMn {β, RT}	11:44
MNPHCY02[84]	PcMn {β, 295K}	11:42
MNPHCY03[86,291]	PcMn {β, N, 5.8K}	11:43
MNPHCY11[83–85]	PcMn {β, 116K}	11:41
MNPHCY21[85]	PcMn {β, 116K, 1.Refinement}	11:39
MPCOPC10[326]	[PcCo(4-mepy)$_2$][2mepy]	15:2
NAKXUS[197]	[Pc_2Nd][NBu_4]	18:77–78
NAKYAZ[197]	[Pc_2Gd][NBu_4]	18:25–26
NAKYED[197]	[Pc_2Ho][NBu_4][H_2O]	18:33–34
NAKYIH[197]	[Pc_2Lu][NBu_4][H_2O]	18:65–66
NALSOI01[174]	PcGa-OH {XI, R}	
NALSOI02[167]	PcGa-OH {V, R}	12:19
NAPHIV[355]	[F-(Pc)Fe-C-Fe(Pc)-F][NBu_4]$_2$[3H_2O]	15:43
NAPJIX[178]	[PcIn(F)$_2$][NBu_4][H_2O]	14:4–5
NAPPAV[383]	PcRe(Pϕ_3)$_2$	15:85
NAPRAX[179]	[($PcTaF_3$)$_{1-x}$($PcTaF_4$)$_x$][NBu_4]	16:28–29,17:1
NEKLUK[356]	[N_3-(Pc)Fe-N-Fe(Pc)-N_3][PNP]$_2$[I_2][2Et_2O]	15:44
NEKRAW[311]	PcRe-RePc	12:41–42
NELCIQ[180]	[PcTl-I][CH_2Cl_2]	12:57
NELCOW[360]	[PcFe(OMe)$_2$][PNP]$_2$[1Me_2CO, 1.5MeOH]	15:50
NEXSIS[387]	[py-(Pc)Rh-Rh(Pc)-py][2C_6H_6]	15:91
NEXVAN[163]	$PcNbCl_3$	16:9–10
NEYCEZ[392]	[PcSi(-O-C(O)-Fc)$_2$][2$CHCl_3$]	15:96
NILVUZ[183]	PcSb(-I-)$_2$SbPc	14:15–16
NIPHTC[13,14,57,58]	PcNi {β}	11:57
NIPHTI[111]	[PcNi][I_3]$_{1/3}$	11:64
NIRDOH[336]	[PcCo(CN)$_2$][PTMA][MeCN] {295K}	15:14
NIRDUN[336]	[PcFe(CN)$_2$][PTMA][MeCN] {123K]	15:48
NIRDUN01[336]	[PcFe(CN)$_2$][PTMA][MeCN] {295K}	15:49
NIRPAF[336]	[PcCo(CN)$_2$][PTMA][MeCN] {113K]	15:13
NOCTII[243]	(Me_3SiO)$_8$PcNi	11:55–56
NOCVAC[243]	(RO_2)$_2$(RO)$_2$(HO)$_2$PcNi	11:53–54
NOCVEC[243]	(RO_2)$_2$(RO)$_2$(HO)$_2$PcNi	11:51–52
NOCYEJ[231]	[Pc_2Nb][I_3][0.5I_2,3.5ClN]	21:13–14
NOFGAQ[334]	[NC-(Pc)Co-CN-(Pc)-CN][B[c]PT]	15:11
NOFMAW[409]	[Pc_2In][I_2]	18:41–42
NOFMEA[409]	[Pc_2Zr][I_2]	18:131–132
NOFNUR[176,403]	$PcZrI_2$(PN)	16:22–23
NOLXOB[397]	PcSi[OCH_2-C_6H_3($OCH_2\phi$)$_2$]$_2$	15:113–114
NOYYIJ[279]	[PcCo][NBu_4][Me_2CO]	11:10
NUGSUD[435]	[(Me_2N)$_8$PazMn(Et_2dtc)]	22:25–26
NUGTAK[435]	[(Me_2N)$_8$PazMnCl]	22:27
NULSUI[433]	(OETAP)$_2$Zr	22:50
NULSOC[433]	(OETAP)(OEP)Zr	22:51
OBUJIE[236]	[Pc_2Gd][$C_4(CN)_6$][2ClB]	18:29–30
OCOBIR[252]	[2,3-NcEuOEP][4$CHCl_3$]	23:13–14
PAGYAX[417]	[PazFeCl]	22:5
PAZRAJ[353]	THF-(Pc)Fe-C-Fe(tpp)-THF	23:1
PAZREN[353]	THF-(Pc)Fe-C-Fe(Pc)-THF	15:40
PBPHCY[133]	PcPb {monoclinic}	12:37–38
PBPHCY01[132,148]	PcPb {triclinic}	12:39–40
PCPCOS[381]	PcOs(CO,py)	15:80

Appendix 9. Continued

CamRef	Crystal description	Figure
		(Continued)
PHALCU01[202,203]	Pc_2U	18:117–118
PHCPMP10[378,379]	[py-(Pc)Mn-O-Mn(Pc)-py][2py]	15:72–73
PHPRSI[87]	$PcSi_{0.17}$	11:74–75
PHTHCY01[1,57,58]	PcH_2 {β}	10:19
PHTHCY02[102]	PcH_2 {β}	10:23
PHTHCY03[96]	PcH_2 {β}	10:21
PHTHCY04[97]	PcH_2 {α}	10:16
PHTHCY05[99]	PcH_2 {X, R}	10:24
PHTHCY06[101]	PcH_2 {X, R}	10:25
PHTHCY07[29,30]	PcH_2 {β}	10:22
PHTHCY08[98]	PcH_2 {α}	10:17
PHTHFE[346]	[PcFe(DMF,CO)][DMF]	15:29
PHTLBC[138]	SubpcB-Cl	20:1
PIBJIT[219]	$[Pc_2Ti][I_3]_{0.66}$	18:113–114
PIFDEN[53]	Pc_3In_2	18:37–38
PILBAN[376]	$[\text{-(Pc)Mn-TCNE-}]_n$ {β}	15:70
PILBER[376]	$[\text{-(Pc)Mn-NC-C(CN)-CN-}]_n$	15:69
PITROZ[422]	$Et_8PazRhMe$	22:9
POFNON[425]	$[(Bn_2N)_8PazMg(O\text{-}C(OEt)CH_3)$	22:11–12
POFNUT[425]	$[\{(allyl)(Bn)N\}_8PazNi][\phi Cl]$	22:13–14
POGKEB[423]	$[(Me_2N)_8\text{-2secoPazH}_2\text{-2,3-dione}]$	22:10
POGKEB01[424]	$[(Me_2N)_8\text{-2secoPazH}_2\text{-2,3-dione}]$	22:10
POQYID[182]	$[PcHfF_3][NBu_4]$	16:4–5
POWLUI[424]	$[(Me_2N)_8\text{-2secoPazZn-2,3-dione}]$	22:22
POWNEU[439]	$[(Pr_2)_2\{(2\text{-PrO})_2Bz\}_2PazH_2$	22:31
PTCNND[205,206]	$[Pc_2Nd][CH_2Cl_2]$	18:67–68
PTCYFE01[326]	$[PcFe(4\text{-mepy})_2][2Mepy]$	15:28
PTHCZN[82]	PcZn {β}	11:76
PTPOCY02[13,15,57,58]	PcPt	11:70
PTPOCY10[77]	PcPt {α}	11:71
PTPOCY11[77]	PcPt {γ}	11:72
PUNTEX[437]	$[(EtS)_8PazMn(SH)]$	22:29
QASKAW[50]	$Pc(Na\text{-}OP\phi_3)_2$	13:7
QEKBEN[126]	$PcSi(OH)_2$ {II}	15:105
QEKBEN01[126]	$PcSi(OH)_2$ {I}	15:104
QEMMIE[306]	$PcCu\text{-}OReO_3$	12:9
QEMMOK[306]	$PcCu(ReO_4)_2$	15:26
QEMTAD[246]	$[(EtO)_8CpPcNi][PF_6]$	11:59–61
QILWAJ[344]	$PcCu(SbF_6)_2$	15:27
QOCJEX[198]	$[Pc_2Bi][NBu_4]$	18:9–10
QOCJIB[198]	$[Pc_2Ce][NBu_4]$	18:17–18
QOCJOH[198]	$[Pc_2La][NBu_4]$	18:45–46
QOCJUN[198]	$[Pc_2La][NPe_4]$	18:47–48
QOCKAU[198]	$[Pc_2Bi][NBu_4]$	18:11–12
QOCKEY[198]	$[Pc_2Pr][NBu_4][0.75MeOH]$	18:85–86
QOCKIC[198]	$[Pc_2Sm][NBu_4][MeOH]$	18:97–98
QOCKOI[198]	$[Pc_2Y][NBu_4]$	18:125–126
QOCKUO[198]	$[Pc_2Sc][NBu_4]$	18:95–96
QOCLAV[198]	$[Pc_2Pr][NBu_4][2py]$	18:87–88
QOCLEZ[198]	$[Pc_2Sb][NBu_4][2THF]$	18:91–92
QOCLID[198]	$[Pc_2Sb][Br][PNP]_2[2Et_2O]$	18:93–94
QOCLOJ[198]	$[Pc_2Bi][Br][PNP]_2[2Et_2O]$	18:7–8
QOCLUP[198]	$[Pc_2In][I][PNP]_2[2Et_2O]$	18:39–40
QOCMAW[198]	$[Pc_2La][NBu_4][EtOH]$	
QOPCON[373]	$PcMg(HOCH_2CH_2OMe)_2$	15:65
QOPCUT[373]	$[PcMg(OH_2)_2][NMP]_2$	15:66
QOTPIY[443]	$[((Me_2N)_2)_3(benzo)PazH_2]$	22:39
QOTPOE[443]	$[cis((Me_2N)_2)_2(benzo)_2PazMg]$	22:41
QOTPUK[443]	$[(Me_2N)_2(benzo)_3PazMg]$	22:49
QOWHIT[258]	PccBi	21:11–12
QUDQIP[390,391]	$[\text{-(Pc)Si-(OSiMe}_2\{CH_2)_7SiMe_2O\text{-)}]_n$	15:98

Appendix 9. Continued

CamRef	Crystal description	Figure
		(Continued)
QUDQOV[390,391]	$PcSi[OSi(C_6H_{13})_3]_2$	15:118–119
QUDQUB[390,391]	$PcSi[OSiMe_2(C_8H_{17})]_2$	15:120–121
RABDEZ[230]	$[Pc_2Ce][BF_4]_{0.33}$	18:21–22
RADPER[104]	2,3-NcNi	19:2–3
RADPIV[104]	2,3-NcCu	19:1
RADPOB[104]	2,3-NcZn	19:4
RENJID[368]	[PcIr(py,I)][py]	15:60
RERGEA[352]	py-(Pc)Fe-C-Fe(Pc)-py	15:39
RERGIE[352]	py-(Pc)Ru-C-Ru(Pc)-py	15:93
RERGOK[177]	$[PcIn(-O,-O'CO)][NBu_4]$	14:8–9
RESFIE[119]	$PcGeI_2$	15:57
RESFOK[119]	$PcSnI_2$	15:124
RESFOK01[118]	$PcSnI_2$	15:125
RESYAP[115]	$[PcLi][NBu_4][THF]$	11:29–30
RESYET[115]	$[PcLi][F][NBu_4]_2[0.5H_2O]$	11:27–28
RIKSIN[380]	$[PcMoBr_2][PNP]$	15:79
RILVOX[371]	$[PcIr(-NO_2)_2][PNP]$	15:63
ROCCAN,ROCCAN01[241,242]	$(i\text{-}H_{11}C_5O)_4PcNi$	11:49–50
ROPLUD[432]	$[(RO_2)_4PazH_2]$	22:23
ROPTAR[375]	(dppe)(Cp)Fe-(Pc)Mn-O-Mn(Pc)-NC-Fe(dppe)Cp	15:68
RUGJOS[140]	SubpcB-OEt	20:9
RUJQES[388]	$[PcRu(-NO_2)_2][NBu_4]_2$	15:94
SACNOZ[405]	$[(Pc)Ce(T(pMeOP)P)][2CHCl_3,pentane]$	23:4
SACNUF[405]	$[PcCeTPP][SbCl_6]$	23:8
SAFPUK[213]	$[Pc_2Nd][O]$	18:79–80
SAMDUF[124,162]	$[PcNb(-O-,-O-,-Br-)NbPc][Br_3][1\text{-}ClN]$	16:15–16
SAMFAN[124,162]	$[PcNb(-O-,-O-,-Br-)NbPc][Br_3][PN]$	16:17
SAXZEW[185]	PcV-I	12:59
SEWRER[214]	$[Pc_2Ti][1\text{-}ClN]$	21:15–16
SILPEI[331,332]	$[PcCo(CN)_2][2H_2O]$	15:6
SIMPLC10[395]	$PcSi(OSiMe_3)_2$	15:106
SIOPCY[449]	$[PcSiOSiEt_3,Me][0.5C_6H_{12}]$	
SNPHCY01[143]	PcSn	12:50–51
SNPTCY[199,201]	Pc_2Sn	18:103–104
SNPTCY01[223]	Pc_2Sn	18:105–106
SOPQET[338]	$[PcCo(CN)_2][p\text{-}PPYNN]$	15:16
SOPQIX[338]	$[PcCo(CN)_2][p\text{-}BPYNN]$	15:17–18
SOTGAJ[215]	Pc_2Zr	18:129–130
SOZBUE[42]	$(n\text{-}H_{13}C_6)_8PcH_2$	10:10–12
SPCNFE[347]	$[PcFe(-OSMe_2)_2][2DMSO]$	15:31
SUKSUM[41]	$(i\text{-}H_{11}C_5O)_8PcH_2$	10:4–6
SURDIS[427]	$[(Ag\text{-}contng\text{-}crownS_2)_4Ag_4PazNi][8BF_4]$	11:49
TANMUQ[407]	[(TPP)Ce(Pc)Gd(OEP)][2ClB]	23:7
TAYSIV[172]	$[PcMg\text{-}Cl][PNP][Me_2CO]$	12:34
TESFAY[171]	$(H_{11}C_5)_8PcRh\text{-}H\text{-}RePc(H_{11}C_5)_8$	12:44–45
TISDOO[428]	$[(EtS)_8PazMnCl]$	22:18
TISDUU[428]	$[(EtS)_8PazFeCl]$	22:19
TOCLIG[369]	py-(Pc)Ir-Ir(Pc)-py	15:61
TOGPEK[429]	$[(EtS)_8PazMn]$	22:20
TOPVID[343]	$[PcCr(-ONO)_2][PNP]$	15:25
TOYSUV[240]	$(H_{13}C_6)_8PcNi$ {103K}	11:45–46
TOYSUV01[240]	$(H_{13}C_6)_8PcNi$ {295K}	11:47–48
TOZCIU[44]	$[(teeda)PtS_2NorpcNi][0.5CHCl_3]$	22:48
TOZCOA[430]	$[tr((Et_3P)_2PtS_2)_2PazNi][C_6H_4Cl_2]_2$	22:40
TUPWEG[410]	$[PcLaTPyP][OH][CHCl_3,MeOH]$	23:10
UDIHEU[446]	$[((PrO(CO)(CH_2)_3)_2)_2((iPrO)_2benzo)_2PazH_2]$	22:45
VACQEV[436]	$[((4\text{-}tBu\text{-}\phi)_2)_3(Cp_2MoS_2)PazH_2][2.5MeOH]$	22:28
VACXOM[249]	$1,2\text{-}NcFe(NC\text{-}C_6H_{11})_2$	19:9–10
VAFMIY[350]	Br-(Pc)Fe-N-Fe(Pc)-Br	15:35
VENDAT	$[K][PcCo(CN)_2][5CH_3CN]$	
VENDAT10[331,332,335]	$[K][PcCo(CN)_2][5CH_3CN]$	15:12

Appendix 9. Continued

CamRef	Crystal description	Figure
		(Continued)
VIKVEQ[127]	$PcSiCl_2$	15:116
VIKZAQ[127]	$[PcSi(OMe)_2][0.48H_2O]$	15:99
VILGOM[127]	$PcSi(thff)_2$	15:107
VIWMOD[250]	$(tBu)_4$-1,2-NcFe(NC-tBu$)_2$	19:11–14
VIYFEO[278]	$[PcCo][(AsF_6)_{0.5}]$	11:8
VODDIB[247]	$[(H_{13}C_6S)_8Pc]_2Lu$	18:56–58
VPHTHC[150]	PcV=O {II}	12:58
VUMZOS[416]	$[(MeS)_8PazMg(OH_2)]$	22:42–43
WAYSIY[218]	$[Pc_2Pr][Br_{1.5}]$ {α}	18:89–90
WETGAD[420,421]	$[(cr18S_2)_4PazMg(OH_2)]$	22:8
WETGAD10[421]	$[(cr18S_2)_4PazMg(OH_2)]$	22:8
WETJUA[43]	$[(dppf)PdS_2NorpcH_2][CH_2Cl_2]$	22:46–47
WETJUA10[44]	$[(dppf)PdS_2NorpcH_2][CH_2Cl_2]$	22:46–47
WOHDUH[251]	$[2,3\text{-}NcCo(CN)_2][2DCB]$	19:7
WOHTES[314]	$[PcZn\text{-}NC\text{-}Fe(dppe)Cp][0.5CH_2Cl_2,THF]$	12:61
WOHTIW[314]	$[PcFeCN,NC\text{-}Ru(P\phi_3)_2Cp][2CHCl_3]$	15:32
WOHTOC[314]	$[PcFe(NC\text{-}Fe(dppe)Cp)_2][SbF_6]$	15:52
WOHTUI[314]	$[PcFe(NC\text{-}Ru(P\phi_3)_2Cp)_2][SbF_6]$	15:53
WOHVAQ[314]	$[PcFe(NC\text{-}Ru(P\phi_3)_2Cp)_2][1.5CH_2Cl_2]$	15:30
WOHVEU[314]	$[PcFe(CN\text{-}Ru(P\phi_3)_2Cp)(NC\text{-}Ru(P\phi_3)_2Cp)][SbF_6]$	15:51
WOKDAB[45]	$(i\text{-}H_{11}C_5)_8PcH_2$	10:1–3
WOPZOQ[359]	$[PcFe(CN)_2][P\phi_4]_{0.5}$	15:47
WOVREE[193]	$PcSb(\text{-}Cl\text{-})_2SbPc$	14:17–18
WOWDOB[251]	$[2,3\text{-}NcCo(CN)_2][CH_3CN]$	19:8
XAQGEB	SubbpcB-Cl	20:5–6
XENLAD[351]	$PcFe(4mepip)_2$	15:36–37
XETZIF[60,61]	PcMg {273K}	12:32
XETZIF01[59]	PcMg {120K}	12:29
XETZIF02[59]	PcMg {200K}	12:30
XETZIF03[59]	PcMg {260K}	12:31
XEYQIB[377]	[PcMn(NO)(-ONO)][PNP]	15:75
XEYQOH[377]	[PcMn(NO)(-NCO)][PNP]	15:74
XEYQUN[377]	$[PcMn(NO)_2][PNP]$	15:76
XEYRAU[377]	$[PcRe(NO)_2][PNP]$	15:89
XEYREY[377]	$PcMn(NO)(OP\phi_3)$	15:71
XEYRIC[377]	$PcRe(NO)(OP\phi_3)$	15:84
XEYROI[377]	$[PcRe(NO)(OP\phi_3)][I_3][CH_2Cl_2]$	15:90
YAVPEQ[327,328]	$[PcCo(CN)_2][2CHBr_3]$	15:3
YEGTIN[220]	$[Pc_2La][PNP]$	18:49–50
YEGTOT[220]	$[Pc_2Gd][PNP][0.33H_2O]$	18:27–28
YEGTUZ[220]	$[Pc_2Tm][PNP][0.5H_2O]$	18:115–116
YEJJIG[305]	PcCr-I	12:8
YOHMAJ[194]	$[PcSb]_4[Sb_4I_{16}]$	16:18–19
YOKMUG[139]	SubpcB-ϕ	
YOKMUG01[141]	SubpcB-ϕ	20:10
YOKWEA[169]	$[PcZrCl_3][PNP][2CH_2Cl_2]$	16:26–27
YUGKOA[361]	$[H_2O\text{-}(Pc)Fe\text{-}N\text{-}Fe(tpp)\text{-}OH_2][I_5][2THF]$	23:2
YUHSUP[225]	Pc_2Er {α}	18:23–24
ZAKZOA[224]	Pc_2In	18:35–36
ZAPNAF[426]	$[(Me_2N)_8PazH_2][TCNQ]$	22:16
ZAPNEJ[426]	$[(Me_2N)_8PazH_2]_2[C_{60}]$	22:15
ZAZQAS[408]	$[TPPLaPc][2CH_2Cl_2]$	23:11
ZAZQEW[408]	$[(TPP)Gd(Pc)][SbCl_6][2CH_2Cl_2,H_2O]$	23:9
ZEHTUB[226]	$[Pc_2La][CH_2Cl_2]$	18:43–44
ZEMBEY[404]	$[PcIn(\text{-}ONO\text{-})_2][NBu_4]$	17:2–3
ZENGUU[256]	PccGd	21:8
ZEPKIO[289]	$[PcLi][PNP][CH_2Cl_2]$	11:31–32
ZEPKOU[289]	$[PcLi][PNP][H_2O]$	11:33–34
ZNPTCY[144]	$[PcZn\text{-}H_2N\text{-}C_6H_{13}][0.5H_2N\text{-}C_6H_{13}]$	12:60
ZODWAQ[329]	$[PcCo(CN)_2][2CHCl_3]$	15:4
ZOLHUD[257]	PccTl	21:9

Appendix 9. Continued

CamRef	Crystal description	Figure
		(*Continued*)
ZOMWUT[227]	$[Pc_2Sm][Br_{1.45}]$ {α}	18:99–100
ZOMXAA[227]	$[Pc_2Sm][ClO_4]_{0.63}$ {α}	18:101–102
ZUFZOP[173]	$[PcIn(\text{-}NCO)_2][PNP]$	14:6–7
ZUHCIO[170]	$[PcZn\text{-}ONO][PNP][H_2O]$	12:64
ZUQDOE[431]	$[(Et)_8PazFe(OClO_3)]$	22:21
ZUVYOE[324]	$[PcAl(\text{-}ONO)_2][NBu_4]$	15:1
ZUWDAW[228]	$[Pc_2Bi][CH_2Cl_2]$	18:3–4

R: Rietveld analysis
N: Neutron diffraction

Appendix 10. Molecular Structures of CN1 Phthalocyanine Complexes

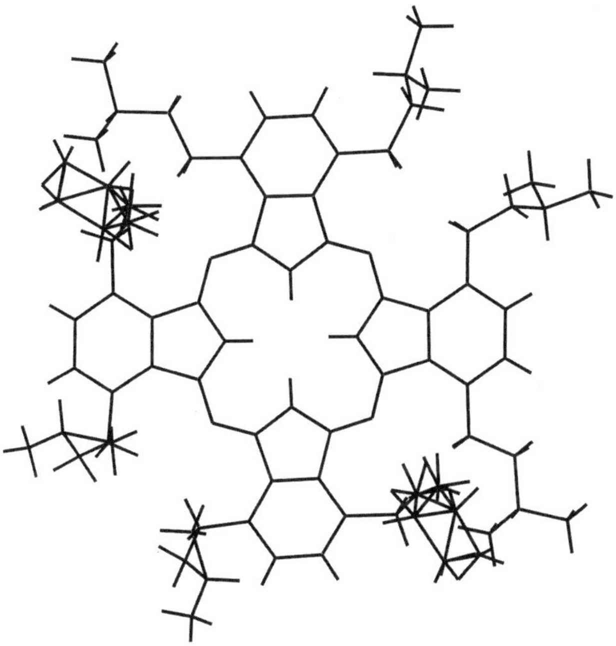

Appendix 10-1. $(i\text{-}H_{11}C_5)_8PcH_2$, WOKDAB[45]

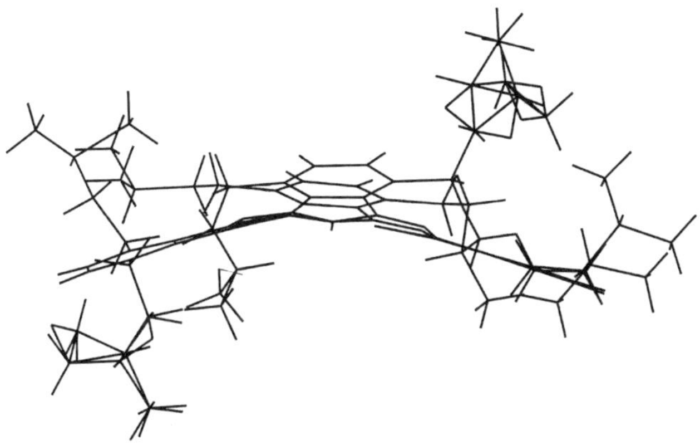

Appendix 10-2. $(i\text{-}H_{11}C_5)_8PcH_2$, WOKDAB[45]

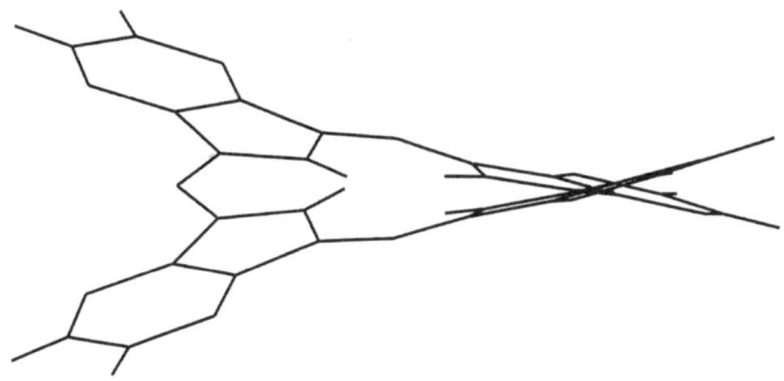

Appendix 10-3. Saddle-shaped phthalocyanine macrocycle of $(i\text{-}H_{11}C_5)_8PcH_2$, WOKDAB[45]

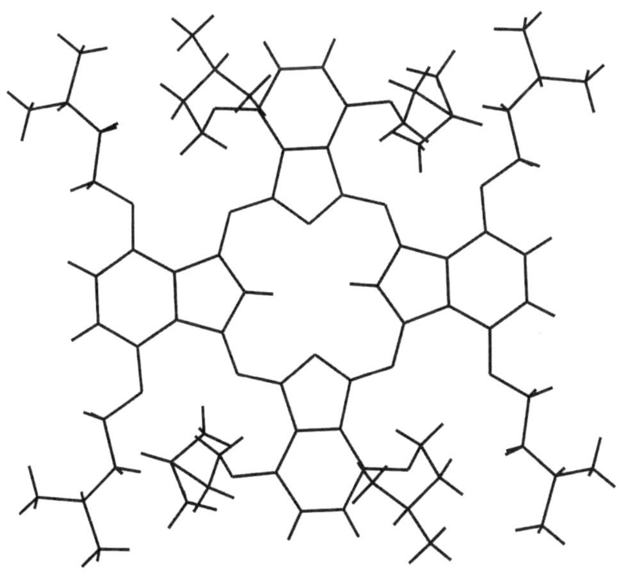

Appendix 10-4. $(i\text{-}H_{11}C_5O)_8PcH_2$, SUKSUM[41]

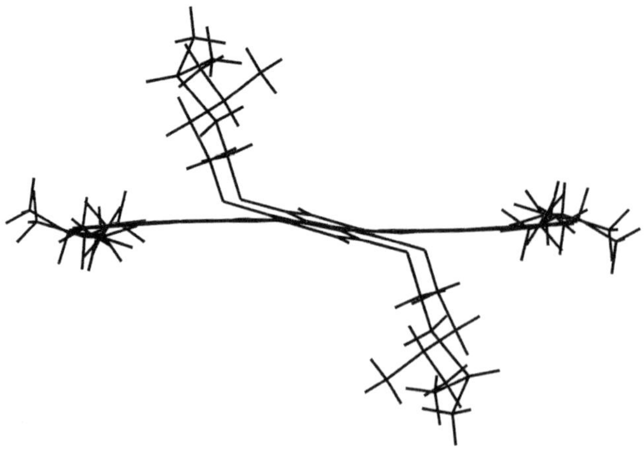

Appendix 10-5. (i-$H_{11}C_5O)_8PcH_2$, SUKSUM[41]

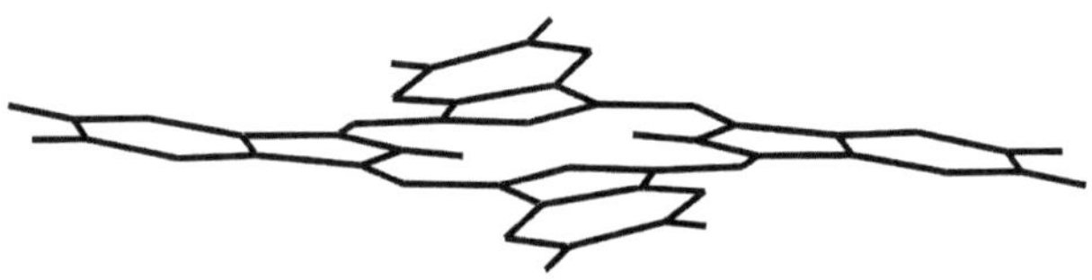

Appendix 10-6. Deformed phthalocyanine macrocycle of (i-$H_{11}C_5O)_8PcH_2$, SUKSUM[41]

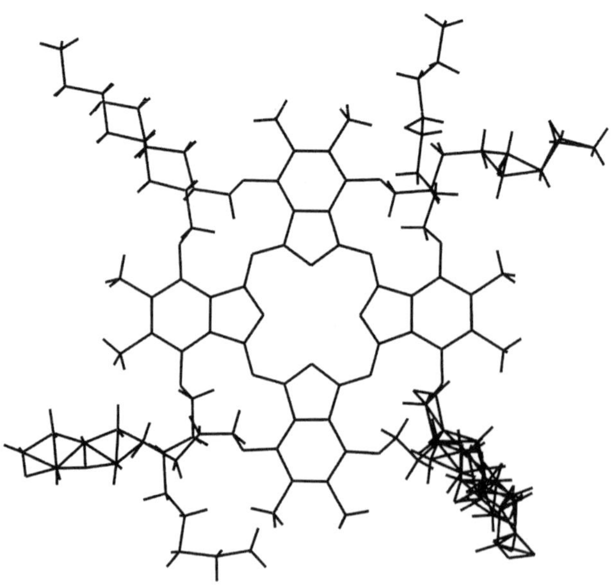

Appendix 10-7. $(Me)_8(H_{17}C_8O)_8PcH_2$, #2002-08[238]

Appendix 10-8. $(Me)_8(H_{17}C_8O)_8PcH_2$, #2002-08[238]

Appendix 10-9. Saddle-shaped phthalocyanine macrocycle of $(Me)_8(H_{17}C_8O)_8PcH_2$, #2002-08[238]

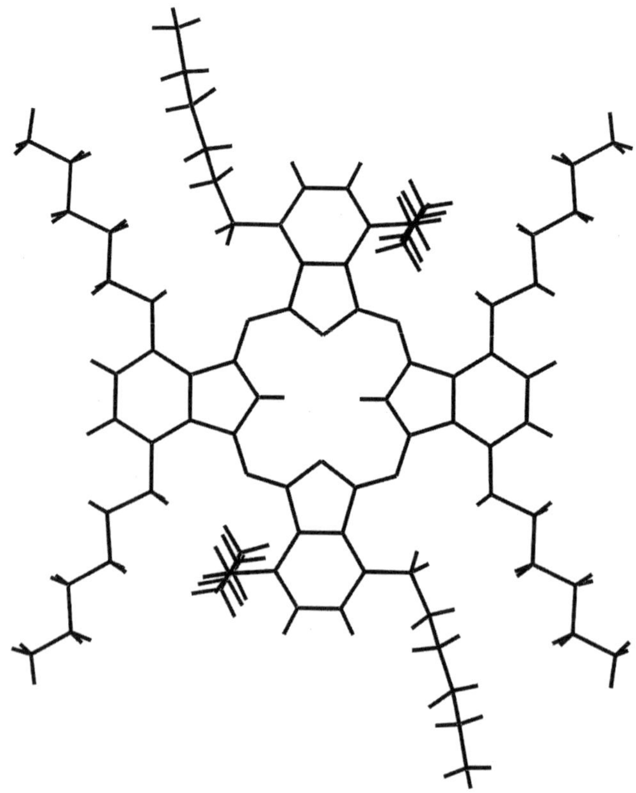

Appendix 10-10. (n-$H_{13}C_6)_8PcH_2$, SOZBUE[42]

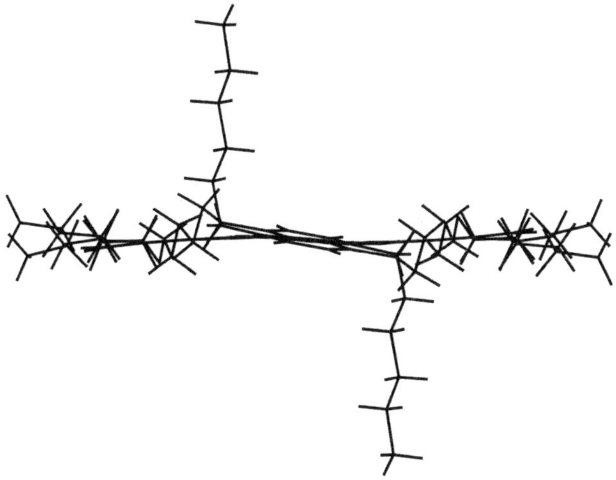

Appendix 10-11. (n-$H_{13}C_6$)$_8$PcH_2, SOZBUE[42]

Appendix 10-12. Nearly undeformed phthalocyanine macrocycle of (n-$H_{13}C_6$)$_8$PcH_2, SOZBUE[42]

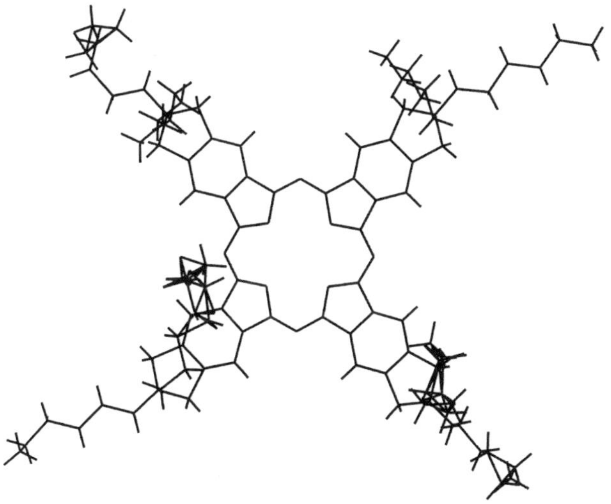

Appendix 10-13. {($H_{13}C_6$)$_2$C}$_4$PcH_2, #2002-09[238]

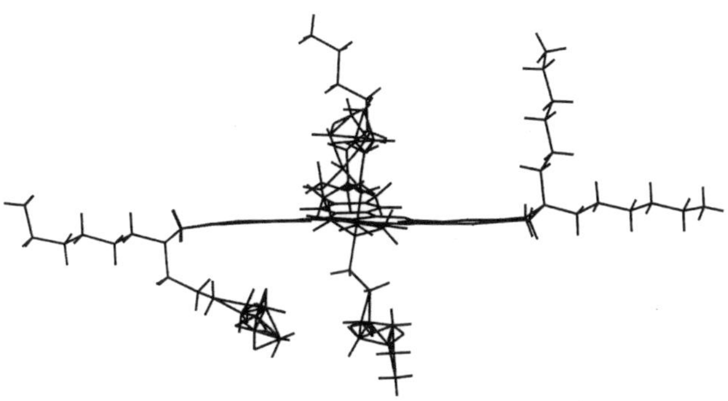

Appendix 10-14. {($H_{13}C_6$)$_2$C}$_4$PcH_2, #2002-09[238]

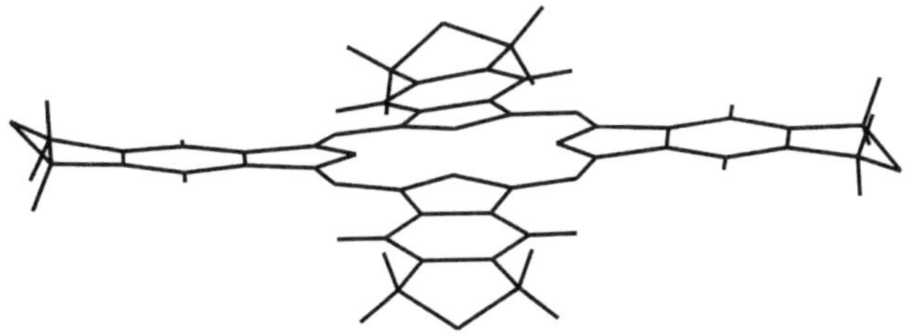

Appendix 10-15. Macrocycle of {($H_{13}C_6$)$_2$C}$_4$PcH_2, #2002-09[238]

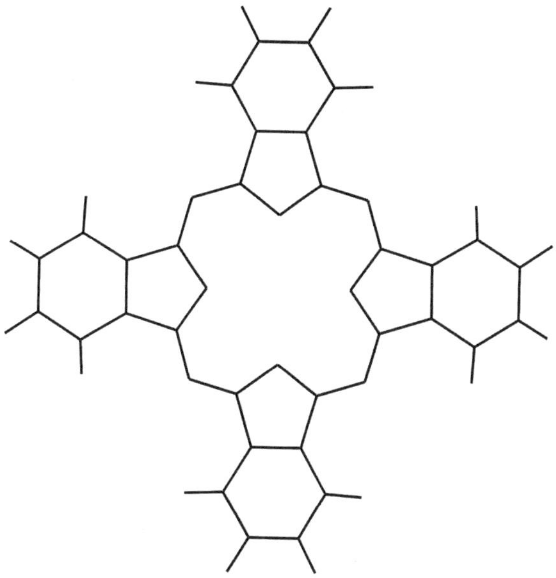

Appendix 10-16. α-PcH_2, PHTHCY04[97]

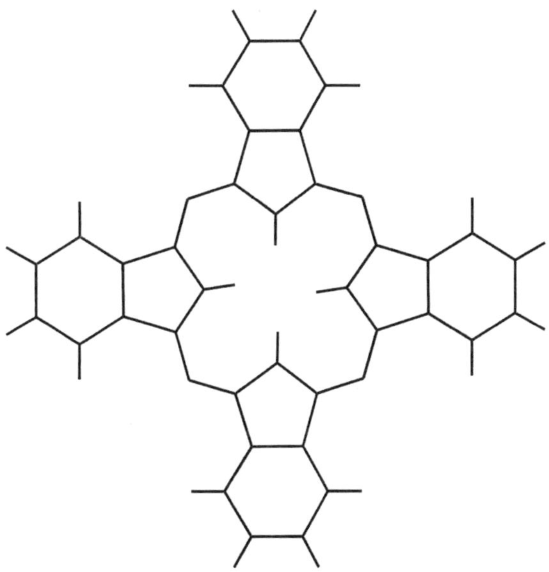

Appendix 10-17. α-PcH_2, PHTHCY08[98]

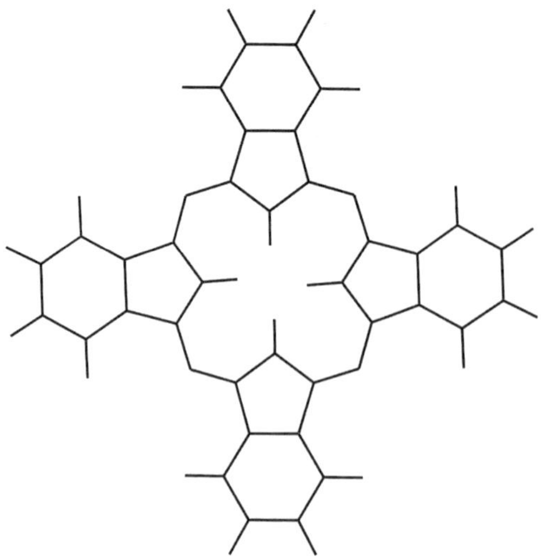

Appendix 10-18. β-PcH_2, #1971-03[31] (App. 26)

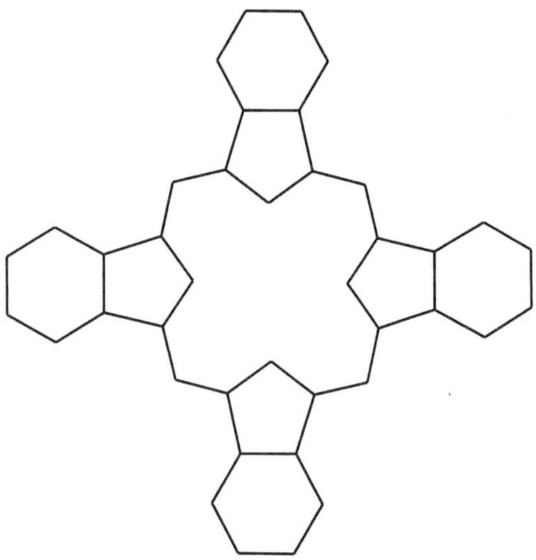

Appendix 10-19. β-PcH_2, PHTHCY01[1,57,58]

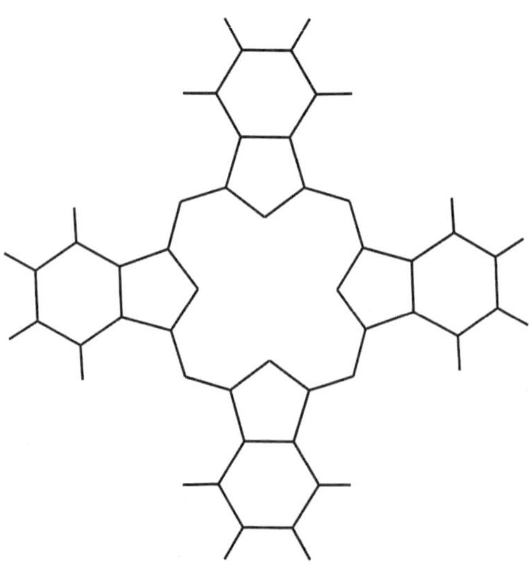

Appendix 10-20. β-PcH_2, #1970-02[40] (App. 25)

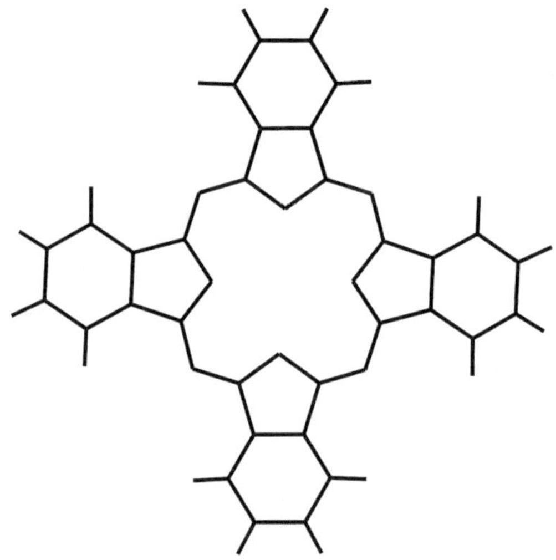

Appendix 10-21. β-PcH_2, PHTHCY03[96]

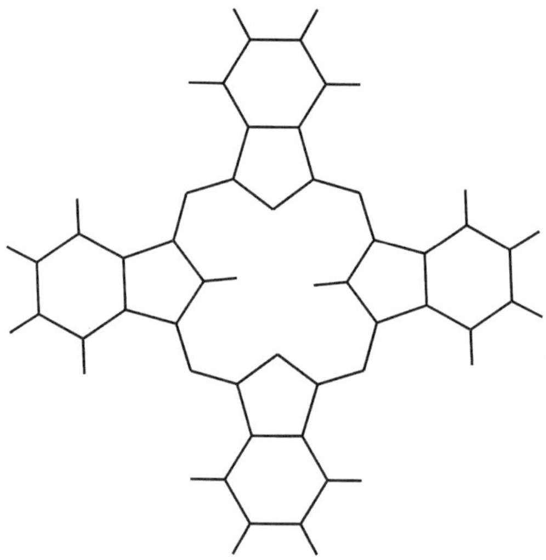

Appendix 10-22. β-PcH_2, PHTHCY07[29,30]

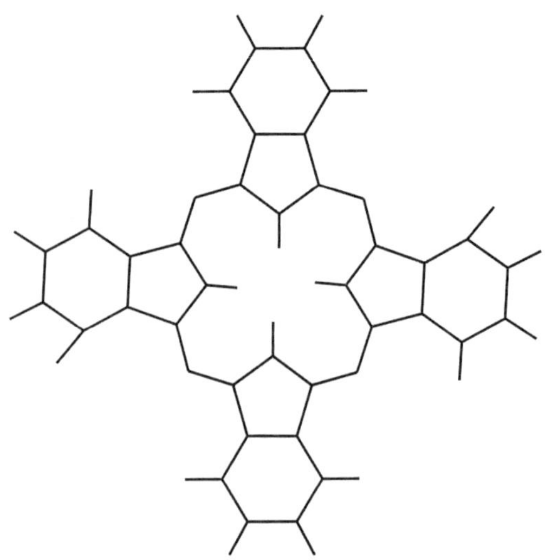

Appendix 10-23. β-PcH_2, PHTHCY02[102]

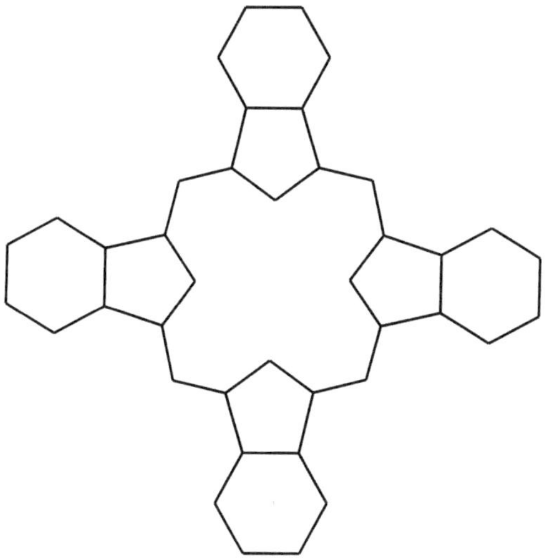

Appendix 10-24. x-PcH_2, PHTHCY05[99]

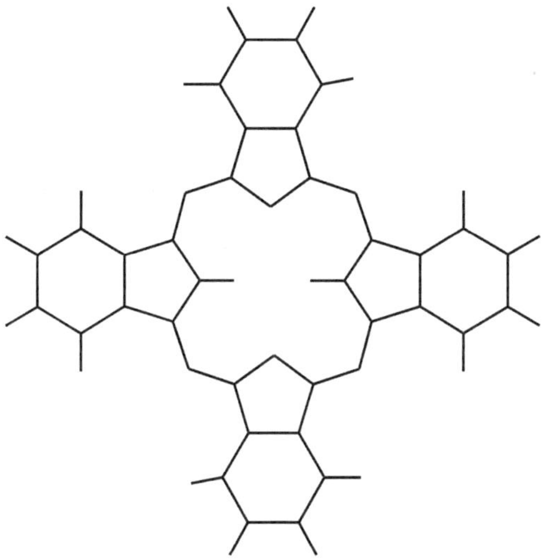

Appendix 10-25. PcH_2 (R), PHTHCY06[101]

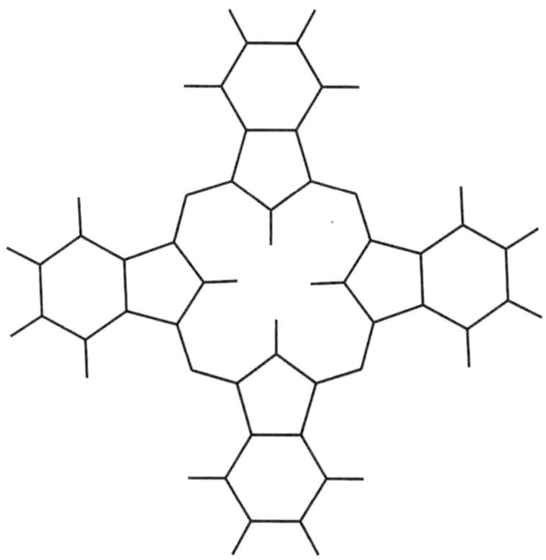

Appendix 10-26. x-PcH_2 (R), #1997-03[102]

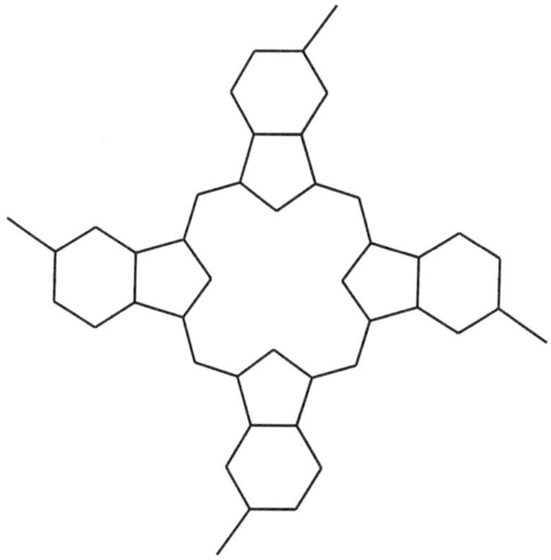

Appendix 10-27. α-PcH_2/Cl_4PcH_2, GAKWUK[93]

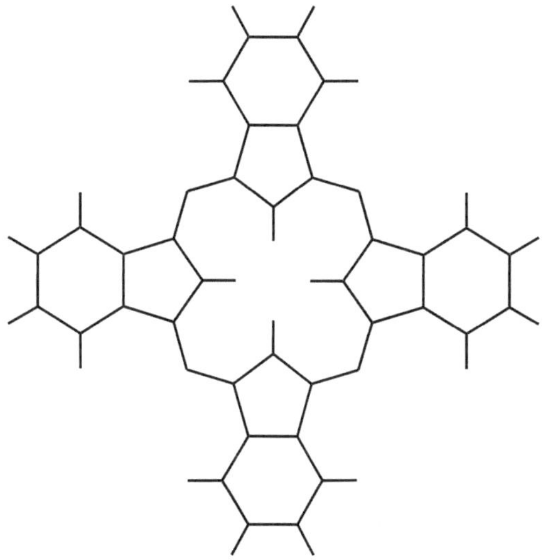

Appendix 10-28. $[PcH_2][I_3]_{1/3}$, LATLIB[273]

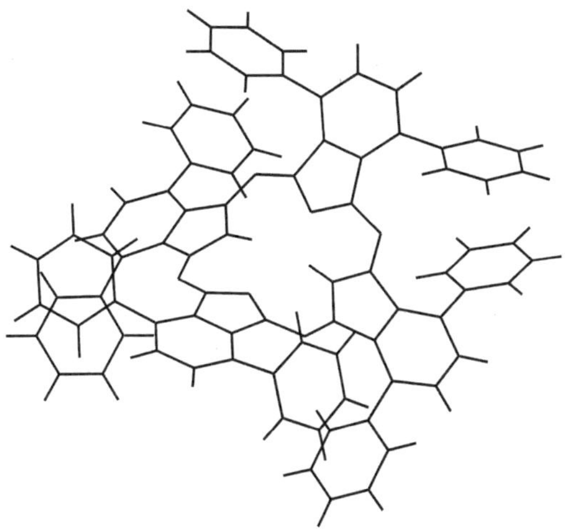

Appendix 10-29. $(\Phi)_8PcH_2$, ECOSOE[239]

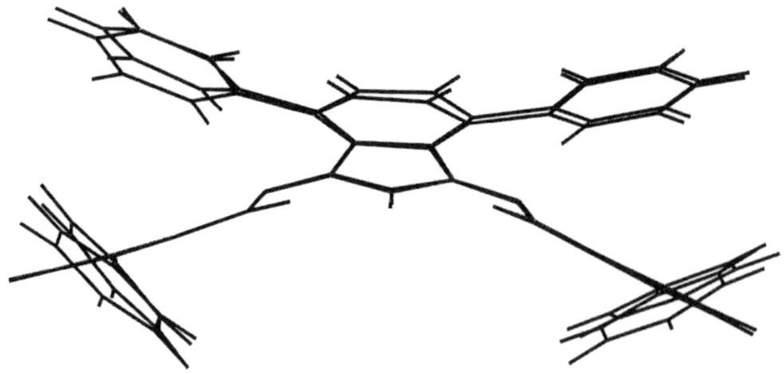

Appendix 10-30. $(\Phi)_8PcH_2$, ECOSOE[239]

Appendix 10-31. $[PcH_2]_3[(AsF_6)_2][1\text{-}ClN]$, CCDC168923[275]

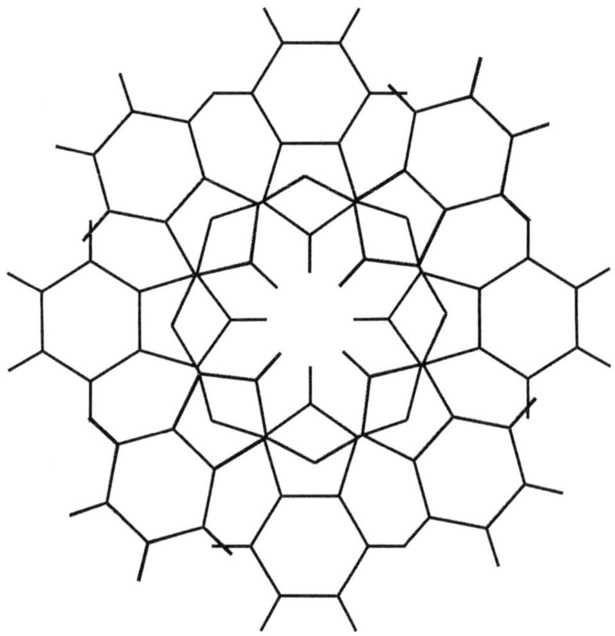

Appendix 10-32. $[PcH_2]_3[(AsF_6)_2][1\text{-}ClN]$, CCDC168923[275]

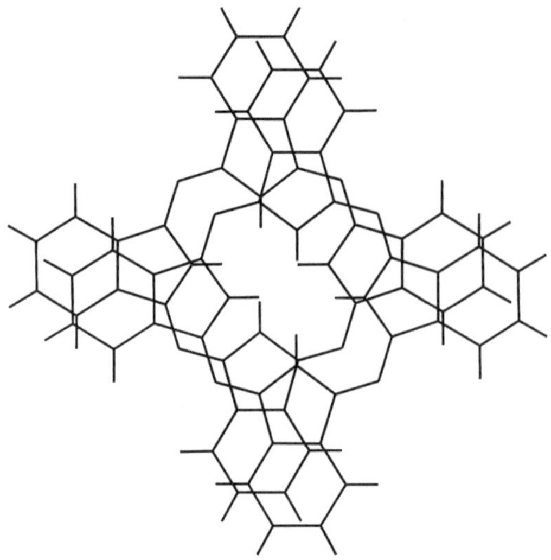

Appendix 10-33. $[PcH_2]_3[(SbF_6)_2][1\text{-}ClN]$, CCDC168921[275]

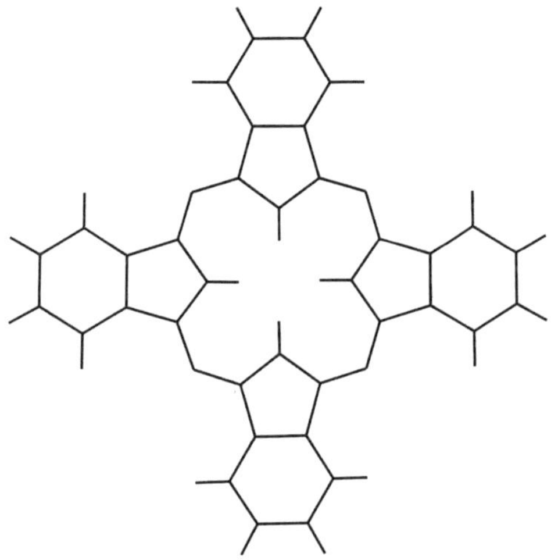

Appendix 10-34. $[PcH_2][IBr_2]$, CCDC184514[114]

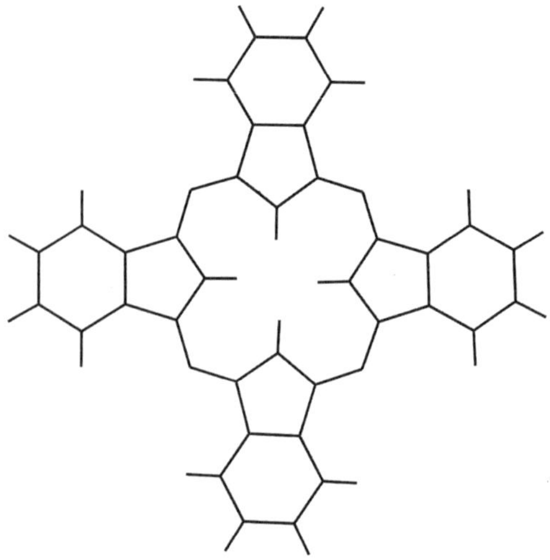

Appendix 10-35. $[PcH_2][IBr_2][Br]$ [1-BrN], CCDC184515[114]

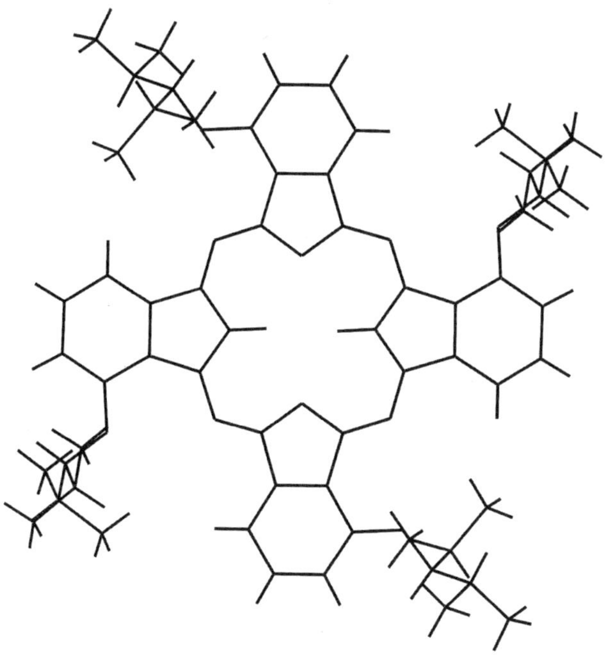

Appendix 10-36. [((iPr)$_2$CHO)$_4$PcH$_2$][(HOOC)$_2$], CCDC175743[248]

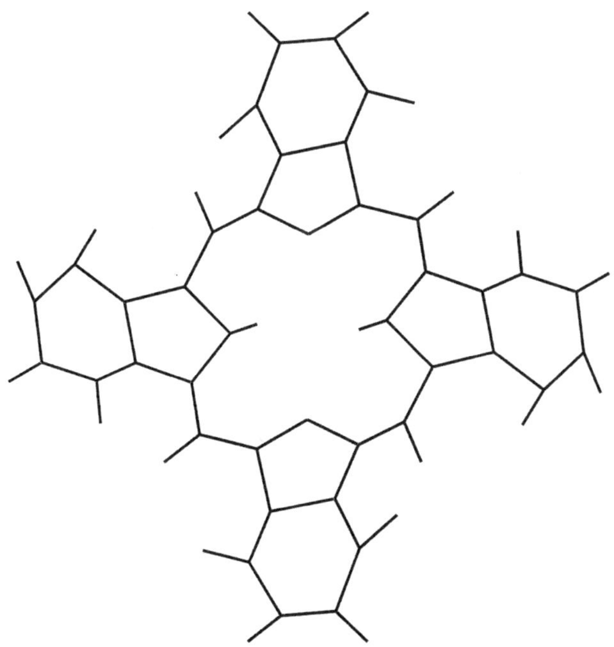

Appendix 10-37. MATBPH$_2$, BZAZPN[68]

Appendix 11. Molecular Structures of CN4 Phthalocyanine Complexes

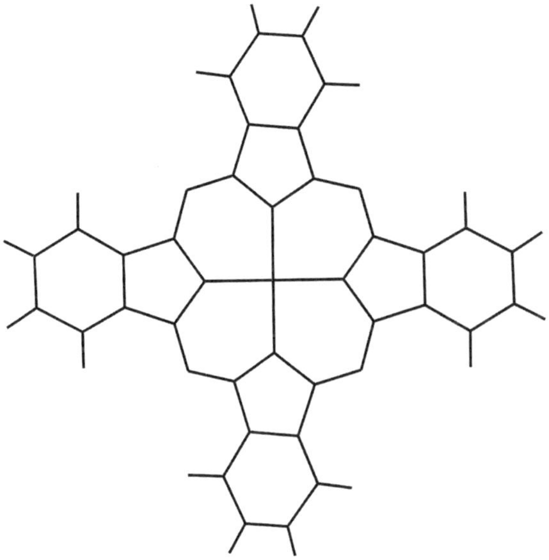

Appendix 11-1. α-PcCo (R), COPTCY07[107]

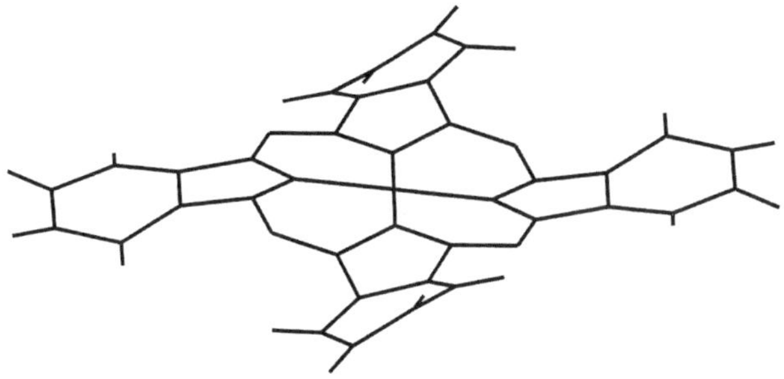

Appendix 11-2. Severely deformed α-PcCo (R), COPTCY07[107]

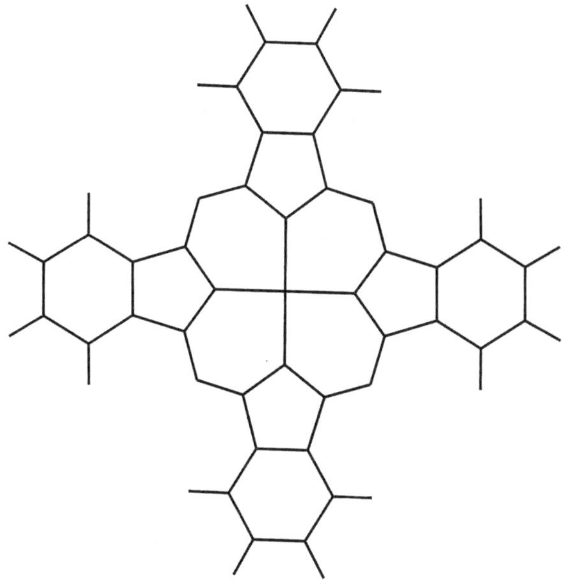

Appendix 11-3. β-PcCo, N115K, COPTCY06[95]

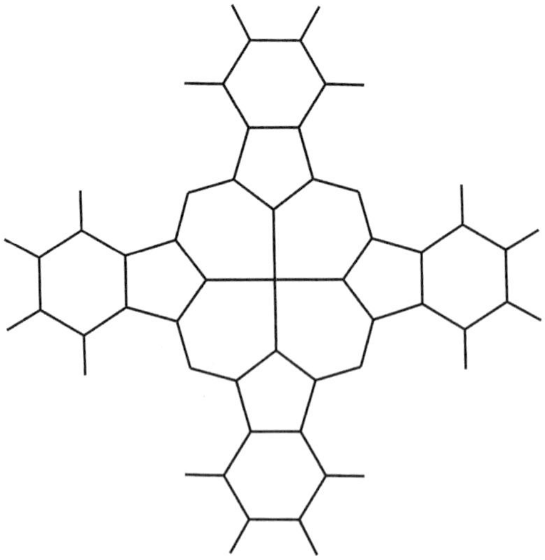

Appendix 11-4. β-PcCo, N4.3K, COPTCY01[27,28]

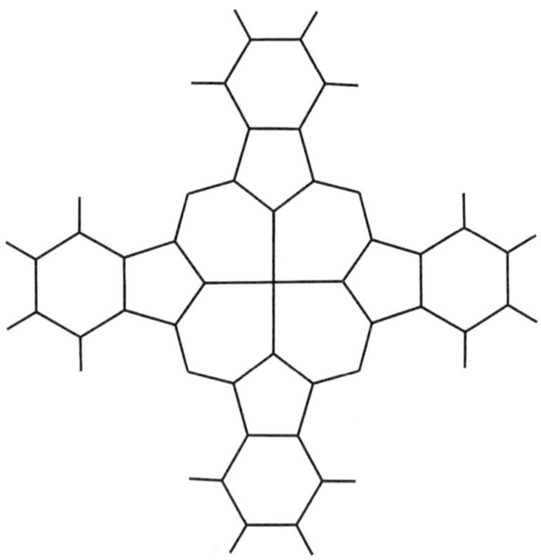

Appendix 11-5. β-PcCo (R), COPTCY08[107]

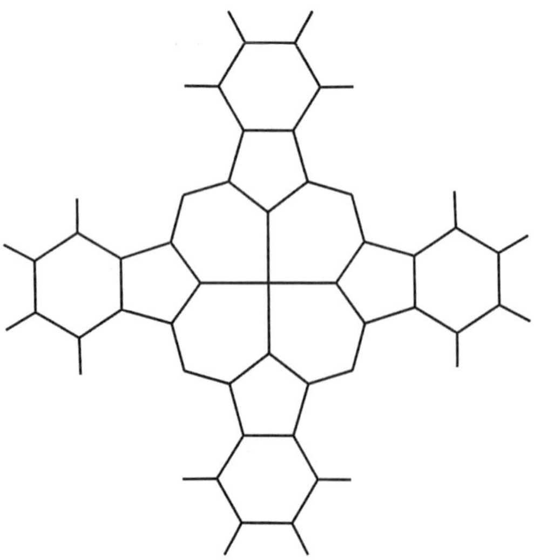

Appendix 11-6. β-PcCo, X115K, COPTCY05[94]

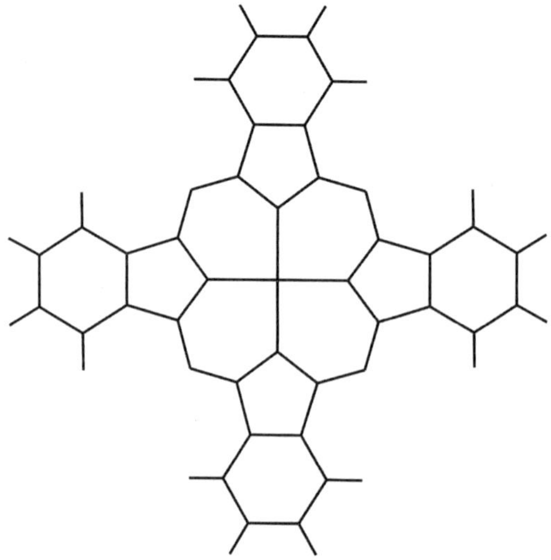

Appendix 11-7. β-PcCo, X295K, COPTCY10[27,83,84]

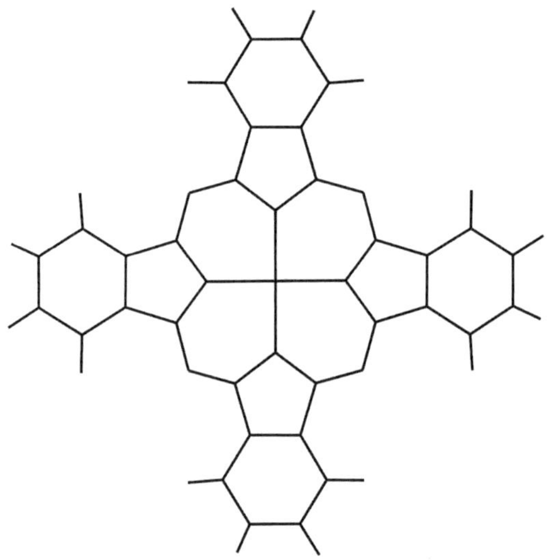

Appendix 11-8. $[PcCo][(AsF_6)_{0.5}]$, VIYFEO[278]

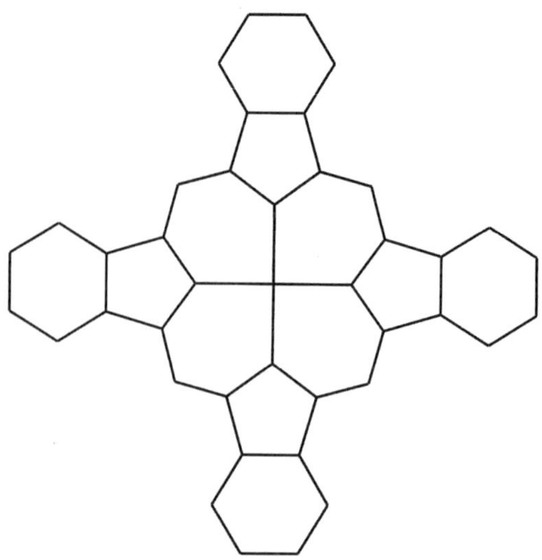

Appendix 11-9. $[PcCo][I_3]_{1/3}$, DIDTEP[112]

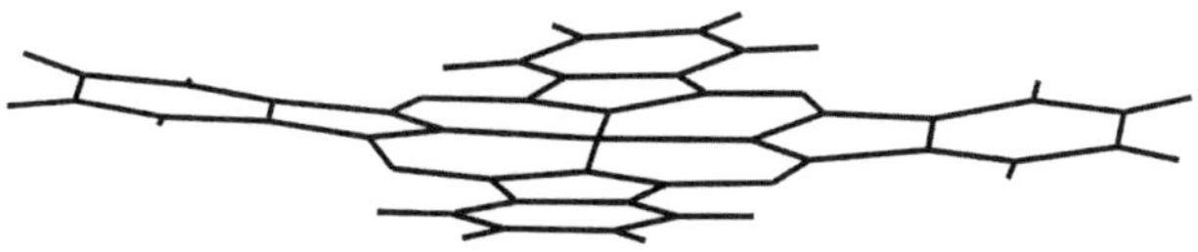

Appendix 11-10. $[PcCo][NBu_4][Me_2CO]$, NOYYIJ[279]

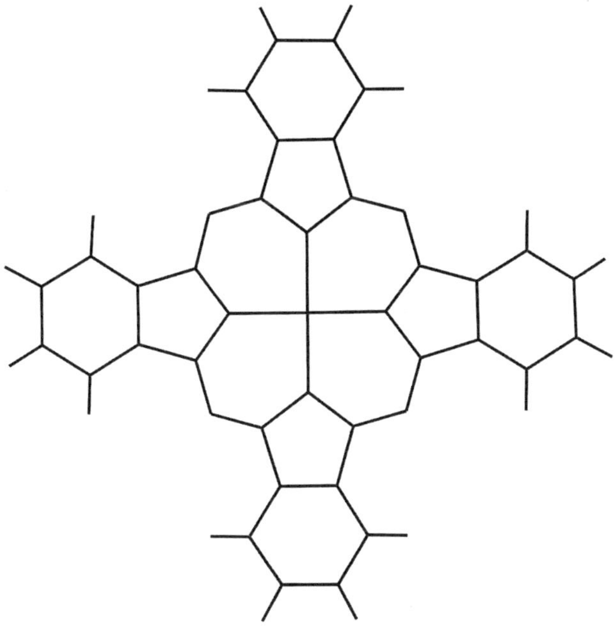

Appendix 11-11. $[PcCo_{0.25}Ni_{0.75}][(AsF_6)_{0.5}]$, CCDC156503[280]

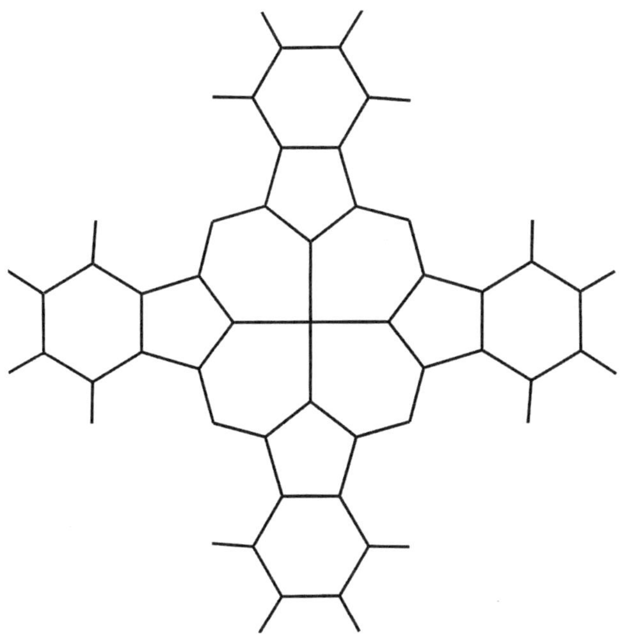

Appendix 11-12. $[PcCo_{0.45}Ni_{0.55}][(AsF_6)_{0.5}]$, CCDC156502[280]

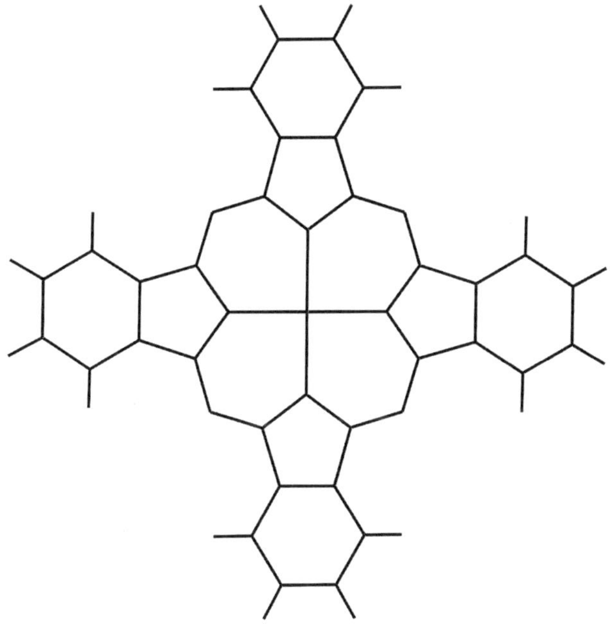

Appendix 11-13. β-PcCr, #1999-01[63]

Appendix 11-14. β-PcCr, #1999-01[63]

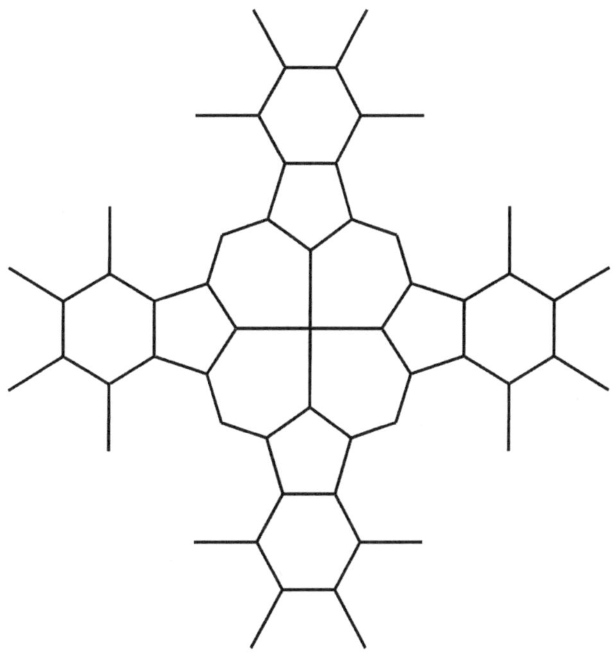

Appendix 11-15. $Cl_{16}PcCu$, #1982-01[88,89,281]

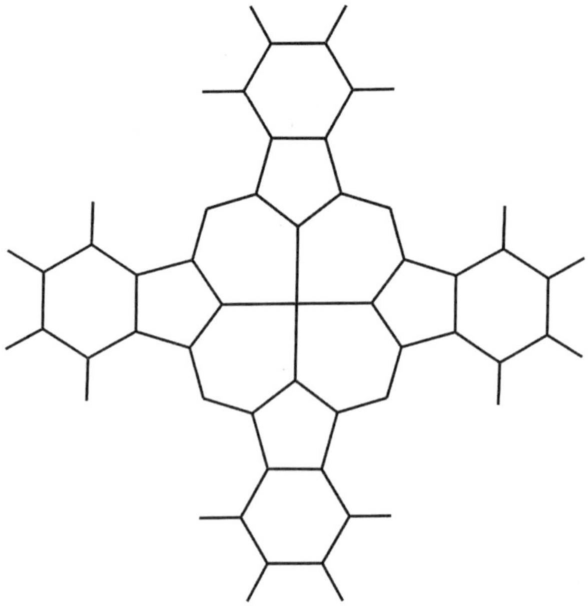

Appendix 11-16. β-PcCu, CUPOCY10[76,286]

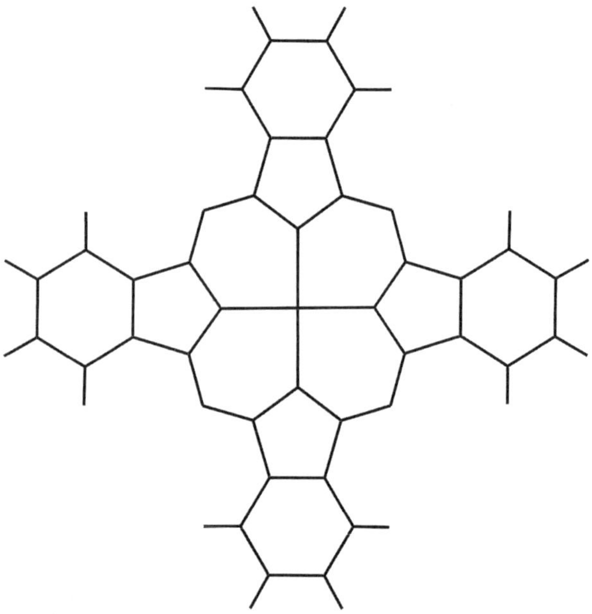

Appendix 11-17. [PcCu][I_3]$_{1/3}$, FEGJOQ[113]

Appendix 11-18. [PcCu]$_3$[(AsF_6)$_2$][1-ClN], CCDC168920[275]

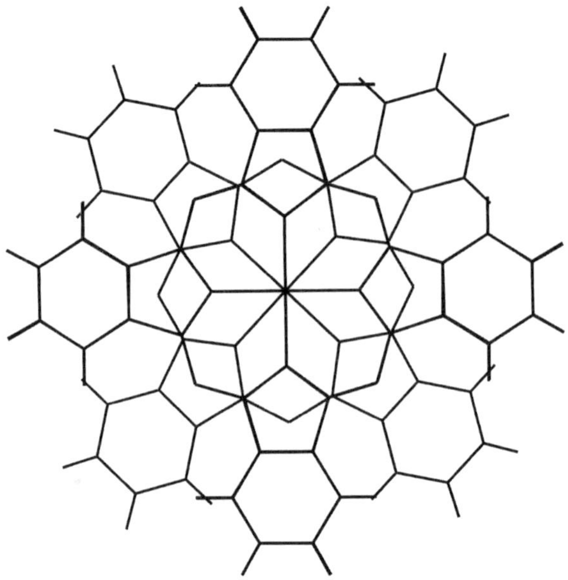

Appendix 11-19. [PcCu]$_3$[(AsF_6)$_2$][1-ClN], CCDC168920[275]

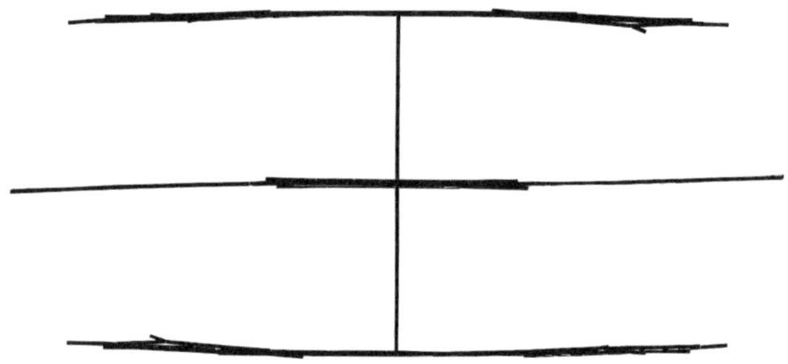

Appendix 11-20. [PcCu]$_3$[(SbF_6)$_2$][1-ClN], CCDC168919[275]

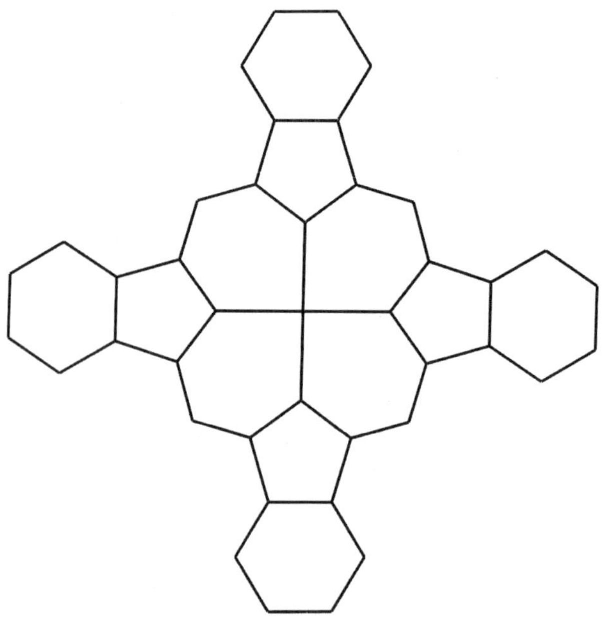

Appendix 11-21. β-PcFe, FEPHCY[81]

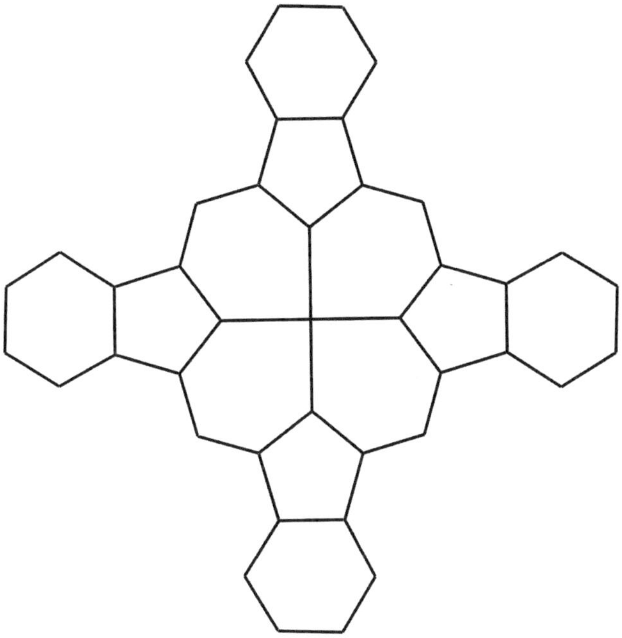

Appendix 11-22. $[PcFe][I_3]_{1/3}$, DAPCEC[157]

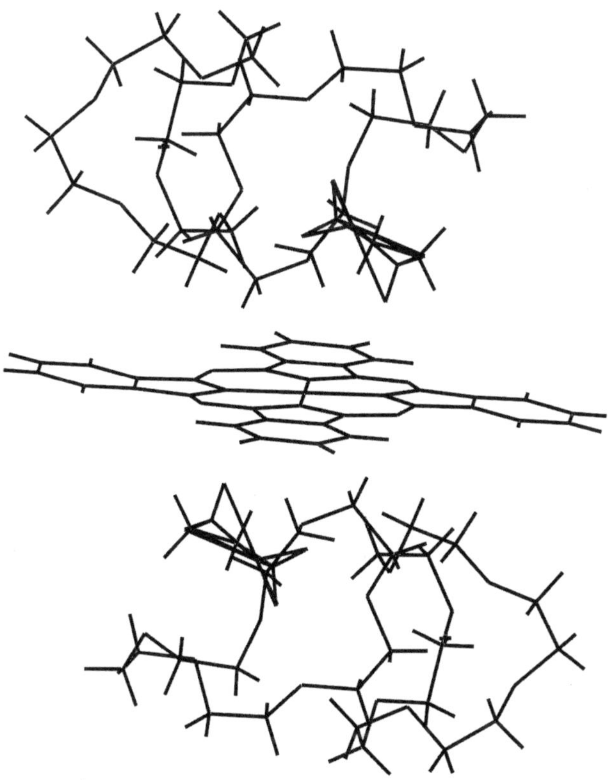

Appendix 11-23. [PcFe][Li.cntnd.18cr6] [Li(THF).cntnd.18c6], FEDXUH[288]

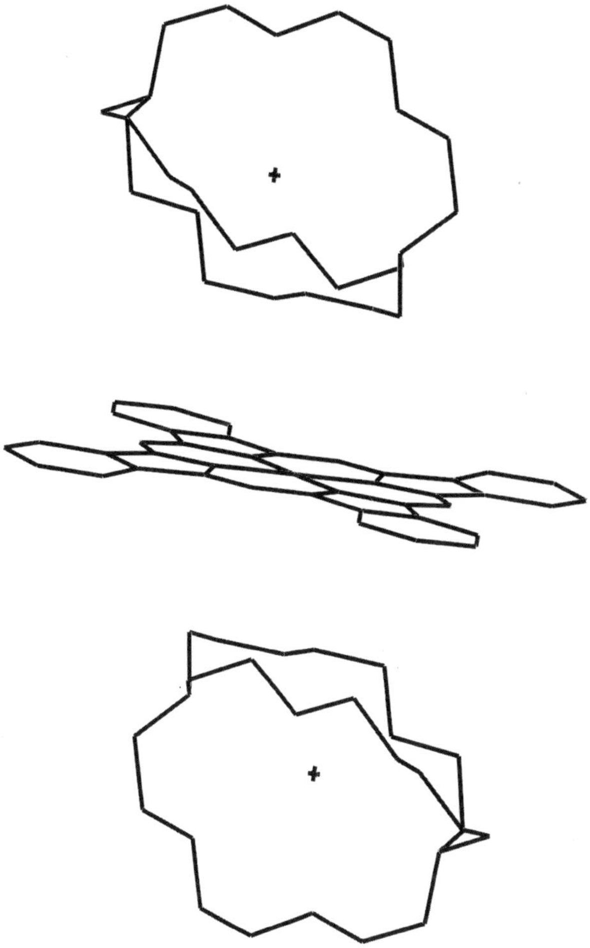

Appendix 11-24. [PcFe][K.cntnd.222], FEDXOB[288]

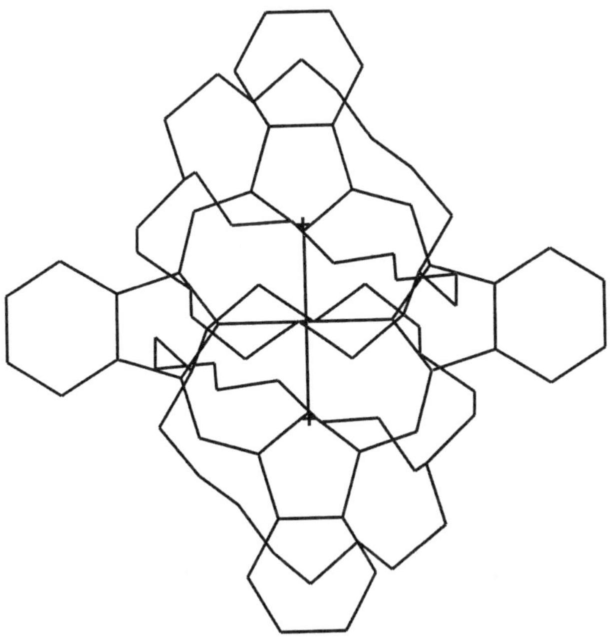

Appendix 11-25. [PcFe][K.cntnd.222], FEDXOB[288]

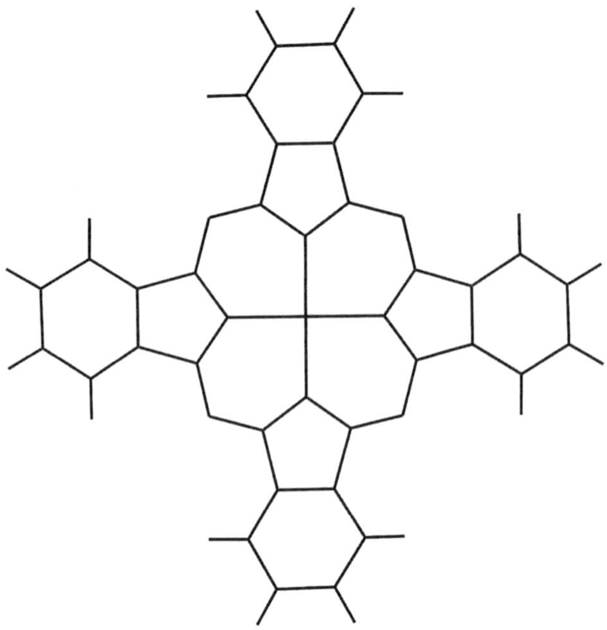

Appendix 11-26. PcLi, DIXNIH02[91]

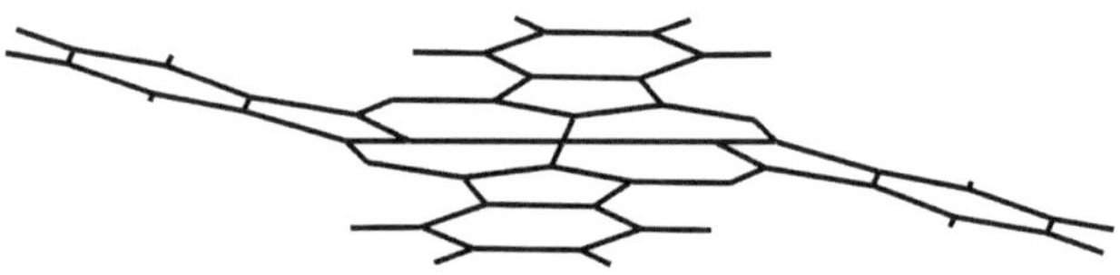

Appendix 11-27. [PcLi][F][NBu_4]$_2$[$0.5H_2O$], RESYET[115]

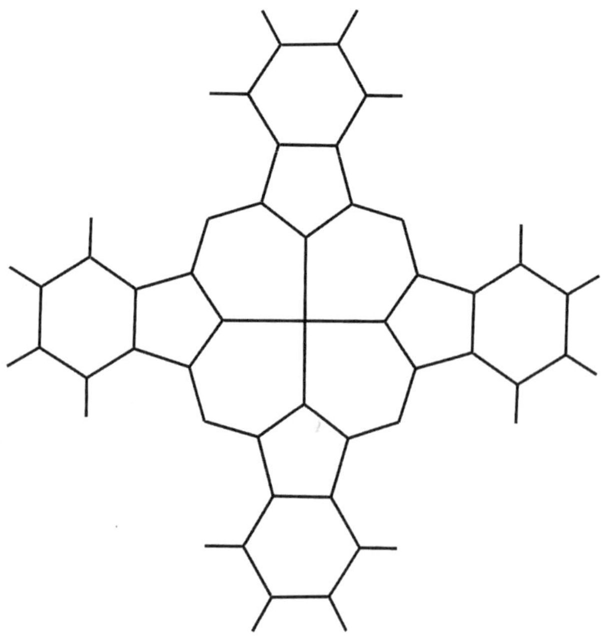

Appendix 11-28. [PcLi][F][NBu_4]$_2$[$0.5H_2O$], RESYET[115]

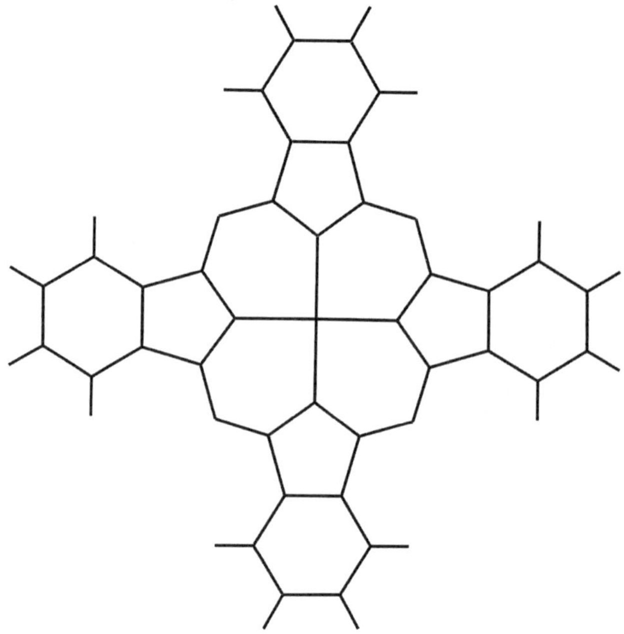

Appendix 11-29. [PcLi][NBu_4][THF], RESYAP[115]

Appendix 11-30. [PcLi][NBu_4][THF], RESYAP[115]

Appendix 11-31. [PcLi][PNP][CH_2Cl_2], ZEPKIO[289]

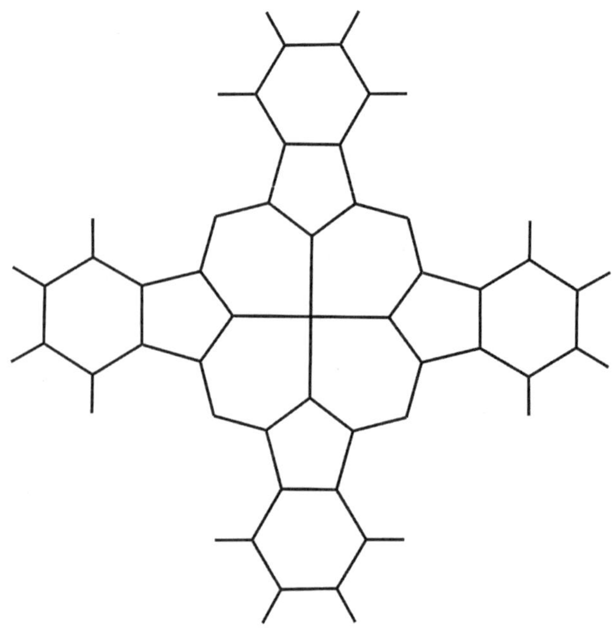

Appendix 11-32. [PcLi][PNP][CH_2Cl_2], ZEPKIO[289]

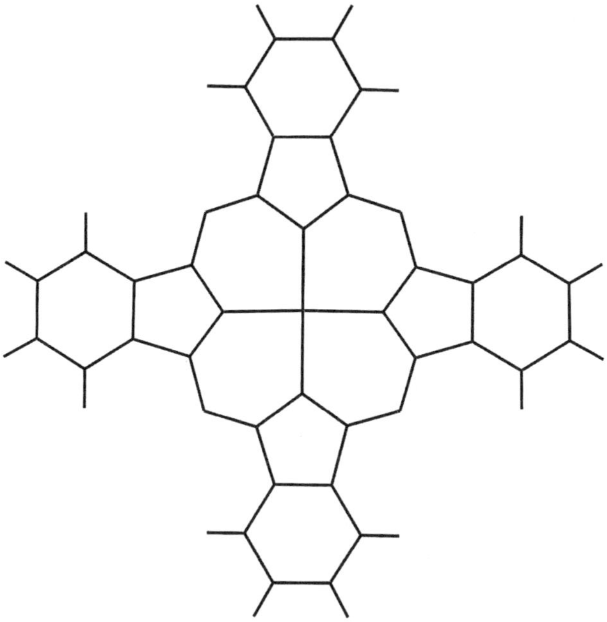

Appendix 11-33. [PcLi][PNP][H_2O], ZEPKOU[289]

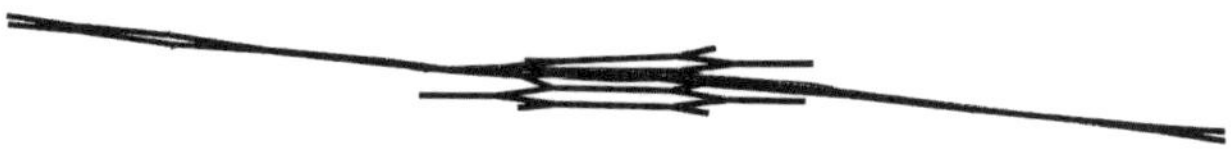

Appendix 11-34. [PcLi][PNP][H_2O], ZEPKOU[289]

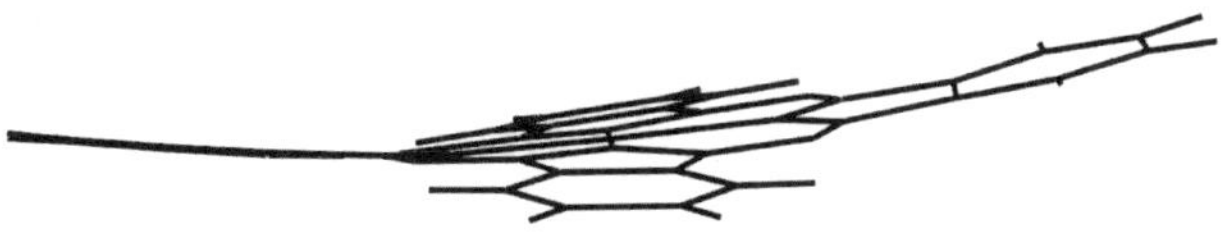

Appendix 11-35. [PcLi][PNP][H_2O], ZEPKOU[289]

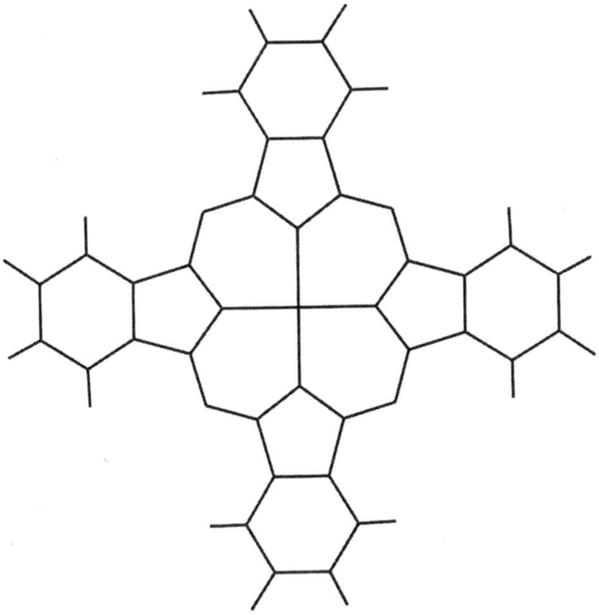

Appendix 11-36. [PcLi][I_3]$_{1/3}$, DIXMUS10[290]

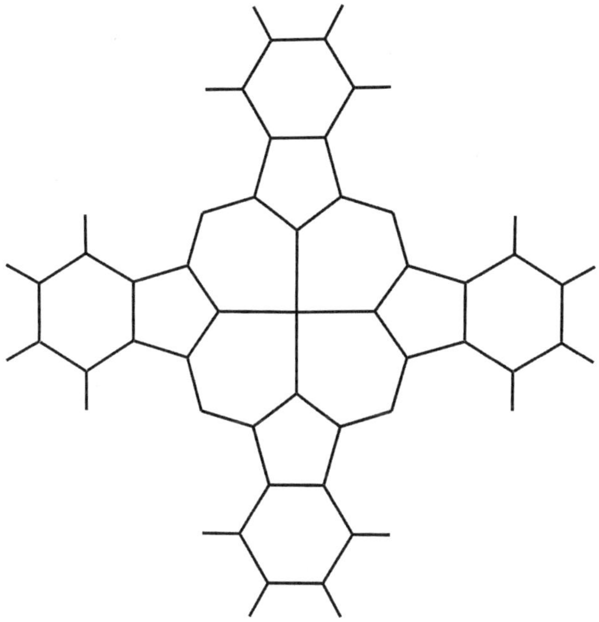

Appendix 11-37. [PcLi][Li($OP\phi_3$)$_4$][EtOH,H_2O], MAKHEL[136]

Appendix 11-38. [PcLi][Li($OP\phi_3$)$_4$][EtOH,H_2O], MAKHEL[136]

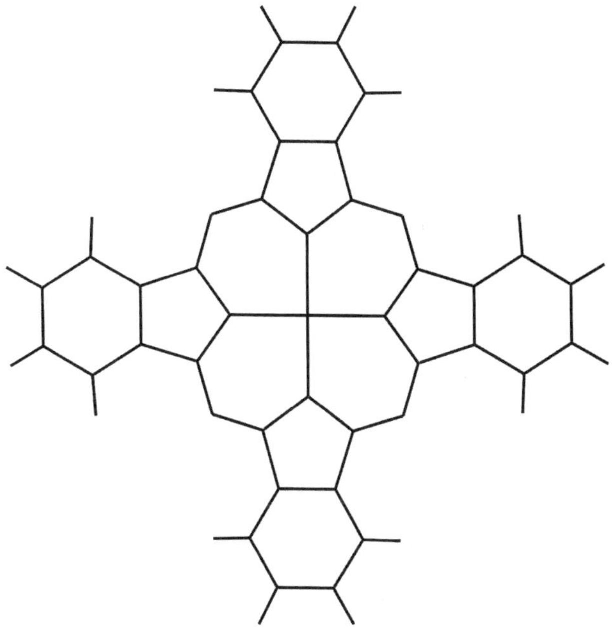

Appendix 11-39. β-PcMn, MNPHCY21[85]

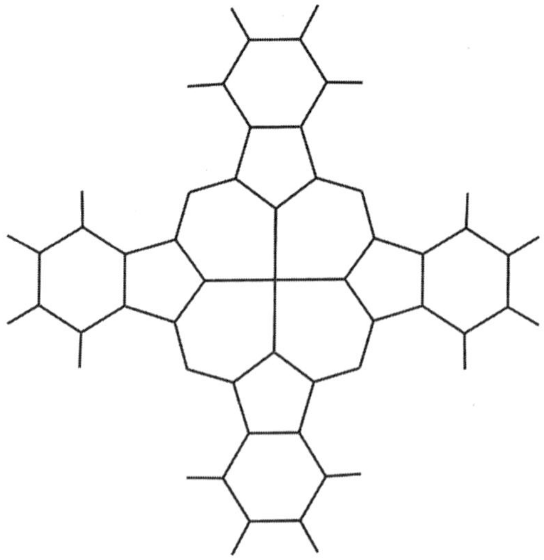

Appendix 11-40. β-PcMn, #1980-01[85]

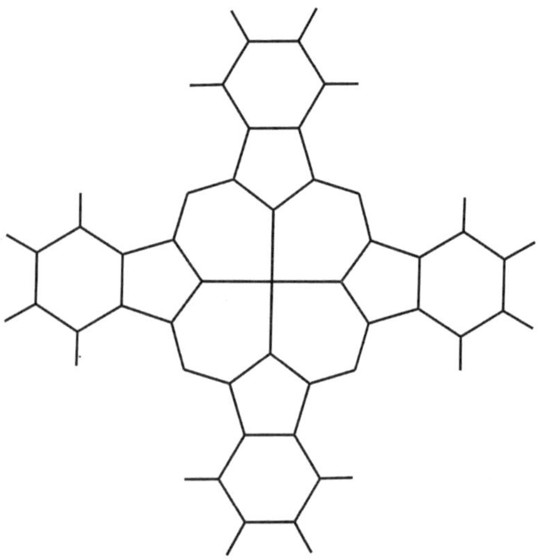

Appendix 11-41. β-PcMn, MNPHCY11[83–85]

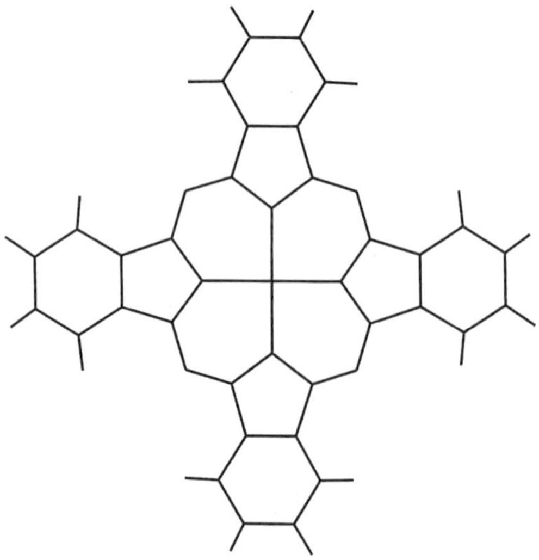

Appendix 11-42. β-PcMn, MNPHCY02[84]

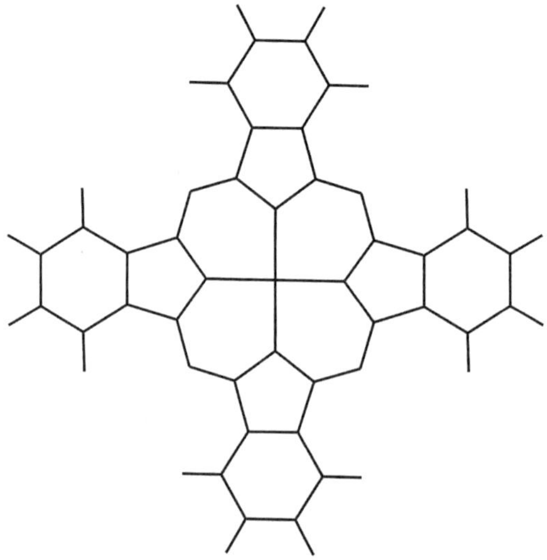

Appendix 11-43. β-PcMn, MNPHCY03[86,291]

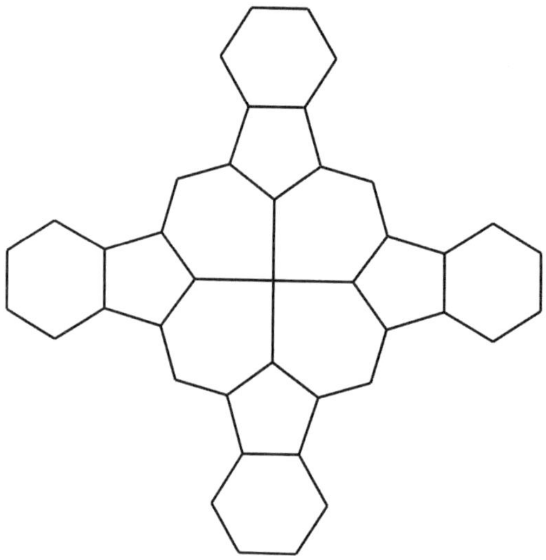

Appendix 11-44. β-PcMn, MNPHCY[81]

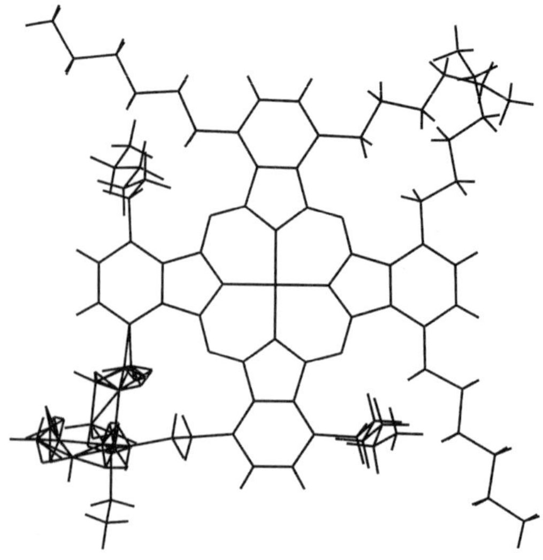

Appendix 11-45. $(H_{13}C_6)_8PcNi$, TOYSUV[240]

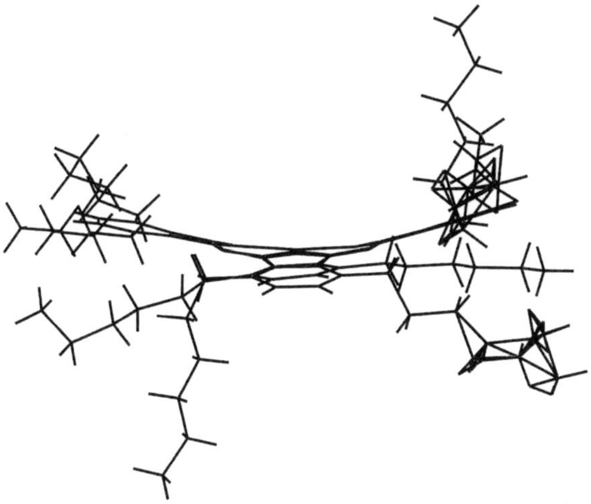

Appendix 11-46. $(H_{13}C_6)_8PcNi$, TOYSUV[240]

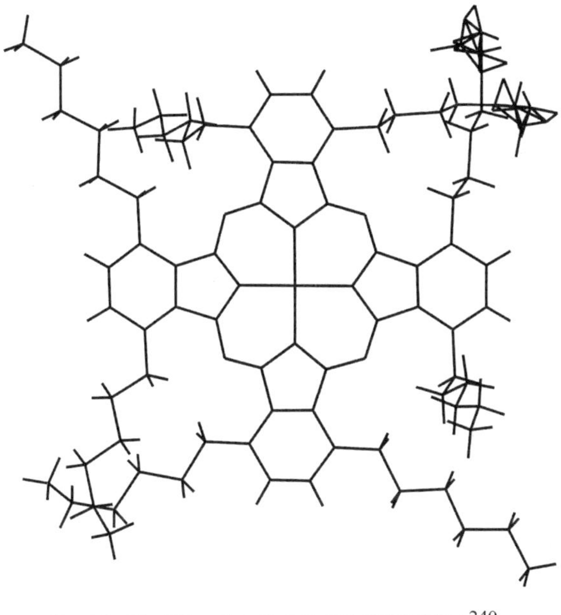

Appendix 11-47. $(H_{13}C_6)_8PcNi$, TOYSUV01[240]

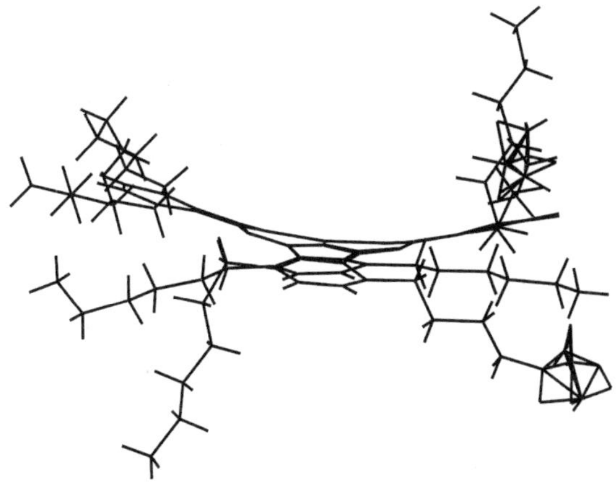

Appendix 11-48. $(H_{13}C_6)_8PcNi$, TOYSUV01[240]

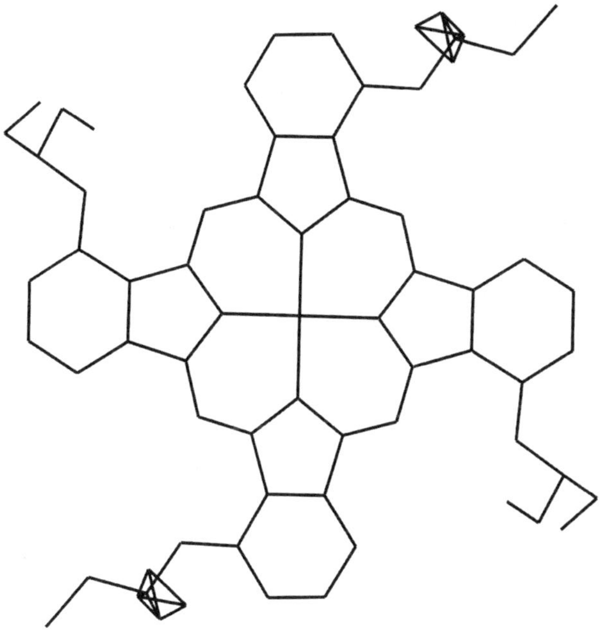

Appendix 11-49. $(i\text{-}H_{11}C_5O)_4PcNi$, ROCCAN[241], ROCCAN01[242]

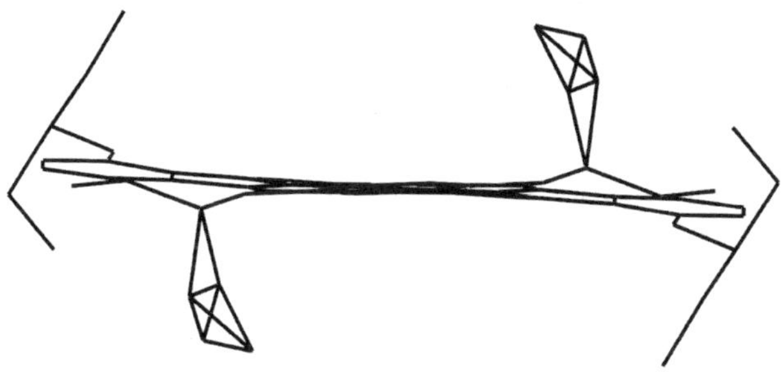

Appendix 11-50. $(i\text{-}H_{11}C_5O)_4PcNi$, ROCCAN[241], ROCCAN01[242]

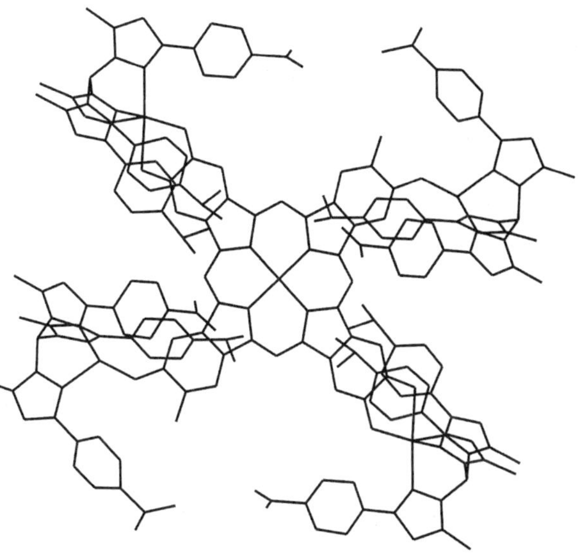

Appendix 11-51. $(RO_2)_2(RO)_2(HO)_2PcNi$, NOCVEC[243]

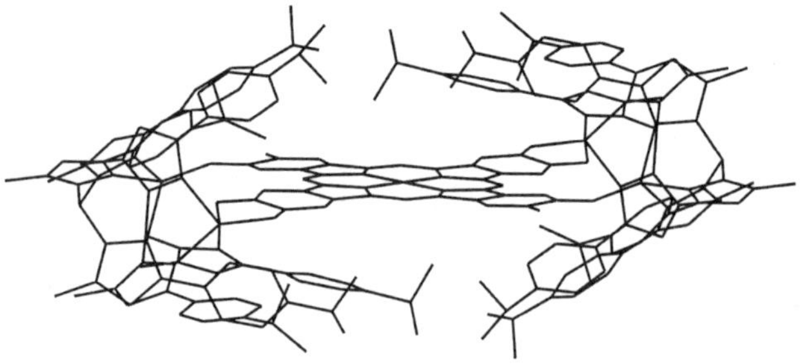

Appendix 11-52. $(RO_2)_2(RO)_2(HO)_2PcNi$, NOCVEC[243]

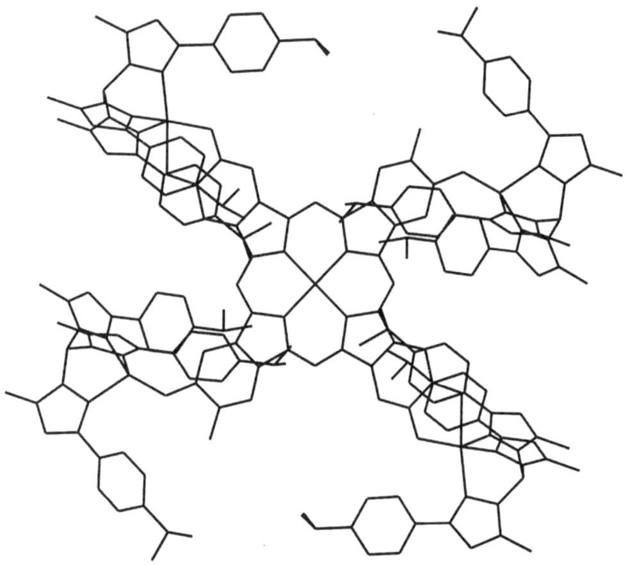

Appendix 11-53. $(RO_2)_2(RO)_2(HO)_2PcNi$, NOCVAC[243]

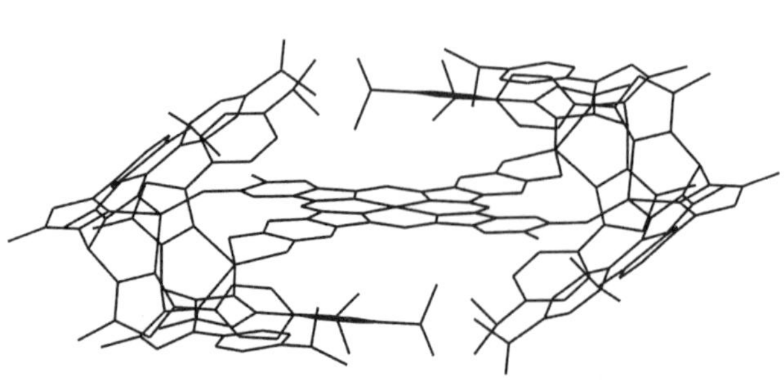

Appendix 11-54. $(RO_2)_2(RO)_2(HO)_2PcNi$, NOCVAC[243]

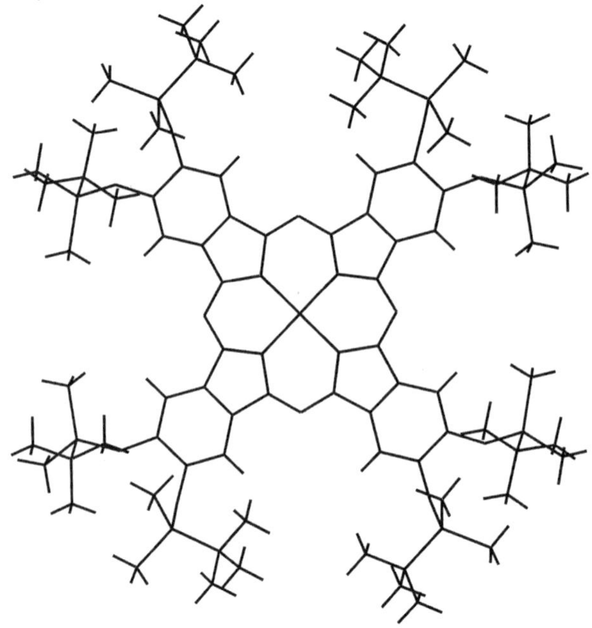

Appendix 11-55. $(tBuMe_2SiO)_8PcNi$, NOCTII[243]

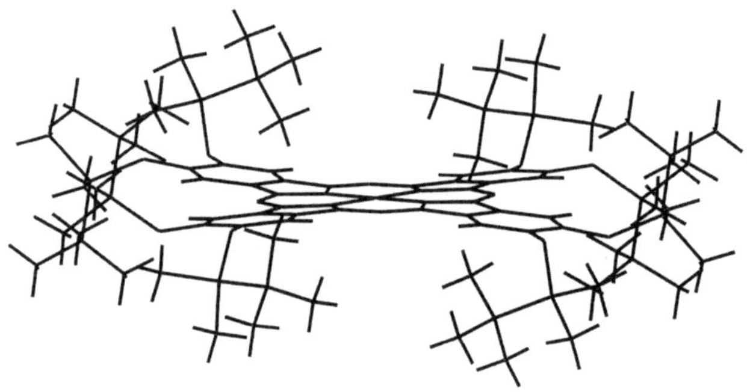

Appendix 11-56. $(tBuMe_2SiO)_8PcNi$, NOCTII[243]

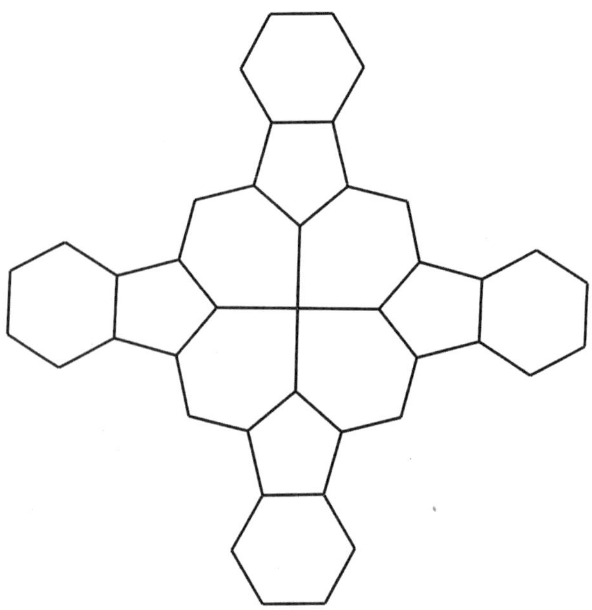

Appendix 11-57. β-PcNi, NIPHTC[13,14,57,58]

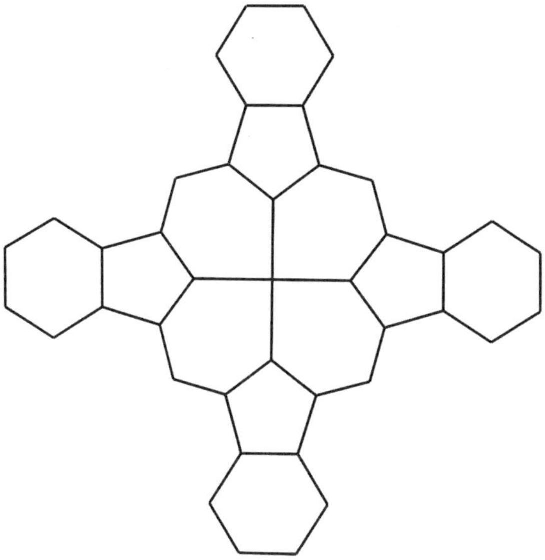

Appendix 11-58. [PcNi][$(AsF_6)_{0.5}$], KARDAI[293,295]

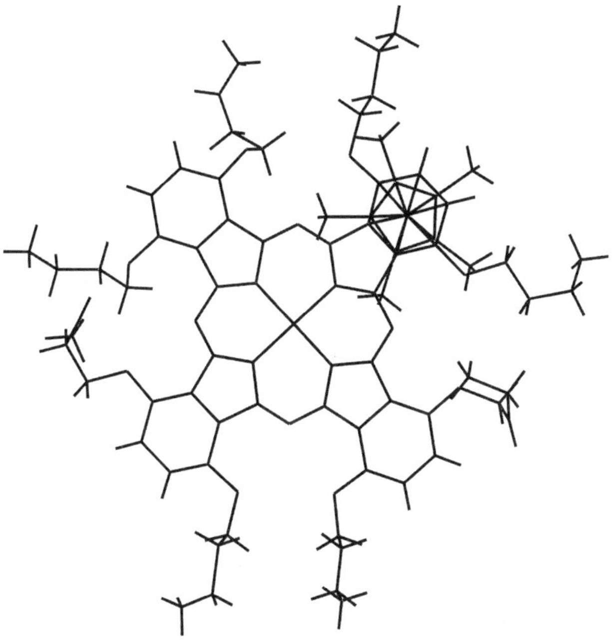

Appendix 11-59. [$(EtO)_8$CpPcNi][PF_6], QEMTAD[246]

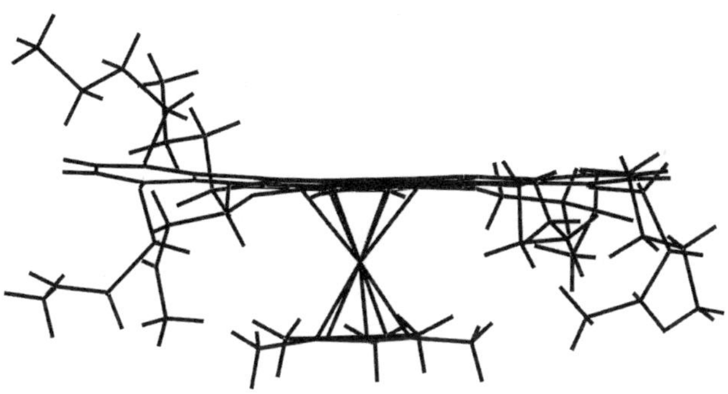

Appendix 11-60. [$(EtO)_8$CpPcNi][PF_6], QEMTAD[246]

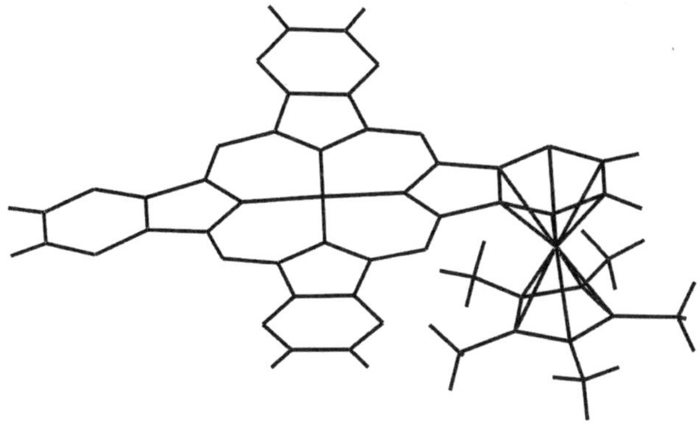

Appendix 11-61. Macrocycle of [$(EtO)_8$CpPcNi][PF_6], QEMTAD[246]

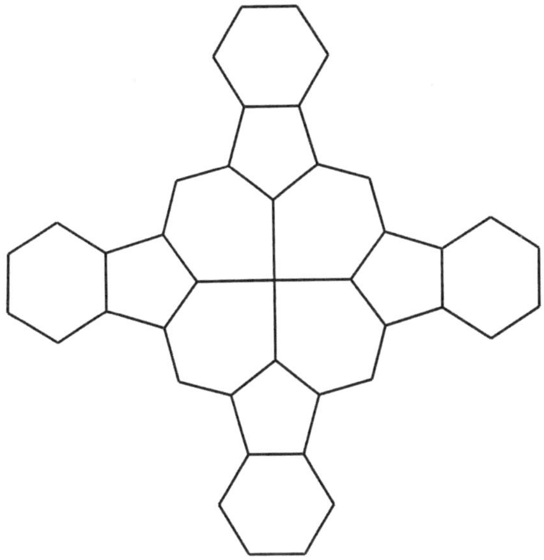

Appendix 11-62. [PcNi][$(SbF_6)_{0.5}$], FANZUP, FANZUP10[292–295]

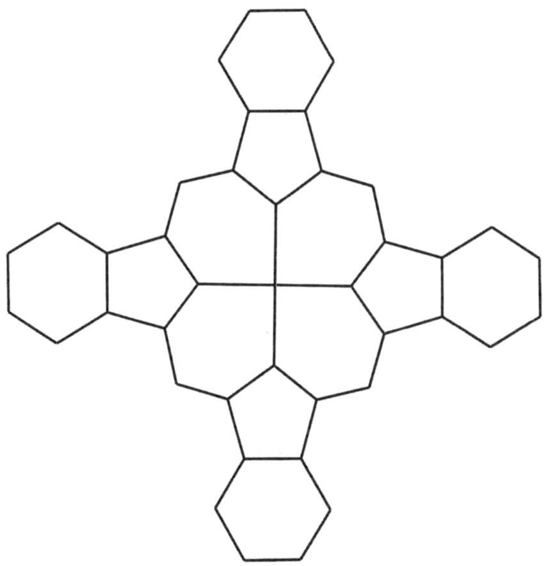

Appendix 11-63. [PcNi][Br_3]$_{1/3}$, DOXCOI[300]

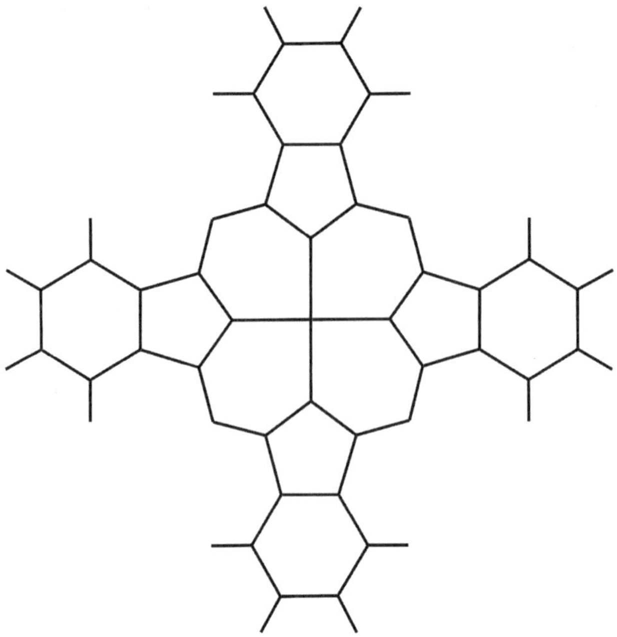

Appendix 11-64. $[PcNi][I_3]_{1/3}$, NIPHTI[111]

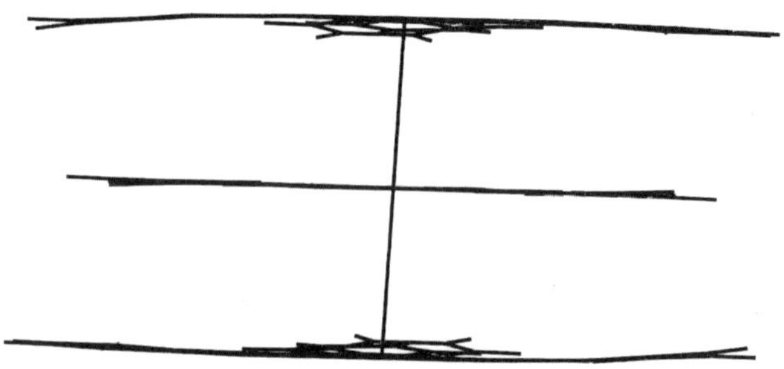

Appendix 11-65. $[PcNi]_3[(ReO_4)_2][1\text{-}ClN]$, CCDC168918[275]

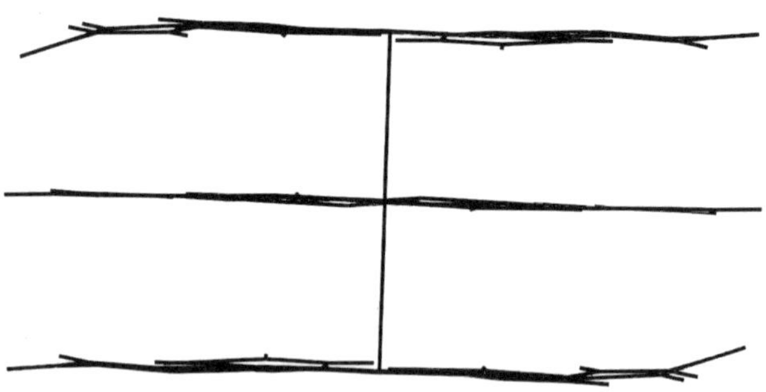

Appendix 11-66. $[PcNi]_3[(SbF_6)_2][1\text{-}ClN]$, CCDC168922[275]

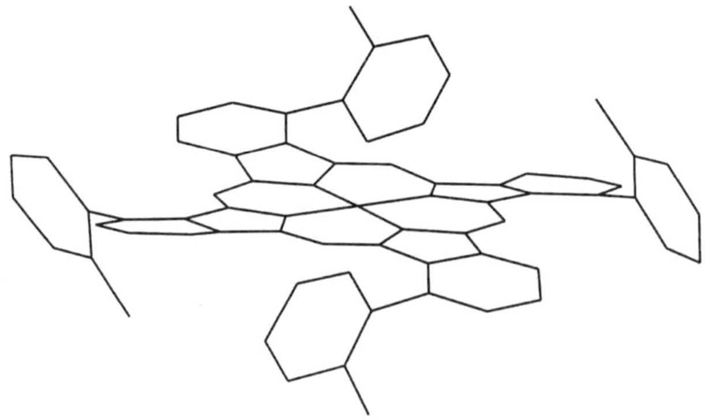

Appendix 11-67. $(o\text{-}tol)_4PcNi$, #2002-03[244]

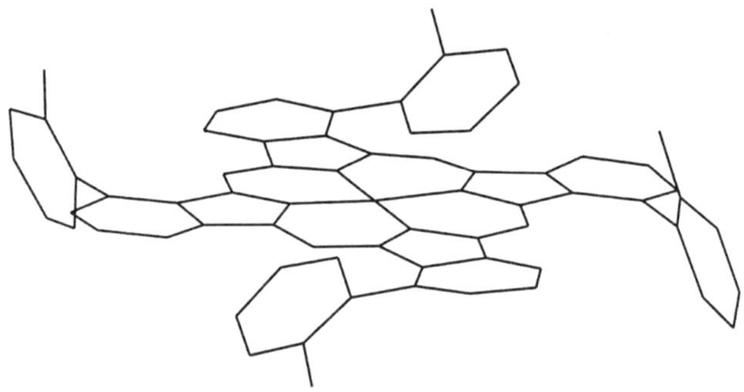

Appendix 11-68. $(o\text{-}tol)_4PcNi$, 2nd molecule #2002-03[244]

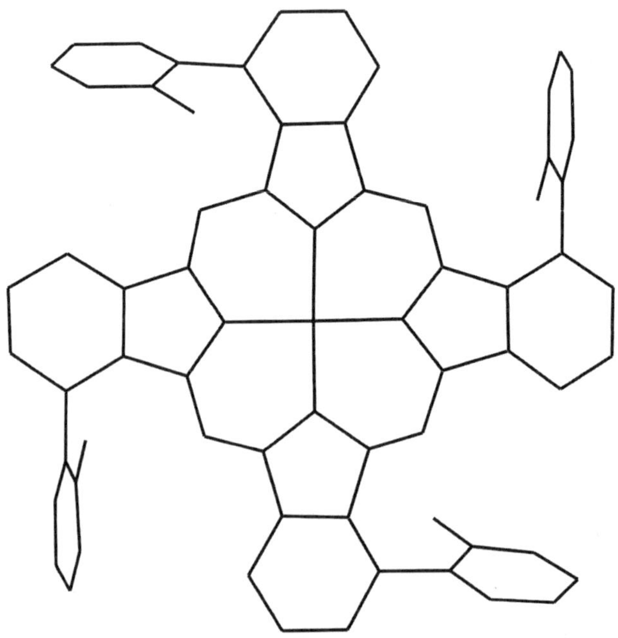

Appendix 11-69. $(o\text{-}tol)_4PcNi$, #2002-03[244]

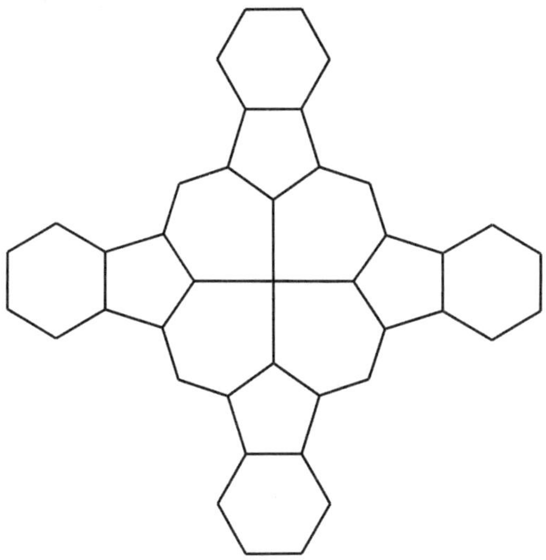

Appendix 11-70. PcPt, PTPOCY02[13,15,57,58]

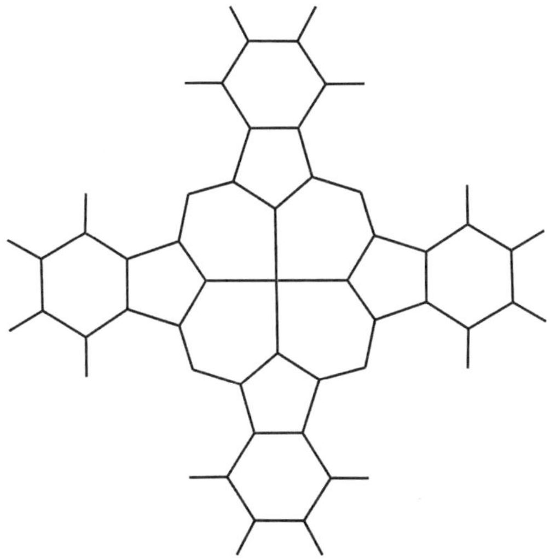

Appendix 11-71. α-PcPt, PTPOCY10[77]

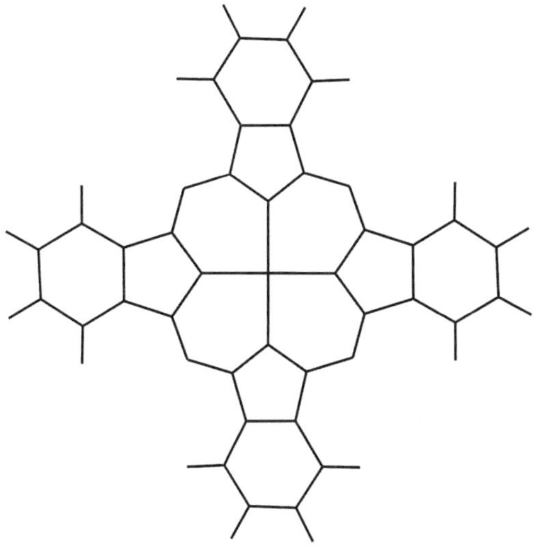

Appendix 11-72. γ-PcPt, PTPOCY11[77]

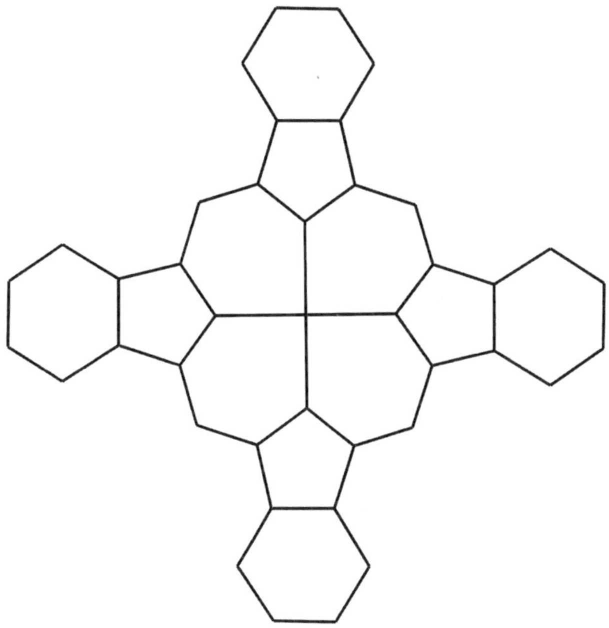

Appendix 11-73. $[PcPt][(ClO_4)_{0.5}]$, JECGAZ[294,301]

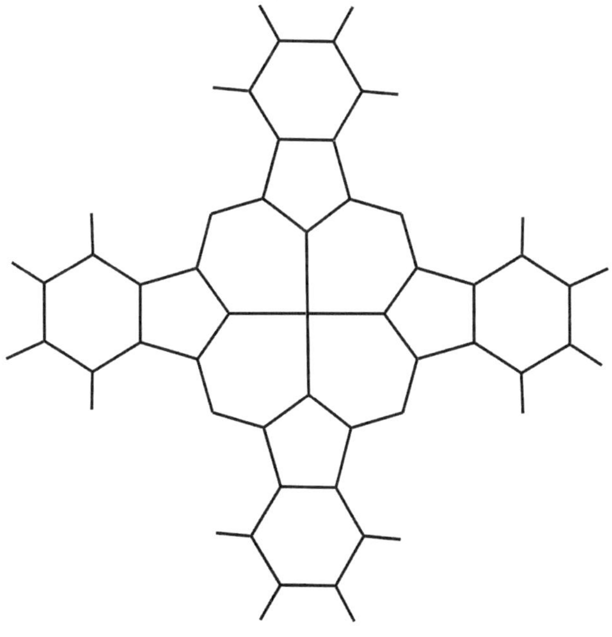

Appendix 11-74. $PcSi_{0.17}$, PHPRSI[87]

Appendix 11-75. $PcSi_{0.17}$, PHPRSI[87]

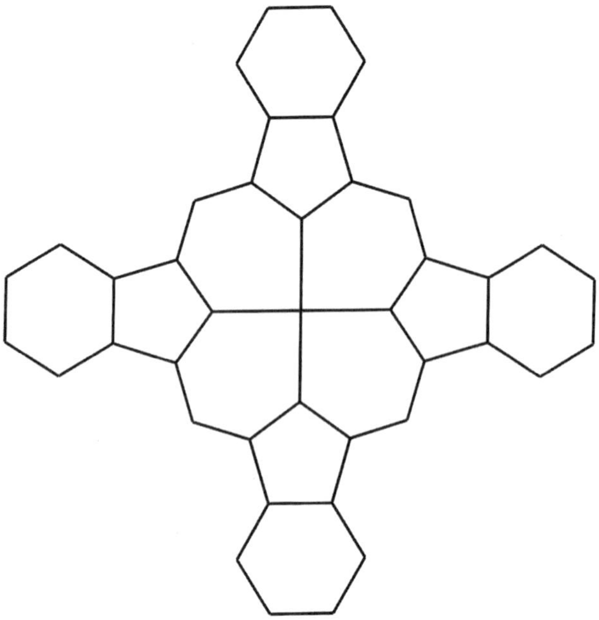

Appendix 11-76. β-PcZn, PTHCZN[82]

Appendix 12. Molecular Structures of CN5 Phthalocyanine Complexes

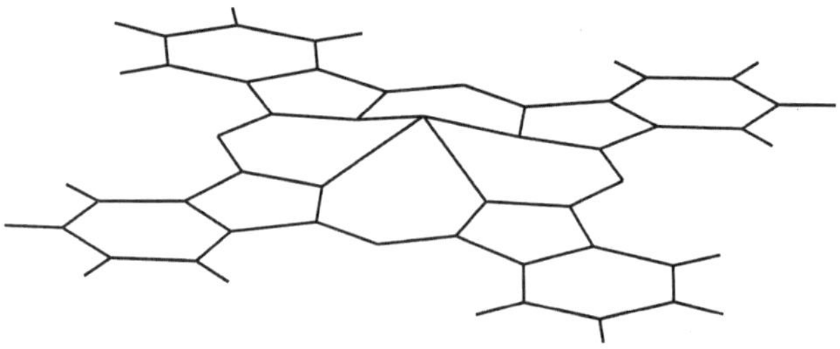

Appendix 12-1. $[PcAs][I_3]$, CCDC180524[303]

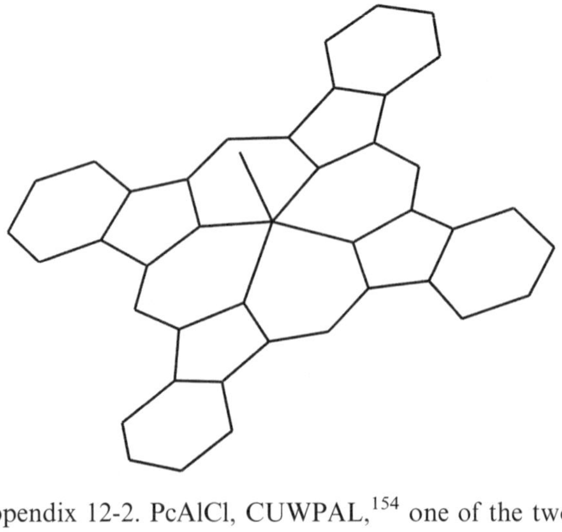

Appendix 12-2. PcAlCl, CUWPAL,[154] one of the two crystallographically different molecules of the unit cell

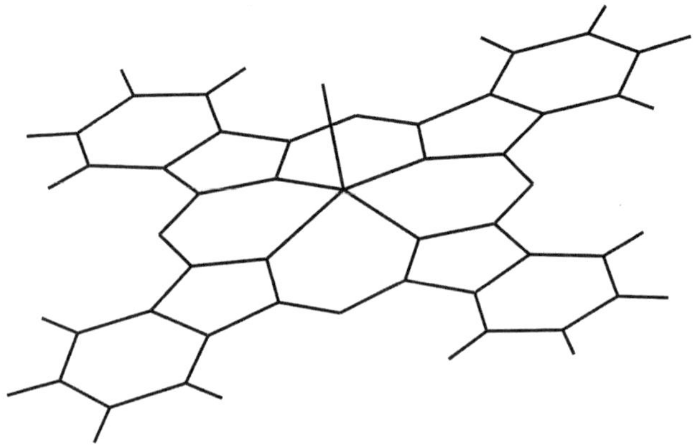

Appendix 12-3. PcAlCl, #1995-01[155,156] (App. 28)

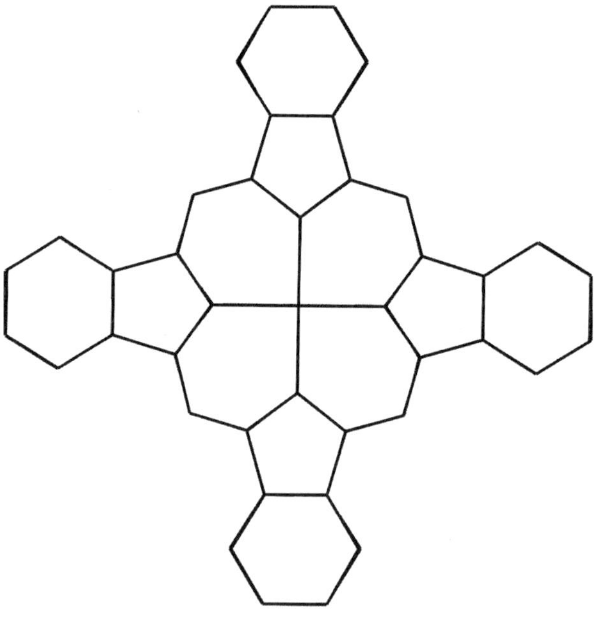

Appendix 12-4. Perpendicular view of PcAl-O-AlPc, DAYYOR[122]

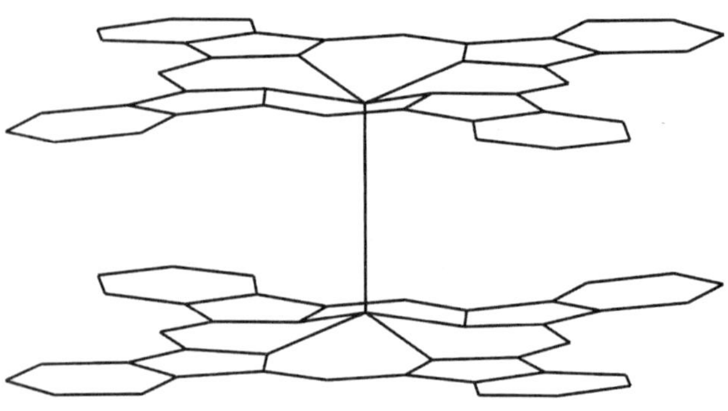

Appendix 12-5. PcAl-O-AlPc, DAYYOR[122]

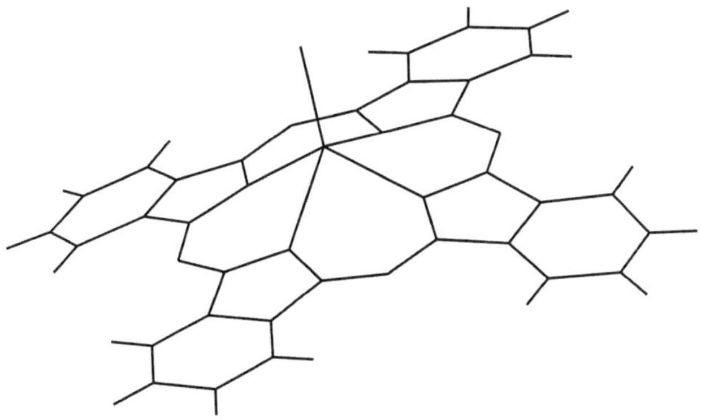

Appendix 12-6. [PcCd-Cl][NEt$_4$][py], #1997-01[304]

Appendix 12-7. [PcCd-Cl][NEt$_4$][py], #1997-01[304]

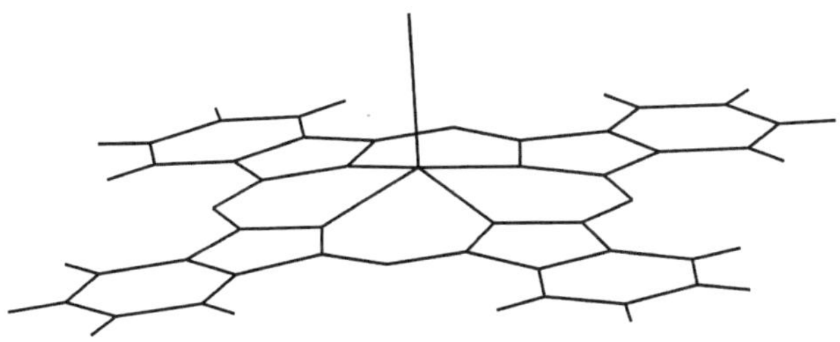

Appendix 12-8. PcCrI, YEJJIG[305]

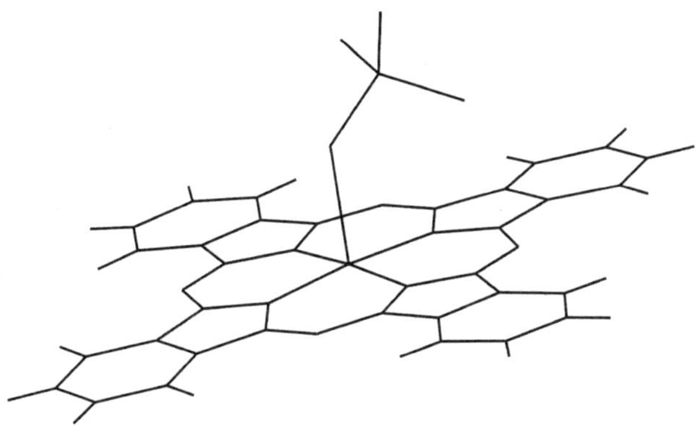

Appendix 12-9. PcCu–OReO$_3$, QEMMIE[306]

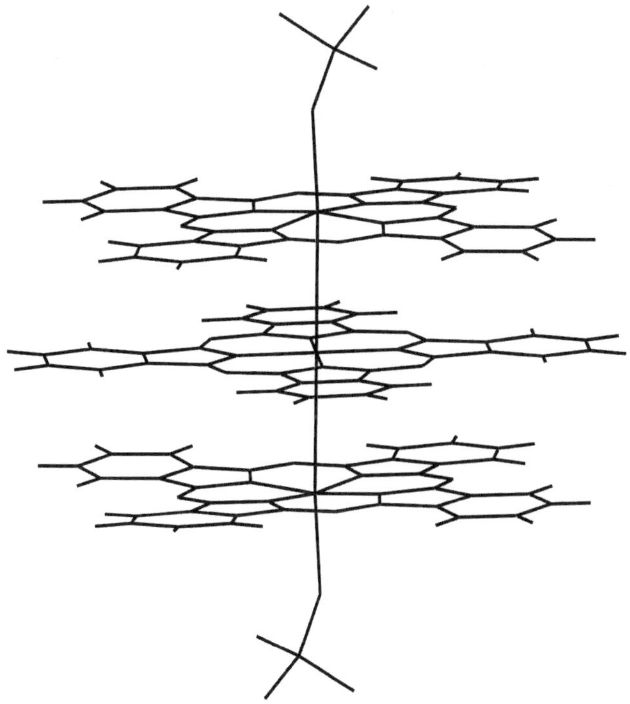

Appendix 12-10. [(O$_3$ReO–CuPc)$_2$PcCu], #2002-01[302]

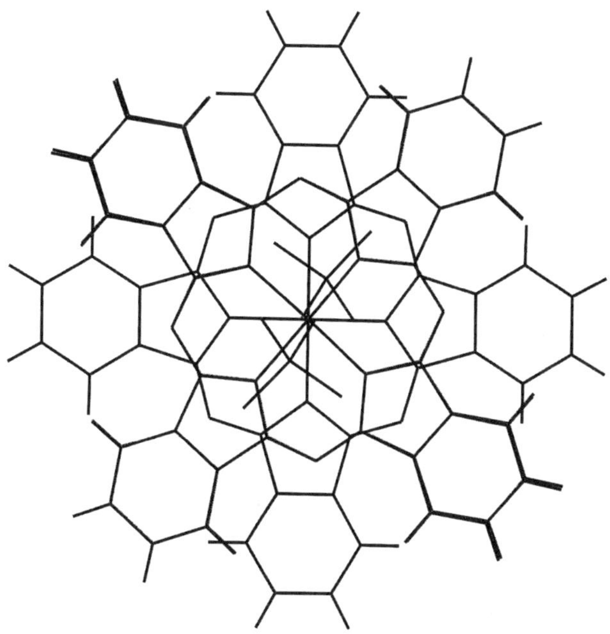

Appendix 12-11. [(O$_3$ReO–CuPc)$_2$PcCu], #2002-01[302]

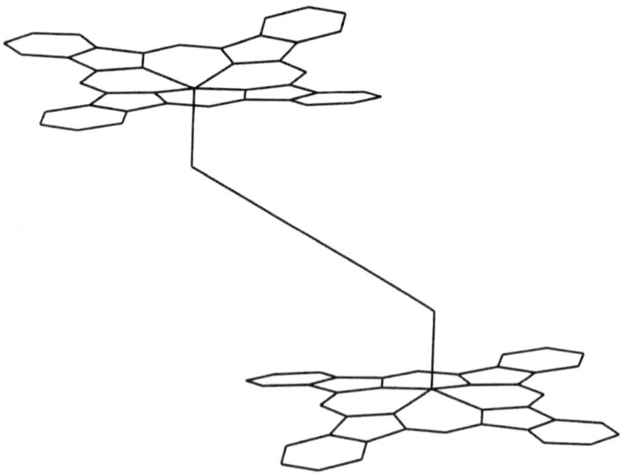

Appendix 12-12. PcFe-Cl-I-I-Cl-FePc, DAPCIG[157]

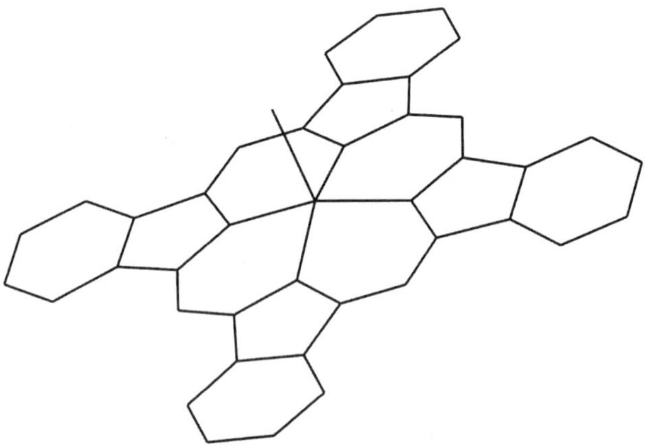

Appendix 12-13. One macrocycle of PcFe-Cl-I-I-Cl-FePc, DAPCIG[157]

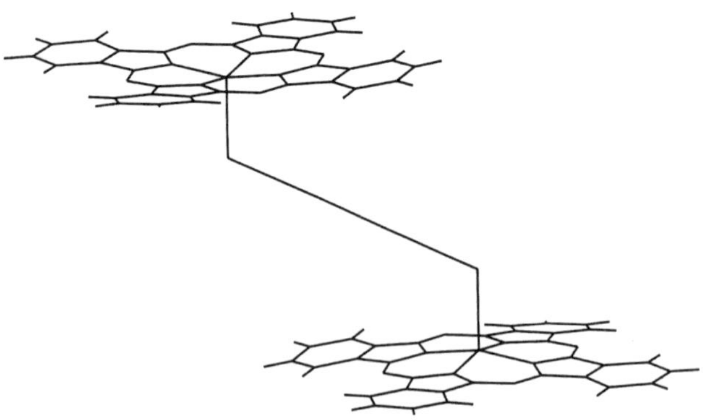

Appendix 12-14. PcFe-I-I-I-I-FePc, FIQFUG[189]

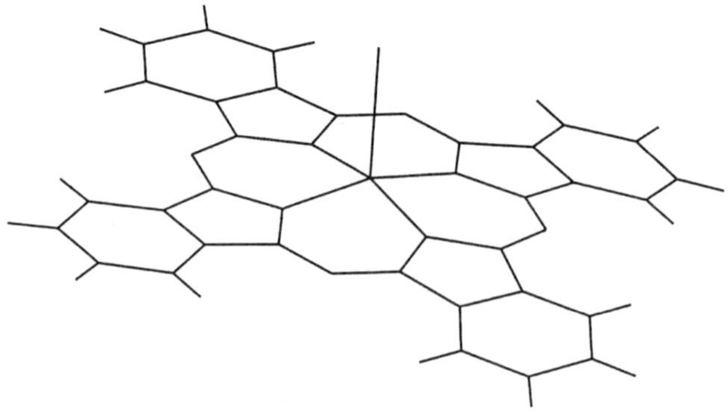

Appendix 12-15. One macrocycle of PcFe-I-I-I-I-FePc, FIQFUG[189]

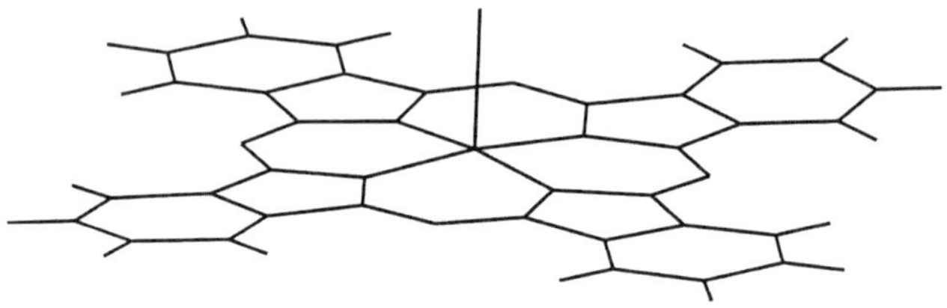

Appendix 12-16. [PcFe-Me][Na.cntnd.18cr6][THF,ϕCl], GEDKAB[161]

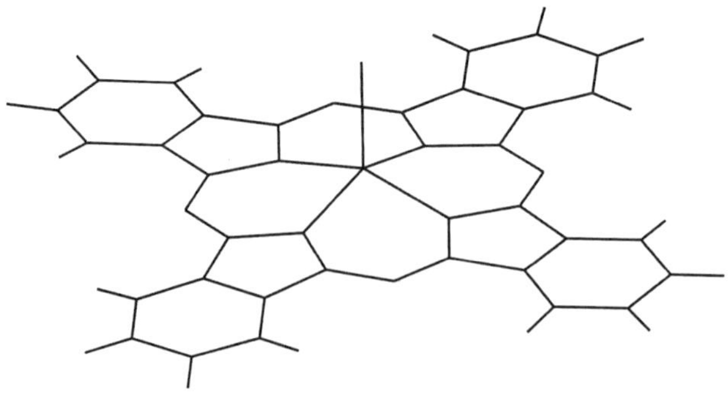

Appendix 12-17. PcGaCl, CUWNUD[154]

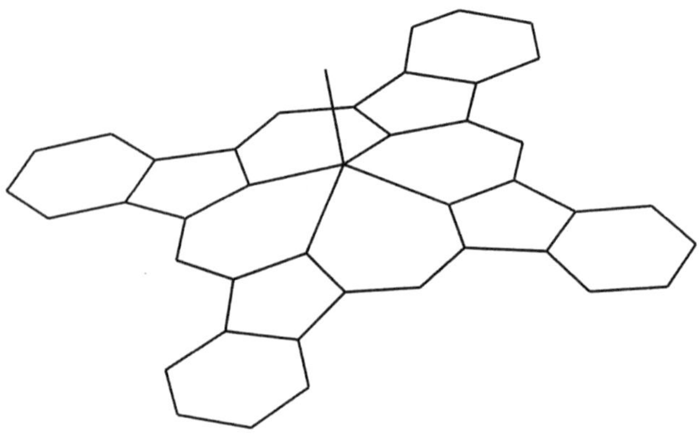

Appendix 12-18. PcGaCl, CUWNUD01[174]

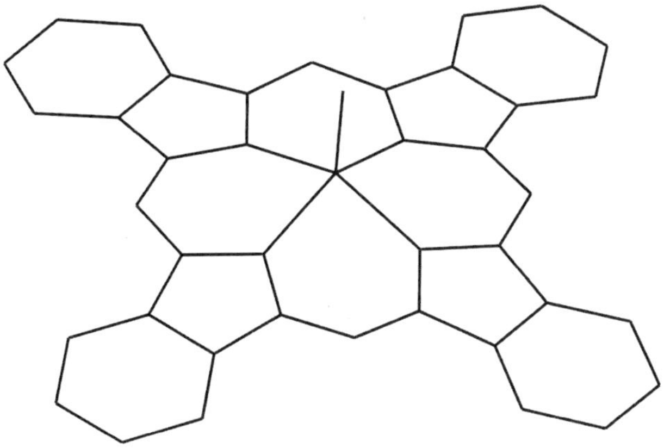

Appendix 12-19. PcGaOH (R), NALSOI02[167]

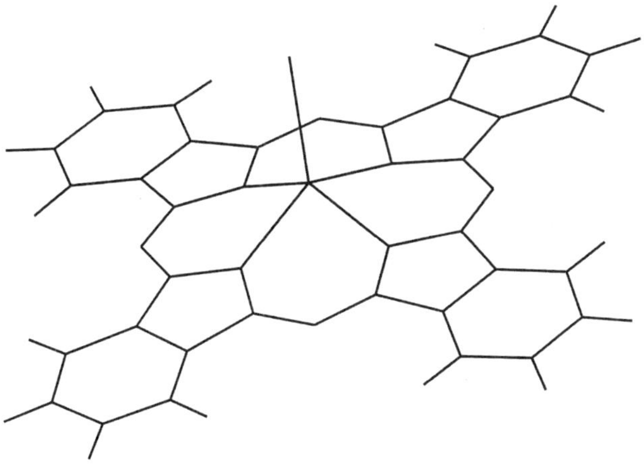

Appendix 12-20. PcInI, BIJQOA[187]

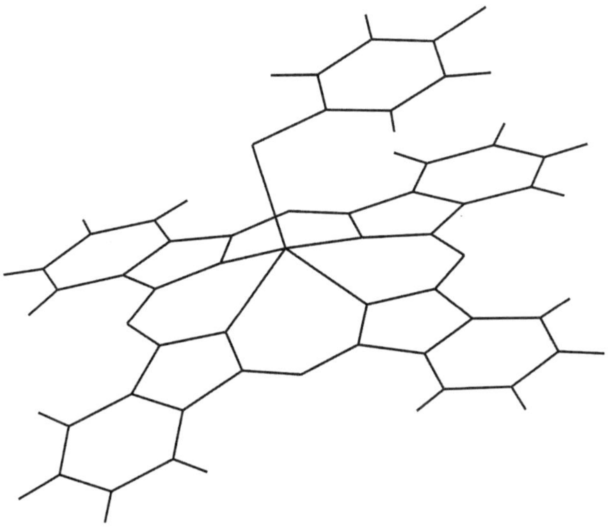

Appendix 12-21. PcIn–Sϕ, #1998-03[308]

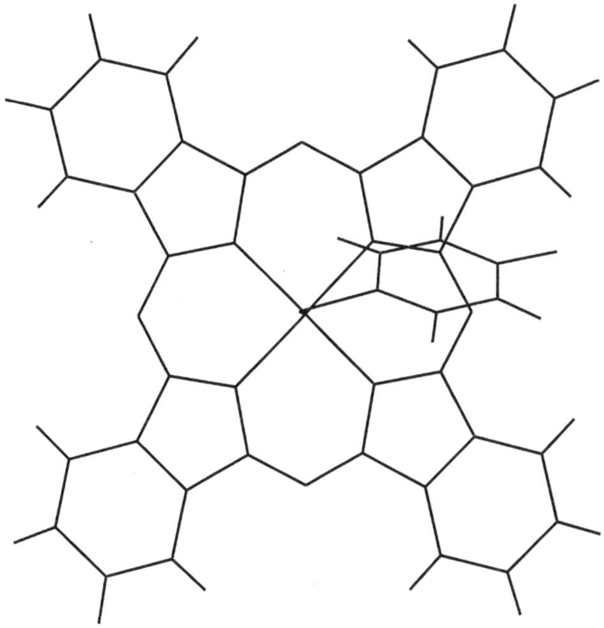

Appendix 12-22. PcIn–Sϕ, #1998-03[308]

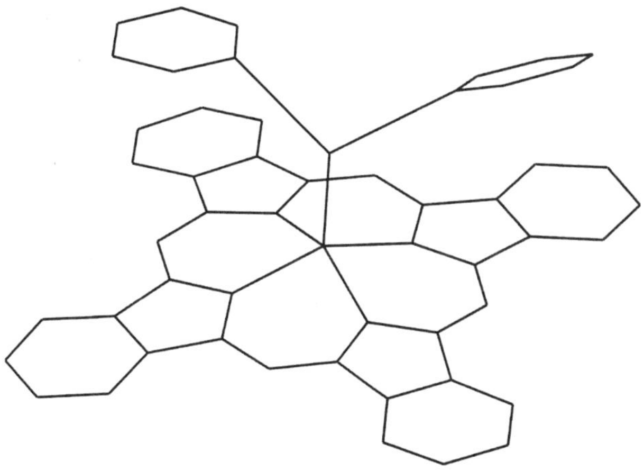

Appendix 12-23. [PcMg-OH_2][2py], MGPCPH[145]

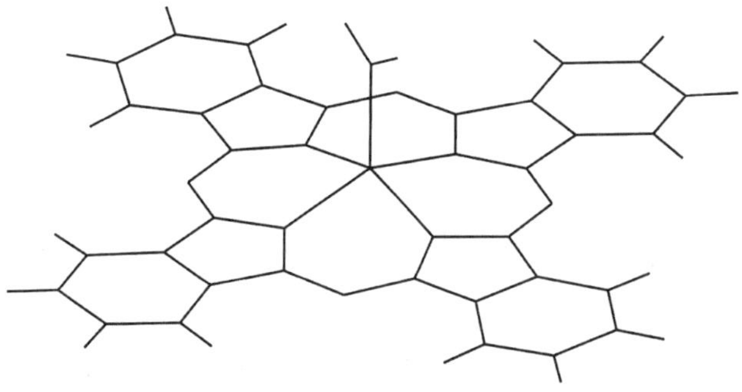

Appendix 12-24. PcMg(OH_2), #2002-06[309]

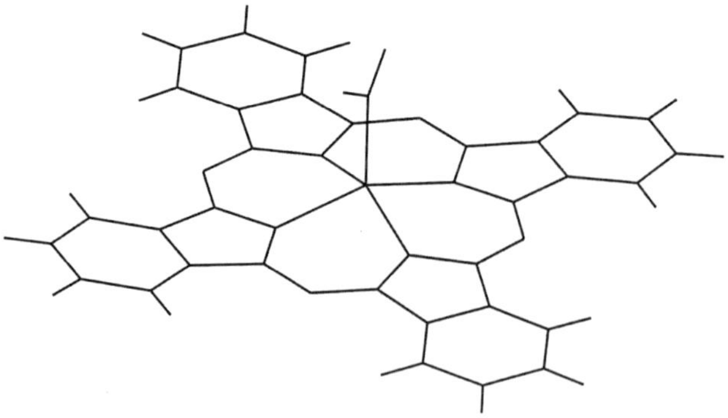

Appendix 12-25. $[PcMg(OH_2py)]_2[py]$, #2002-07[309]

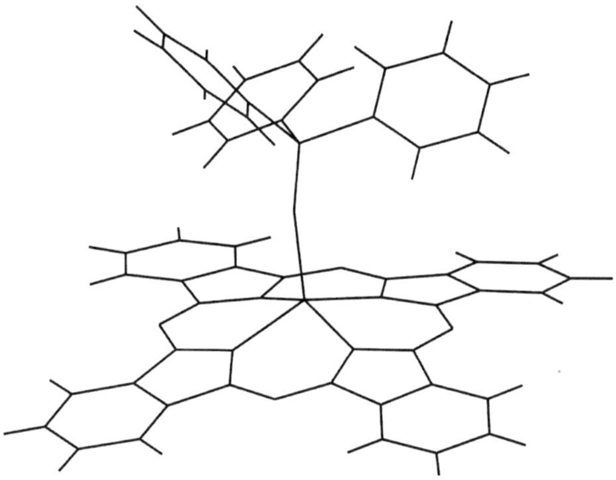

Appendix 12-26. $[PcMg\text{-}OP\phi_3][2py]$, MAKHOV[136]

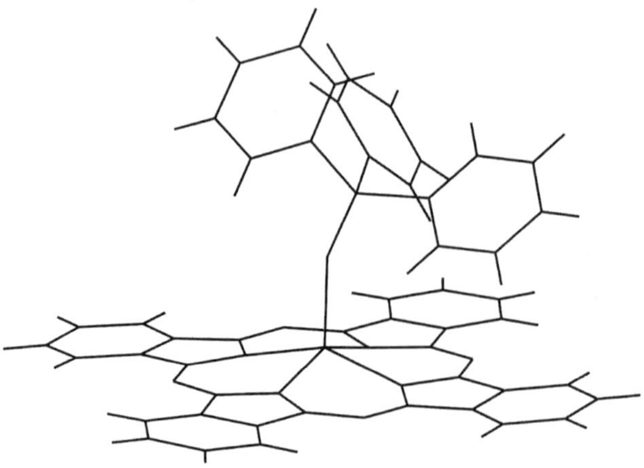

Appendix 12-27. $[PcMg\text{-}OP\phi_3][CH_2Cl_2]$, MAKHUB[136]

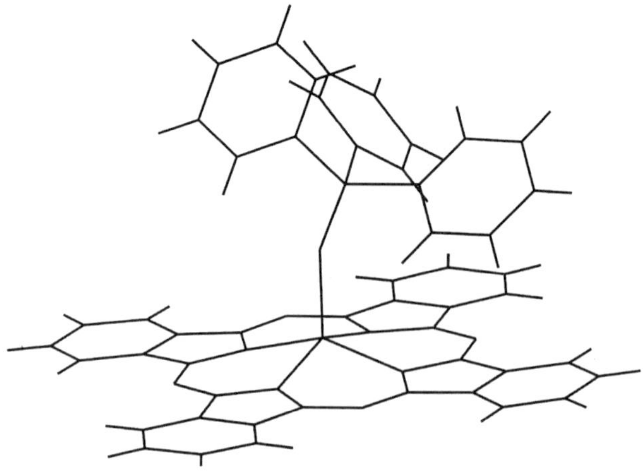

Appendix 12-28. $[PcMg\text{-}OP\phi_3][THF]$, MAKJAJ[136]

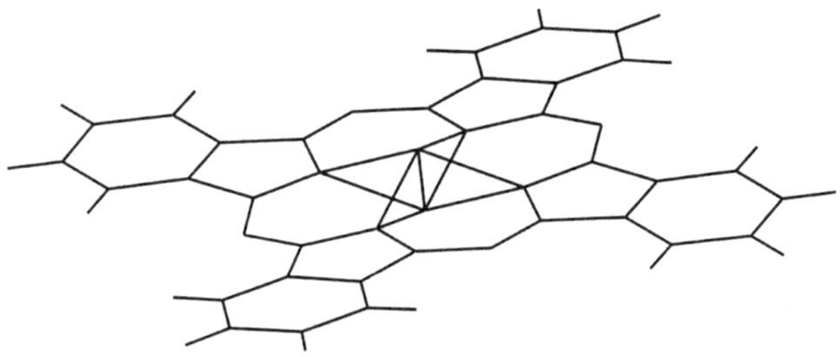

Appendix 12-29. PcMg, XETZIF01[59]

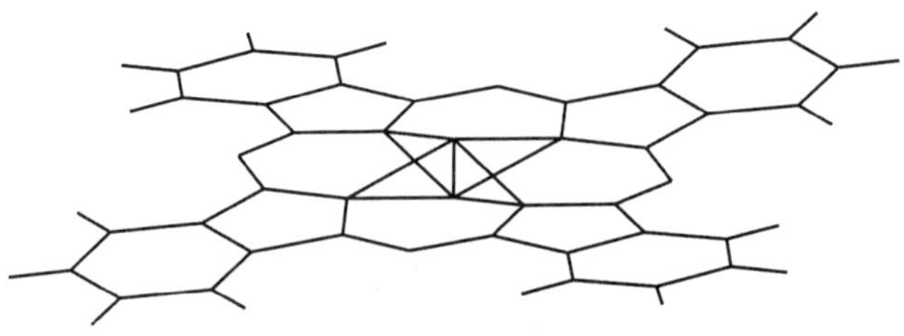

Appendix 12-30. PcMg, XETZIF02[59]

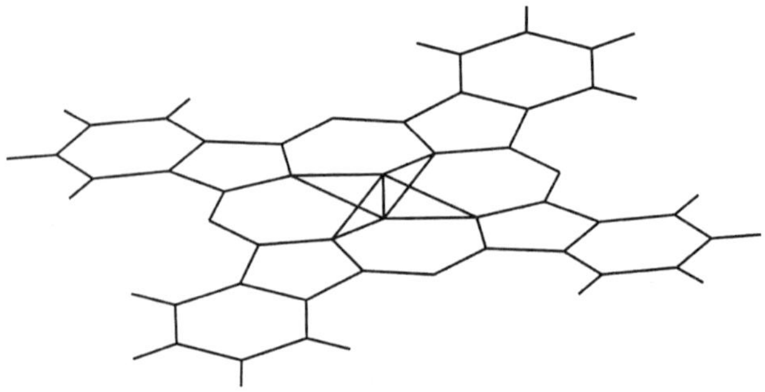

Appendix 12-31. PcMg, XETZIF03[59]

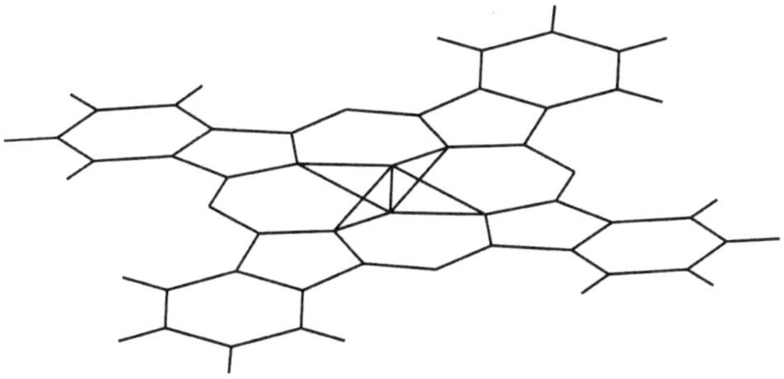

Appendix 12-32. PcMg, XETZIF[60,61]

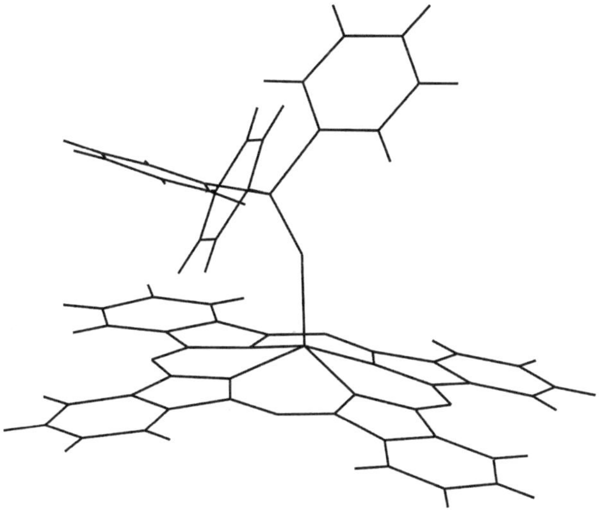

Appendix 12-33. PcMg-OPϕ_3, MAKHIP[136]

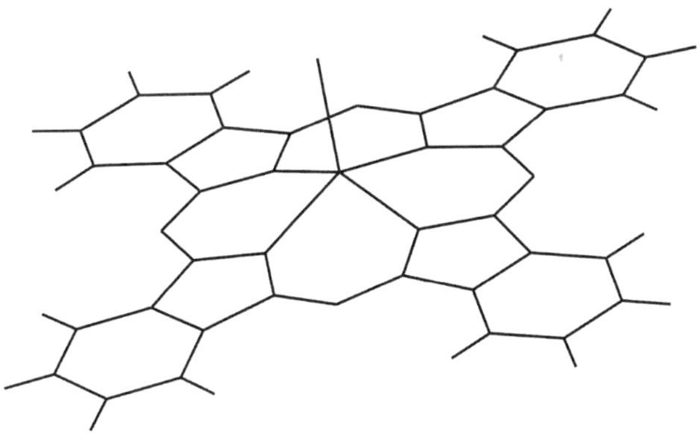

Appendix 12-34. [PcMg–Cl][PNP][Me_2CO], TAYSIV[172]

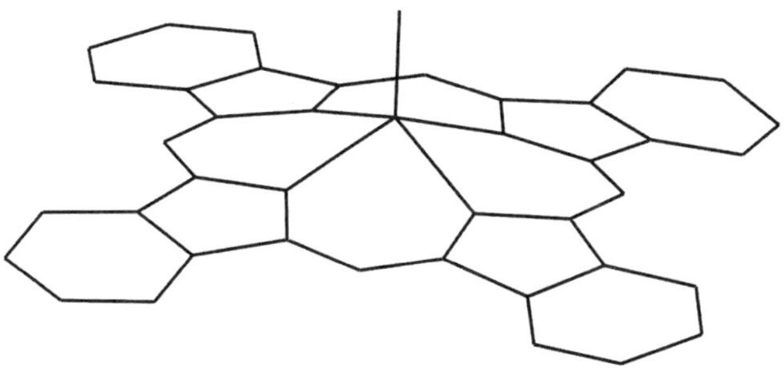

Appendix 12-35. PcM=O, CUTGIH[153,310]

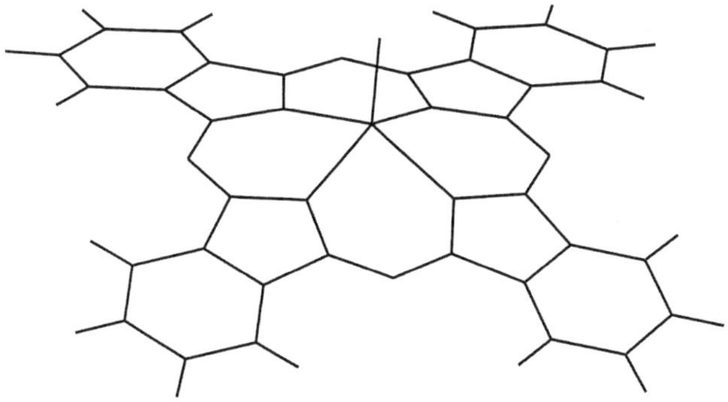

Appendix 12-36. [PcMo=O][I_3], FEWPEC[186]

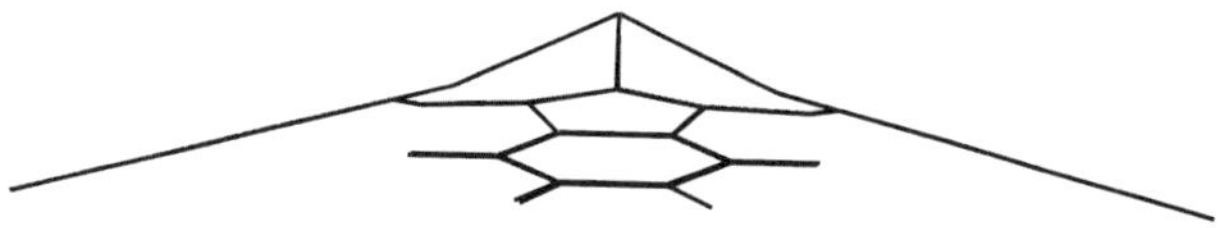

Appendix 12-37. PcPb, PBPHCY[133]

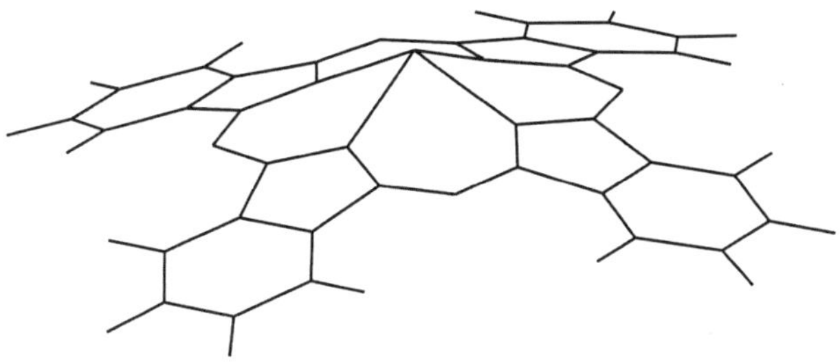

Appendix 12-38. PcPb, PBPHCY[133]

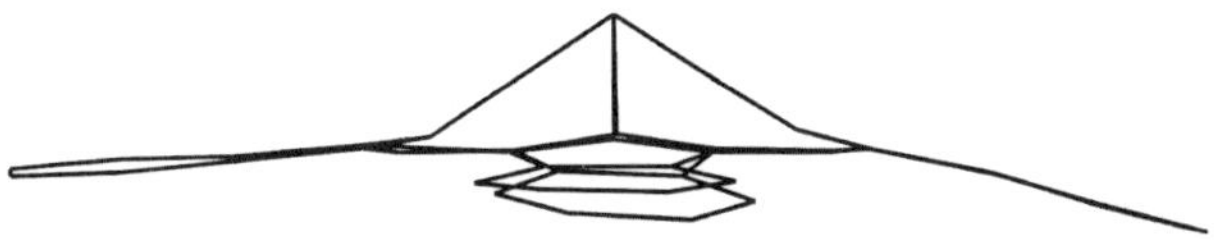

Appendix 12-39. PcPb, PBPHCY01[132]

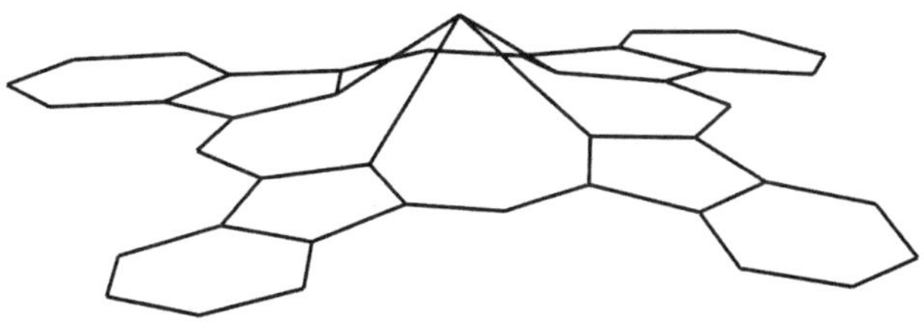

Appendix 12-40. PcPb, PBPHCY01[132]

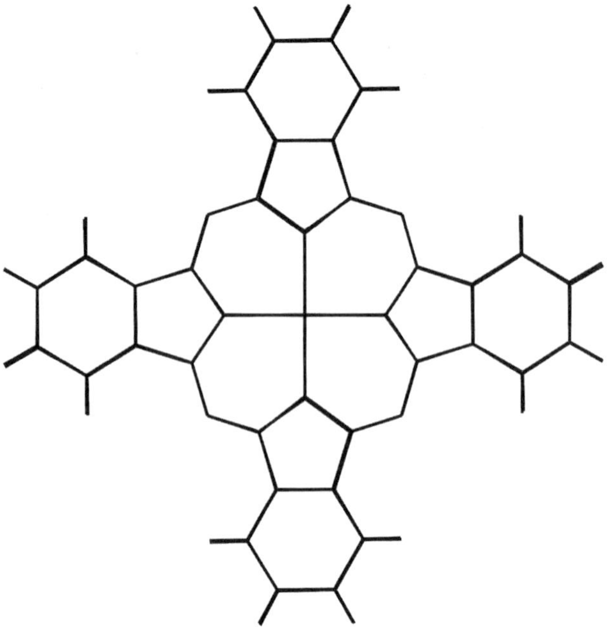

Appendix 12-41. PcRe–RePc, NEKRAW[181]

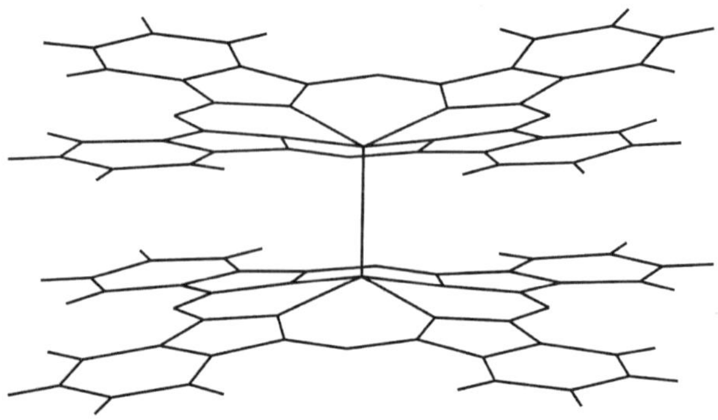

Appendix 12-42. PcRe–RePc, NEKRAW[181]

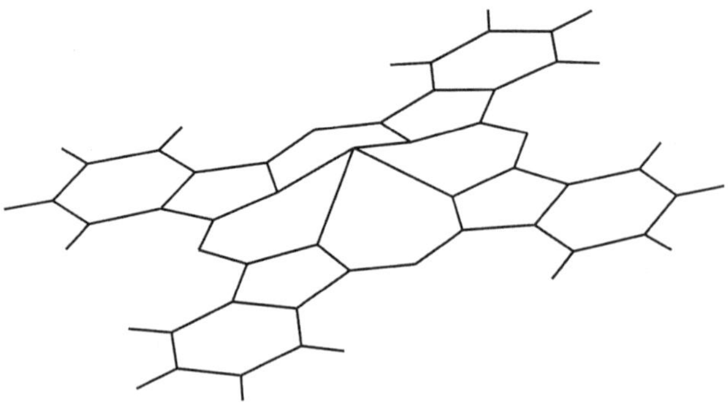

Appendix 12-43. [PcSb][I_3], HOVCOK[188]

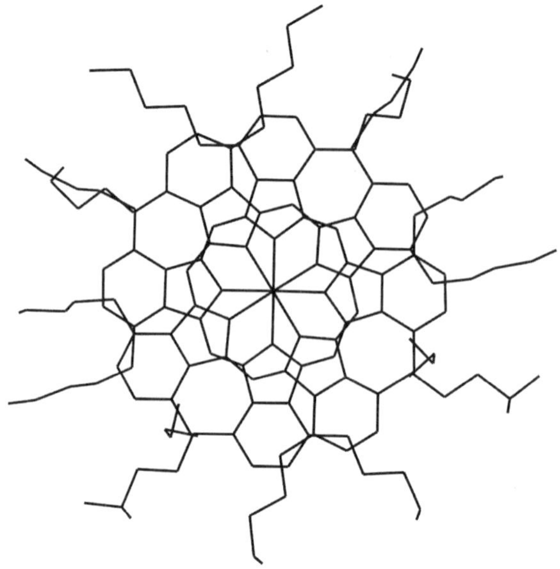

Appendix 12-44. [$(H_{11}C_5)_8$PcRh]$_2$-H, TESFAY[171]

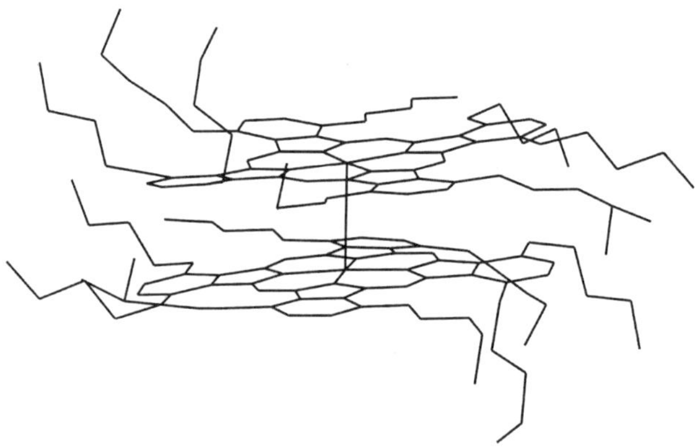

Appendix 12-45. [$(H_{11}C_5)_8$PcRh]$_2$-H, TESFAY[171]

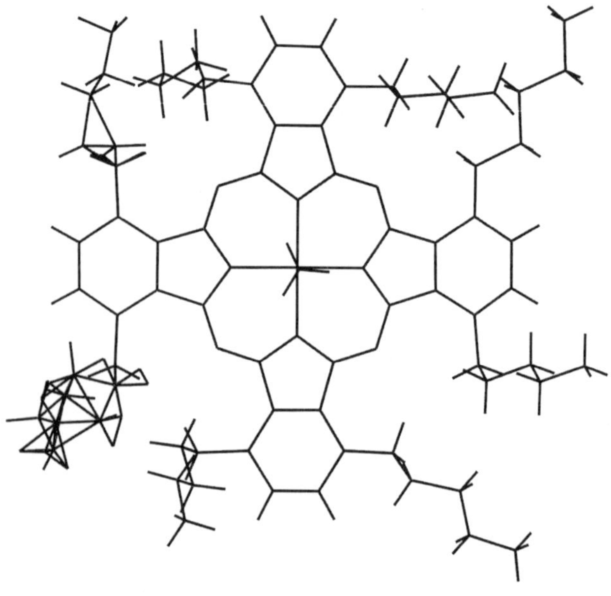

Appendix 12-46. $(H_{11}C_5)_8$PcRh-Me, HASZEG[166]

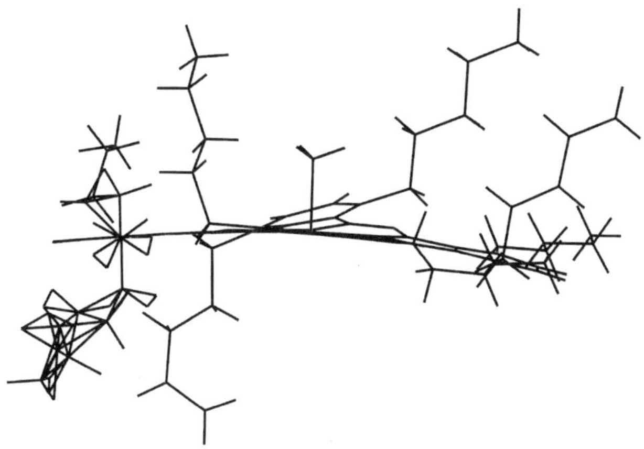

Appendix 12-47. $(H_{11}C_5)_8$PcRh-Me, HASZEG[166]

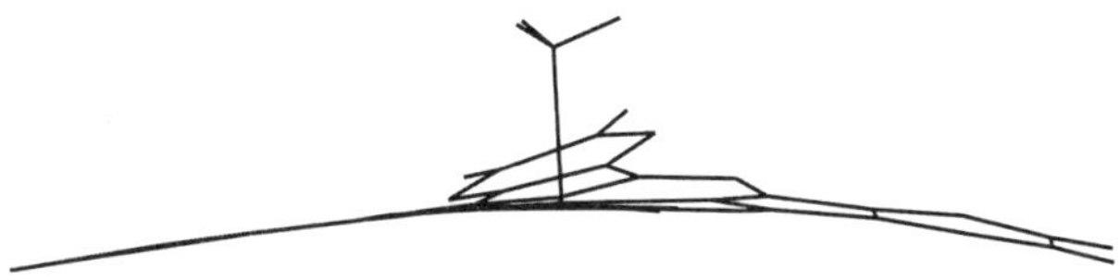

Appendix 12-48. Macrocycle of $(H_{11}C_5)_8$PcRh-Me, HASZEG[166]

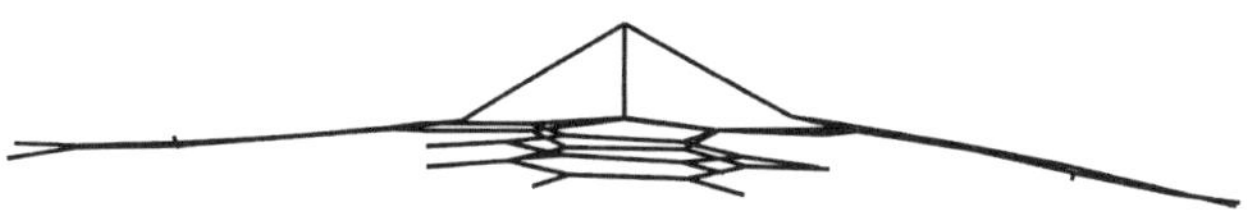

Appendix 12-49. PcSn, #1971-02,[31] SNPHCY[142] (App. 27)

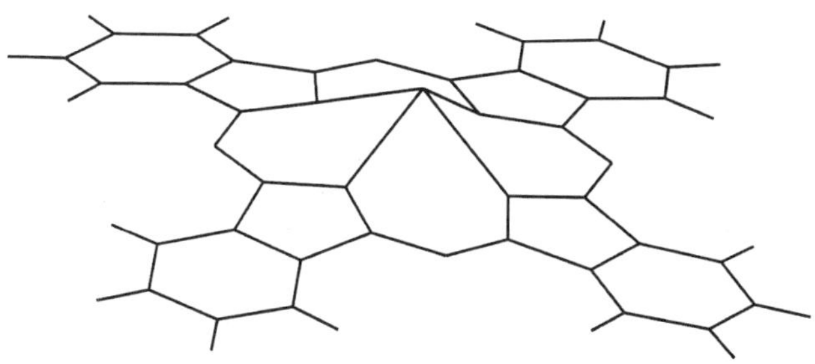

Appendix 12-50. PcSn, SNPHCY01[143]

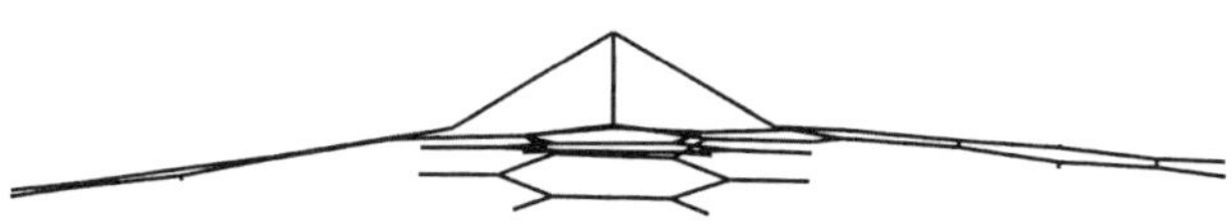

Appendix 12-51. PcSn, SNPHCY01[143]

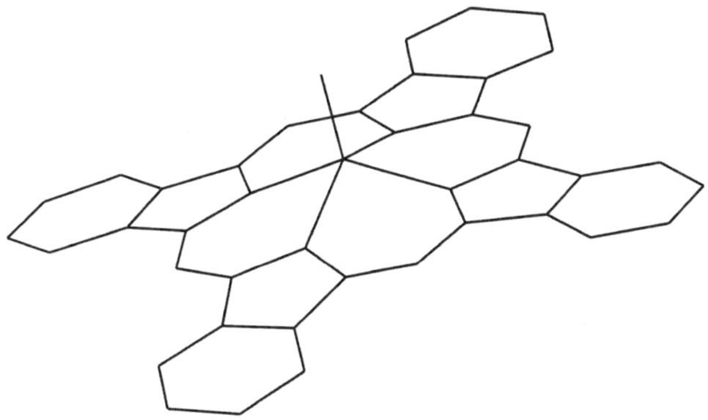

Appendix 12-52. C-PcTiO (R), BITSAY04[3]

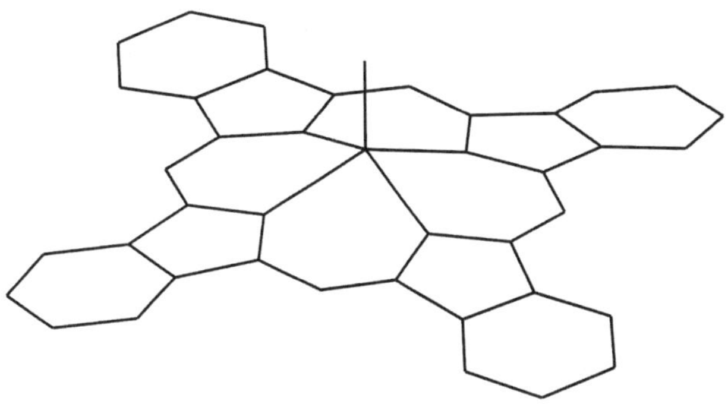

Appendix 12-53. I-PcTiO, BITSAY[2]

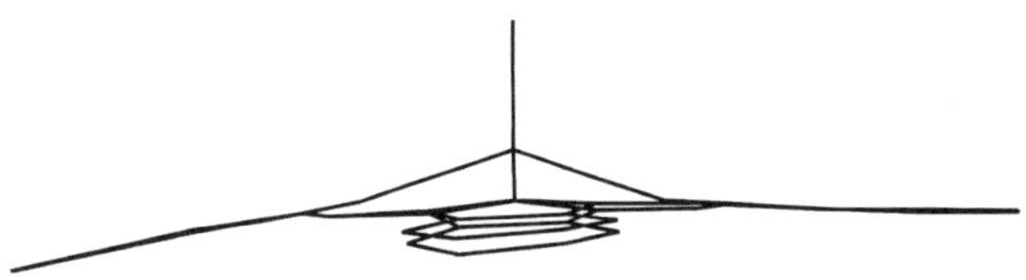

Appendix 12-54. II-PcTiO, BITSAY01[2]

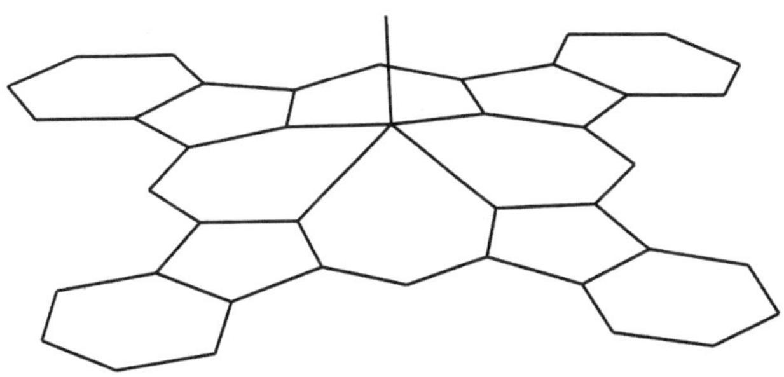

Appendix 12-55. IV-PcTiO, R, BITSAY03[164]

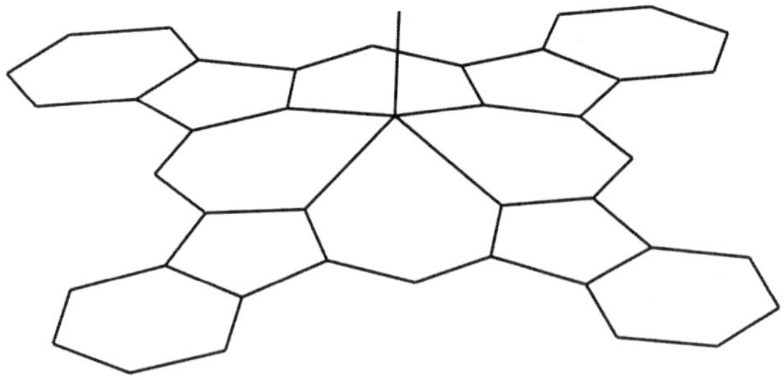

Appendix 12-56. Y-PcTiO, R, BITSAY02[165]

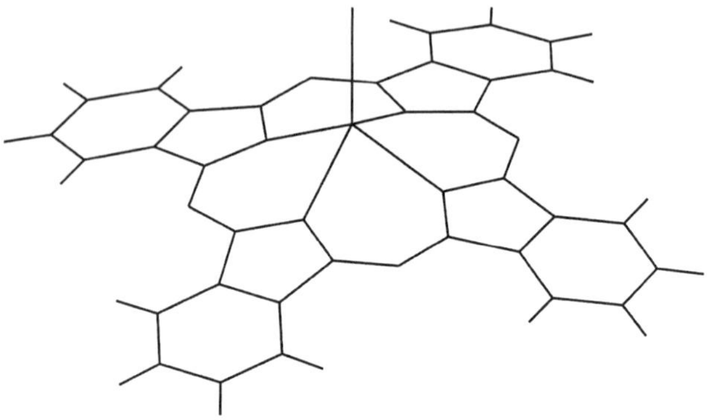

Appendix 12-57. [PcTl-I][CH_2Cl_2], NELCIQ[180]

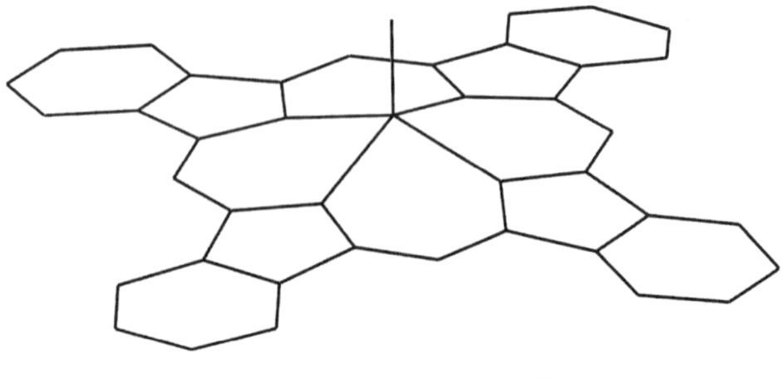

Appendix 12-58. PcV=O, VPHTHC[150]

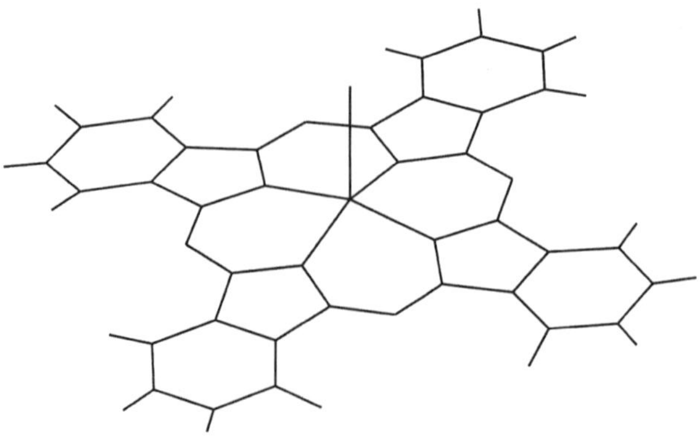

Appendix 12-59. PcVI, SAXZEW[185]

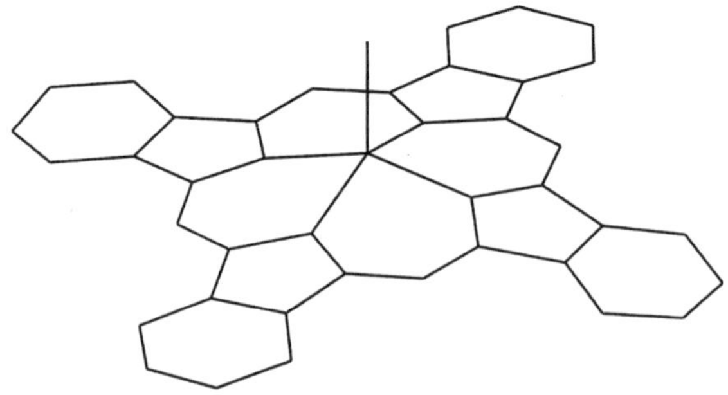

Appendix 12-60. [PcZn-H_2N-C_6H_{13}] [0.5H_2N-C_6H_{13}], ZNPTCY;[144] C_6H_{13} carbon atoms disordered

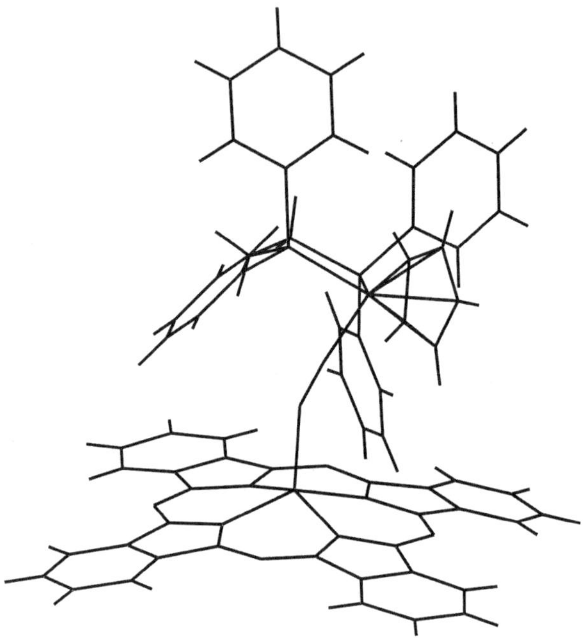

Appendix 12-61. [PcZn-NC-Fe(dppe)Cp][0.5CH_2Cl_2,THF], WOHTES[314]

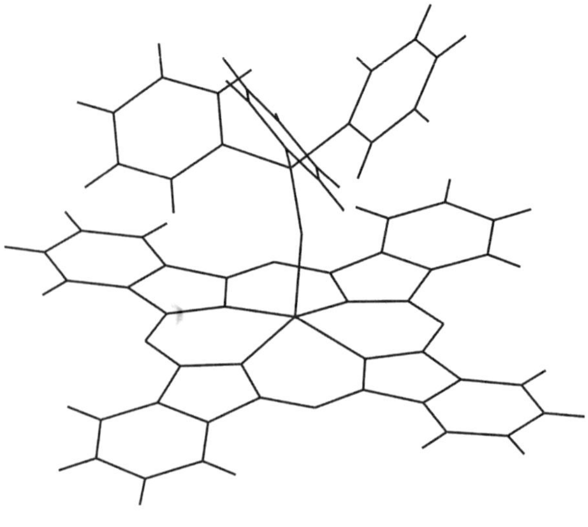

Appendix 12-62. [PcZn–OPϕ_3][Et_2O,1-ClN], MAKHAH[136]

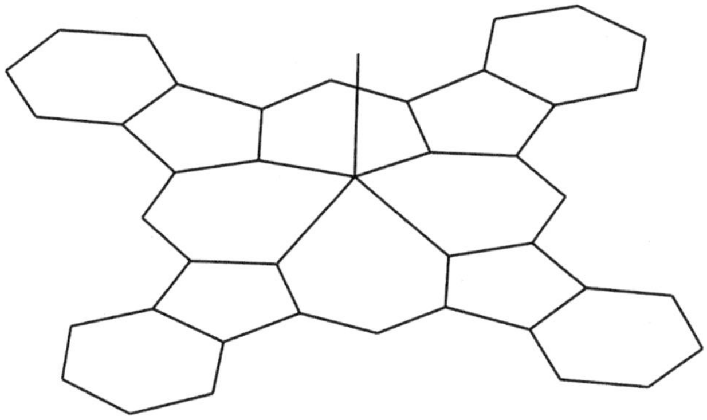

Appendix 12-63. PcZnCl, DAMTEQ[315]

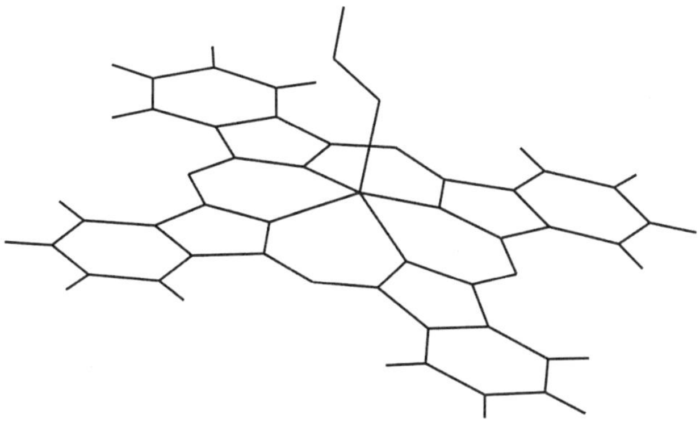

Appendix 12-64. [PcZn–ONO][PNP][H_2O], ZUHCIO[170]

Appendix 13. Molecular Structures of CN5* Phthalocyanine Complexes

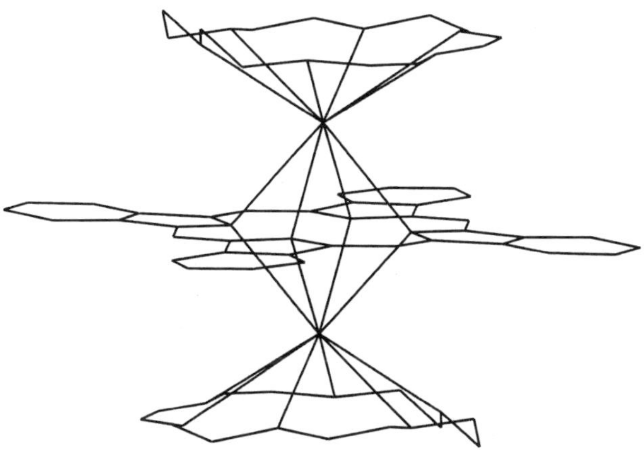

Appendix 13-1. [Pc(K(18cr6))$_2$][1.5C_6H_6], Mol.1, BANCOI[316]

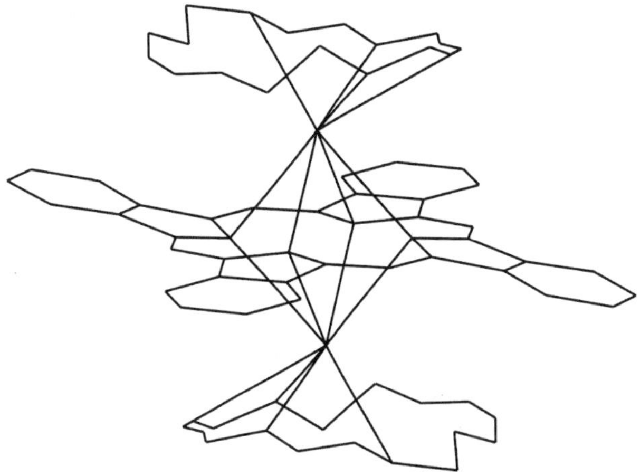

Appendix 13-2. [Pc(K(18cr6))$_2$][1.5C_6H_6], Mol.2, BANCOI[316]

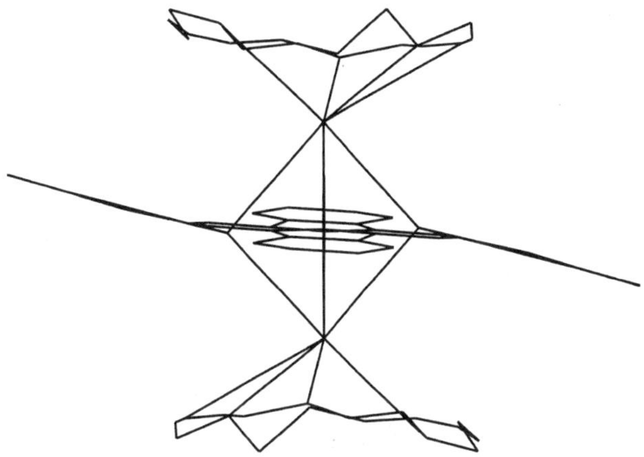

Appendix 13-3. [Pc(K(18cr6))$_2$][1.5C_6H_6], Mol.2, BANCOI[316]

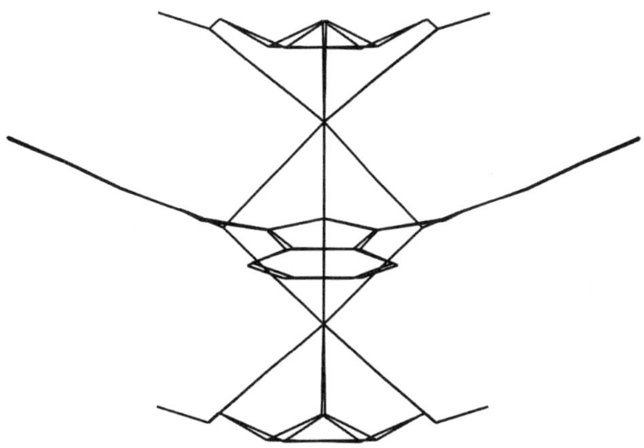

Appendix 13-4. Pc[K(diglyme)]$_2$, BANCUO[316]

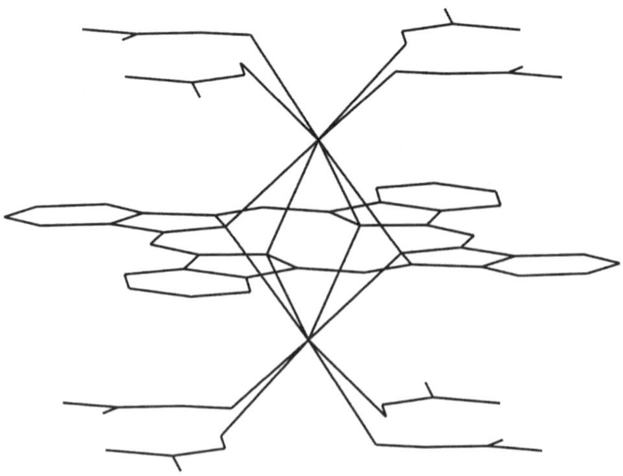

Appendix 13-5. Pc[K(DMF)$_2$]$_2$, BARMAI[317]

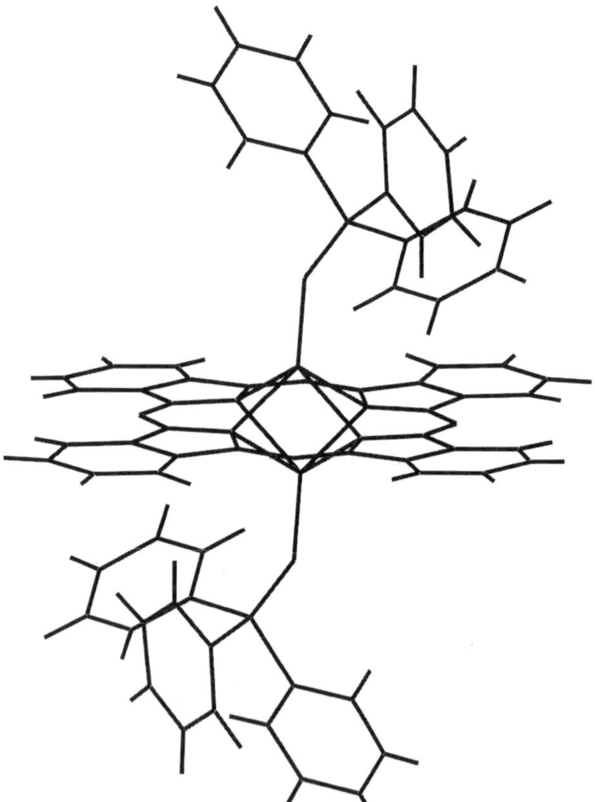

Appendix 13-6. Pc(Li-OPϕ_3)$_2$, MAKJEN[136]

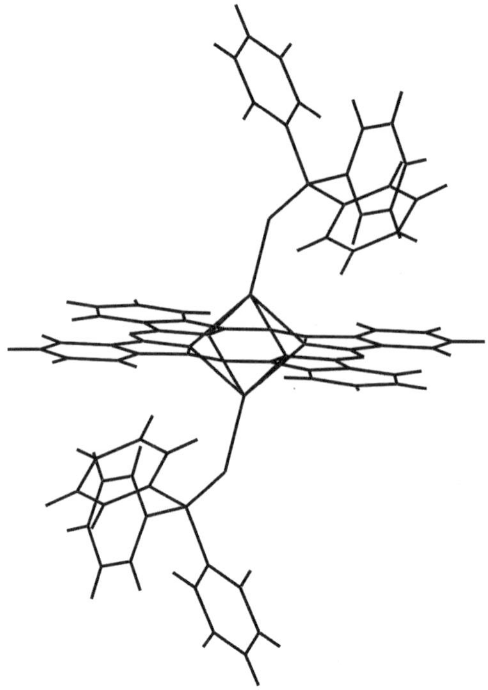

Appendix 13-7. Pc(Na-OPϕ_3)$_2$, QASKAW[50]

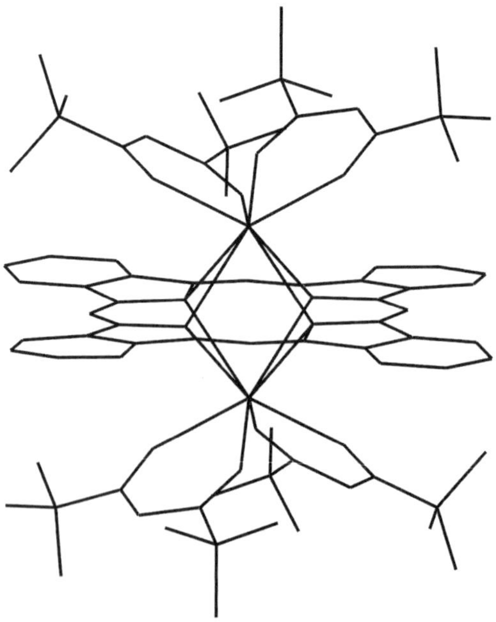

Appendix 13-8. Pc(SmL$_2$)$_2$, CALWIV[318,319]

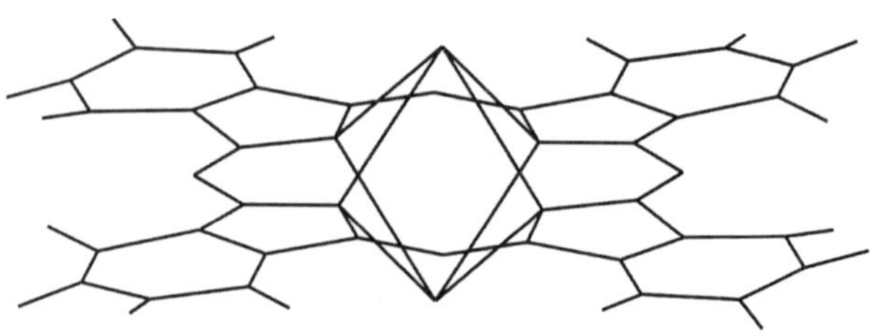

Appendix 13-9. PcTl$_2$, HEBHOL[49]

Appendix 14. Molecular Structures of CN6*cis* Phthalocyanine Complexes

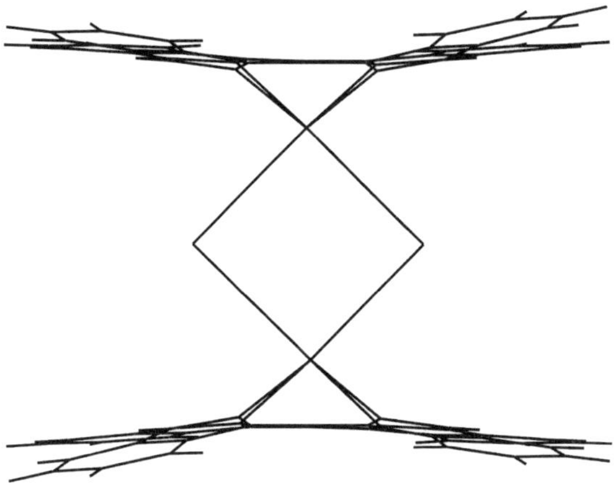

Appendix 14-1. $PcBi(\text{-}I\text{-})_2BiPc$, HOXCIG[184]

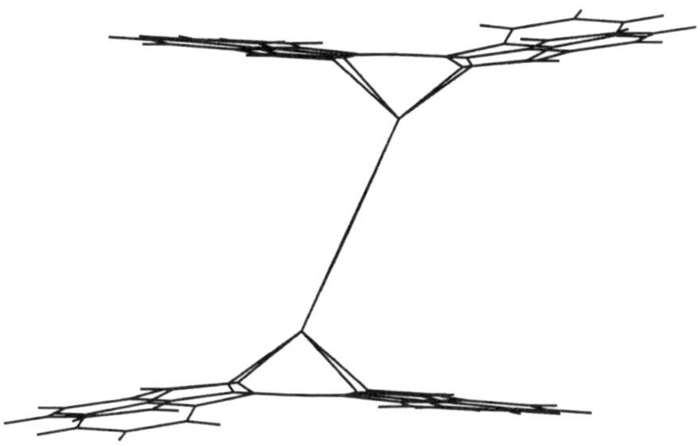

Appendix 14-2. $PcBi(\text{-}I\text{-})_2BiPc$, HOXCIG[184]

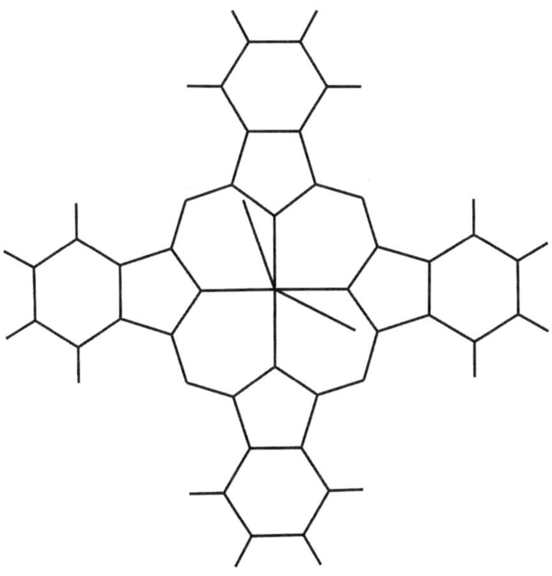

Appendix 14-3. One macrocycle of $PcBi(\text{-}I\text{-})_2BiPc$, HOXCIG[184]

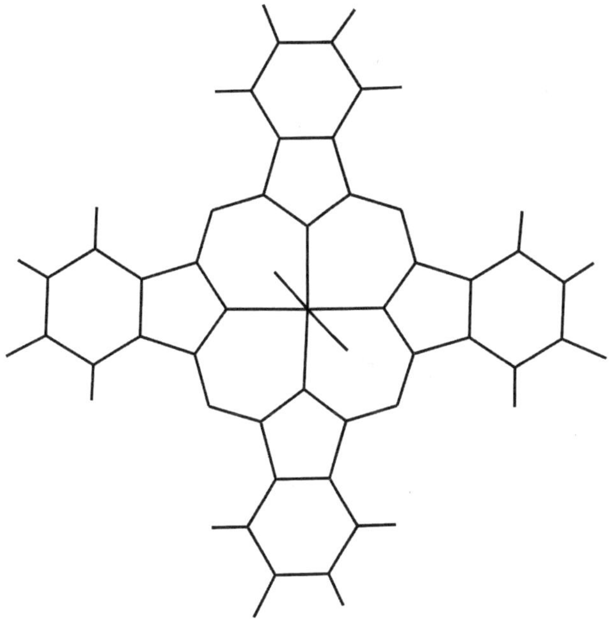

Appendix 14-4. $[PcInF_2][NBu_4][H_2O]$, NAPJIX[178]

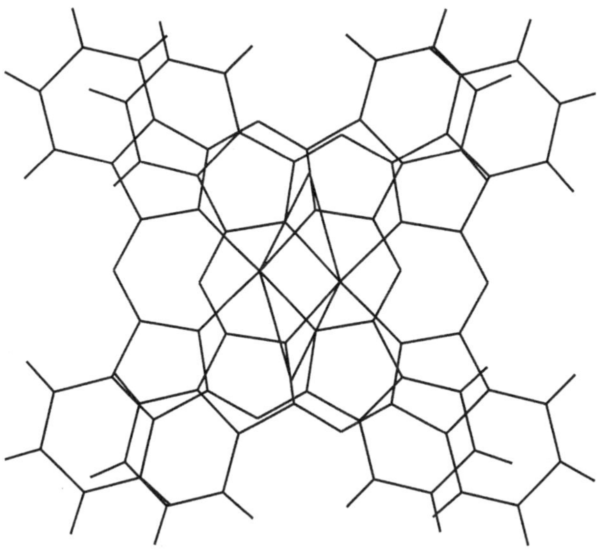

Appendix 14-5. $[PcInF_2][NBu_4][H_2O]$, NAPJIX[178]

Appendix 14-6. $[PcIn(\text{-}NCO)_2][PNP]$, ZUFZOP[173]

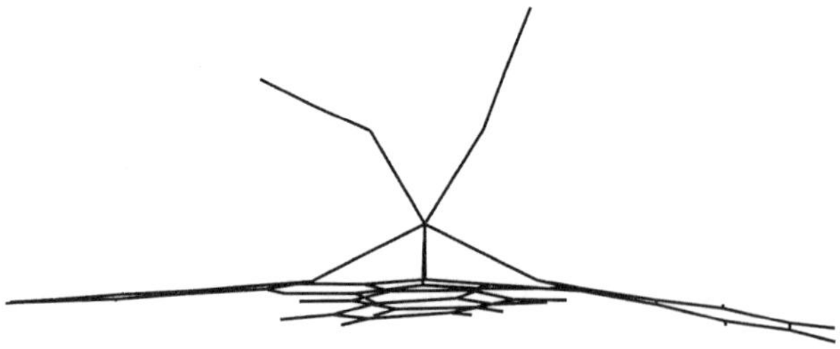

Appendix 14-7. $[PcIn(\text{-}NCO)_2][PNP]$, ZUFZOP[173]

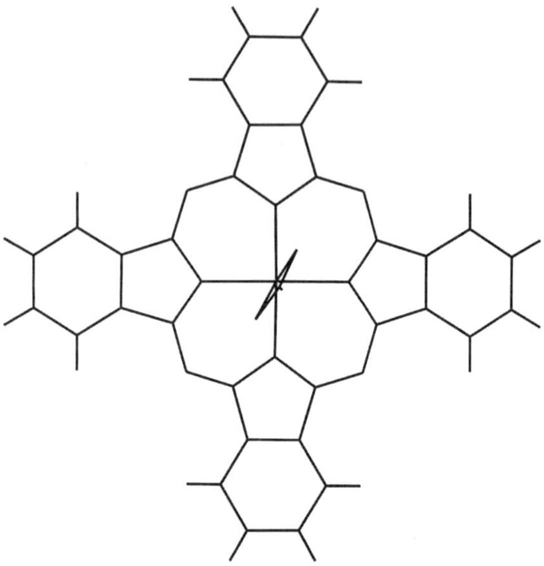

Appendix 14-8. $[PcIn(\text{-}O,\text{-}O'CO)][NBu_4]$, RERGOK[177]

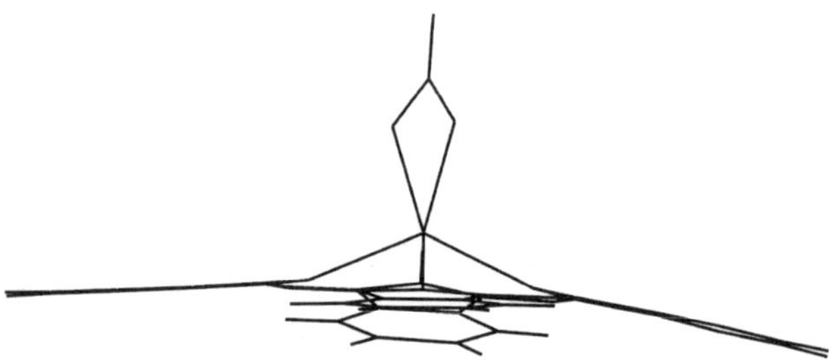

Appendix 14-9. $[PcIn(\text{-}O,\text{-}O'CO)][NBu_4]$, RERGOK[177]

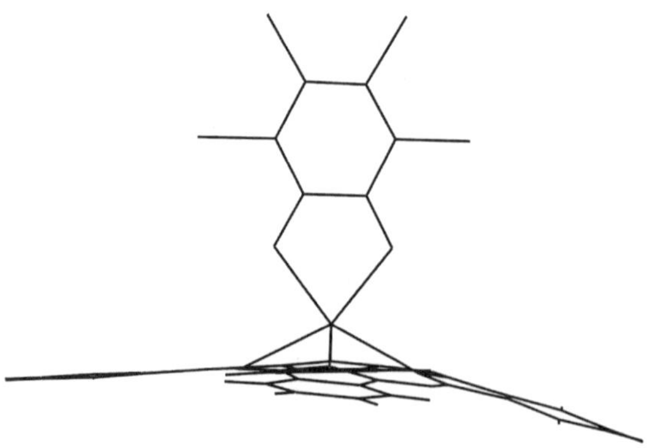

Appendix 14-10. $PcIn(O_2catCl_4)$, #1998-02[308]

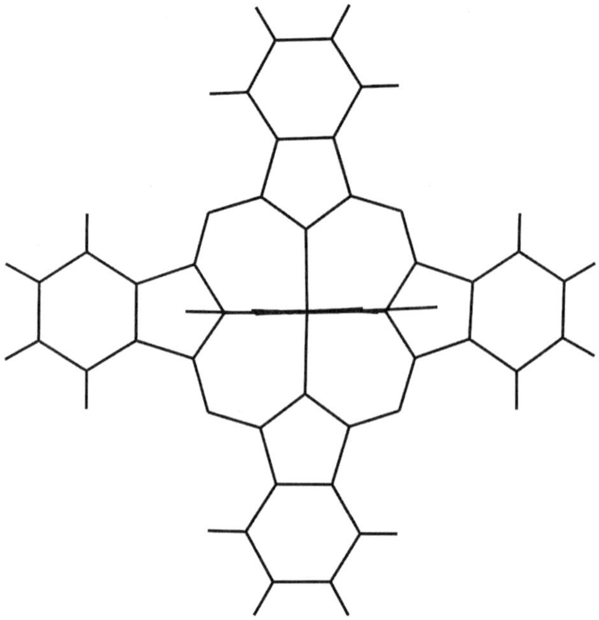

Appendix 14-11. $PcIn(O_2catCl_4)$, #1998-02[308]

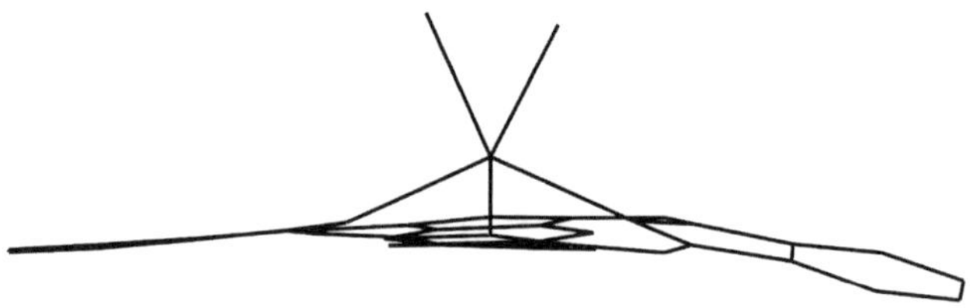

Appendix 14-12. $PcNbCl_2$, BEJTIT[152]

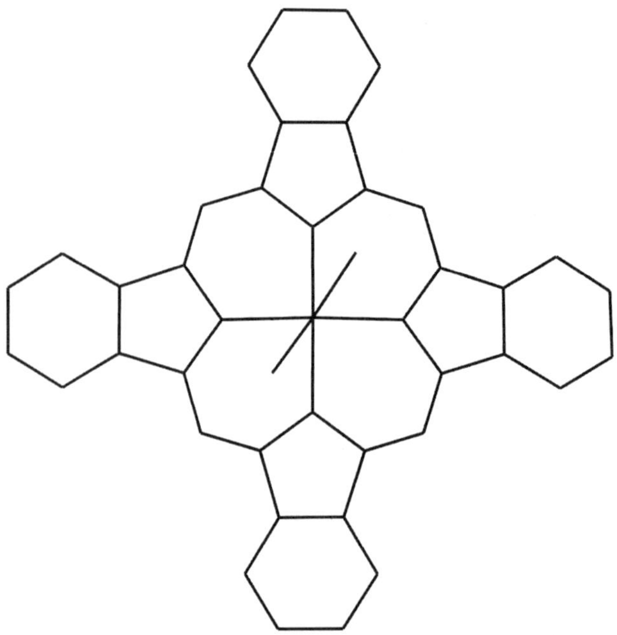

Appendix 14-13. $PcNbCl_2$, BEJTIT[152]

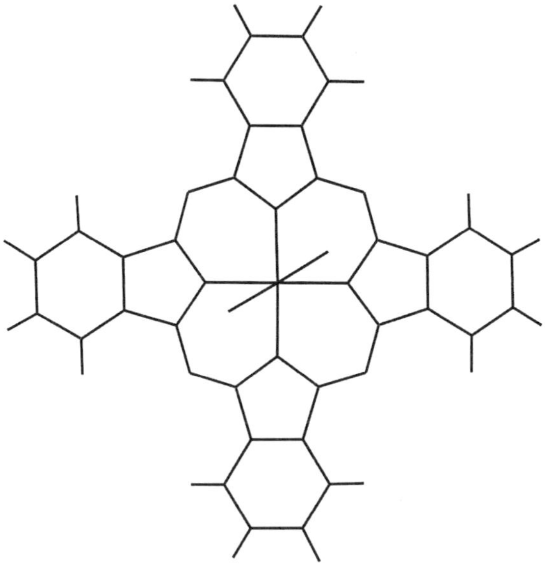

Appendix 14-14. $PcNbCl_2$, BEJTIT01[163]

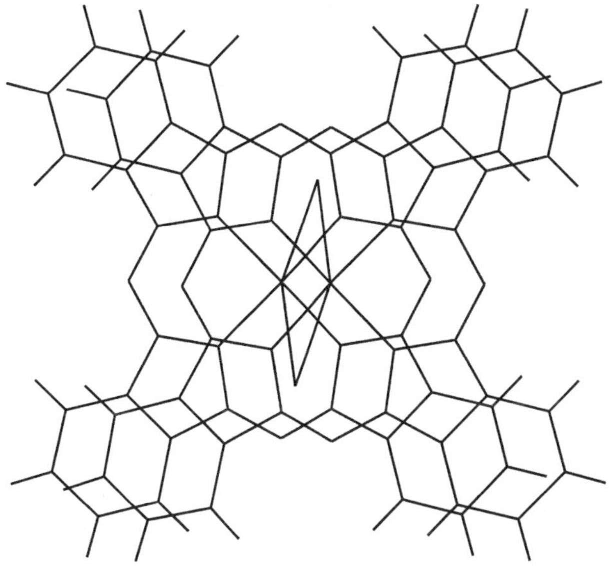

Appendix 14-17. $PcSb(\text{-}Cl\text{-})_2SbPc$, WOVREE[193]

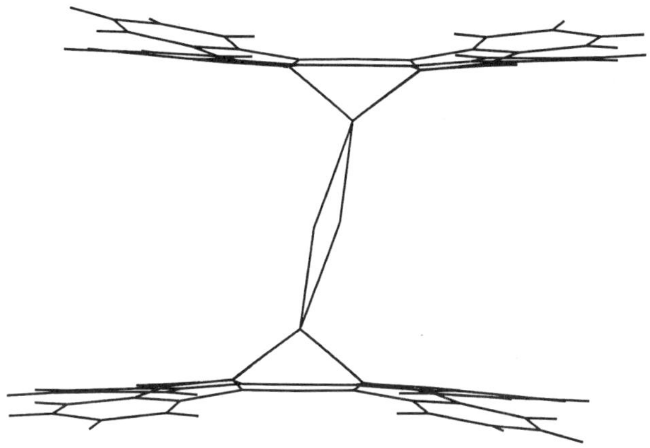

Appendix 14-18. $PcSb(\text{-}Cl\text{-})_2SbPc$, WOVREE[193]

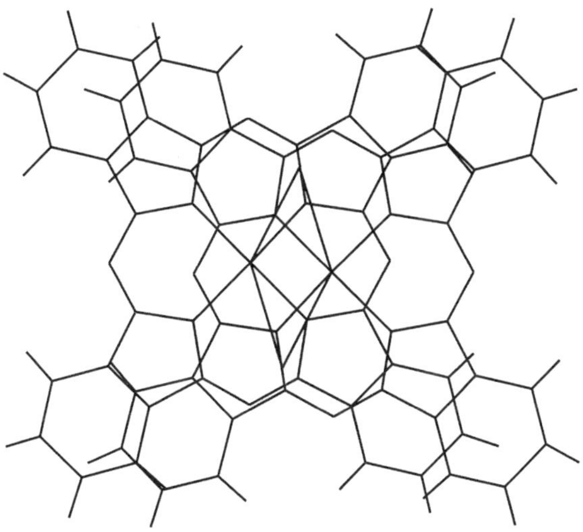

Appendix 14-15. $PcSb(\text{-}I\text{-})_2SbPc$, NILVUZ[183]

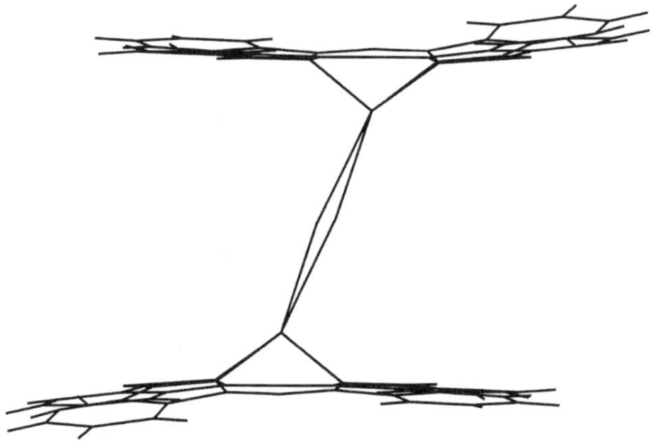

Appendix 14-16. $PcSb(\text{-}I\text{-})_2SbPc$, NILVUZ[183]

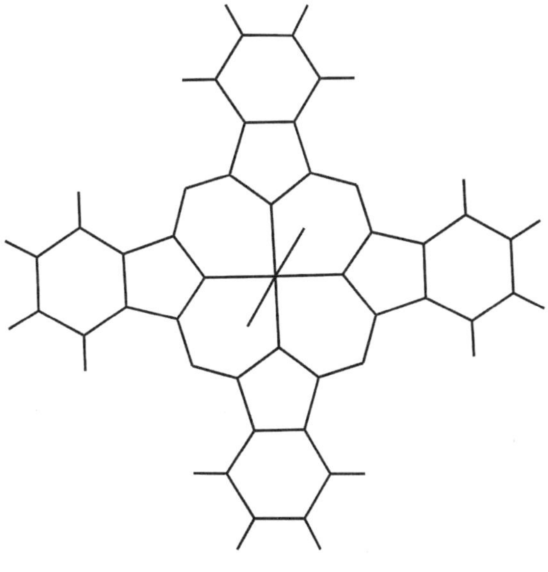

Appendix 14-19. $PcTiCl_2$, DAGXUE[158]

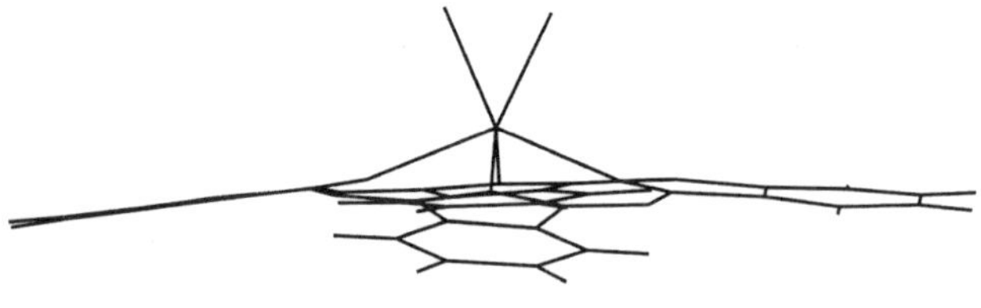

Appendix 14-20. $PcTiCl_2$, DAGXUE[158]

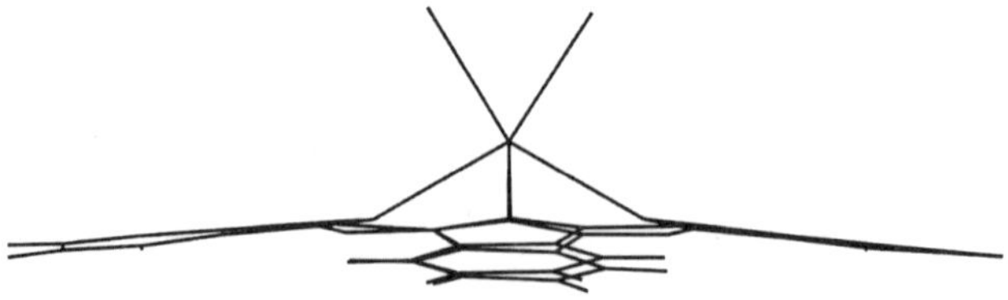

Appendix 14-21. $[PcTlCl_2][NBu_4][0.5Et_2O]$, BOFQES[191]

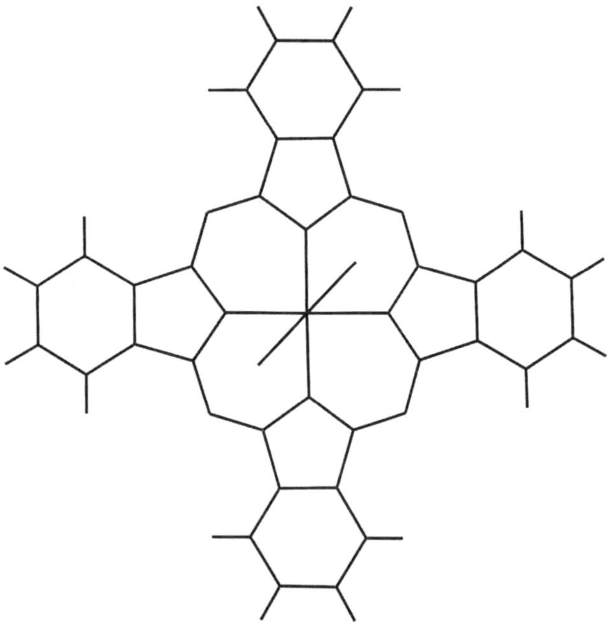

Appendix 14-22. $[PcTlCl_2][NBu_4][0.5Et_2O]$, BOFQES[191]

Appendix 15. Molecular Structures of CN6*tr* Phthalocyanine Complexes

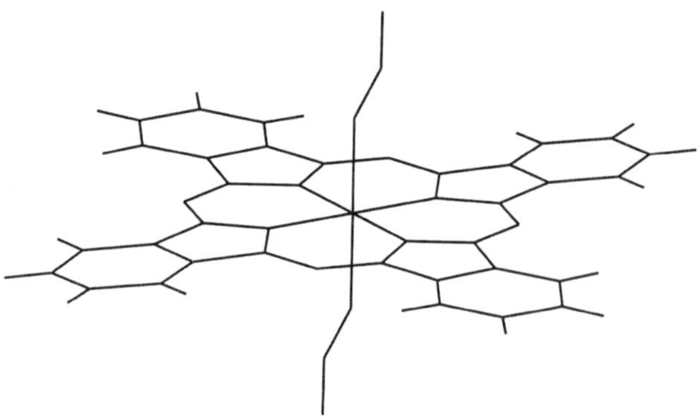

Appendix 15-1. $[PcAl(\text{-}ONO)_2][NBu_4]$, ZUVYOE[324]

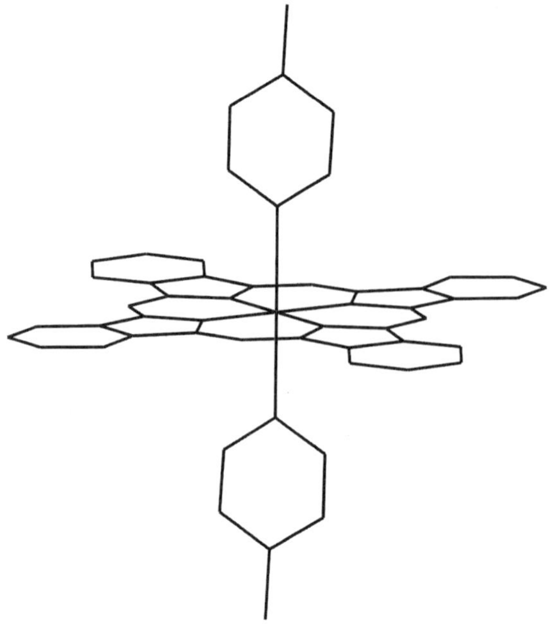

Appendix 15-2. $[PcCo(4\text{-}mepy)_2][2Mepy]$, MPCOPC10[326]

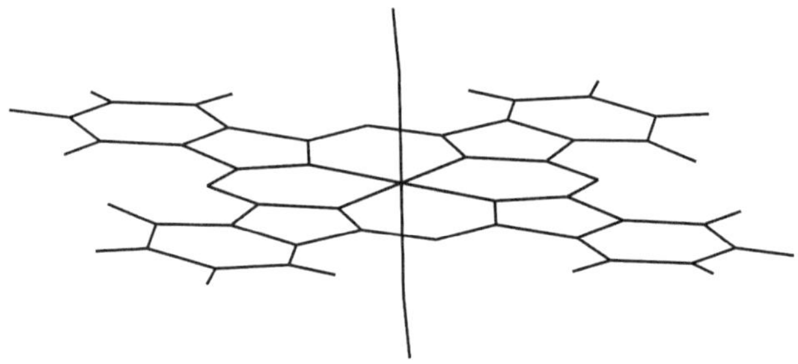

Appendix 15-3. $[PcCo(CN)_2][2CHBr_3]$, YAVPEQ[327,328]

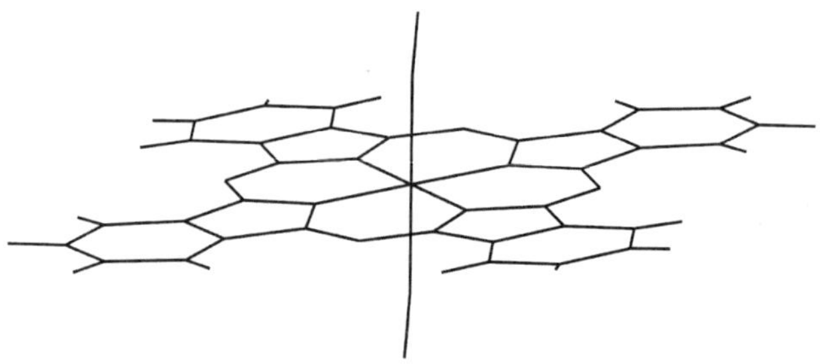

Appendix 15-4. $[PcCo(CN)_2][2CHCl_3]$, ZODWAQ[329]

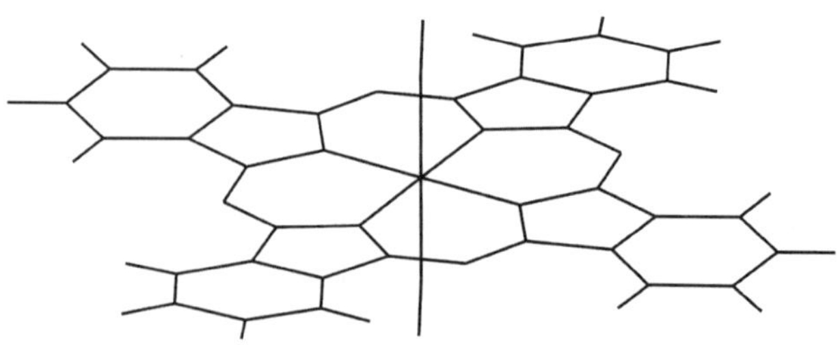

Appendix 15-5. $[PcCo(CN)_2][2DMSO]$, DOVYUI[330]

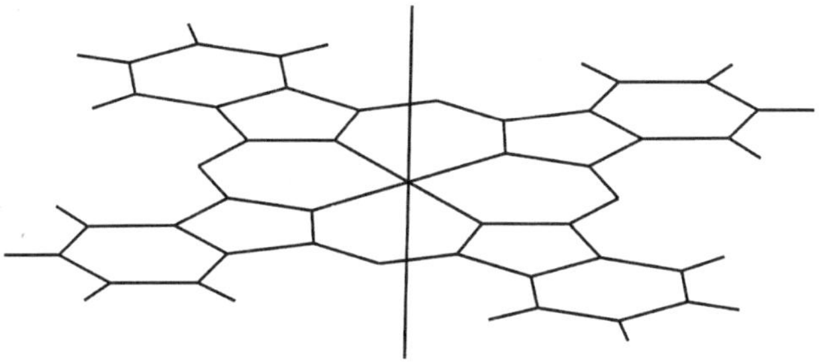

Appendix 15-6. $[PcCo(CN)_2][2H_2O]$, SILPEI[331,332]

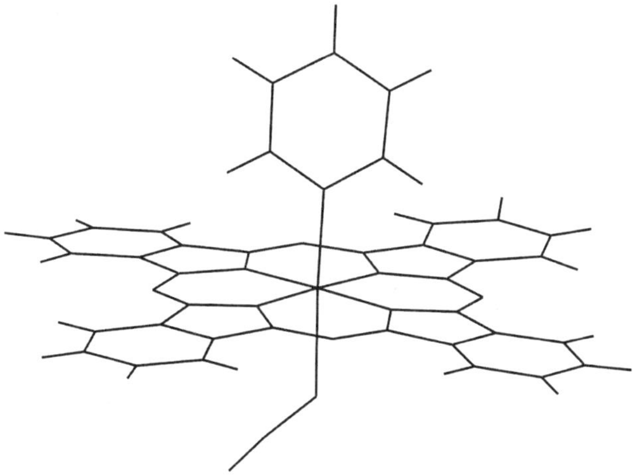

Appendix 15-7. [PcCo(py,-SCN)]$[CHCl_3]$, FODLOZ[333]

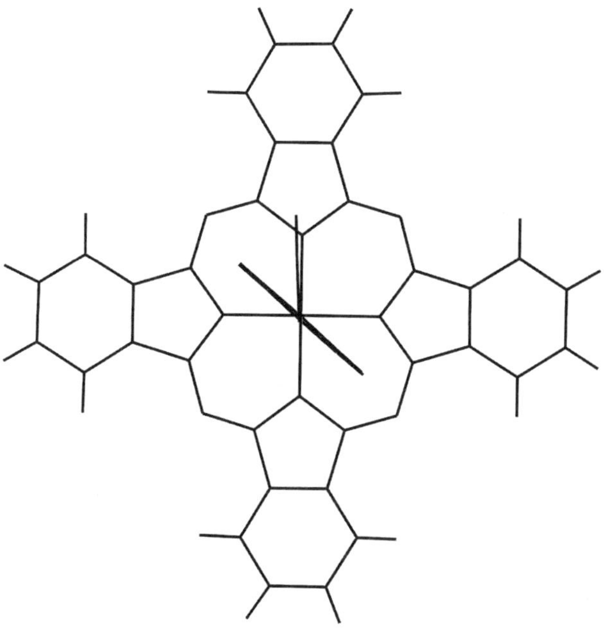

Appendix 15-8. [PcCo(py,-SCN)]$[CHCl_3]$, FODLOZ[333]

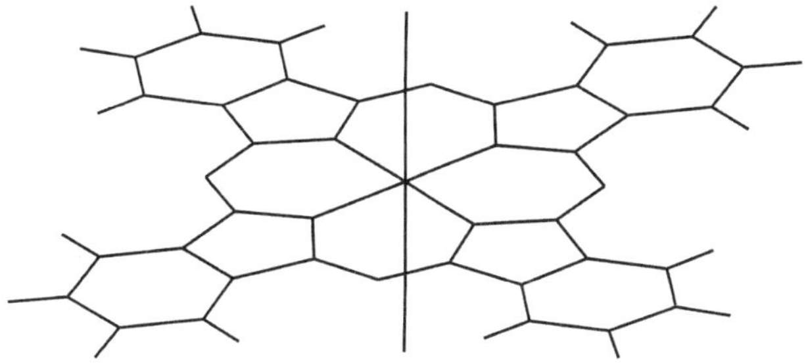

Appendix 15-9. $PcCo(CN)_2$, DOVZAP[330]

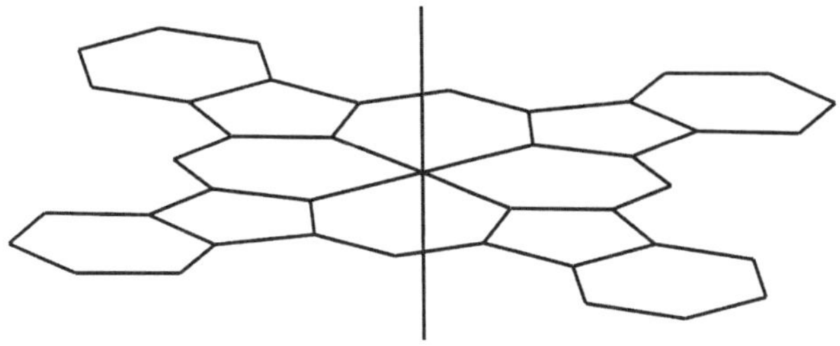

Appendix 15-10. $PcCoCl_2$, KEHXEA[123]

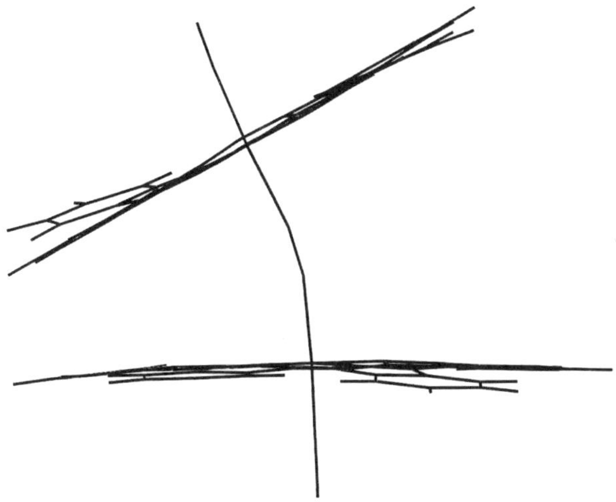

Appendix 15-11. [NC-(Pc)Co-CN-Co(Pc)-CN][B[c]PT], NOFGAQ[334]

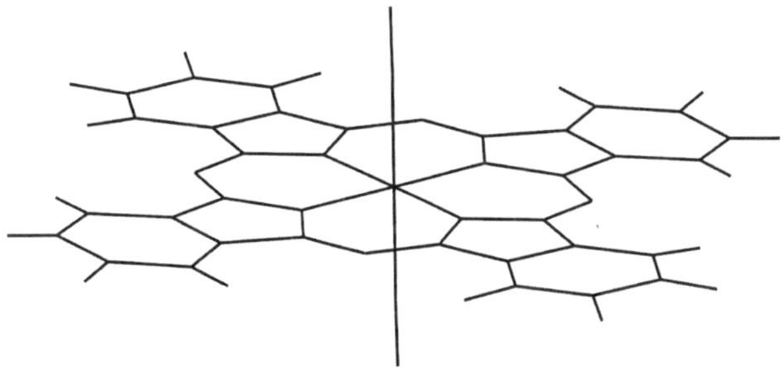

Appendix 15-12. [K]$[PcCo(CN)_2][5CH_3CN]$, VENDAT10[331,332,335]

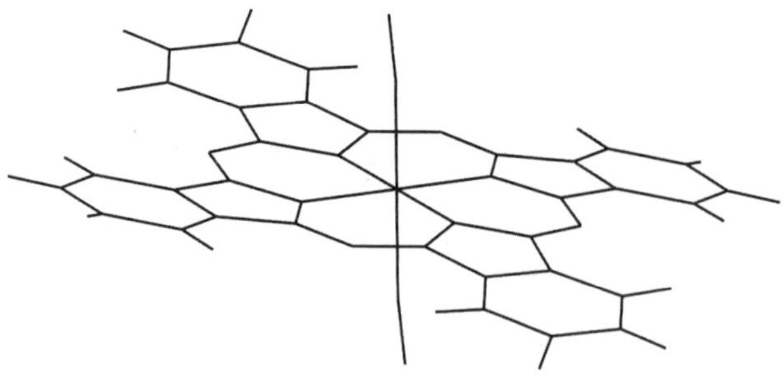

Appendix 15-13. $[PcCo(CN)_2][PTMA][MeCN]$, NIRPAF[336]

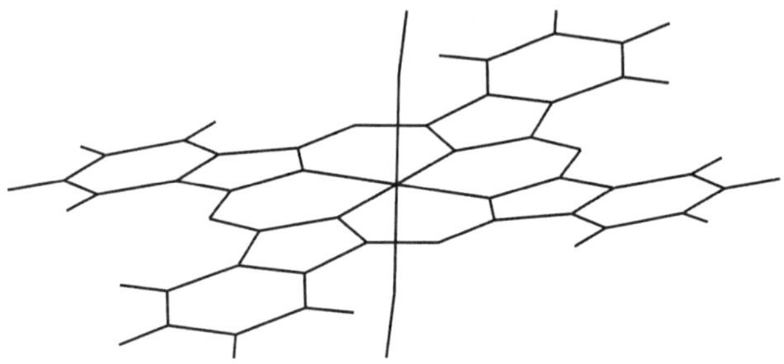

Appendix 15-14. $[PcCo(CN)_2][PTMA][MeCN]$, NIRDOH[336]

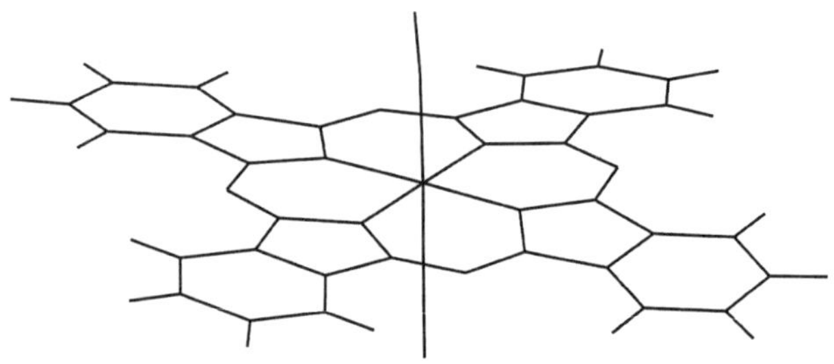

Appendix 15-15. $[PcCo(CN)_2][PXX]_2$, BAPZOH[130,337]

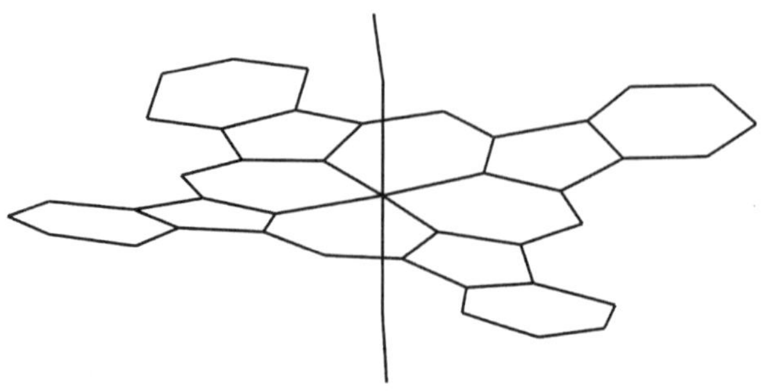

Appendix 15-16. $[PcCo(CN)_2]$[p-PPYNN], SOPQET[338]

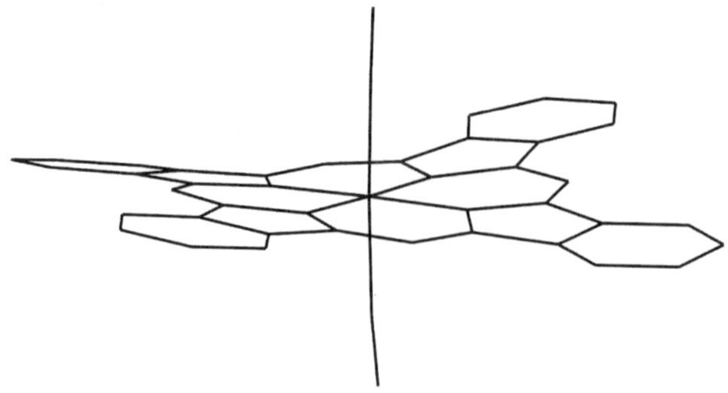

Appendix 15-17. $[PcCo(CN)_2]$[p-BPYNN], SOPQIX[338]

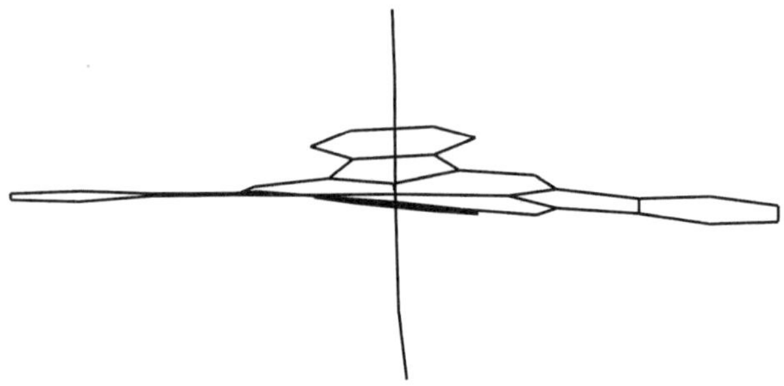

Appendix 15-18. $[PcCo(CN)_2]$[p-BPYNN], SOPQIX[338]

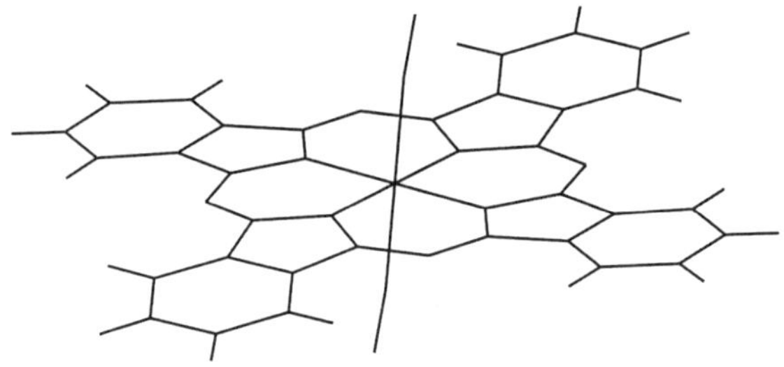

Appendix 15-19. $[PcCo(CN)_2][P\phi_4]_{0.5}$, JAMRAQ[339,340]

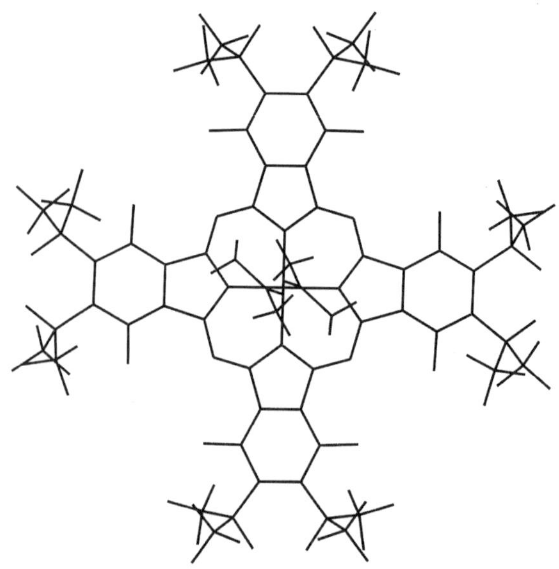

Appendix 15-20. $[F_8((F_3C)_2FC)_8PcCo(OCMe_2)_2]$, CCDC174676[341]

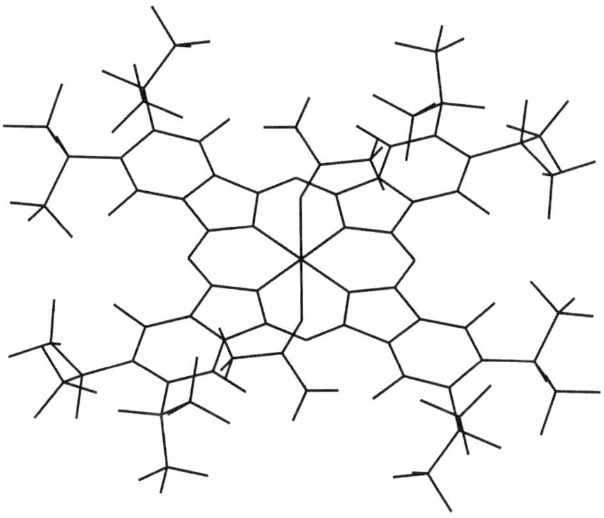

Appendix 15-21. $[F_8((F_3C)_2FC)_8PcCo(OCMe_2)_2]$, CCDC174676[341]

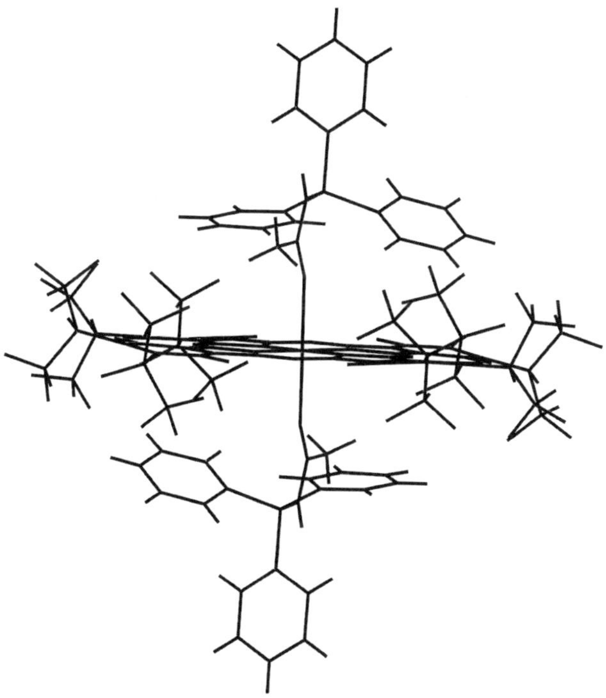

Appendix 15-22. $[F_8((F_3C)_2FC)_8PcCo(OC(Me)(CH_2P\phi_3)_2)$, CCDC174677[341]

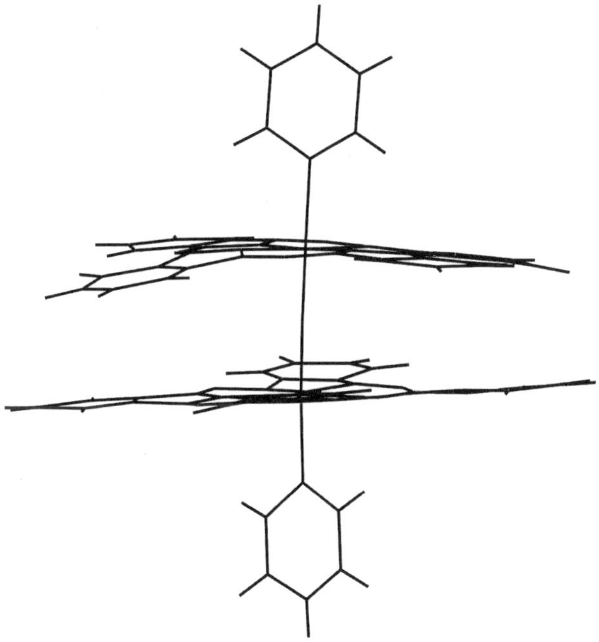

Appendix 15-23. [py-(Pc)Cr-O-Cr(Pc)-py], #2001-08[342]

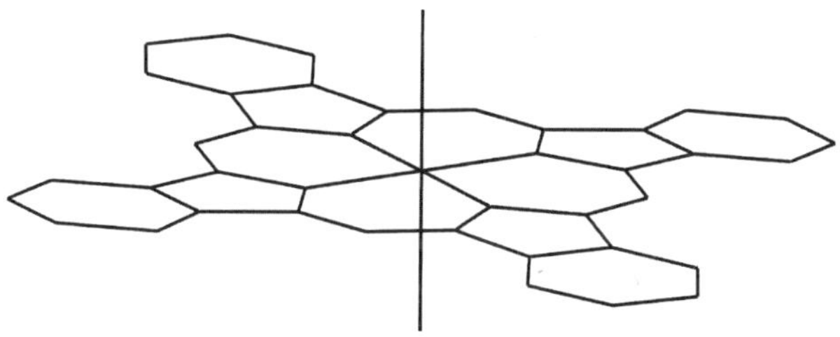

Appendix 15-24. $PcCrCl_2$, KEHWUB[123]

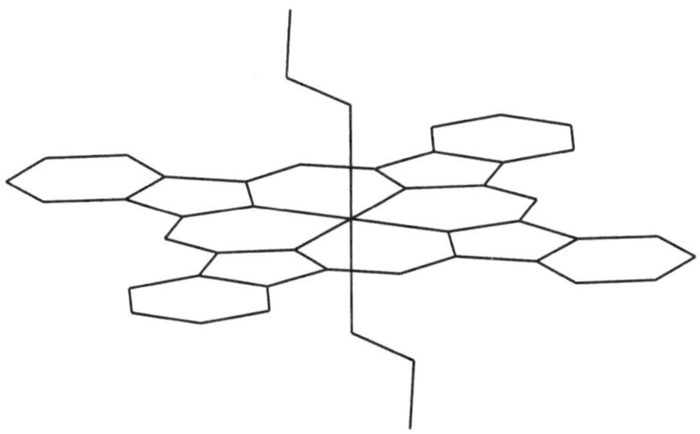

Appendix 15-25. $[PcCr(-ONO)_2][PNP]$, TOPVID[343]

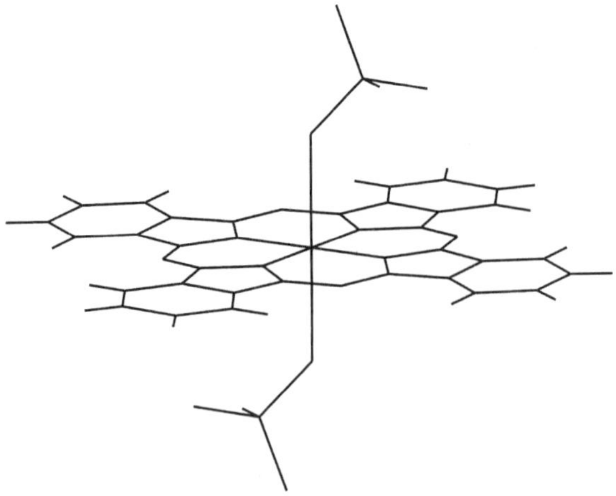

Appendix 15-26. $PcCu(ReO_4)_2$, QEMMOK[306]

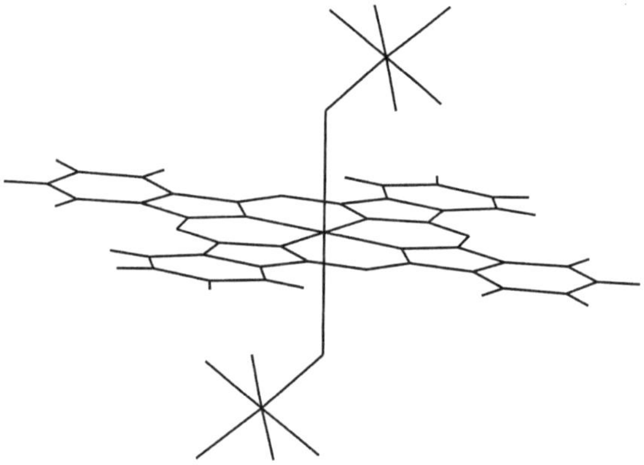

Appendix 15-27. $PcCu(SbF_6)_2$, QILWAJ[344]

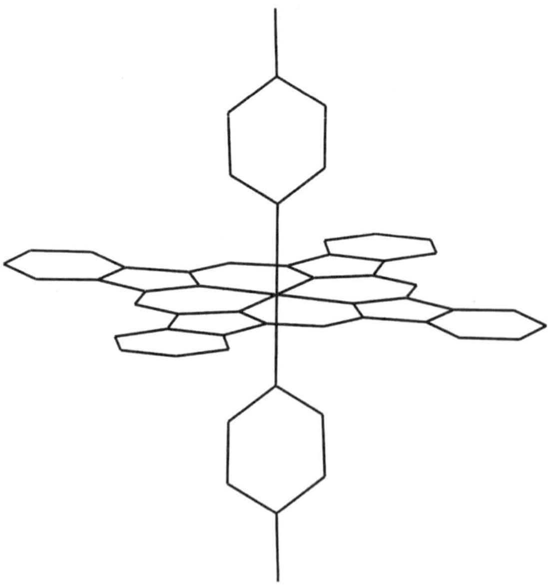

Appendix 15-28. [PcFe(4-mepy)$_2$][2mepy], PTCYFE01[326]

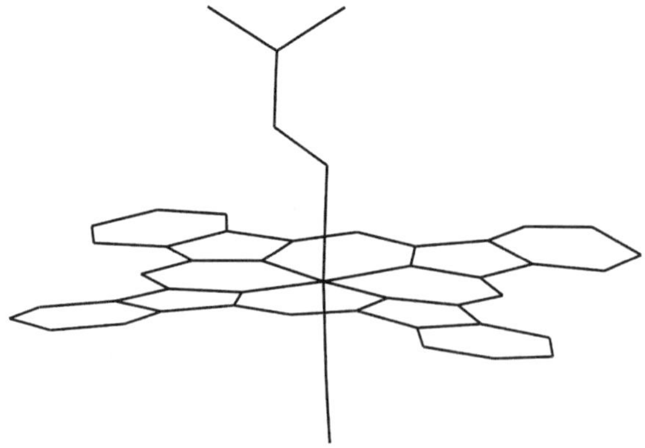

Appendix 15-29. [PcFe(DMF,CO)][DMF], PHTHFE[346]

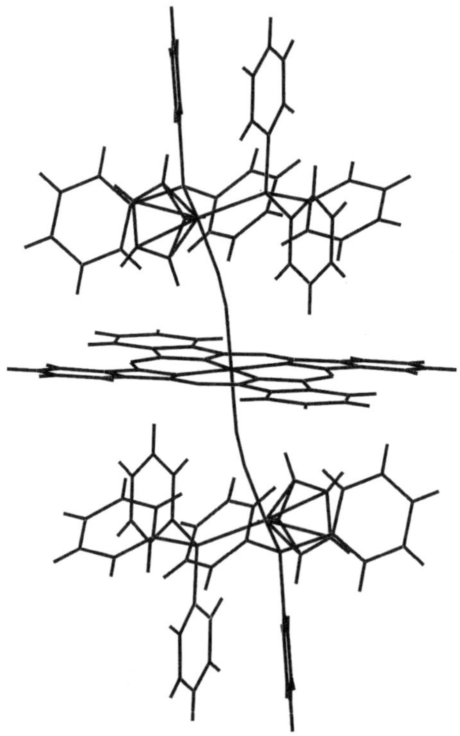

Appendix 15-30. [PcFe(NC-Ru($P\phi_3$)$_2$Cp)$_2$][1.5CH_2Cl_2], WOHVAQ[314]

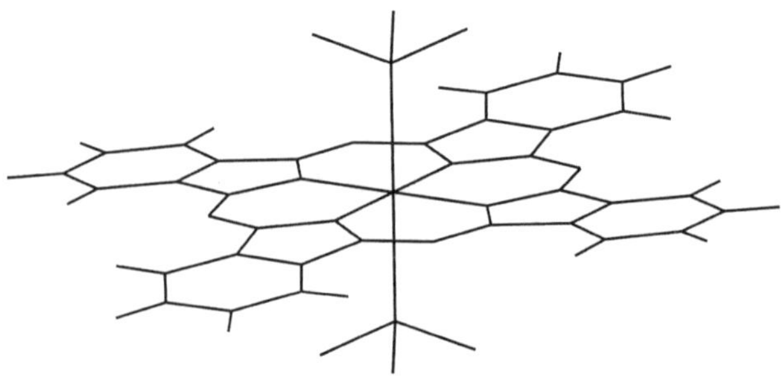

Appendix 15-31. [PcFe(-$OSMe_2$)$_2$][2DMSO], SPCNFE[347]

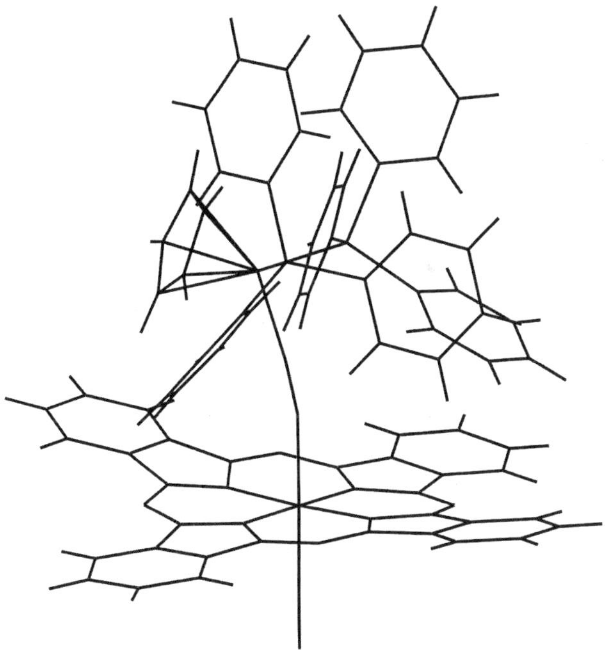

Appendix 15-32. [PcFeCN,NC-Ru(Pϕ_3)$_2$Cp][2CHCl$_3$], WOHTIW[314]

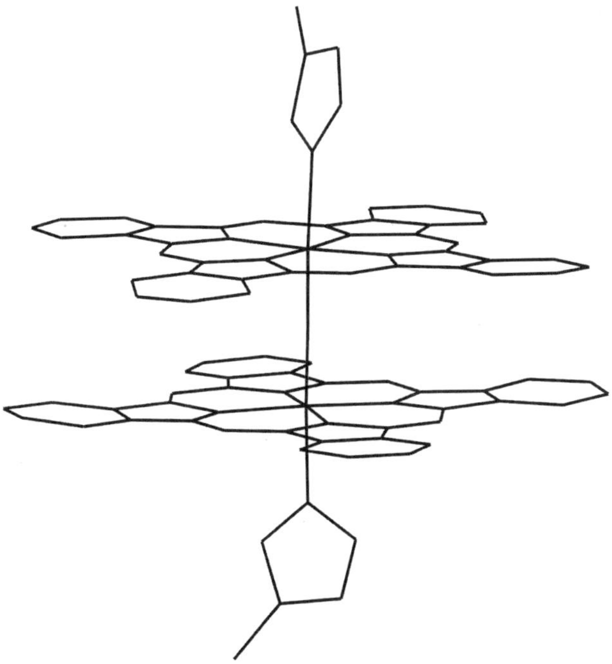

Appendix 15-33. [R-(Pc)Fe-C-Fe(Pc)-R][3Me$_2$CO] {+Monomer}, FUNGEA[348]

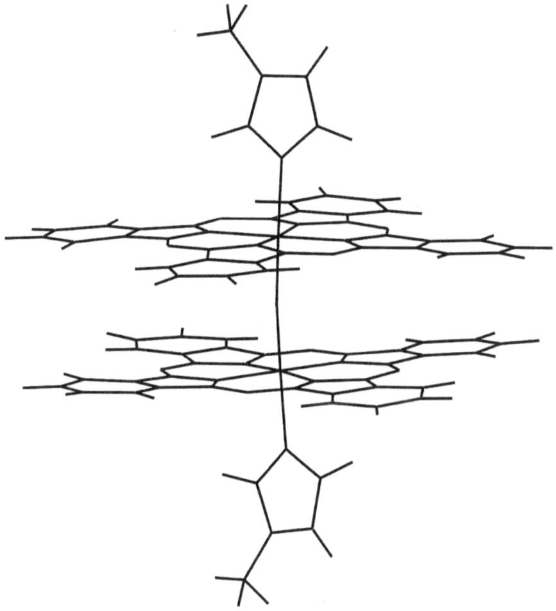

Appendix 15-34. [R-(Pc)Fe-O-Fe(Pc)-R][3Me$_2$CO] {+Monomer}, KITPIM[349]

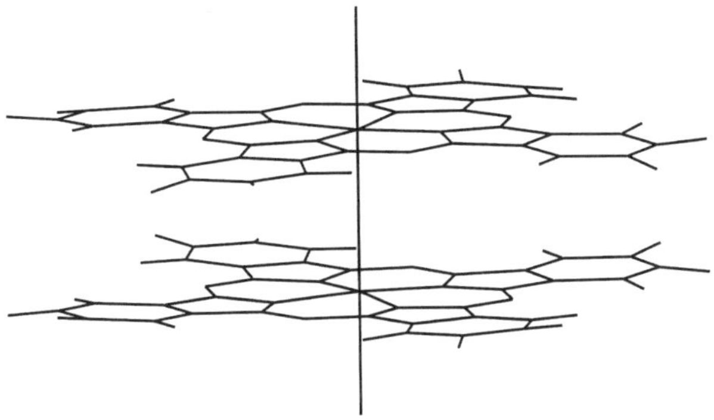

Appendix 15-35. Br-(Pc)Fe-N-Fe(Pc)-Br, VAFMIY[350]

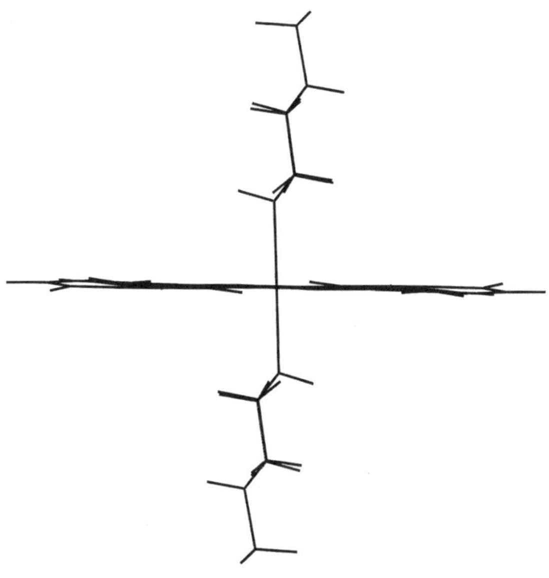

Appendix 15-36. PcFe(4mepip)$_2$, XENLAD[351]

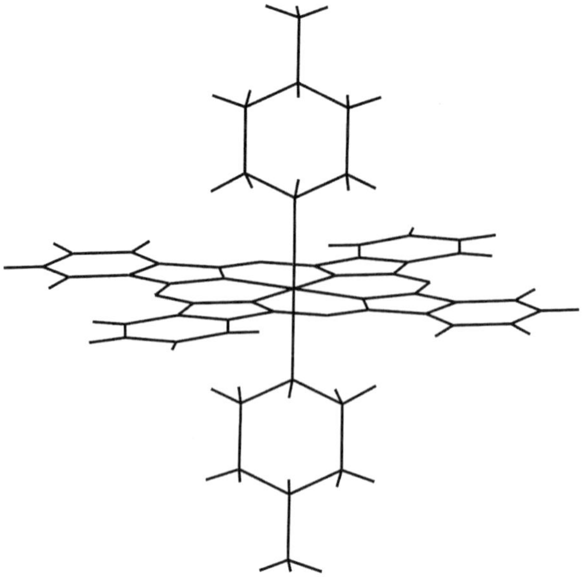

Appendix 15-37. PcFe(4mepip)$_2$, XENLAD[351]

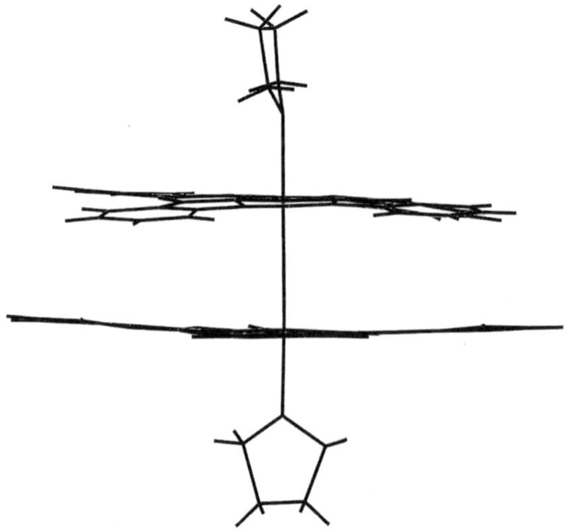

Appendix 15-40. THF-(Pc)Fe-C-Fe(Pc)-THF, PAZREN[353]

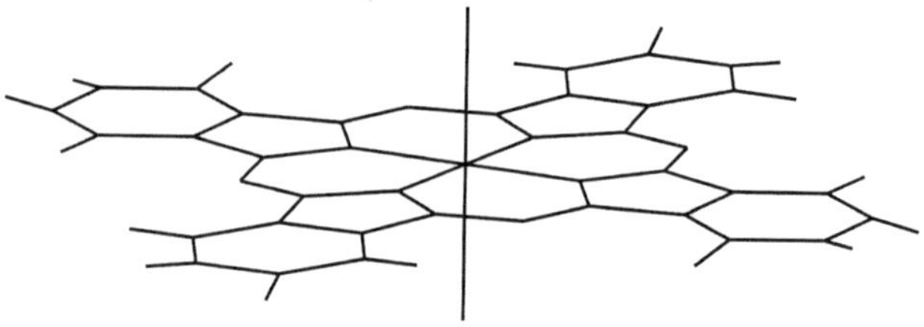

Appendix 15-38. PcFeCl$_2$, KEHXAW[123]

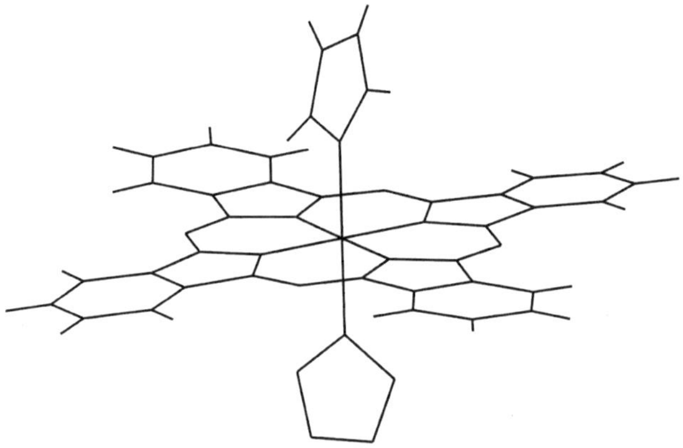

Appendix 15-41. [PcFe(im)$_2$][PNP], #2001-09[354]

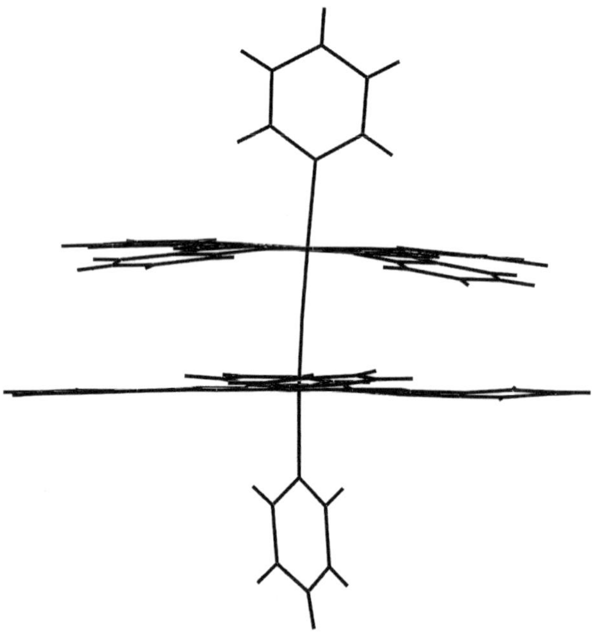

Appendix 15-39. py-(Pc)Fe-C-Fe(Pc)-py, RERGEA[352]

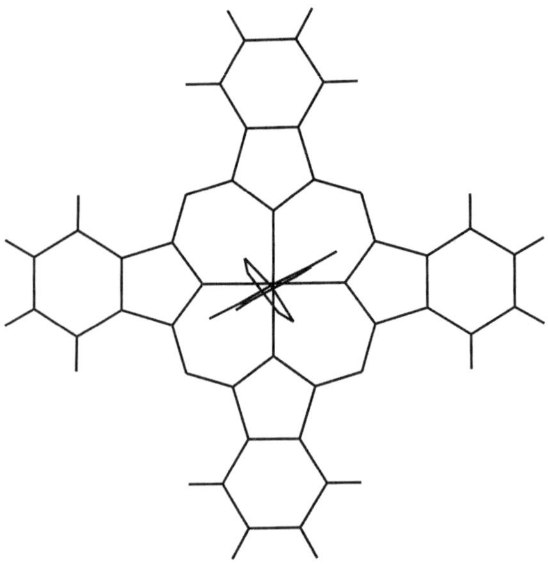

Appendix 15-42. [PcFe(im)$_2$][PNP], #2001-09[354]

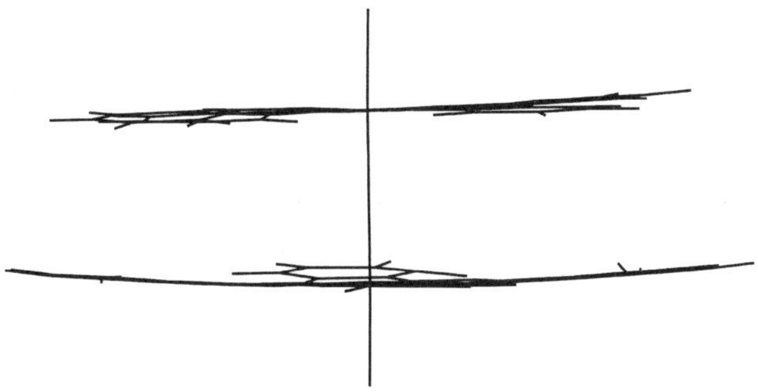

Appendix 15-43. [F-(Pc)Fe-C-Fe(Pc)-F][NBu_4]$_2$[$3H_2O$], NAPHIV[355]

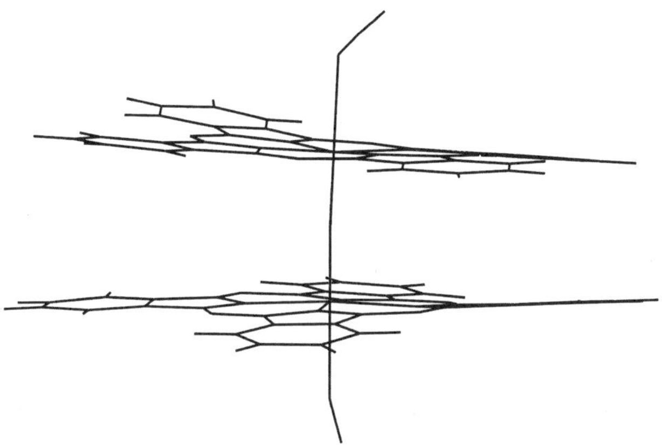

Appendix 15-44. [N_3-(Pc)Fe-N-Fe(Pc)-N_3][PNP]$_2$[I_2][$2Et_2O$], NEKLUK[356]

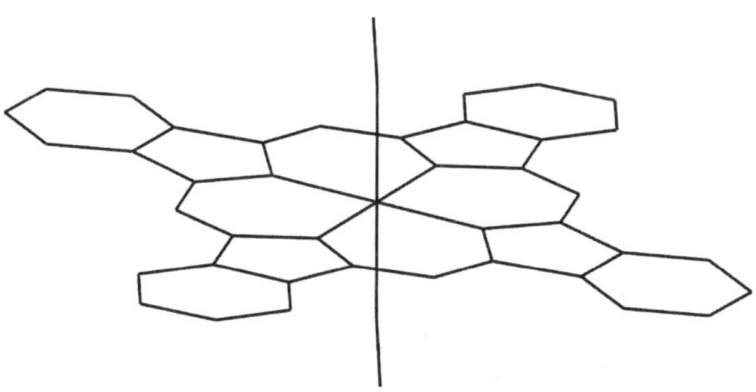

Appendix 15-45. [PcFe(CN)$_2$][PNP], DEFNUX[129,357]

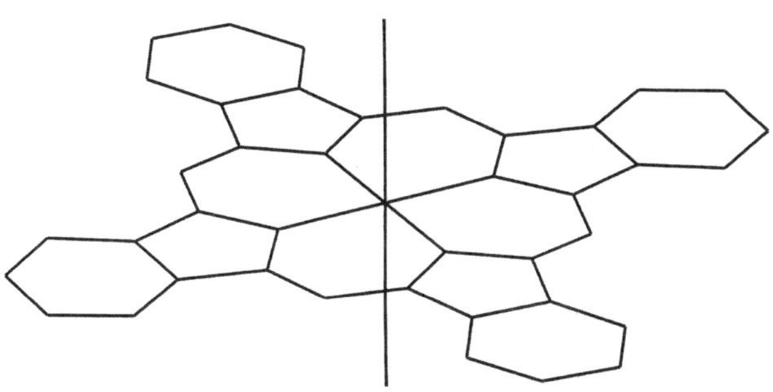

Appendix 15-46. [PcFe(CN)$_2$][PNP]$_2$[CH_2Cl_2], DODXUP[358]

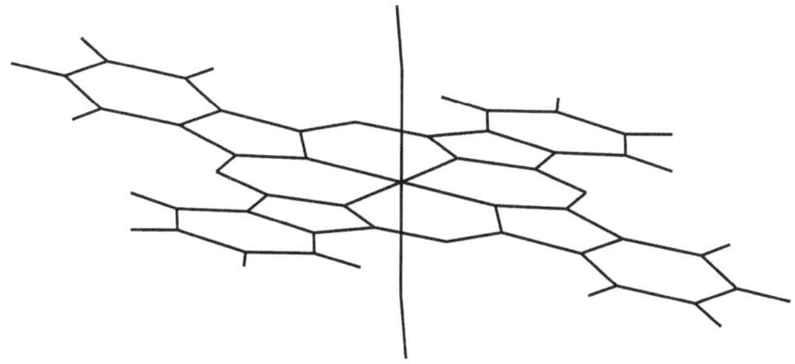

Appendix 15-47. [PcFe(CN)$_2$][Pϕ_4]$_{0.5}$, WOPZOQ[359]

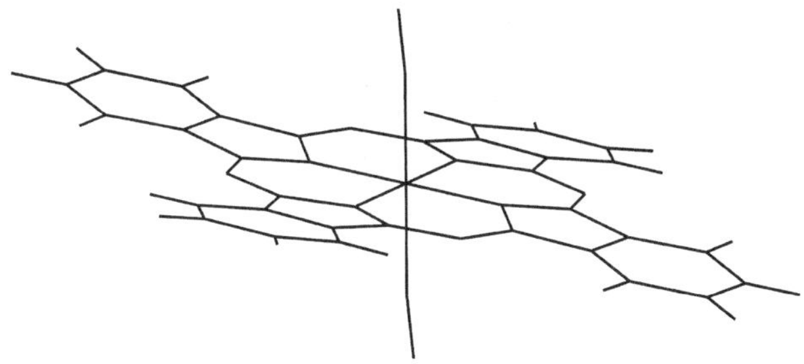

Appendix 15-48. [PcFe(CN)$_2$][PTMA][MeCN], NIRDUN[336]

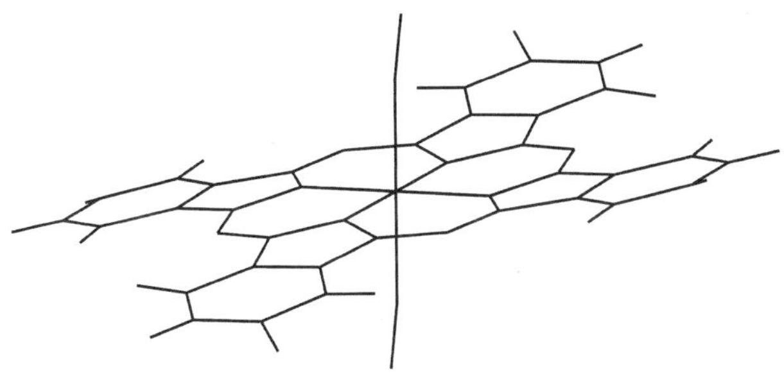

Appendix 15-49. [PcFe(CN)$_2$][PTMA][MeCN], NIRDUN01[336]

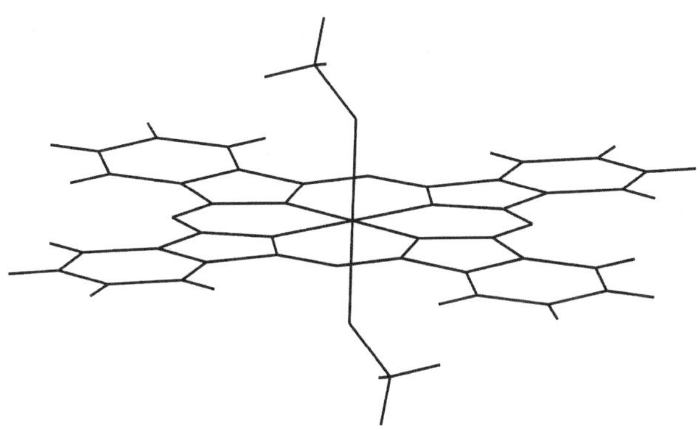

Appendix 15-50. [PcFe(OMe)$_2$][PNP]$_2$[1Me_2CO, 1.5MeOH], NELCOW[360]

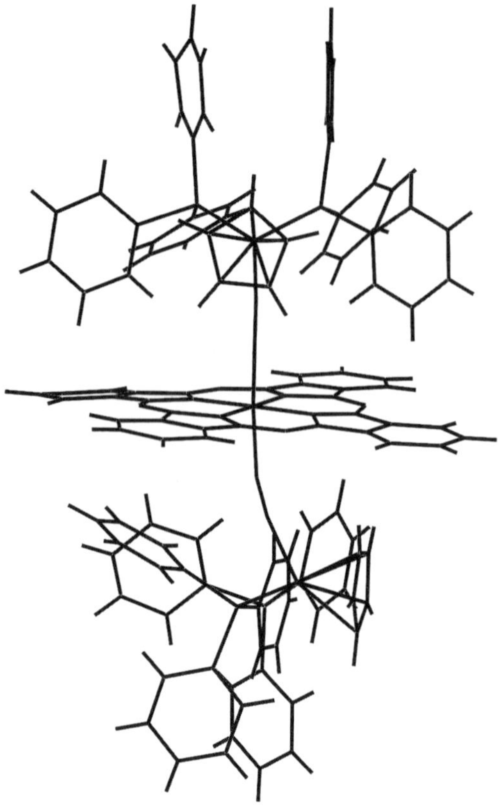

Appendix 15-51. [PcFe(CN-Ru(Pϕ_3)$_2$Cp)(NC-Ru(Pϕ_3)$_2$Cp)][SbF$_6$], WOHVEU[314]

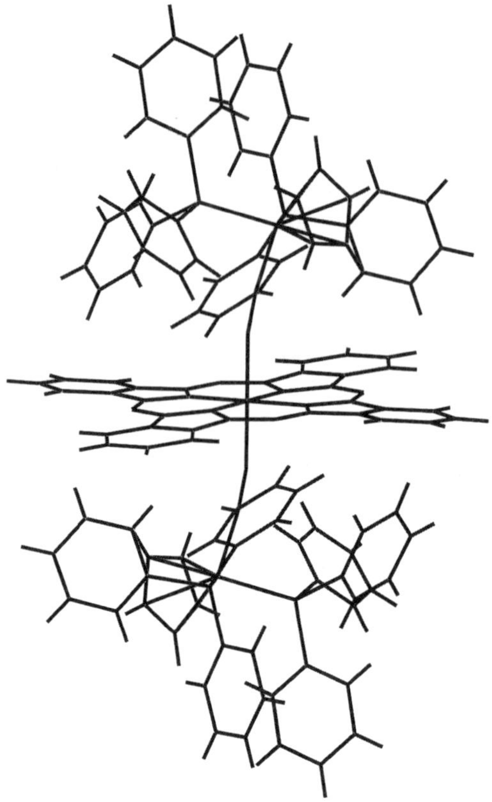

Appendix 15-53. [PcFe(NC-Ru(Pϕ_3)$_2$Cp)$_2$][SbF$_6$], WOHTUI[314]

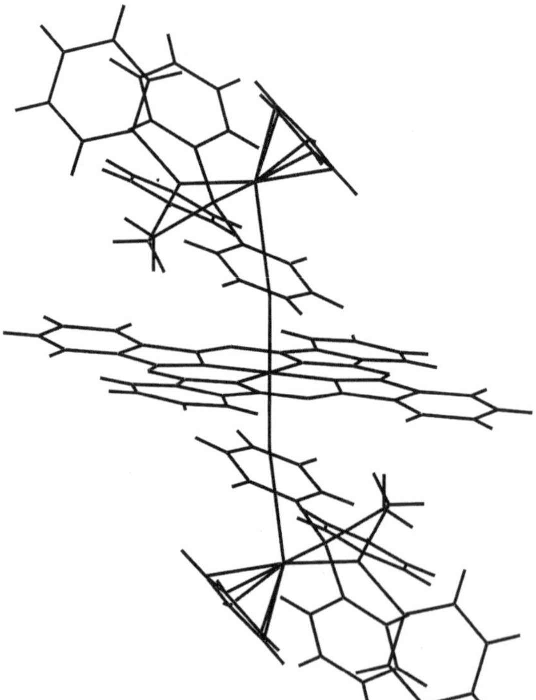

Appendix 15-52. [PcFe(NC-Fe(dppe)Cp)$_2$][SbF$_6$], WOHTOC[314]

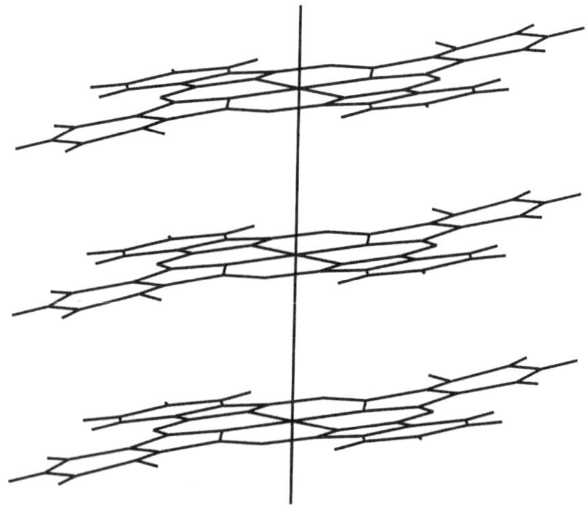

Appendix 15-54. [-Ga(Pc)F-]$_n$, BAMSEN10[122]

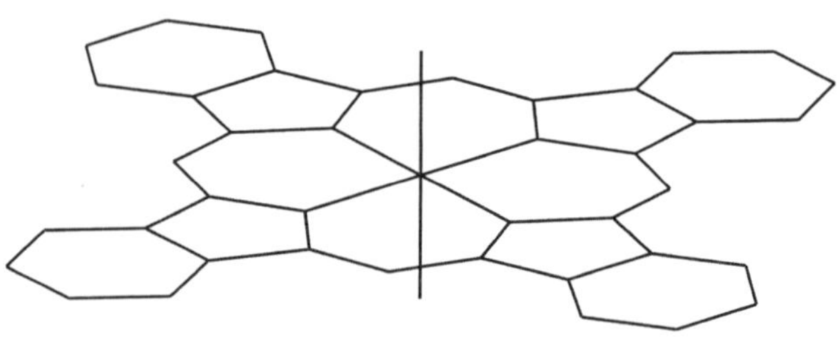

Appendix 15-55. α-PcGe(OH)$_2$, #1985-01[120] (App. 29)

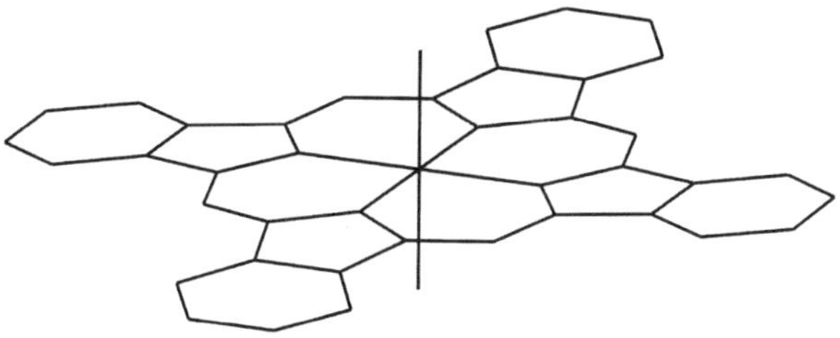

Appendix 15-56. β-$PcGe(OH)_2$, #1985-02[120] (App. 30)

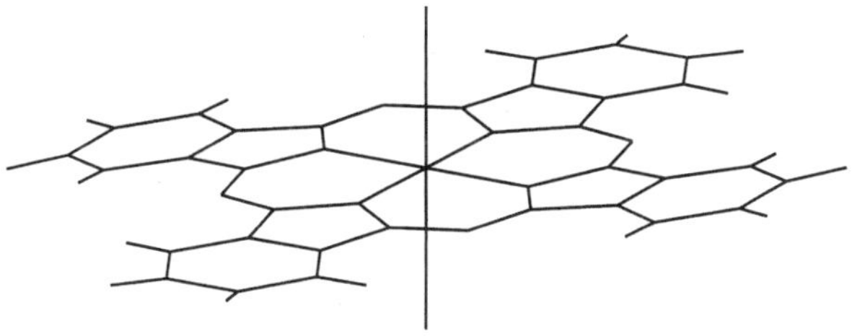

Appendix 15-57. $PcGeI_2$, RESFIE[119]

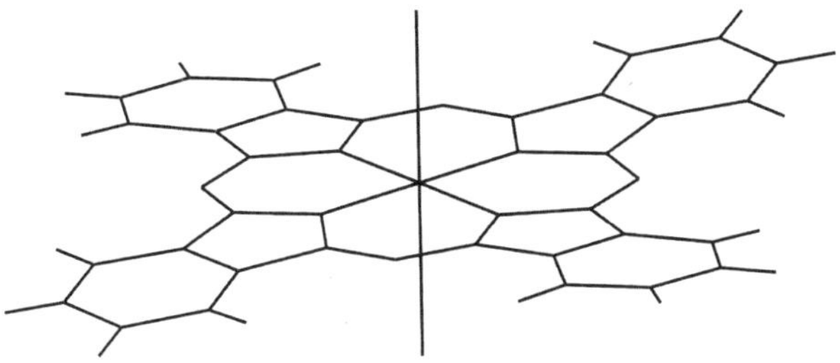

Appendix 15-58. $[PcGeI_2][I_2]$, GODLEQ[366]

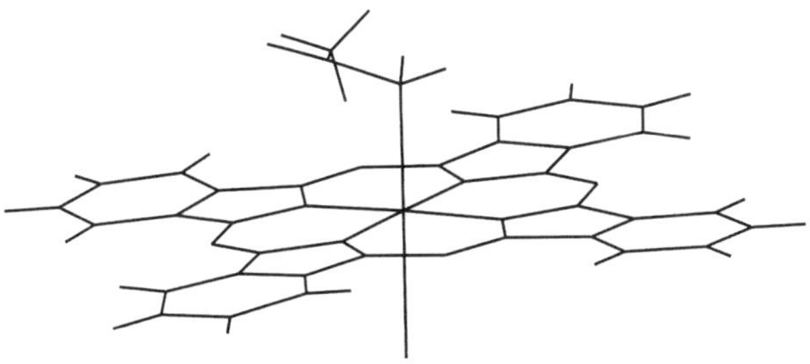

Appendix 15-59. $[PcIr(Cl,\text{-}CC(O)C)][H_2O]$, GUDZUA[367]

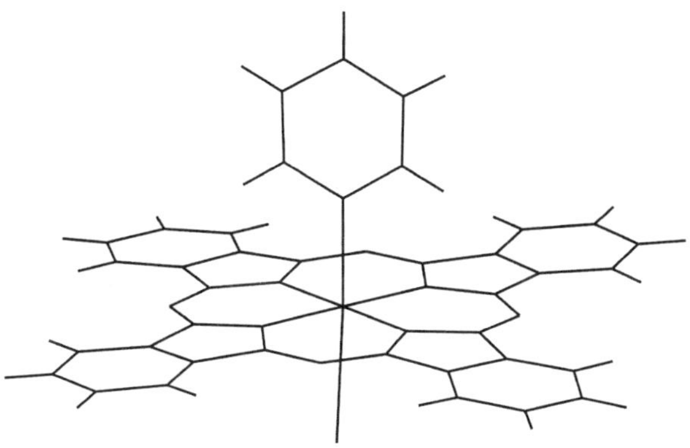

Appendix 15-60. [PcIr(py, I)][py], RENJID[368]

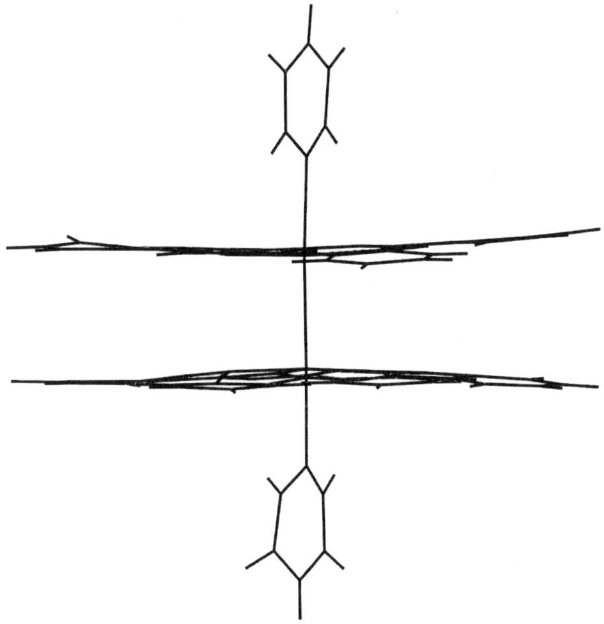

Appendix 15-61. py-(Pc)Ir-Ir(Pc)-py, TOCLIG[369]

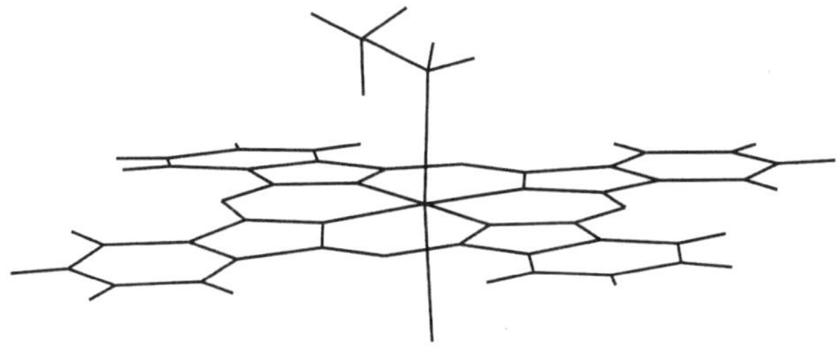

Appendix 15-62. [PcIr-(acido,N_3)]$[NBu_4][0.25Et_2O,0.5CH_2Cl_2]$, JAHSOA[370]

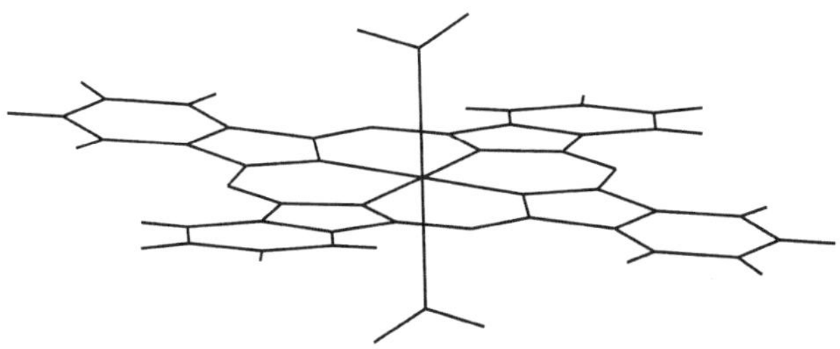

Appendix 15-63. $[PcIr(\text{-}NO_2)_2][PNP]$, RILVOX[371]

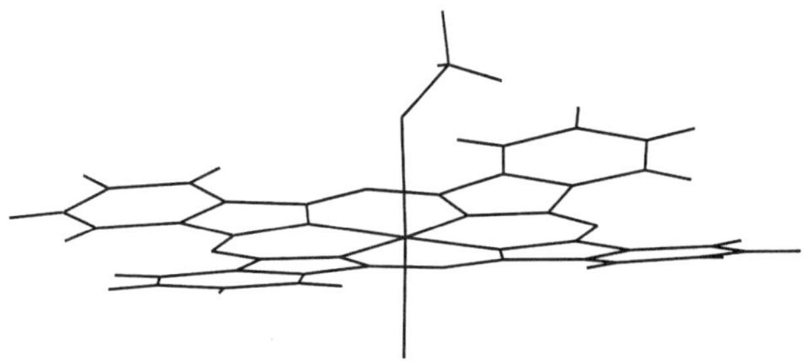

Appendix 15-64. $[PcIr(OH,OMe)][NBu_4]$, LIQVAI[372]

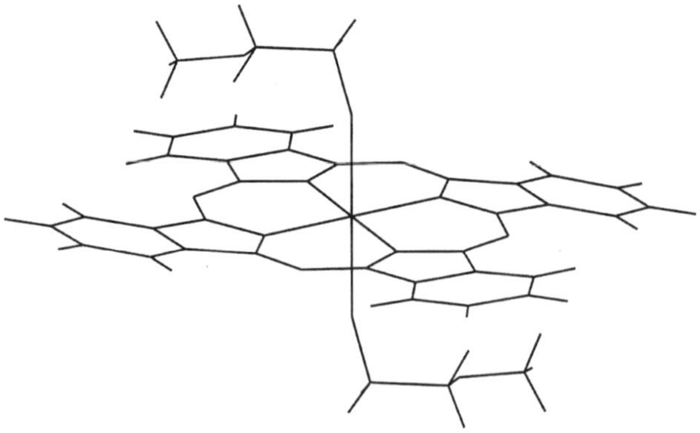

Appendix 15-65. PcMg($HOCH_2CH_2OMe)_2$, QOPCON[373]

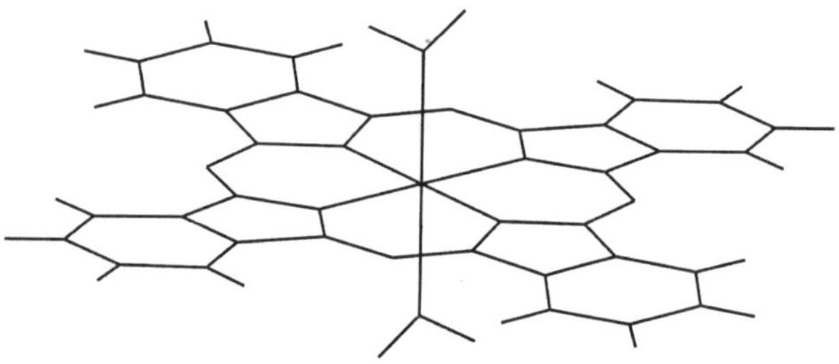

Appendix 15-66. PcMg($OH_2)_2$(NMP)$_2$, QOPCUT[373]

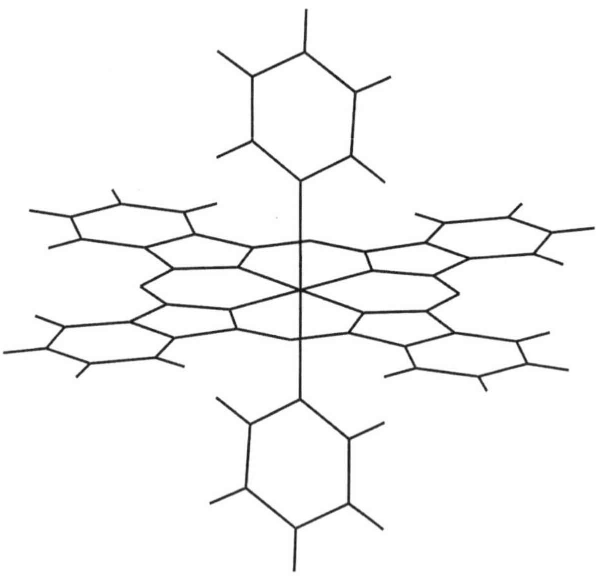

Appendix 15-67. [PcMg(py)$_2$], CCDC166120[374]

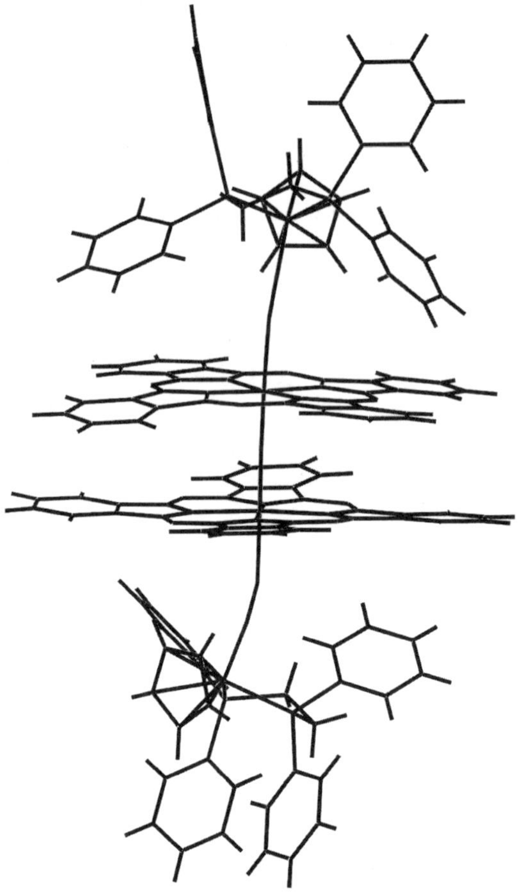

Appendix 15-68. (dppe)(Cp)Fe-(Pc)Mn-O-Mn(Pc)-NC-Fe(dppe)Cp, ROPTAR[375]

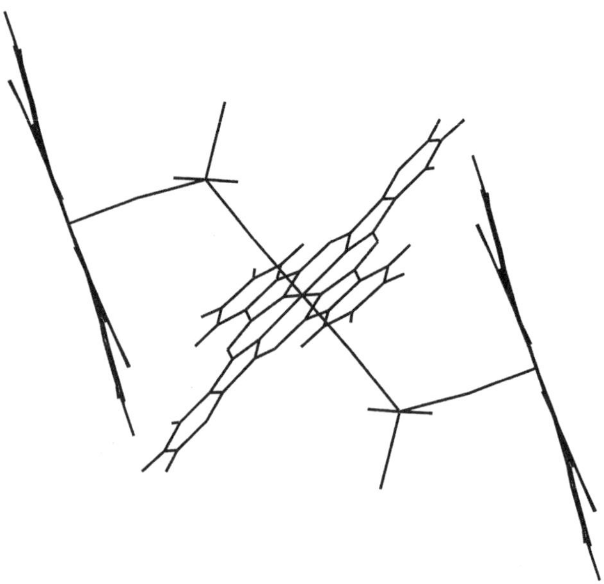

Appendix 15-69. [-(Pc)Mn-NC-C(CN)-CN-]$_n$, PILBER[376]

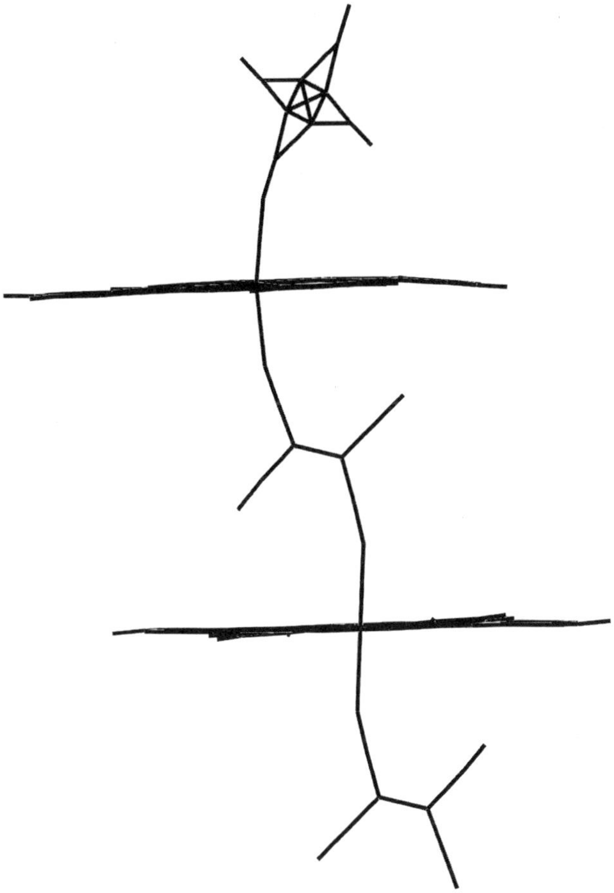

Appendix 15-70. [-(Pc)Mn-TCNE-]$_n$, PILBAN[376]

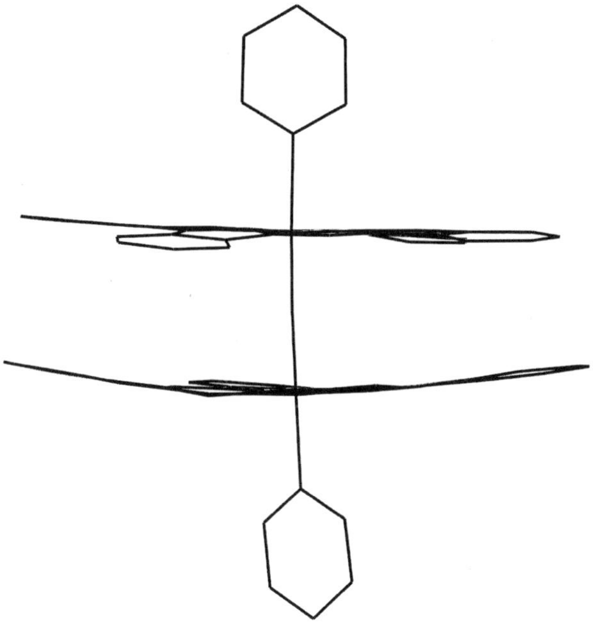

Appendix 15-72. [py-(Pc)Mn-O-Mn(Pc)-py][2py], PHCPMP10[378,379]

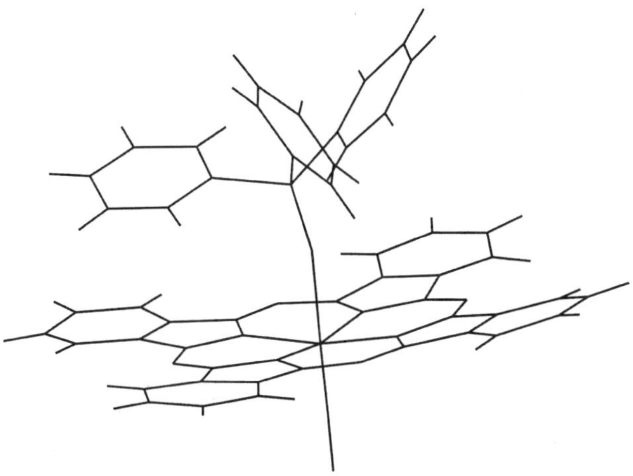

Appendix 15-71. [PcMn(NO)(OPϕ_3)], XEYREY[377]

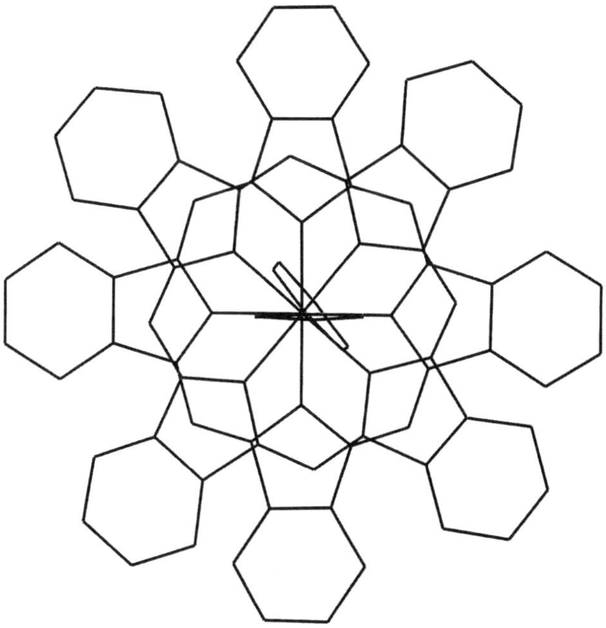

Appendix 15-73. [py-(Pc)Mn-O-Mn(Pc)-py][2py], PHCPMP10[378,379]

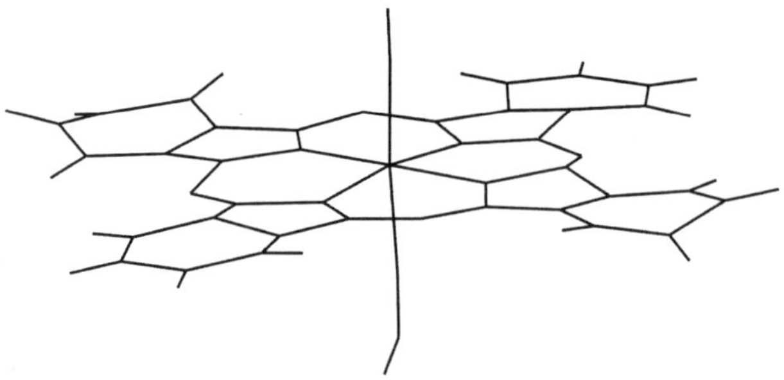

Appendix 15-74. [PcMn(NO)(-NCO)][PNP], XEYQOH[377]

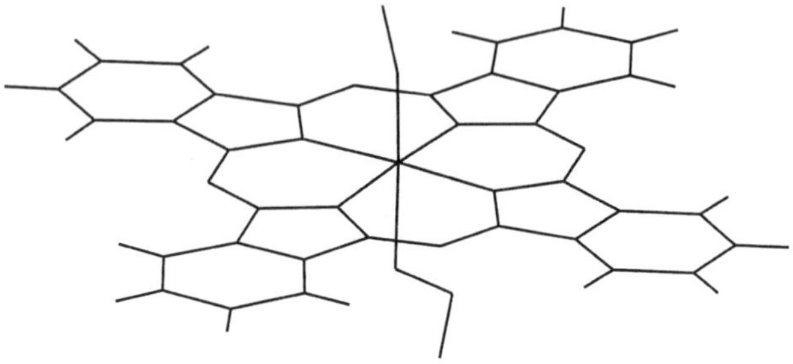

Appendix 15-75. [PcMn(NO)(-ONO)][PNP], XEYGIB[377]

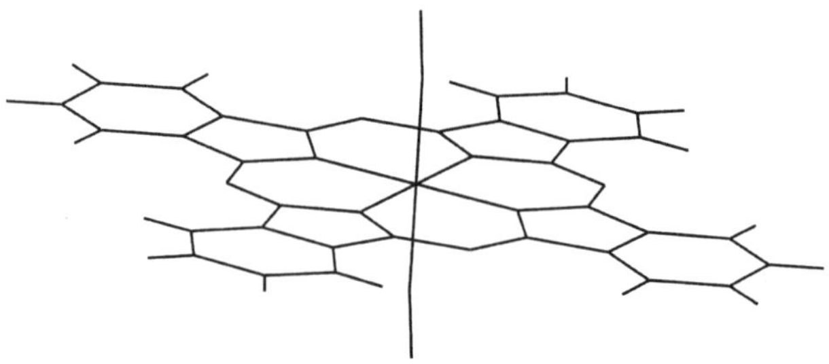

Appendix 15-76. [PcMn$(NO)_2$][PNP], XEYQUN[377]

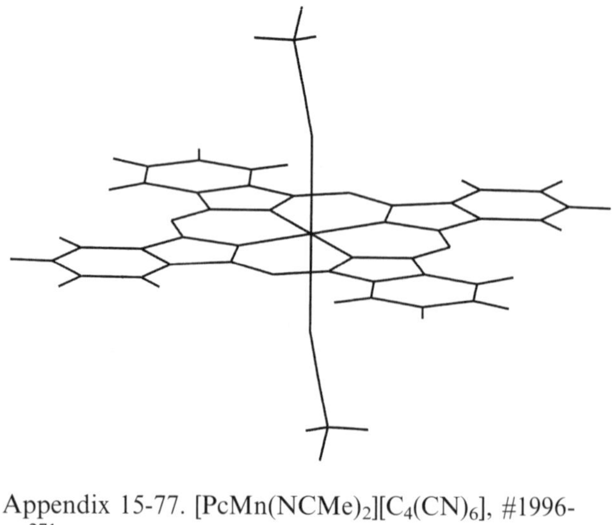

Appendix 15-77. [PcMn$(NCMe)_2$][$C_4(CN)_6$], #1996-02[271] (App. 31)

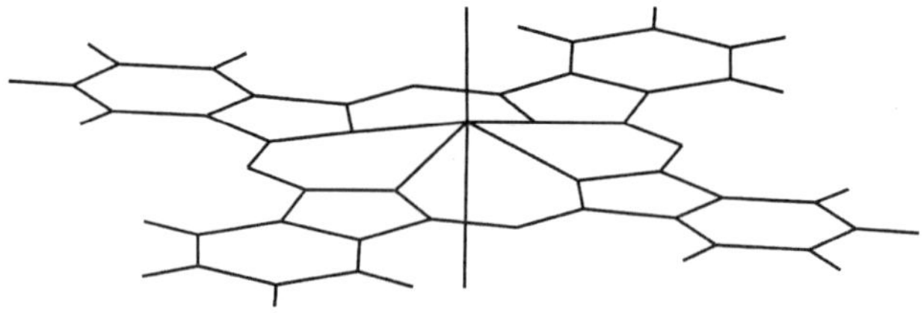

Appendix 15-78. PcMo(=O,-Cl), #1990-03[124] (App. 32)

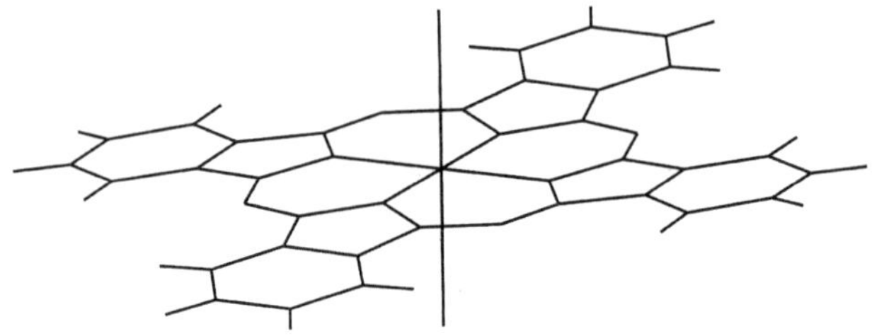

Appendix 15-79. [PcMoBr_2][PNP], RIKSIN[380]

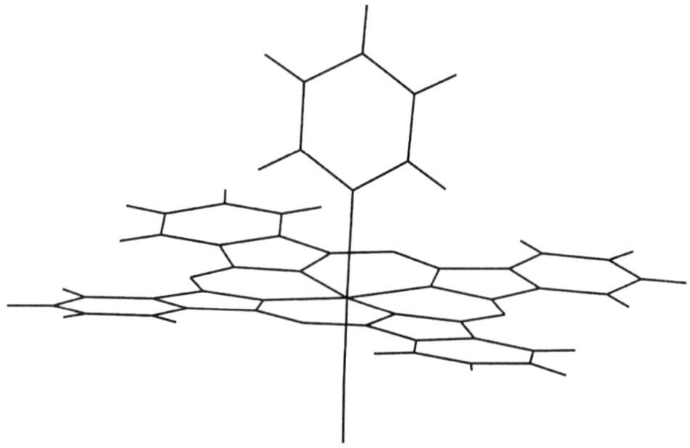

Appendix 15-80. PcOs(CO,py), PCPCOS[381]

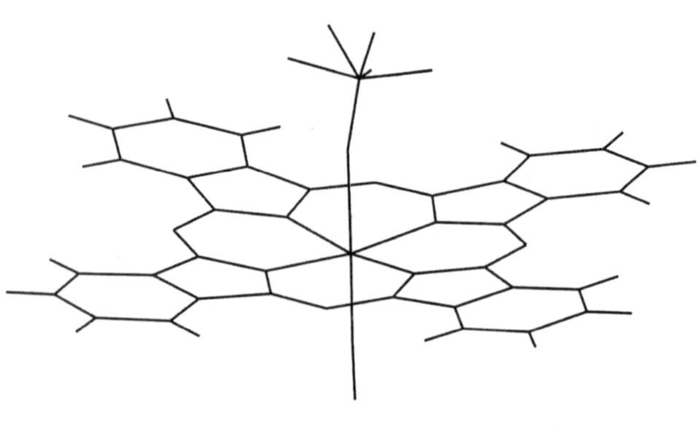

Appendix 15-81. [PcOs(NO,$C_6H_{12}O_2$)][BF_4], #1996-01[382]

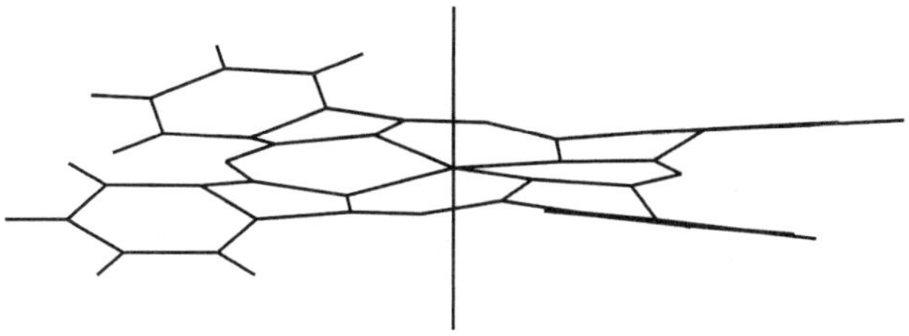

Appendix 15-82. $[PcOsCl_2][PNP][0.33PNPOH]$, #1997-04[131]

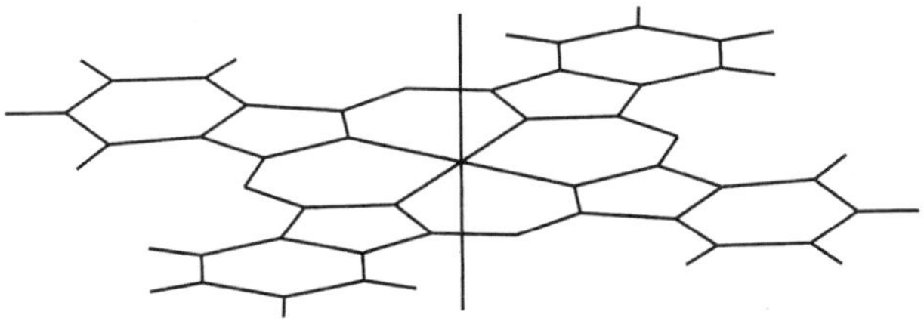

Appendix 15-83. $[PcOsCl_2][PNP]$, #1997-05[131]

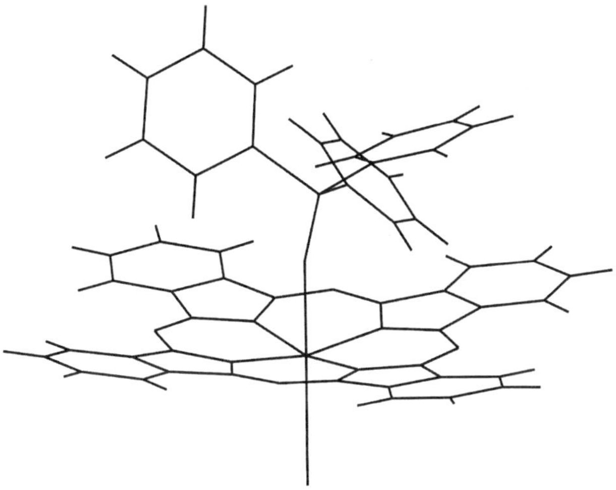

Appendix 15-84. $PcRe(NO)(OP\phi_3)$, XEYRIC[377]

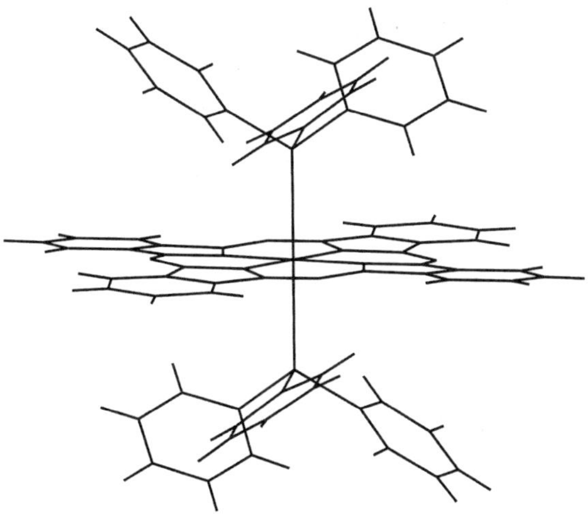

Appendix 15-85. $PcRe(P\phi_3)_2$, NAPPAV[383]

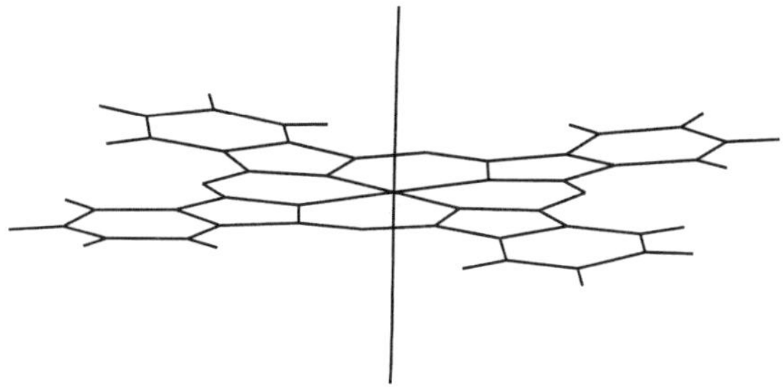

Appendix 15-86. $[PcRe(CN)_2][NBu_4][0.5Me_2CO]$, DOSWIR[384]

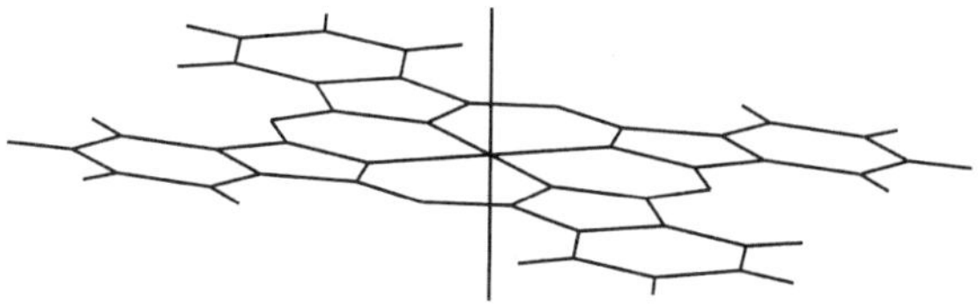

Appendix 15-87. $[PcRe(CN)_2][PNP]$, DOSWOX[384]

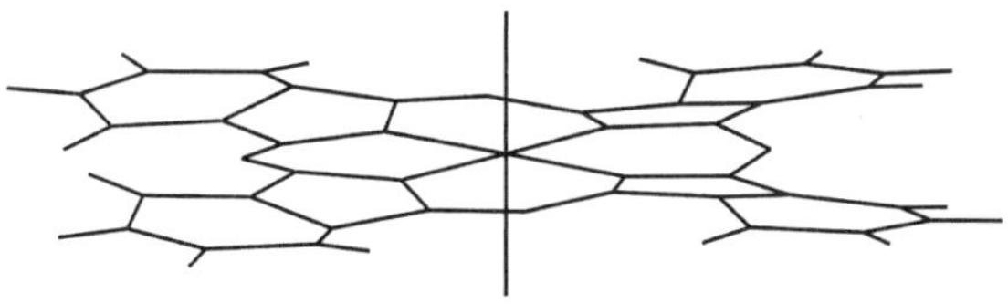

Appendix 15-88. $[PcReF_2][0.33(PNP)F][2H_2O]$, GEQCOU[385]

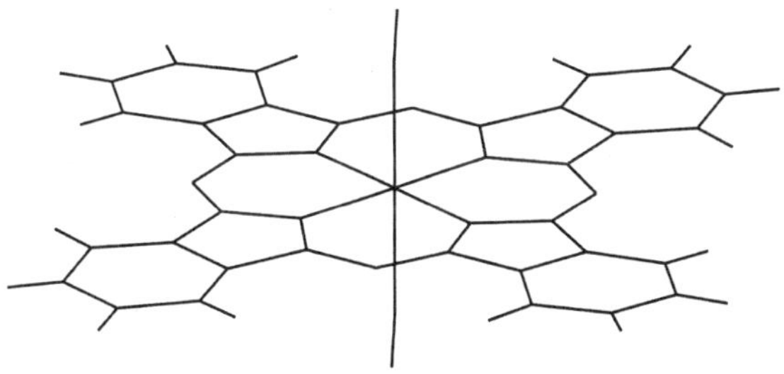

Appendix 15-89. $[PcRe(NO)_2][PNP]$, XEYRAU[377]

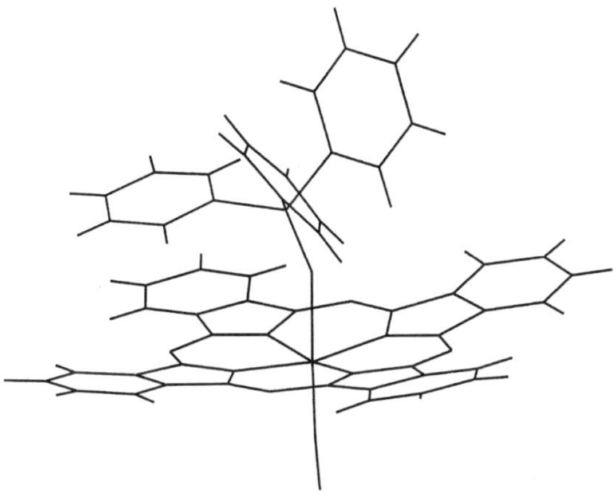

Appendix 15-90. $[PcRe(NO)(OP\phi_3)][I_3][CH_2Cl_2]$, XEYROI[377]

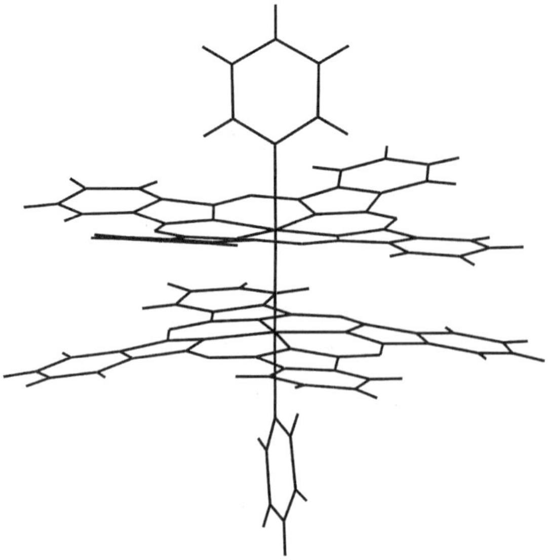

Appendix 15-91. [py-(Pc)Rh-Rh(Pc)-py][$2C_6H_6$], NEXSIS[387]

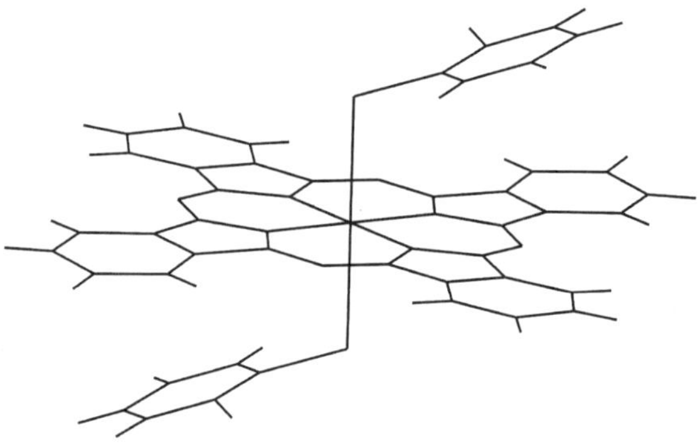

Appendix 15-92. [PcRh(OϕCl)$_2$][NBu_4], #1997-02[386]

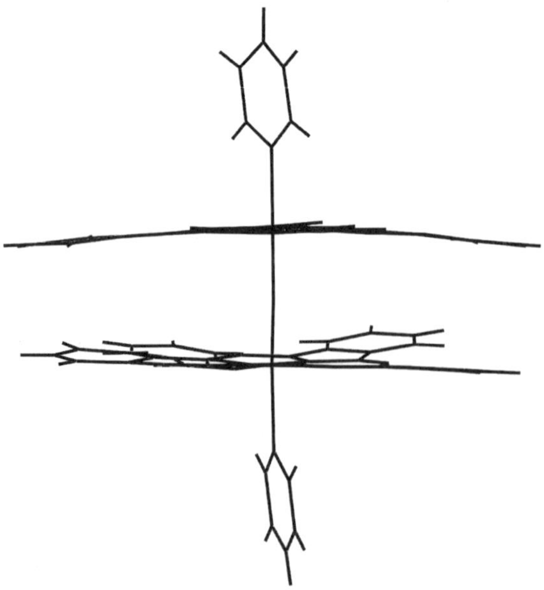

Appendix 15-93. py-(Pc)Ru-C-Ru(Pc)-py, RERGIE[352]

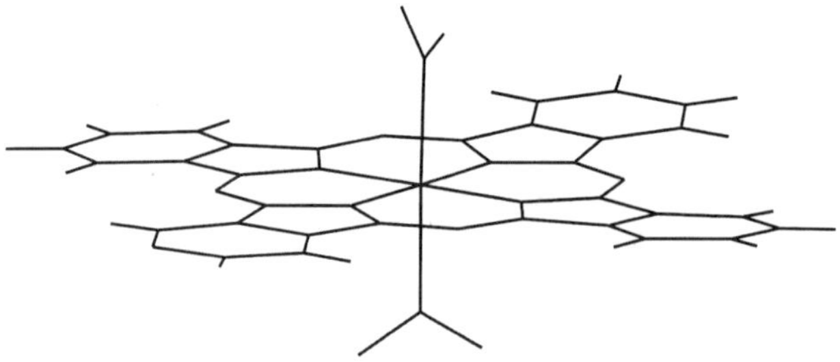

Appendix 15-94. [PcRu(-NO_2)$_2$][NBu_4]$_2$, RUJQES[388]

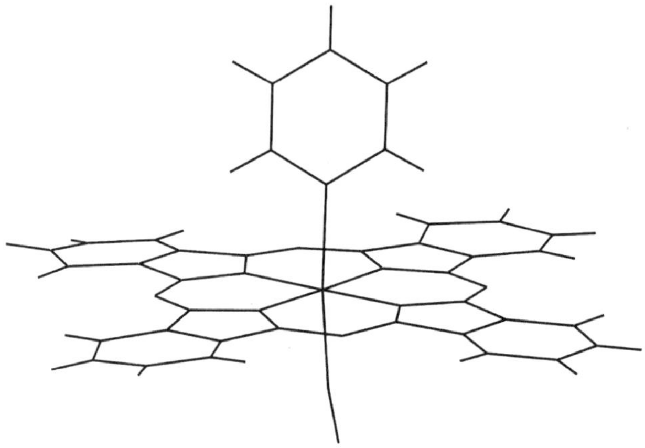

Appendix 15-95. [PcRu(py,CN)][NBu_4], GEQDIP[389]

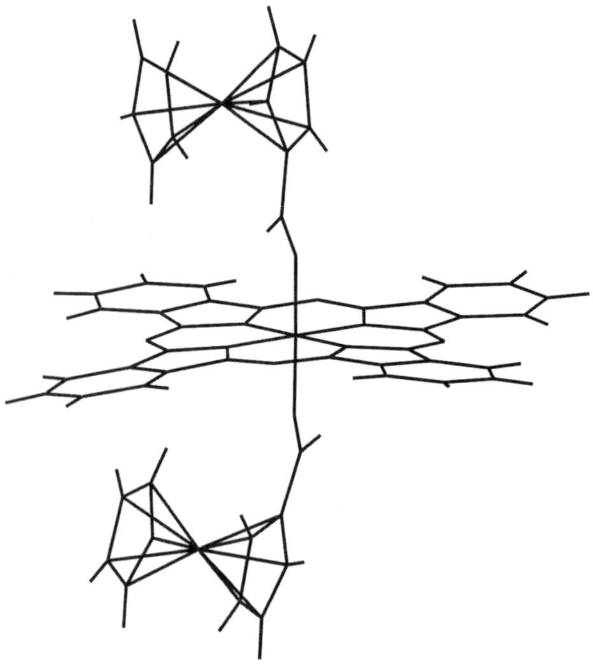

Appendix 15-96. [PcSi(-O-C(O)-Fc)$_2$][$2CHCl_3$], NEYCEZ[392]

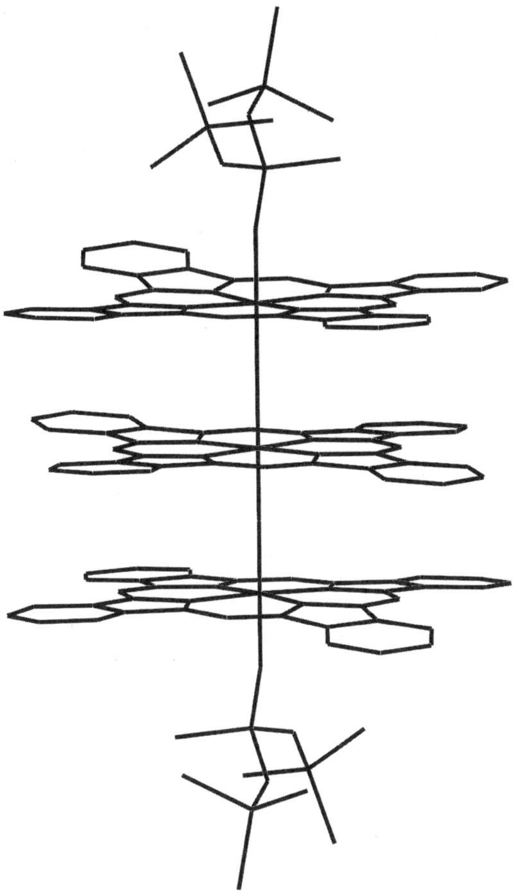

Appendix 15-97. $(Me_3SiO)_2MeSiO$-((Pc)Si-O$)_3$-SiMe$(OSiMe_3)_2$, #1970-01[269] (App. 34)

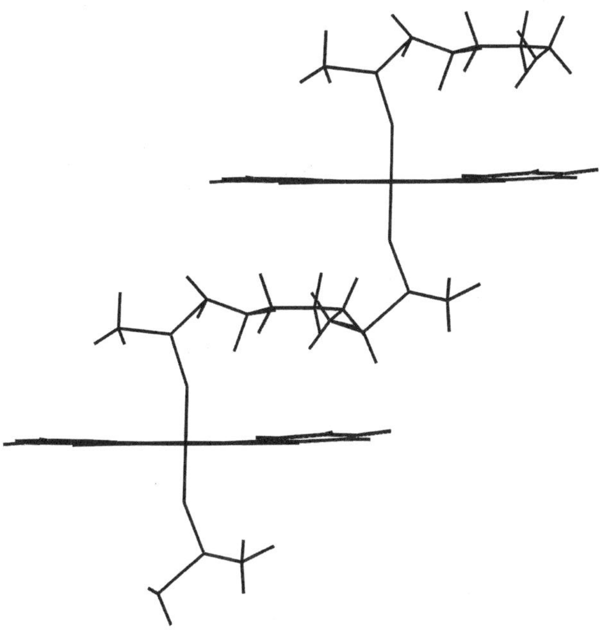

Appendix 15-98. [-(Pc)Si-(OSiMe$_2${CH$_2$}$_7$SiMe$_2$O-)]$_n$, QUDQIP[390,391]

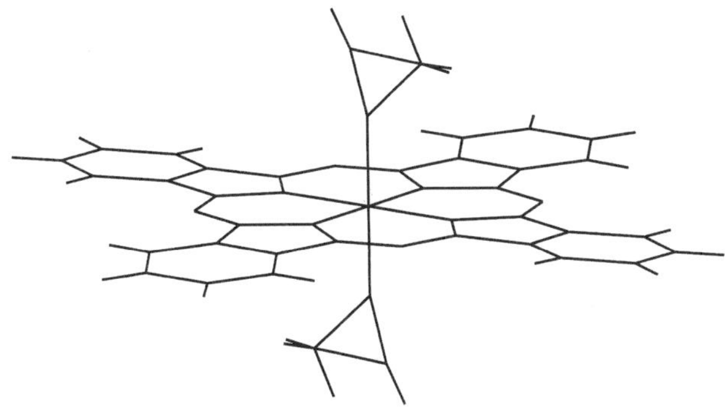

Appendix 15-99. [PcSi(OMe)$_2$][0.48H_2O], VIKZAQ[127]

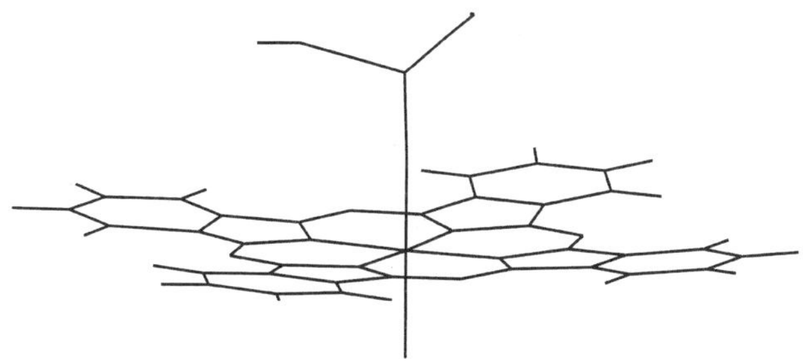

Appendix 15-100. [PcSiOSiEt$_3$,Me][0.5C_6H_{12}], SIOPCY,[449] #1971-01[270] (App. 33)

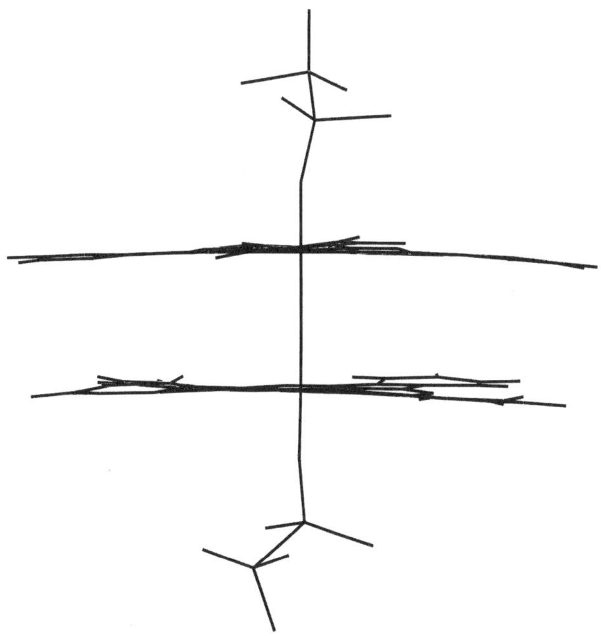

Appendix 15-101. Me$_2$(tBu)SiO-(Pc)Si-O-Si(Pc)-OSi(tBu)Me$_2$, DOHYAA[393]

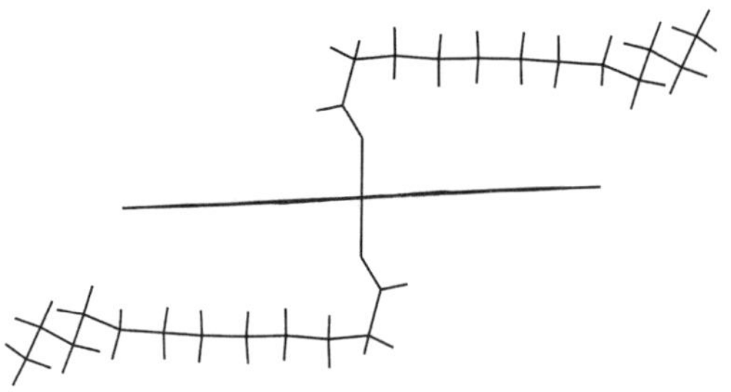

Appendix 15-102. PcSi(O-C(O)-$C_{10}H_{21})_2$, CCDC173623[128]

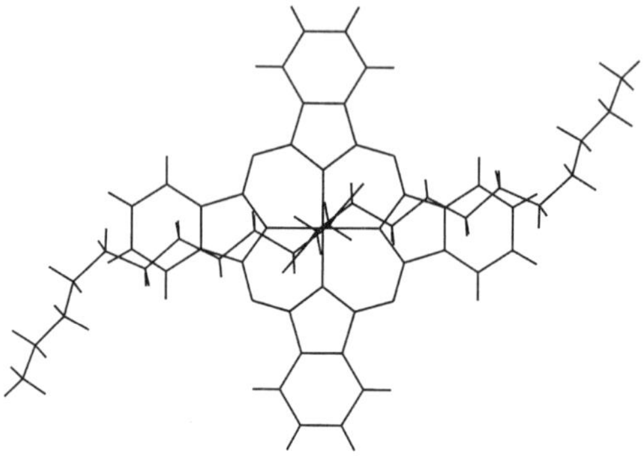

Appendix 15-103. PcSi(O-C(O)-$C_{10}H_{21})_2$, CCDC173623[128]

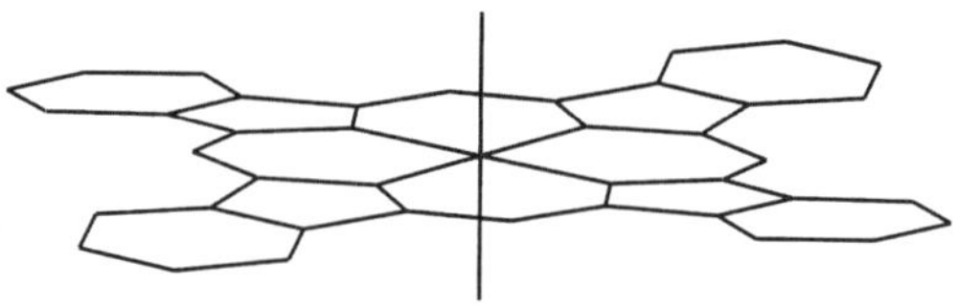

Appendix 15-104. I-PcSi$(OH)_2$, QEKBEN01[126]

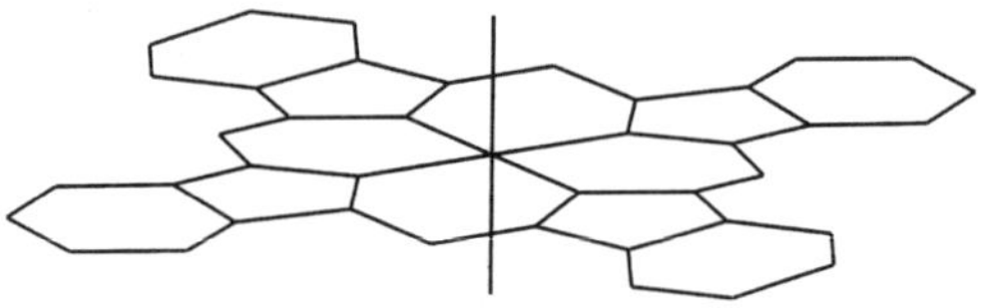

Appendix 15-105. II-PcSi$(OH)_2$, QEKBEN[126]

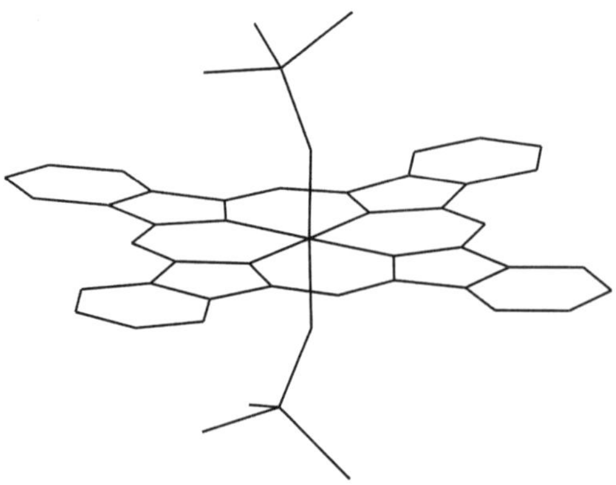

Appendix 15-106. PcSi$(OSiMe_3)_2$, SIMPLC10[395]

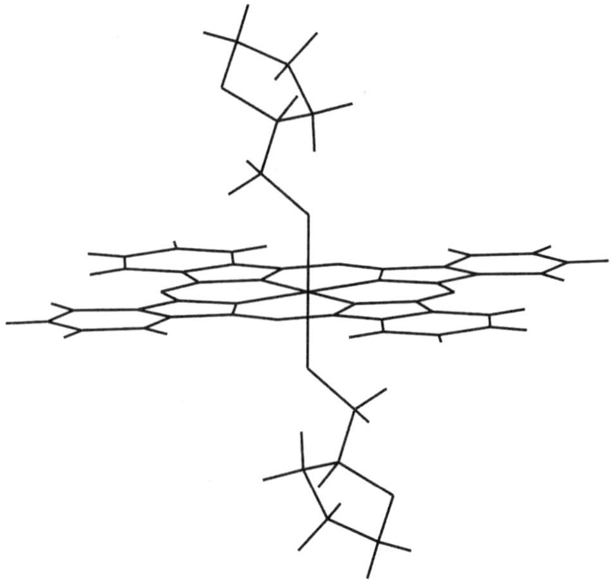

Appendix 15-107. PcSi$(thff)_2$, VILGOM[127]

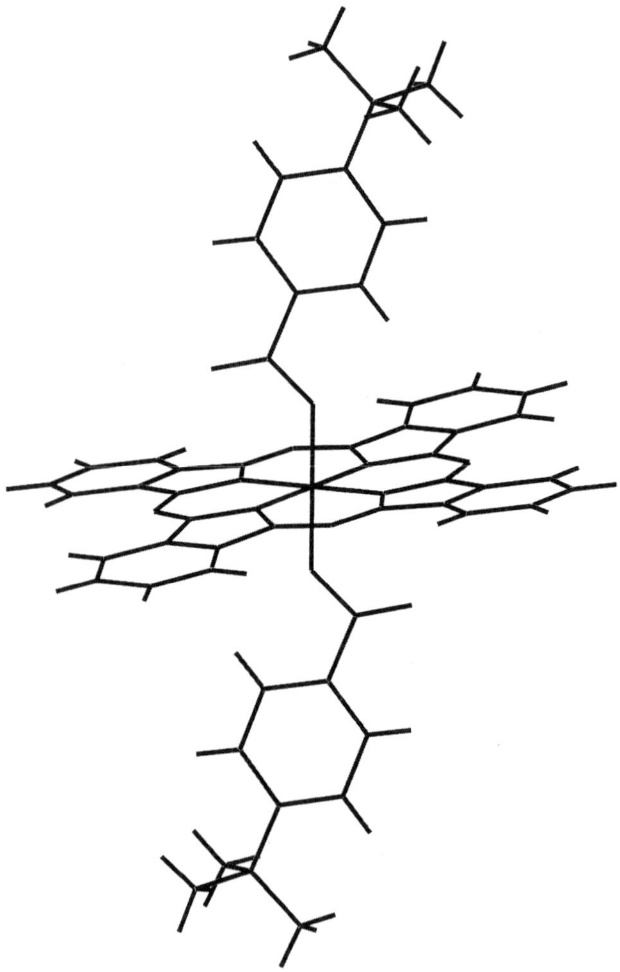

Appendix 15-108. PcSi[O(CO)ϕ-4tBu$]_2$, ACUMEQ[399]

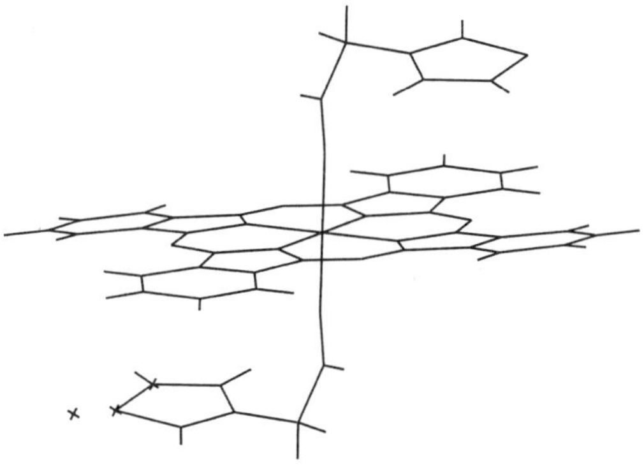

Appendix 15-109. PcSi[O(CO)-(3-thienyl)]$_2$, ACUMIU[399]

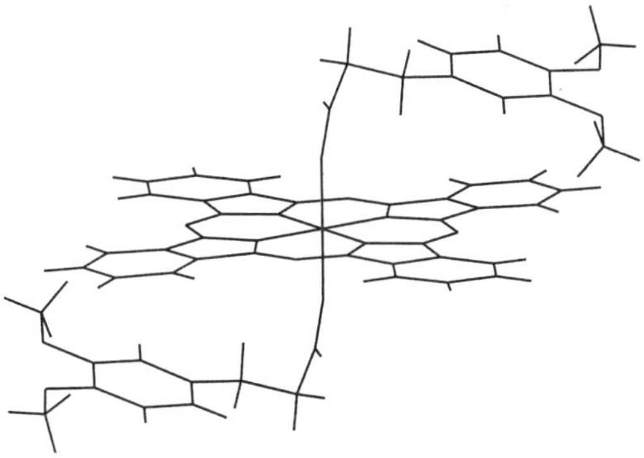

Appendix 15-110. PcSi[O(CO)(CH$_2$)$_2\phi$(3,4-(OMe)$_2$)$_2$, ACUMOA[399]

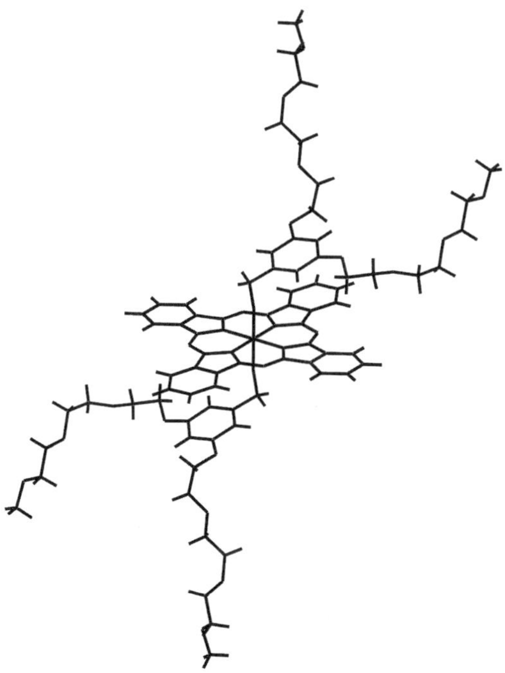

Appendix 15-111. PcSi[OCH$_2$-C$_6$H$_3$((OCH$_2$CH$_2$O)$_3$Me)$_2$]$_2$, #2001-02[396]

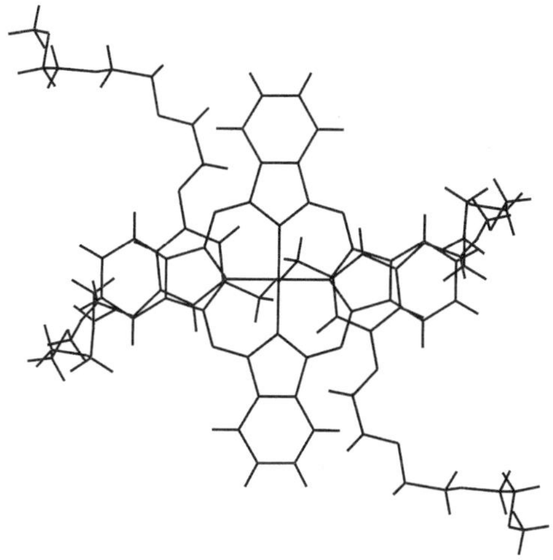

Appendix 15-112. PcSi[OCH$_2$-C$_6$H$_3$((OCH$_2$CH$_2$O)$_3$Me)$_2$]$_2$, #2001-02[396]

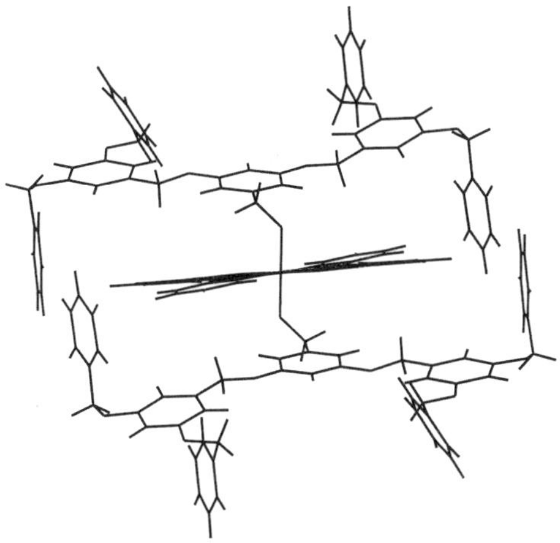

Appendix 15-113. PcSi[OCH$_2$-C$_6$H$_3$(OCH$_2\phi$)$_2$]$_2$, NOLXOB[397]

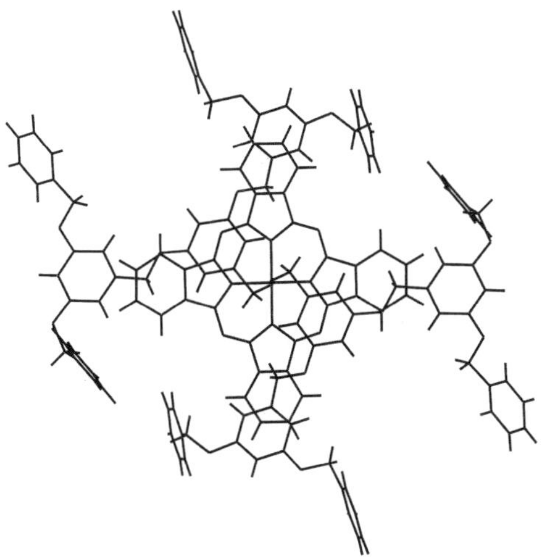

Appendix 15-114. PcSi[OCH$_2$-C$_6$H$_3$(OCH$_2\phi$)$_2$]$_2$, NOLXOB[397]

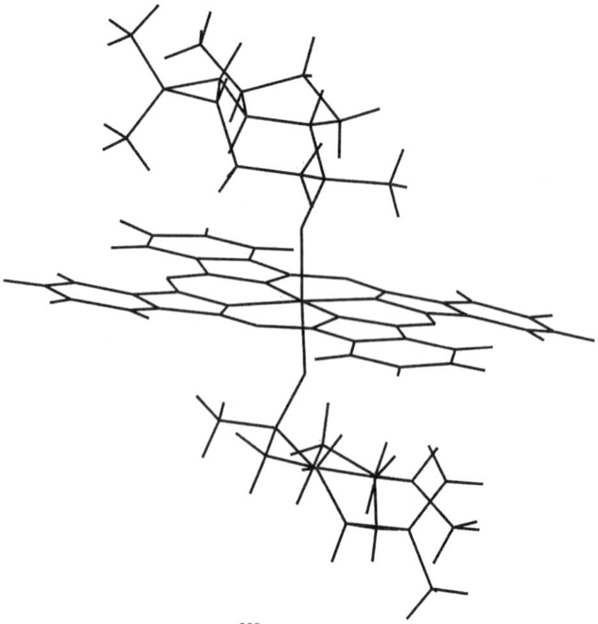

Appendix 15-115. PcSi[Oepiglobulyl]$_2$, CAVZAA[398]

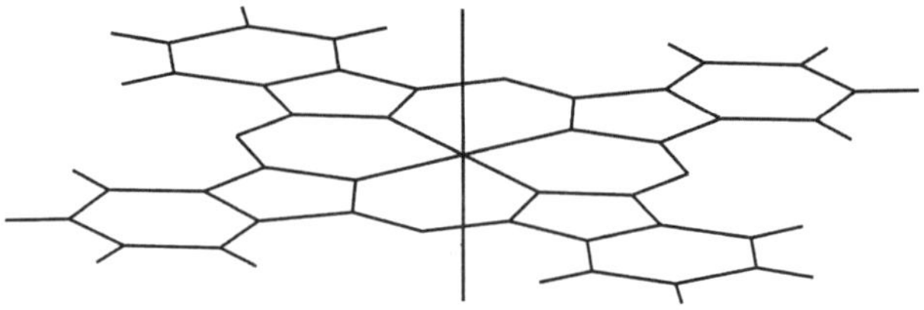

Appendix 15-116. PcSiCl$_2$, VIKVEQ[127]

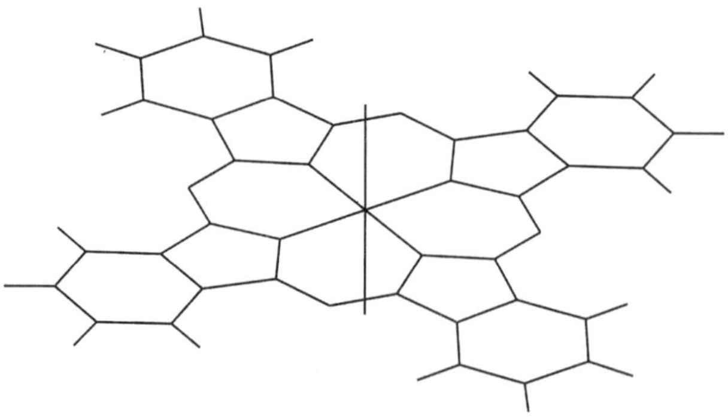

Appendix 15-117. PcSiCl$_2$, CCDC173621[128]

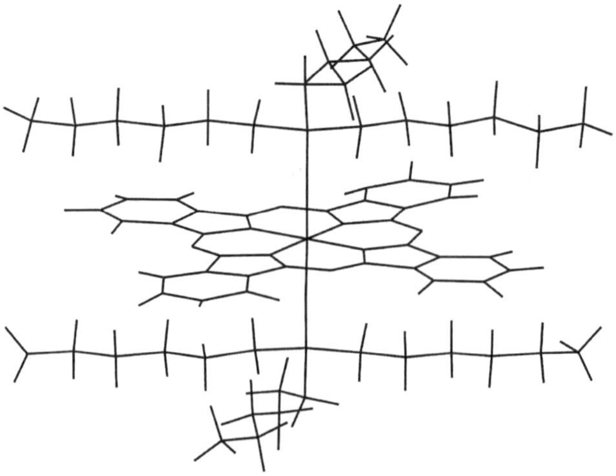

Appendix 15-118. PcSi[OSi(C$_6$H$_{13}$)$_3$]$_2$, QUDQOV[390,391]

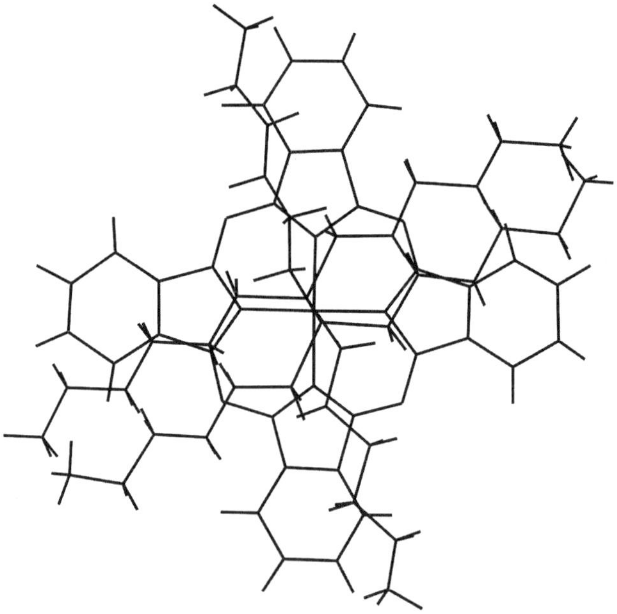

Appendix 15-119. PcSi[OSi(C$_6$H$_{13}$)$_3$]$_2$, QUDQOV[390,391]

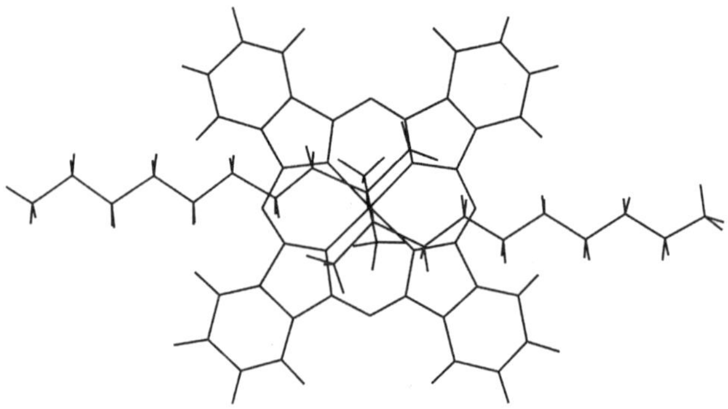

Appendix 15-120. PcSi[OSiMe$_2$(C$_8$H$_{17}$)]$_2$, QUDQUB[390,391]

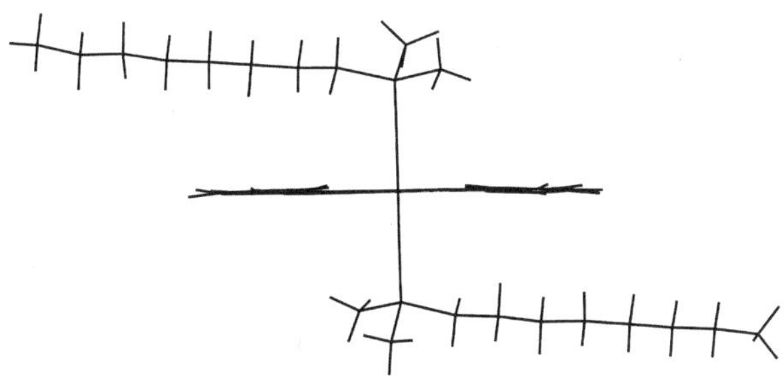

Appendix 15-121. $PcSi[OSiMe_2(C_8H_{17})]_2$, QUDQUB[390,391]

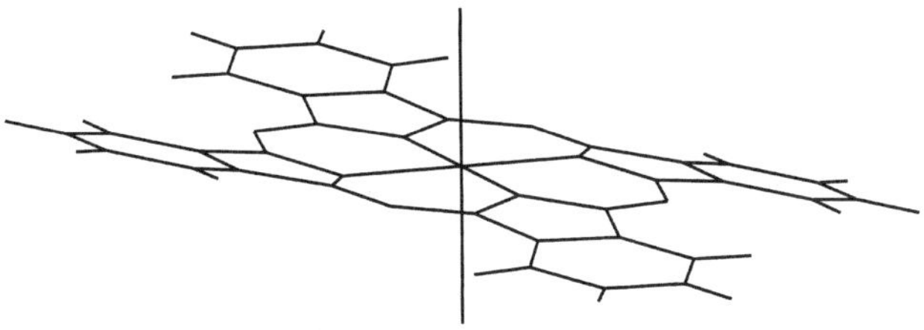

Appendix 15-122. $PcSnCl_2$, #1972-01[117] (App. 35)

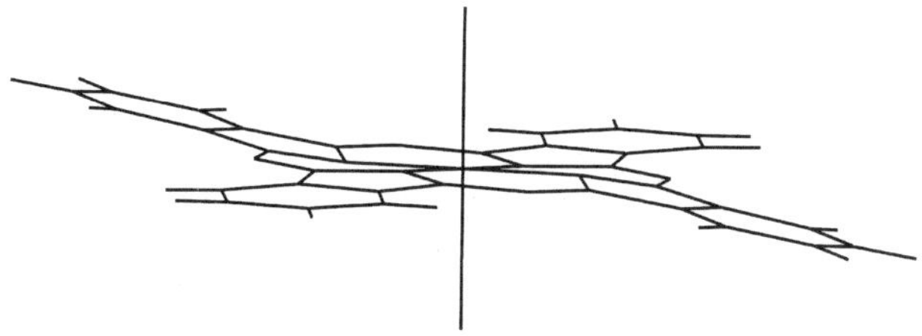

Appendix 15-123. $PcSnCl_2$, #1972-01[117] (App. 35)

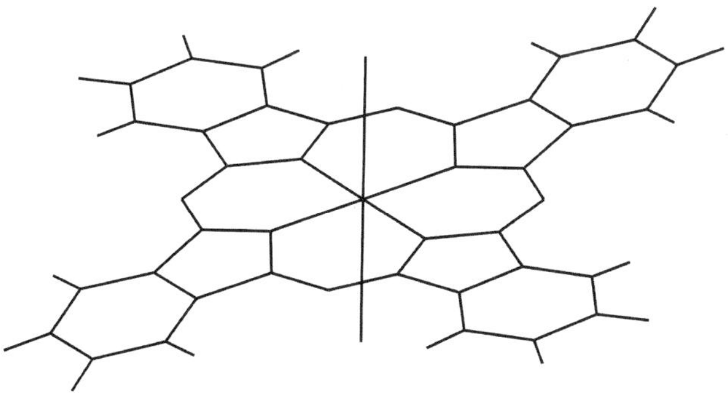

Appendix 15-124. $PcSnI_2$, RESFOK[119]

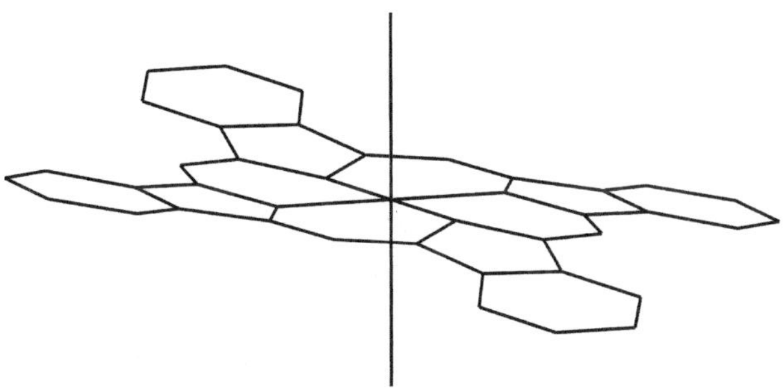

Appendix 15-125. $PcSnI_2$, RESFOK01[118]

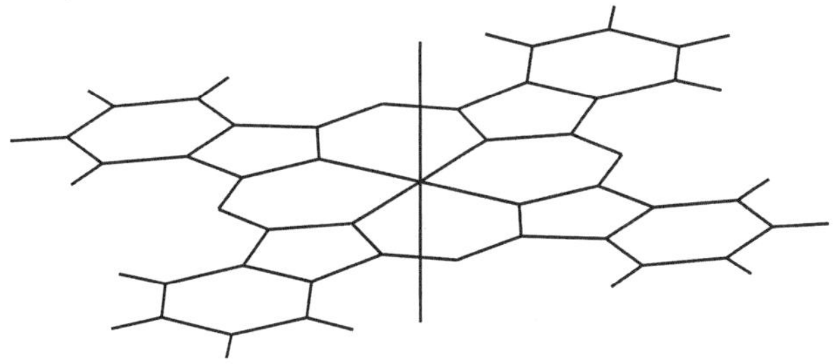

Appendix 15-126. $[PcWCl_2][PNP]$, #2001-05[342]

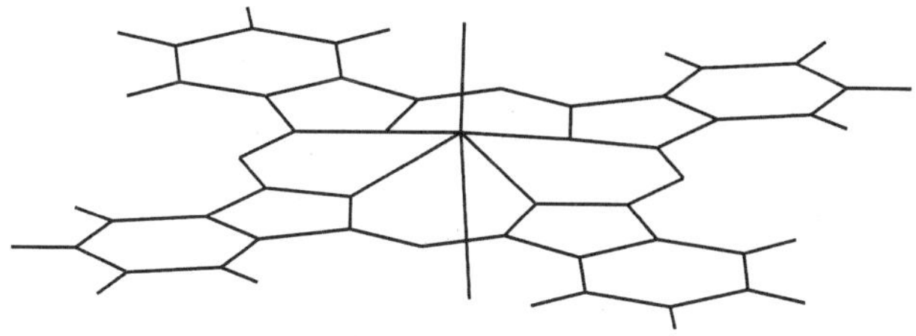

Appendix 15-127. PcW(=O,-Cl), #1990-02[124] (App. 36)

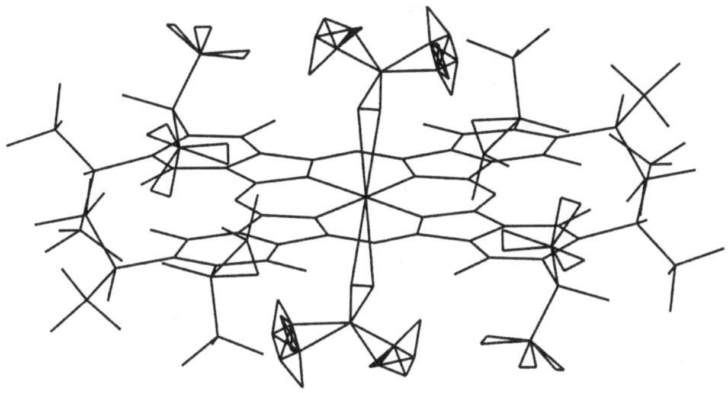

Appendix 15-128. $[F_8((F_3C)_2FC)_8PcZn(OCMe_2)_2]$, CCDC169755[341]

Appendix 16. Molecular Structures of CN7*cis* Phthalocyanine Complexes

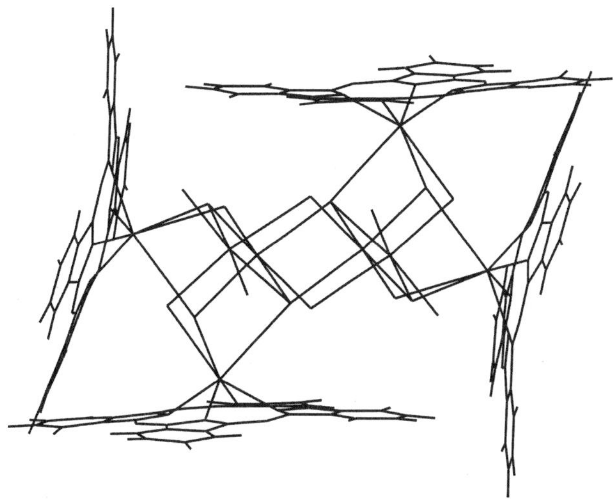

Appendix 16-1. $(PcBi)_4(Bi_4I_{16})$, LOKSAF[190]

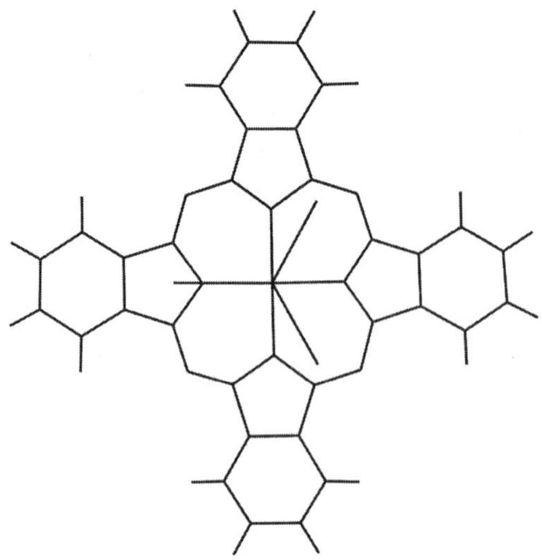

Appendix 16-2. One macrocycle of $(PcBi)_4(Bi_4I_{16})$, LOKSAF[190]

Appendix 16-3. One macrocycle of $(PcBi)_4(Bi_4I_{16})$, LOKSAF[190]

Appendix 16-4. $[PcHfF_3][NBu_4]$, POQYID[182]

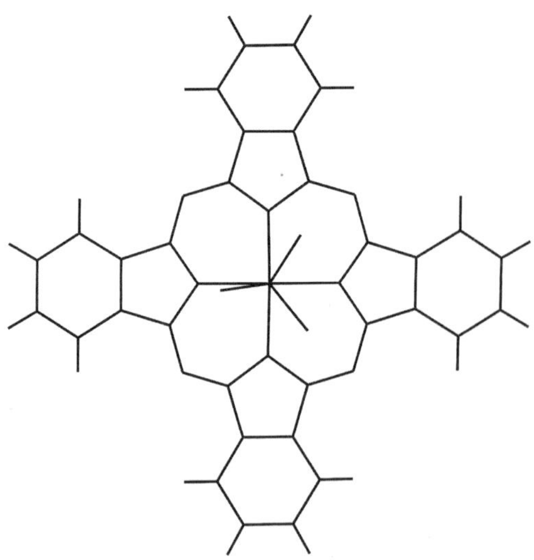

Appendix 16-5. $[PcHfF_3][NBu_4]$, POQYID[182]

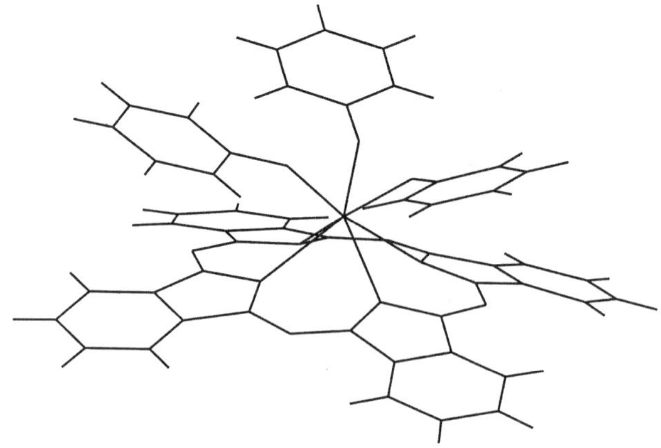

Appendix 16-6. $[PcHf(O\phi)_3][NBu_4][0.5Et_2O]$, CCDC165833[400]

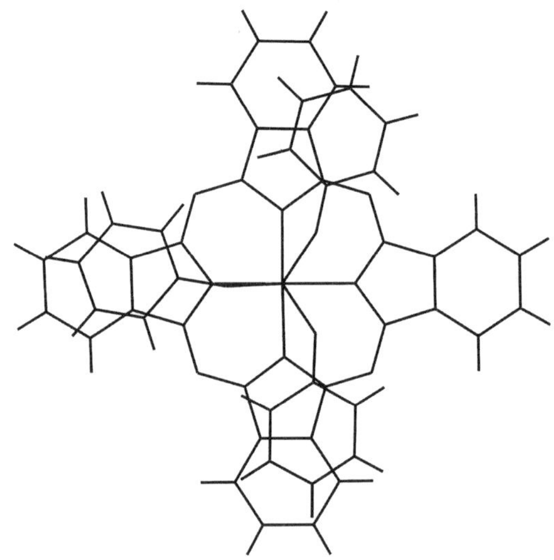

Appendix 16-7. $[PcHf(O\phi)_3][NBu_4][0.5Et_2O]$, CCDC165833[400]

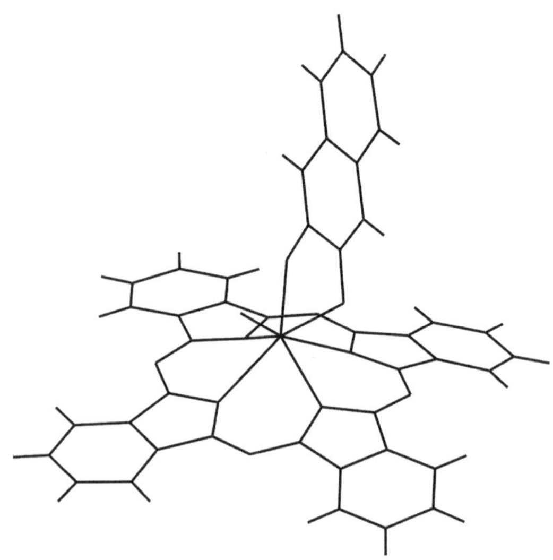

Appendix 16-8. $[PcNb(O,O_2Naphth)][CH_2Cl_2]_{0.5}$, #1998-01[308]

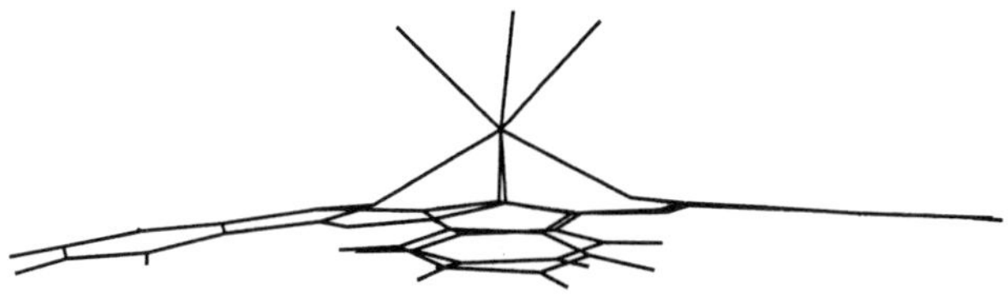

Appendix 16-9. $PcNbCl_3$, NEXVAN[163]

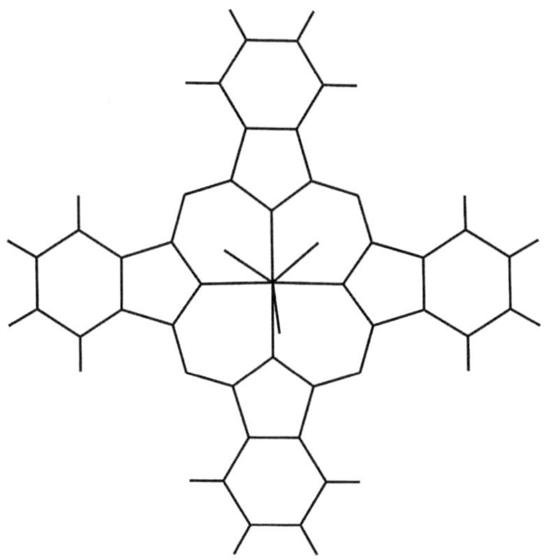

Appendix 16-10. $PcNbCl_3$, NEXVAN[163]

Appendix 16-11. $PcNbCl_3$, #1990-01[124] (App. 37)

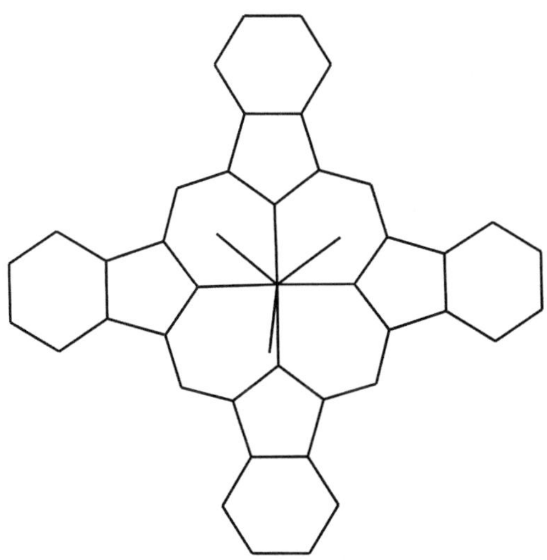

Appendix 16-12. $PcNbCl_3$, #1990-01[124] (App. 37)

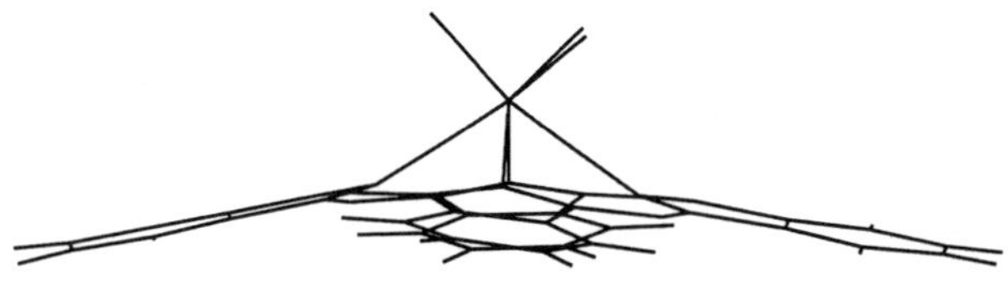

Appendix 16-13. $[PcNb(O,F_2)][NBu_4]$, CAPQOZ[401]

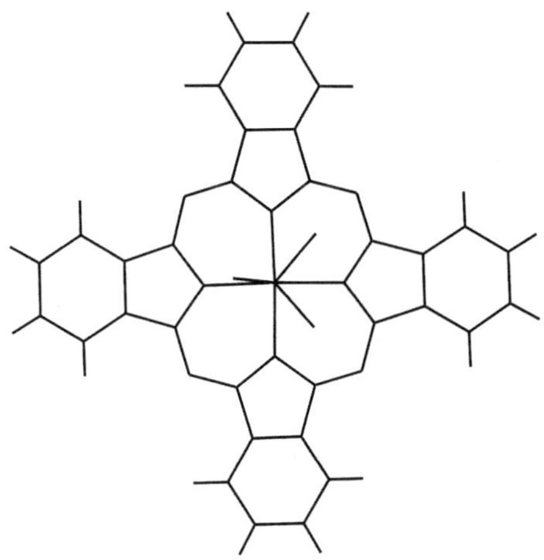

Appendix 16-14. $[PcNb(O,F_2)][NBu_4]$, CAPQOZ[401]

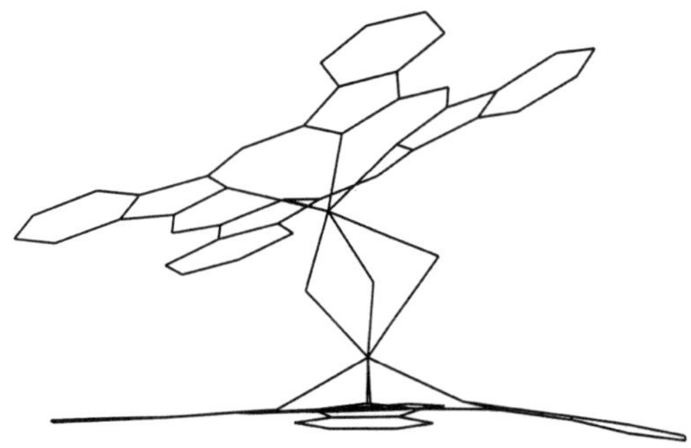

Appendix 16-15. $[PcNb(\text{-O-,-O-,-Br-})NbPc][Br_3][1\text{-ClN}]$, SAMDUF[124,162]

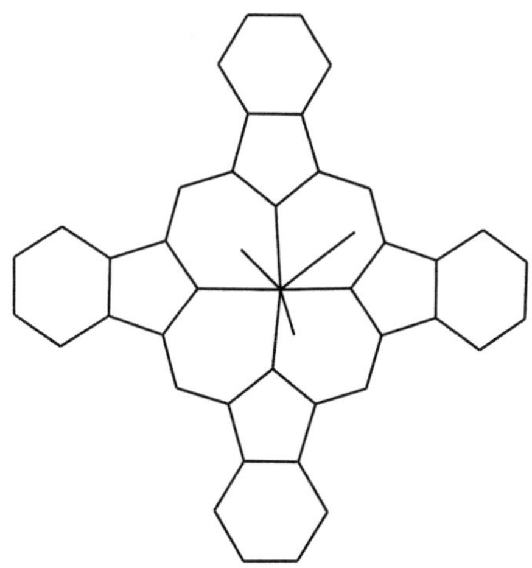

Appendix 16-16. One macrocycle of $[PcNb(\text{-O-,-O-,-Br-})NbPc][Br_3][1\text{-ClN}]$, SAMDUF[124,162]

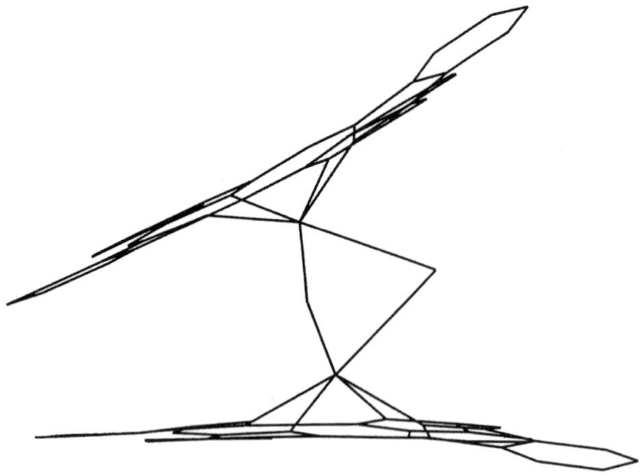

Appendix 16-17. [PcNb(-O-,-O-,-Br-)NbPc][Br_3][PN], SAMFAN[124,162]

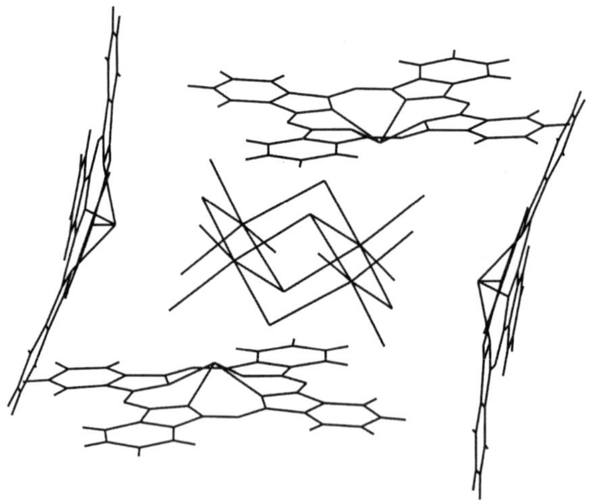

Appendix 16-18. [PcSb]$_4$[Sb_4I_{16}], YOHMAJ[194]

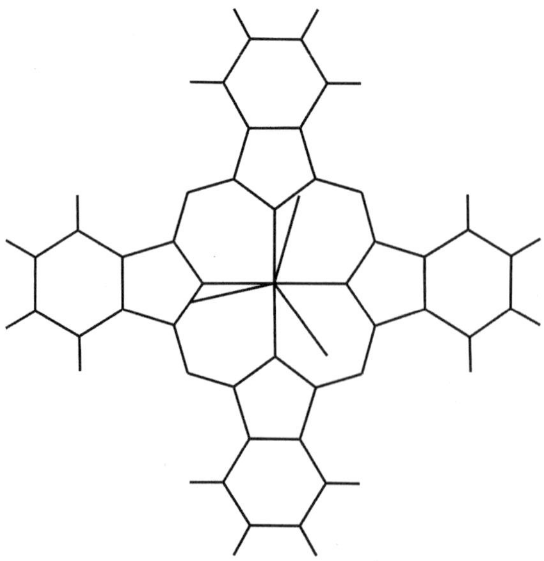

Appendix 16-19. [PcSb]$_4$[Sb_4I_{16}], YOHMAJ[194]

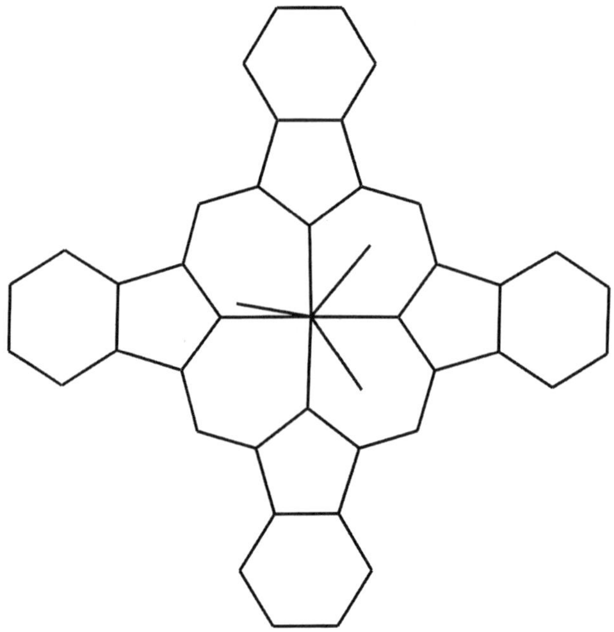

Appendix 16-20. $PcTaCl_3$, GIFZUQ[124,160]

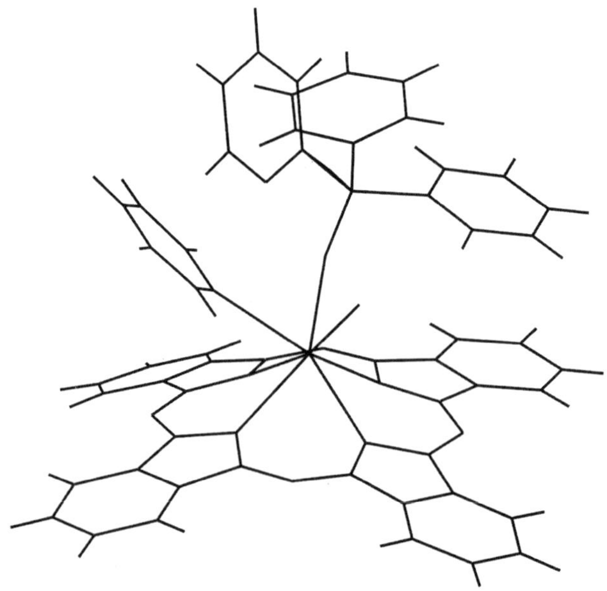

Appendix 16-21. [PcY(Cl)(py)(OPϕ_3)][0.5Et_2O], BORQUU[192]

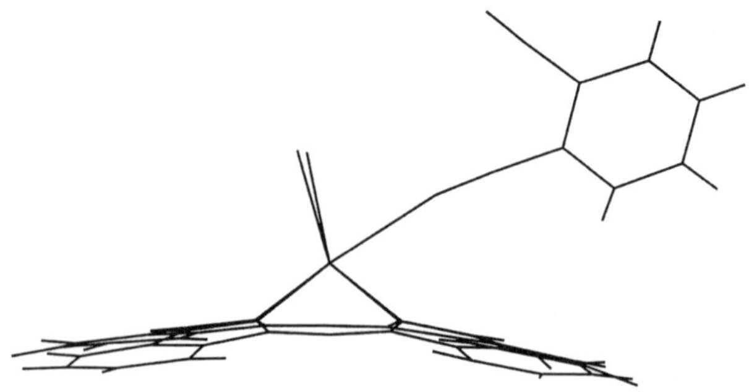

Appendix 16-22. $PcZrI_2(PN)_1$, NOFNUR[176,403]

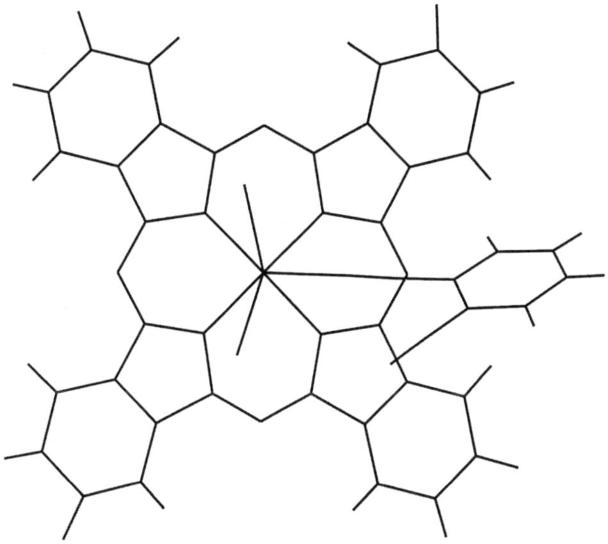

Appendix 16-23. $PcZrI_2(PN)$, NOFNUR[176,403]

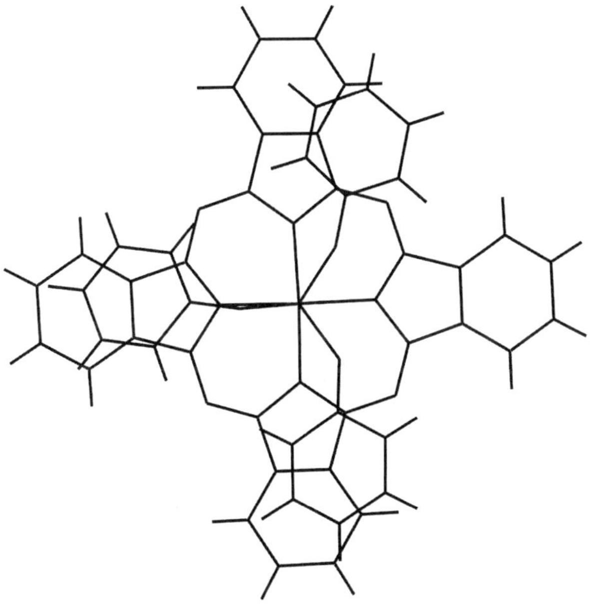

Appendix 16-24. $[PcZr(O\phi)_3][NBu_4][Et_2O]$, CCDC165838[400]

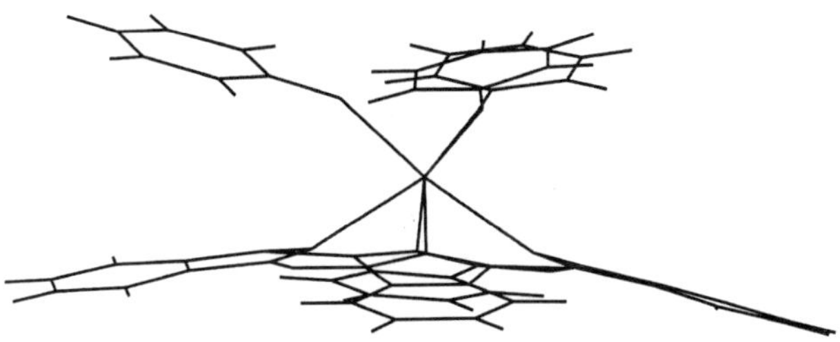

Appendix 16-25. $[PcZr(O\phi)_3][NBu_4][Et_2O]$, CCDC165838[400]

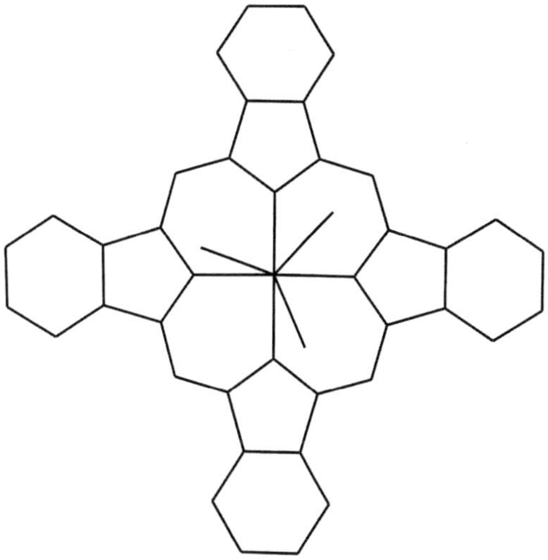

Appendix 16-26. $[PcZrCl_3][PNP]$,YOKWEA[169]

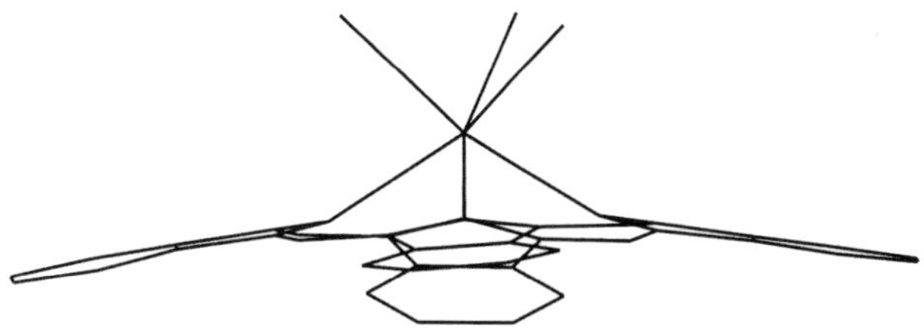

Appendix 16-27. $[PcZrCl_3][PNP]$, YOKWEA[169]

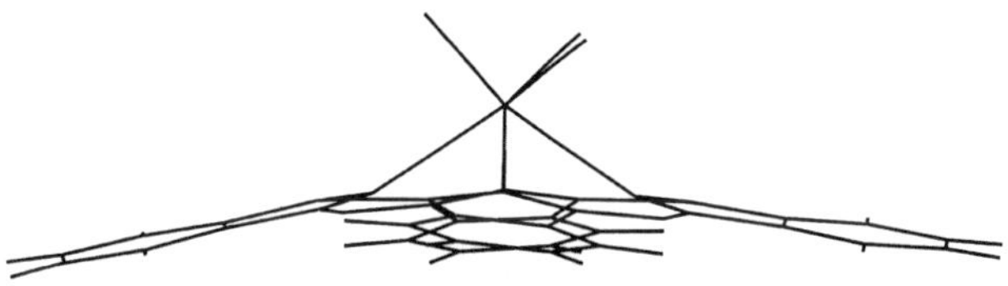

Appendix 16-28. $[(PcTaF_3)_{1-x}(PcTaF_4)_x][NBu_4]$, x=0, NAPPRAX[179]

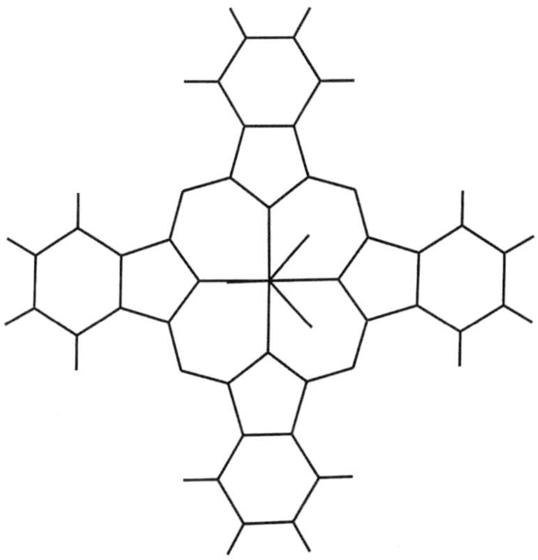

Appendix 16-29. $[(PcTaF_3)_{1-x}(PcTaF_4)_x][NBu_4]$, x=0, NAPRAX[179]

Appendix 17. Molecular Structures of CN8*cis* Phthalocyanine Complexes

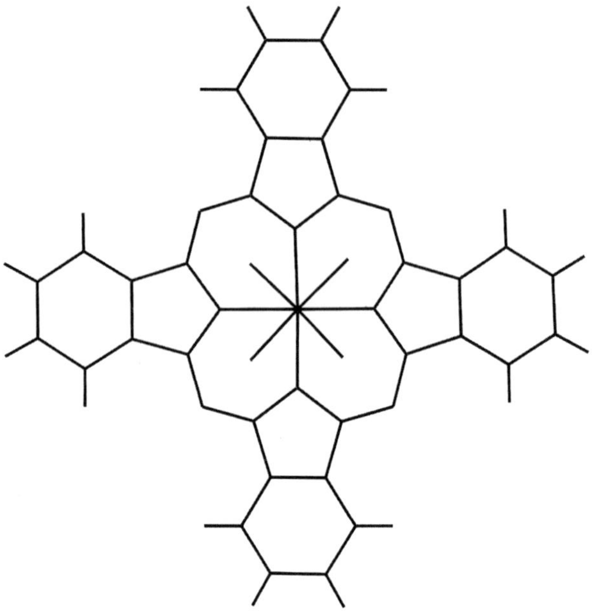

Appendix 17-1. $[(PcTaF_3)_{1-x}(PcTaF_4)_x][NBu_4]$, x=1, NAPRAX[179]

Appendix 17-2. $[PcIn(\text{-ONO-})_2][NBu_4]$, ZEMBEY[404]

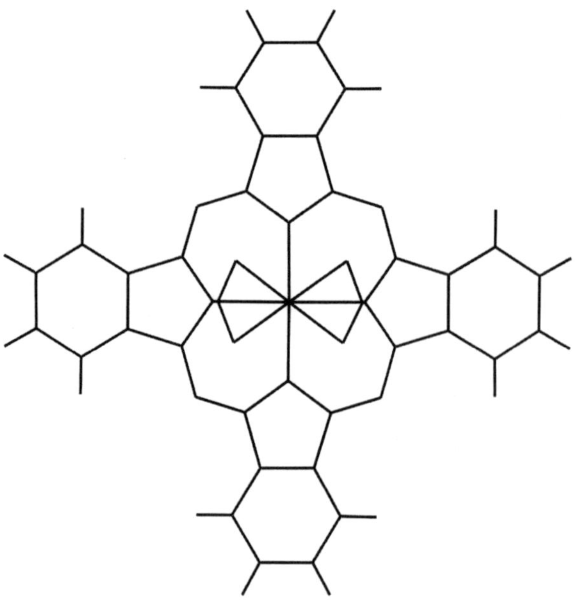

Appendix 17-3. $[PcIn(\text{-ONO-})_2][NBu_4]$, ZEMBEY[404]

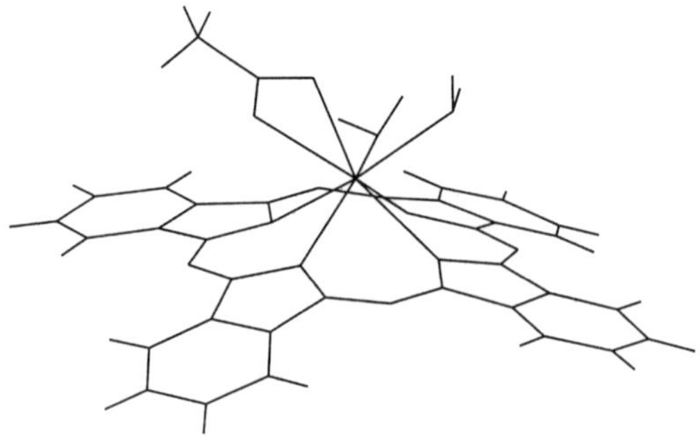

Appendix 17-4. $PcLu(OAc,H_2O,H_2O)$, DICCAT[159]

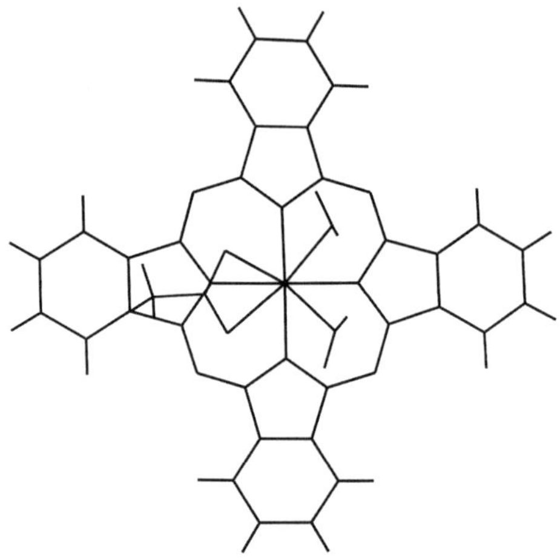

Appendix 17-5. $PcLu(OAc,H_2O,H_2O)$, DICCAT[159]

Appendix 17-6. $PcSn[O\text{-}C(O)\text{-}C_{10}H_{21}]_2$, CCDC173622[128]

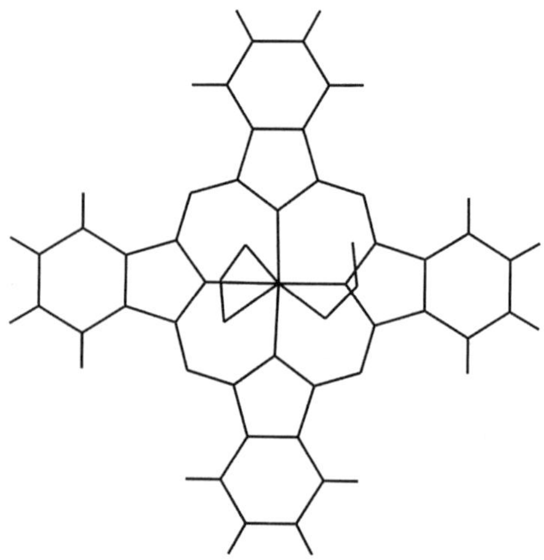

Appendix 17-7. $PcSn[O\text{-}C(O)\text{-}C_{10}H_{21}]_2$, CCDC173622[128]

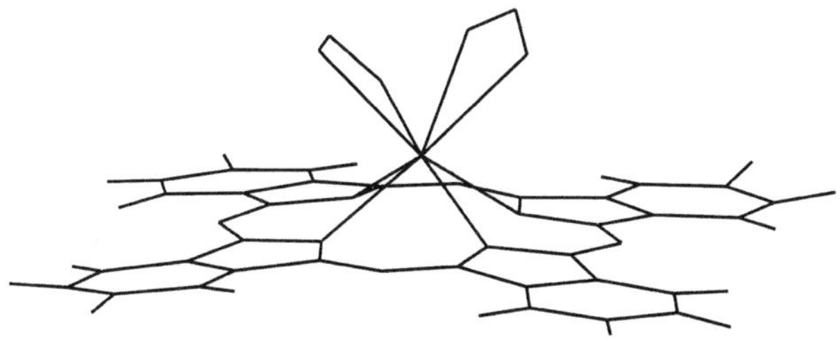

Appendix 17-8. [PcTl(-ONO-)$_2$][NBu$_4$], BOFQAO[191]

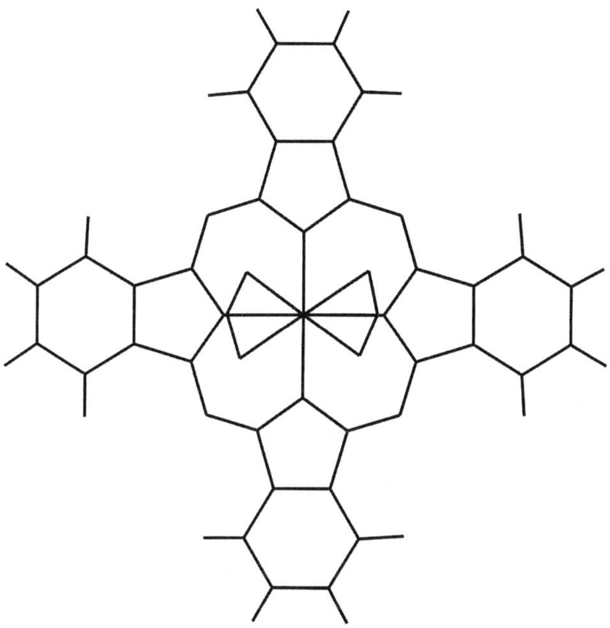

Appendix 17-9. [PcTl(-ONO-)$_2$][NBu$_4$], BOFQAO[191]

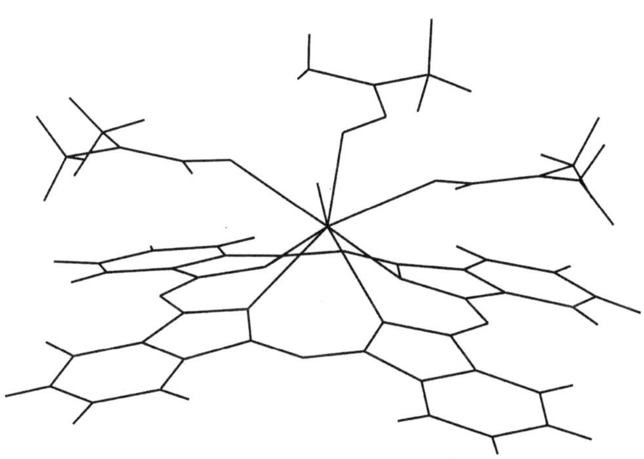

Appendix 17-10. [PcY(Cl)(DMF)$_3$][Et$_2$O], BORQII[192]

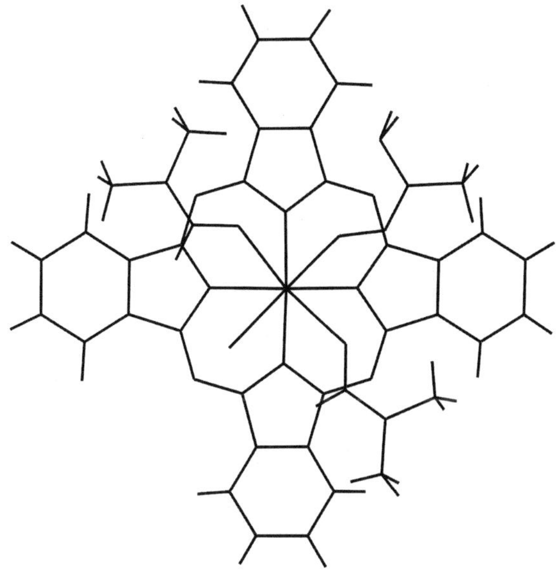

Appendix 17-11. [PcY(Cl)(DMF)$_3$][Et$_2$O], BORQII[192]

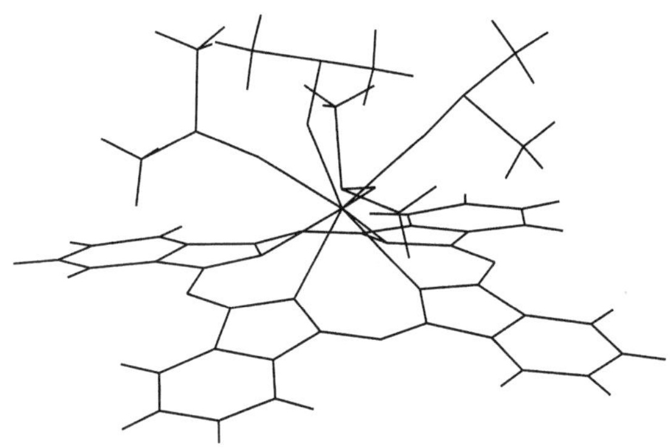

Appendix 17-12. [PcY(Cl)(-OSMe$_2$)$_4$][Et$_2$O,2MeOH], BORQOO[192]

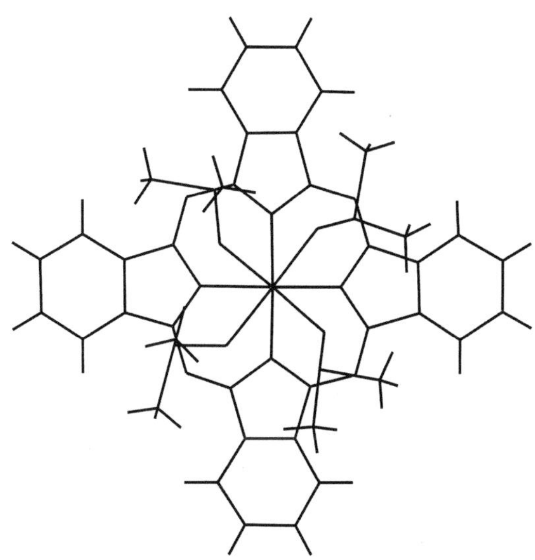

Appendix 17-13. [PcY(Cl)(-OSMe$_2$)$_4$][Et$_2$O,2MeOH], BORQOO[192]

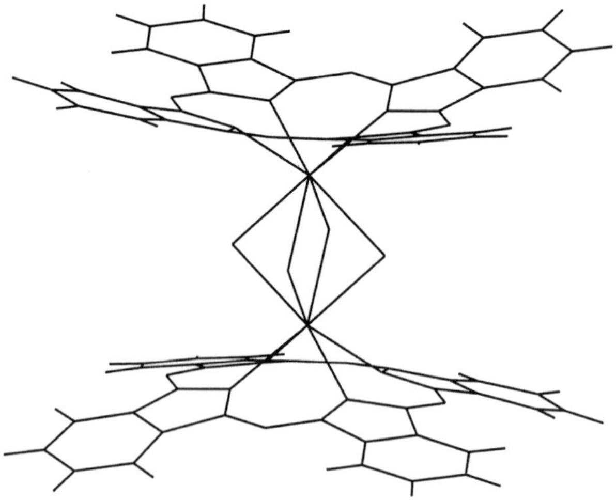

Appendix 17-14. PcY(-(Cl)$_2$-,-(O(Hpy)$_2$)$_2$-)YPc, BOSQAB[192]

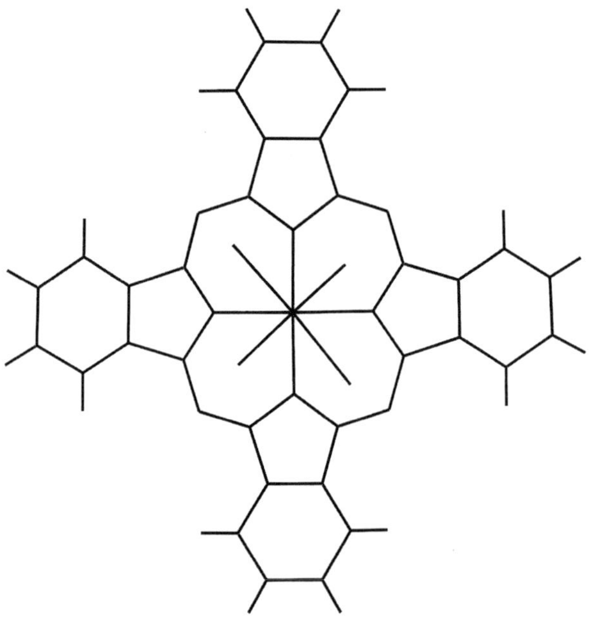

Appendix 17-15. One macrocycle of PcY(-(Cl)$_2$-,-(O(Hpy)$_2$)$_2$-)YPc, BOSQAB[192]

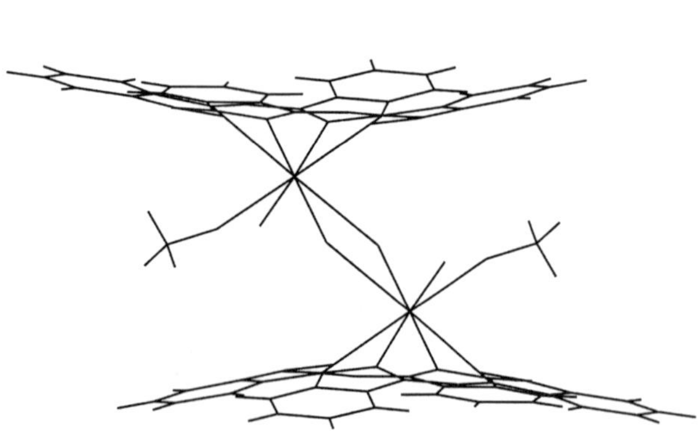

Appendix 17-16. PcY(L,L′,-Cl-,-Cl-)YPc, BOSJAU[192]

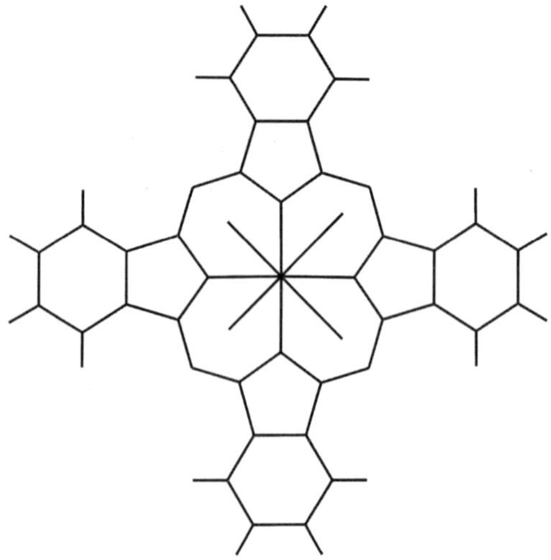

Appendix 17-17. One macrocycle of PcY(L,L′,-Cl-,-Cl-)-YPc, BOSJAU[192]

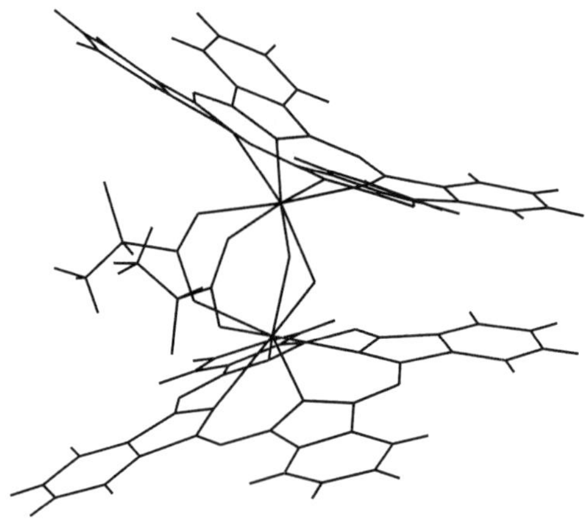

Appendix 17-18. [PcZr(-(OH)$_2$-,-(O,Oala)$_2$-)ZrPc] [DMF$_2$,Et$_2$O,H$_2$O], CCDC165829[400]

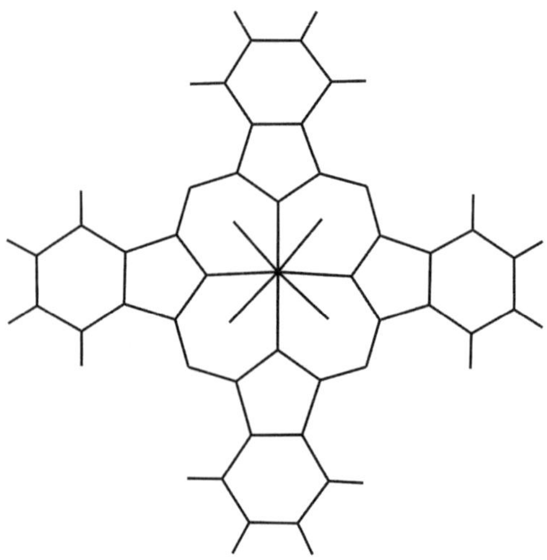

Appendix 17-19. One macrocycle of [PcZr(-(OH)$_2$-,-(O,Oala)$_2$-)ZrPc] [2DMF,Et$_2$O,H$_2$O], CCDC165829[400]

Appendix 17-20. $[PcZr(O_2CatCl_4)_2][NBu_4]_2$, CCDC165834[400]

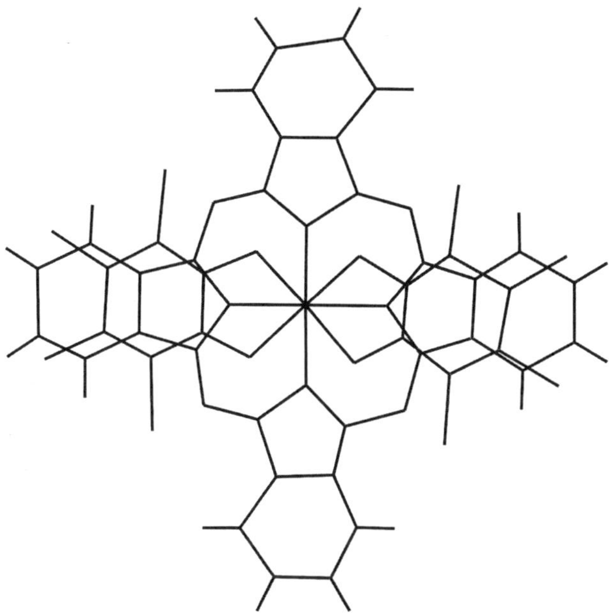

Appendix 17-21. $[PcZr(O_2CatCl_4)_2][NBu_4]_2$, CCDC165834[400]

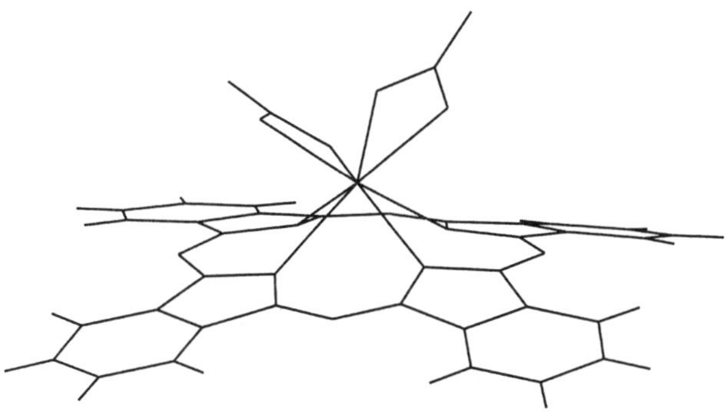

Appendix 17-22. $[PcZr(\text{-}OC(O)C)_2][NEt_4]_2[CH_2Cl_2]_2$, CCDC165836[400]

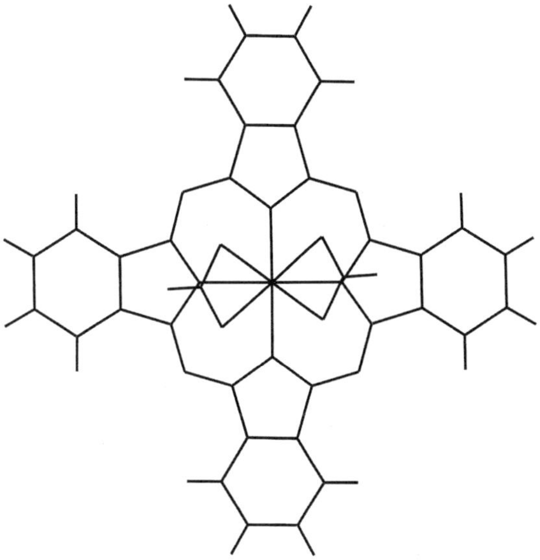

Appendix 17-23. $[PcZr(\text{-}OC(O)C)_2][NEt_4]_2[CH_2Cl_2]_2$, CCDC165836[400]

Appendix 18 Molecular Structures of CN8*bis* Phthalocyanine Complexes

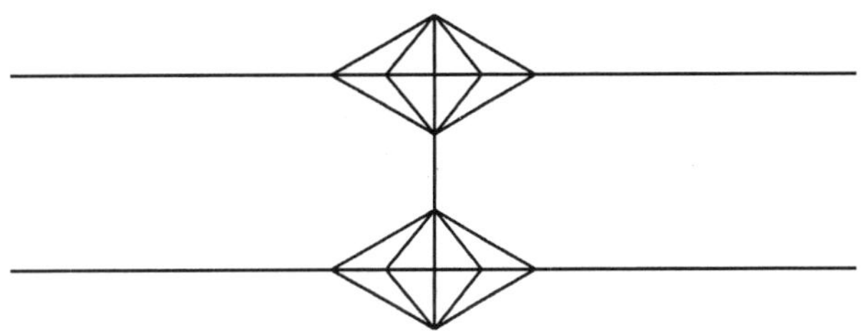

Appendix 18-1. $[Pc_2As][I_2]$, BECTOS[233]

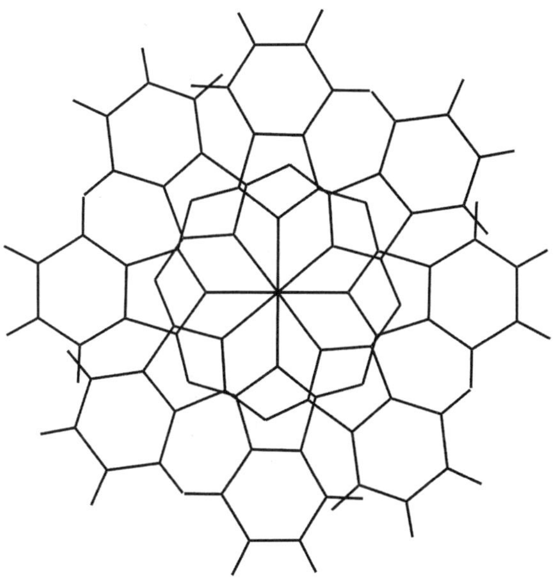

Appendix 18-2. $[Pc_2As][I_2]$, BECTOS[233]

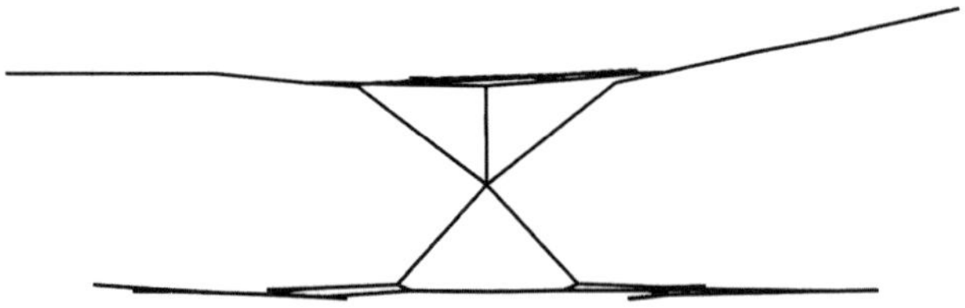

Appendix 18-3. $[Pc_2Bi][CH_2Cl_2]$, ZUWDAW[228]

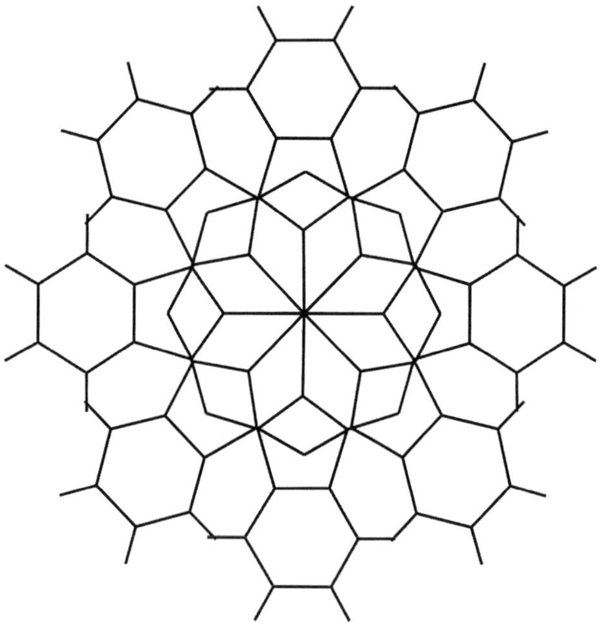

Appendix 18-4. $[Pc_2Bi][CH_2Cl_2]$, ZUWDAW[228]

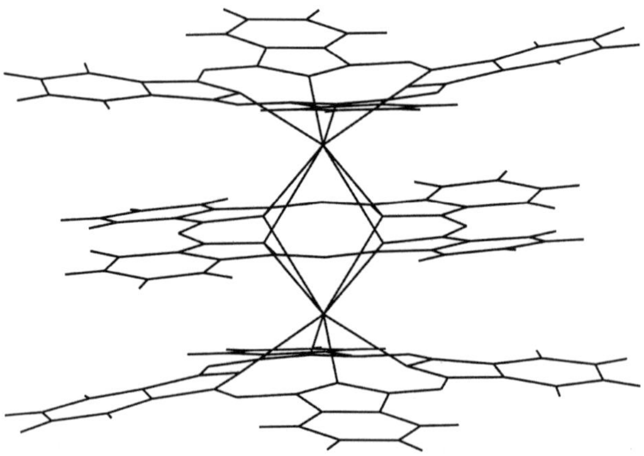

Appendix 18-5. Pc_3Bi_2, DIQPIC[51]

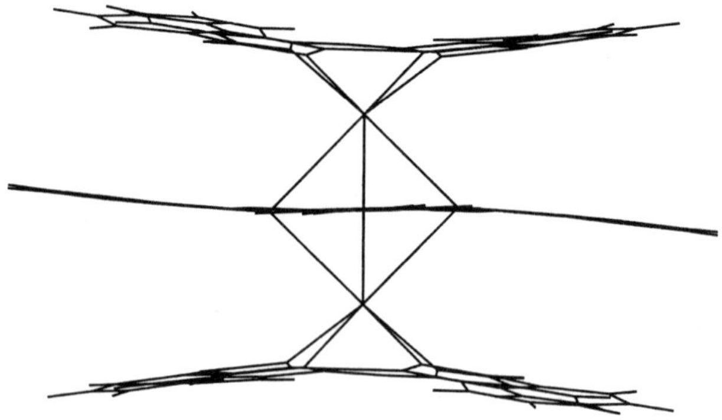

Appendix 18-6. Pc_3Bi_2, DIQPIC01[52]

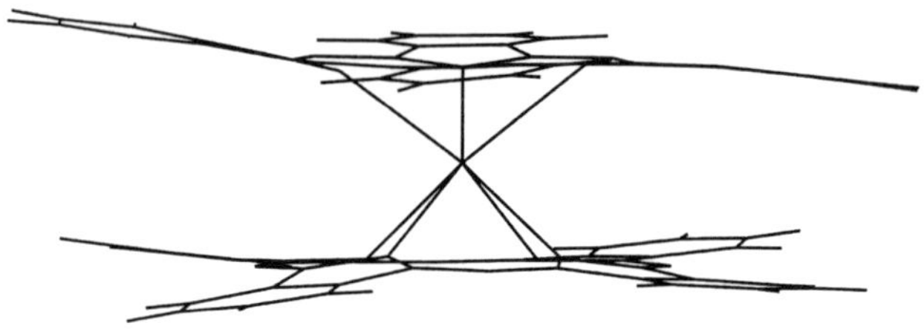

Appendix 18-7. $[Pc_2Bi][Br][PNP]_2[2Et_2O]$, QOCLOJ[198]

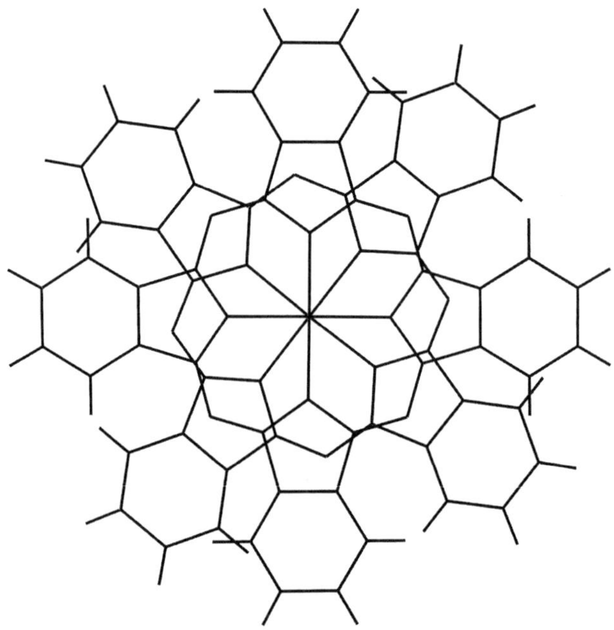

Appendix 18-8. $[Pc_2Bi][Br][PNP]_2[2Et_2O]$, QOCLOJ[198]

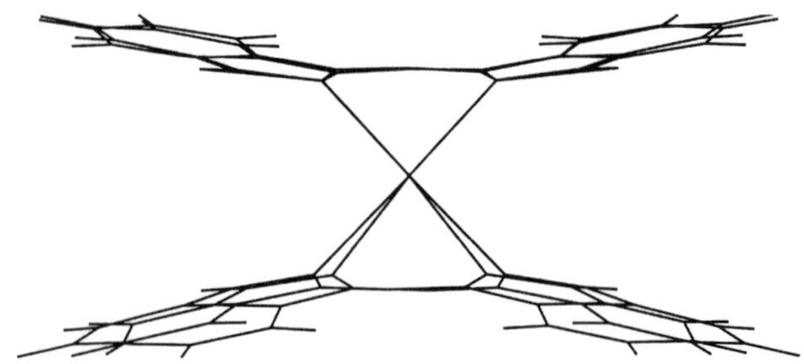

Appendix 18-9. $[Pc_2Bi][NBu_4]$, QOCJEX[198]

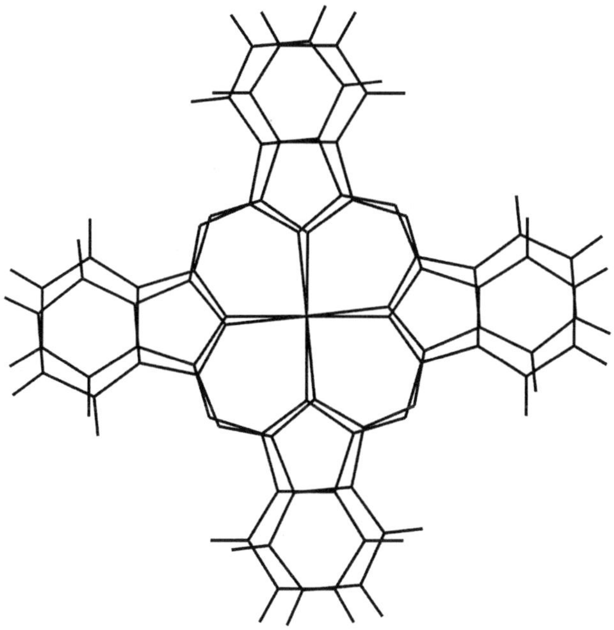

Appendix 18-10. $[Pc_2Bi][NBu_4]$, QOCJEX[198]

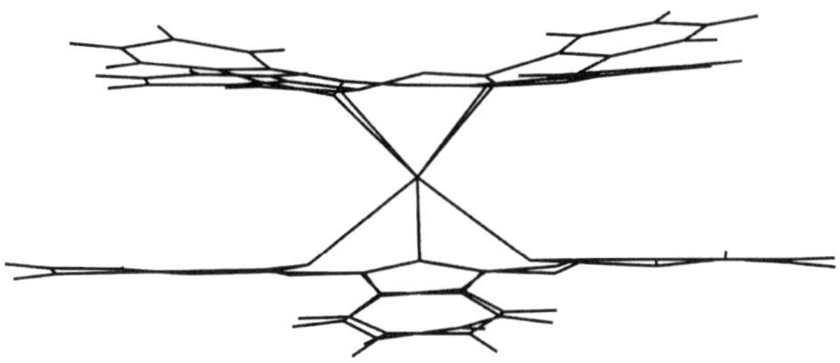

Appendix 18-11. $[Pc_2Bi][NBu_4]$, QOCKAU[198]

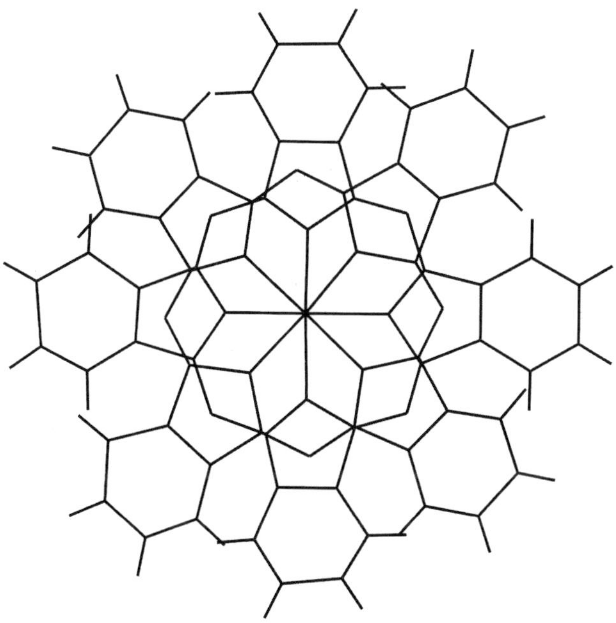

Appendix 18-12. $[Pc_2Bi][NBu_4]$, QOCKAU[198]

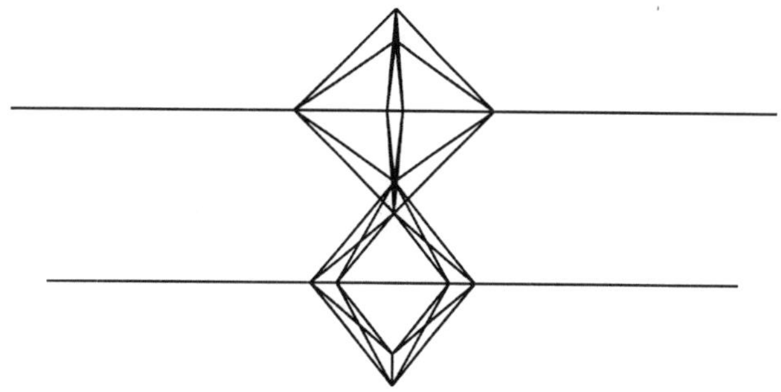

Appendix 18-13. $[Pc_2Bi][I_{1.5}]$, KOJQUV[232]

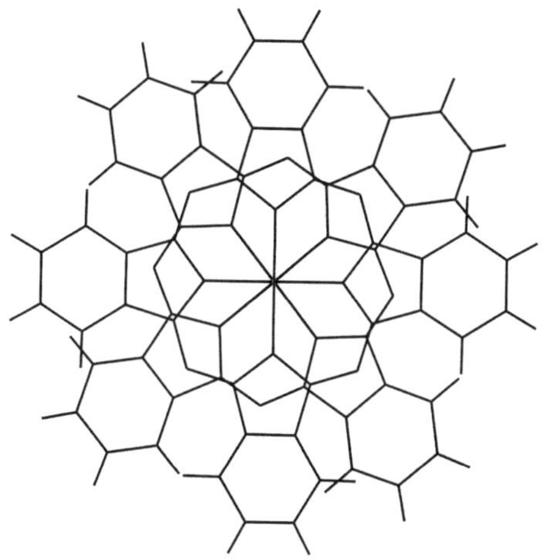

Appendix 18-14. $[Pc_2Bi][I_{1.5}]$, KOJQUV[232]

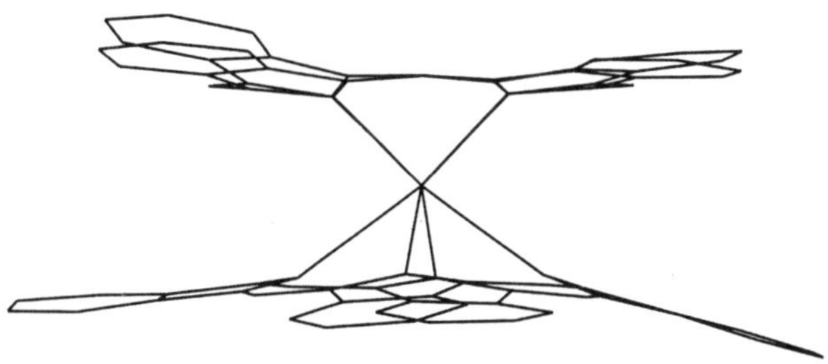

Appendix 18-15. Pc_2Ce, JONZAN[217]

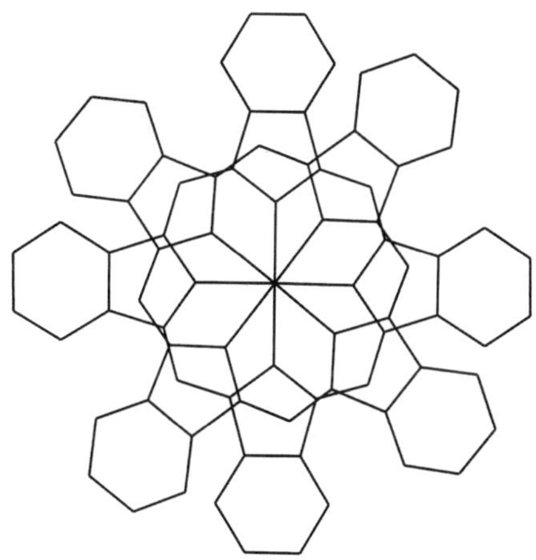

Appendix 18-16. Pc_2Ce, JONZAN[217]

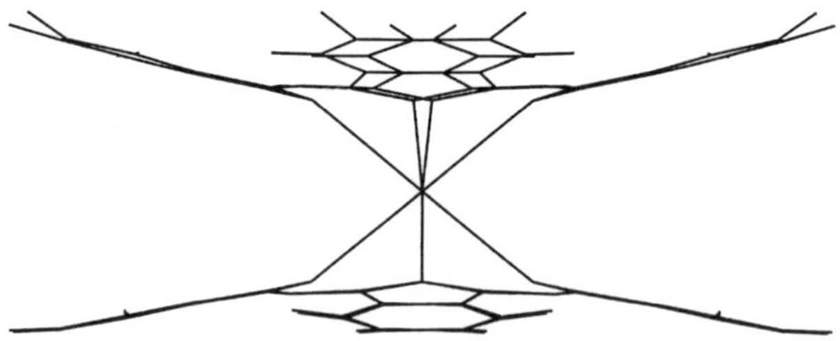

Appendix 18-17. [Pc_2Ce][NBu_4], QOCJIB[198]

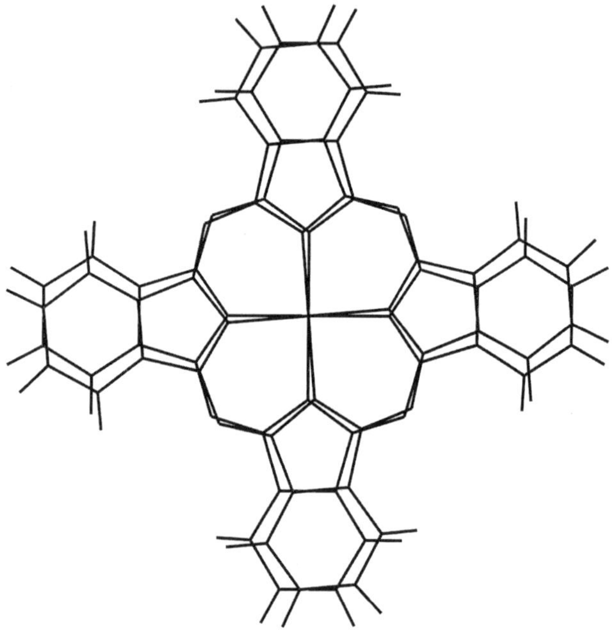

Appendix 18-18. [Pc_2Ce][NBu_4], QOCJIB[198]

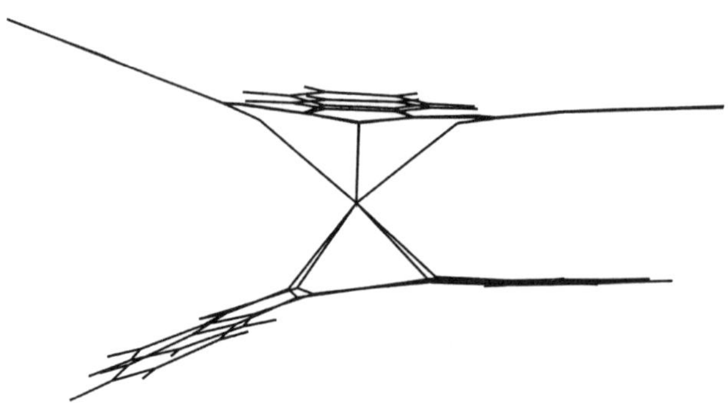

Appendix 18-19. [Pc_2Ce][PNP], #2002-10[221]

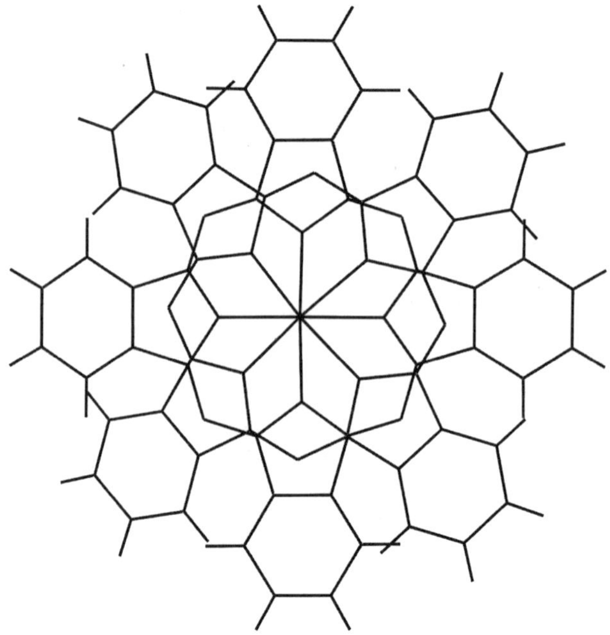

Appendix 18-20. [Pc_2Ce][PNP], #2002-10[221]

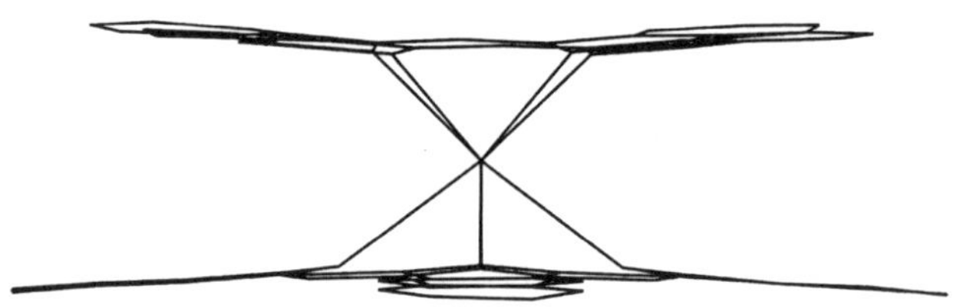

Appendix 18-21. [Pc_2Ce][BF_4]$_{0.33}$, RABDEZ[230]

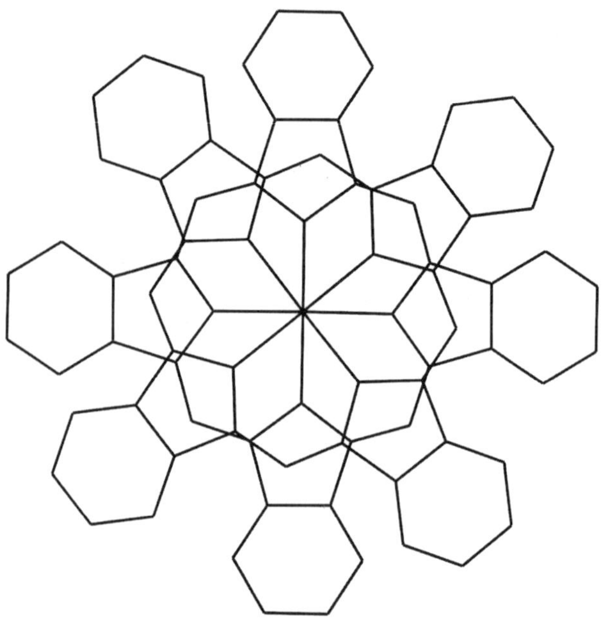

Appendix 18-22. [Pc_2Ce][BF_4]$_{0.33}$, RABDEZ[230]

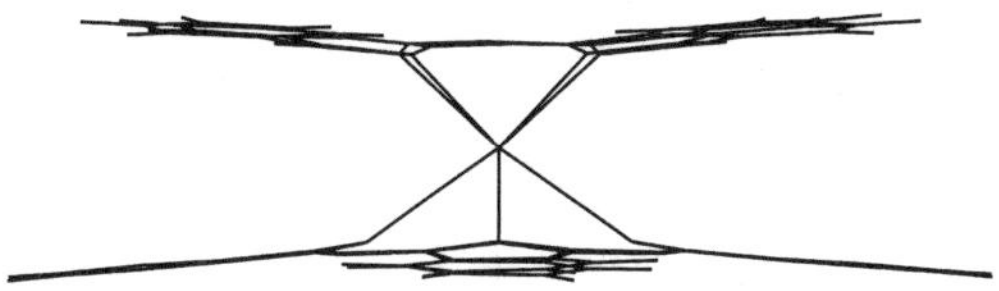

Appendix 18-23. α-Pc_2Er, YUHSUP[225]

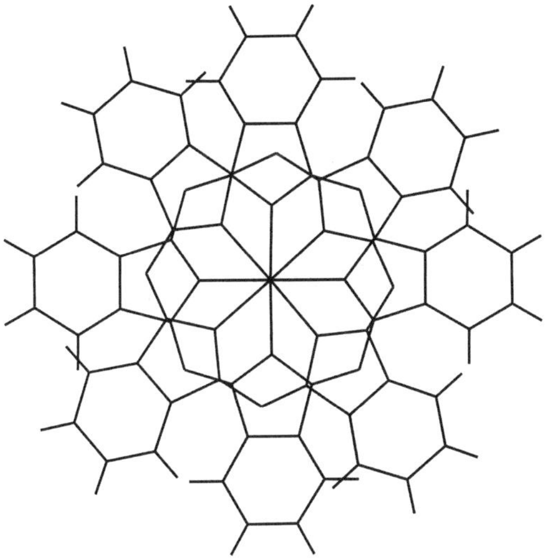

Appendix 18-24. α-Pc_2Er, YUHSUP[225]

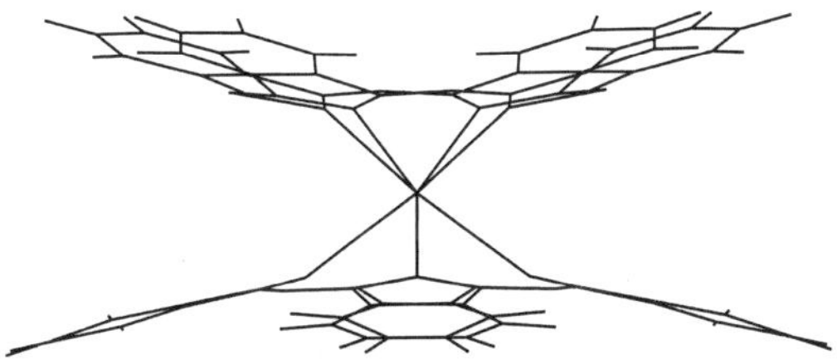

Appendix 18-25. [Pc_2Gd][NBu_4], NAKYAZ[197]

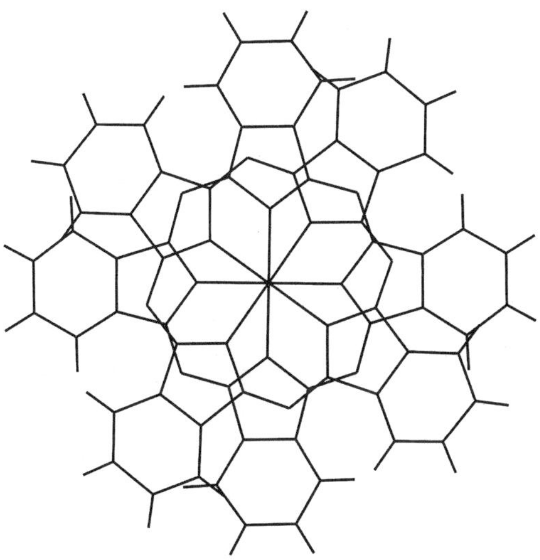

Appendix 18-26. [Pc_2Gd][NBu_4], NAKYAZ[197]

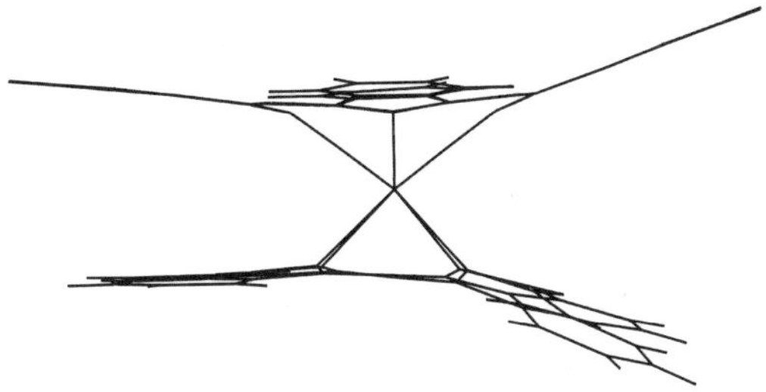

Appendix 18-27. [Pc_2Gd][PNP][0.33H_2O], YEGTOT[220]

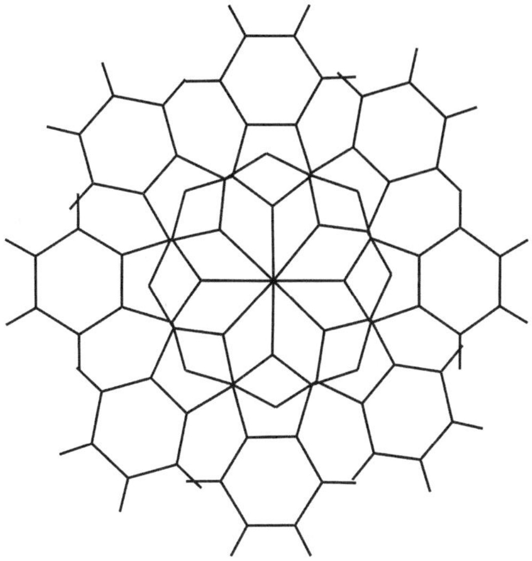

Appendix 18-28. [Pc_2Gd][PNP][0.33H_2O], YEGTOT[220]

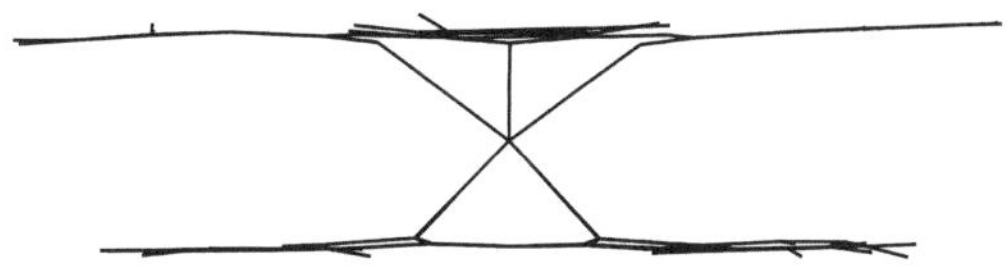

Appendix 18-29. [Pc_2Gd][$C_4(CN)_6$][2ClB], OBUJIE[236]

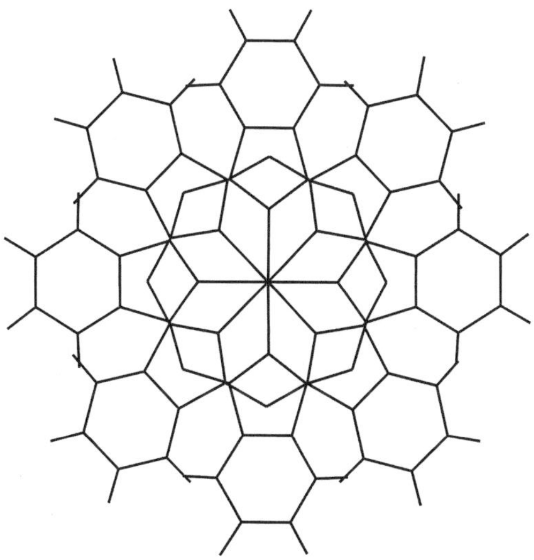

Appendix 18-30. [Pc_2Gd][$C_4(CN)_6$][2ClB], OBUJIE[236]

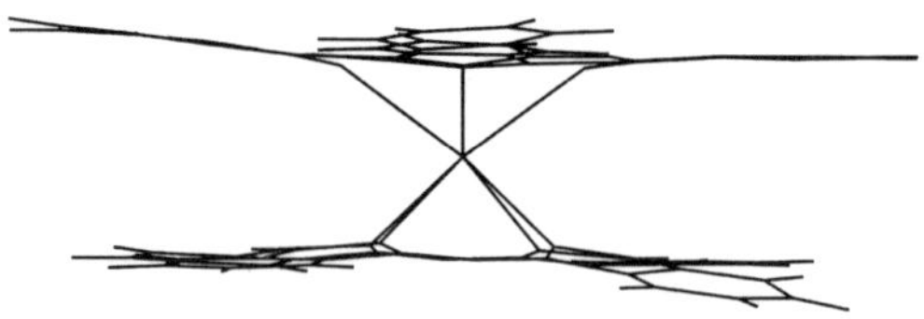

Appendix 18-31. Pc_2Gd, GAWBEL[212]

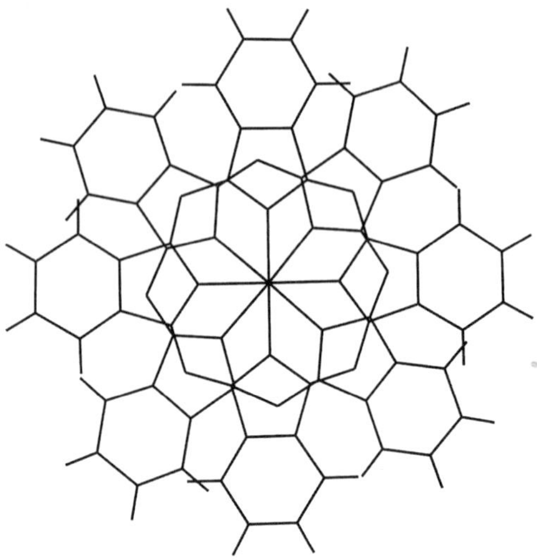

Appendix 18-32. Pc_2Gd, GAWBEL[212]

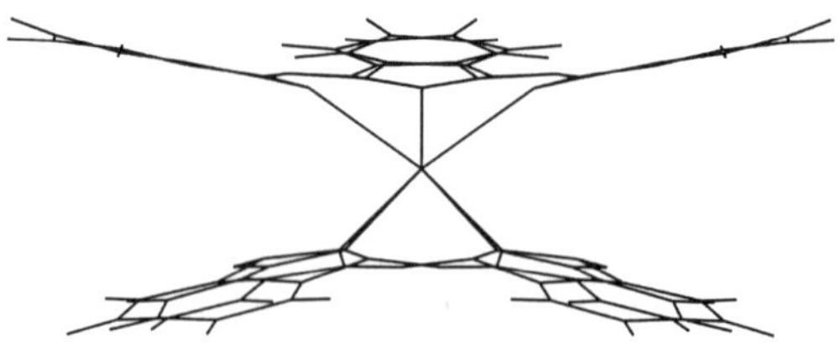

Appendix 18-33. $[Pc_2Ho][NBu_4][H_2O]$, NAKYED[197]

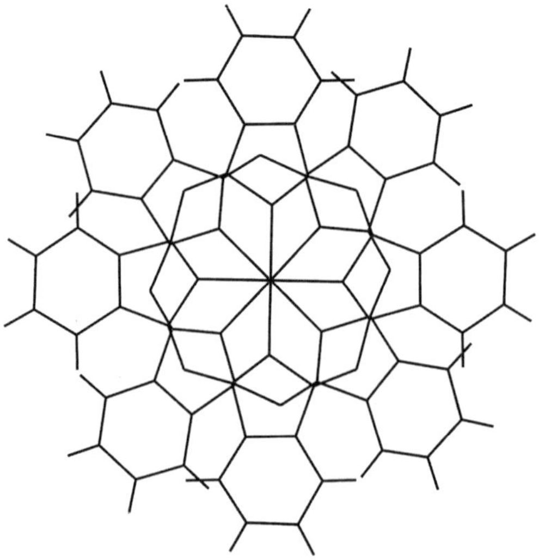

Appendix 18-34. $[Pc_2Ho][NBu_4][H_2O]$, NAKYED[197]

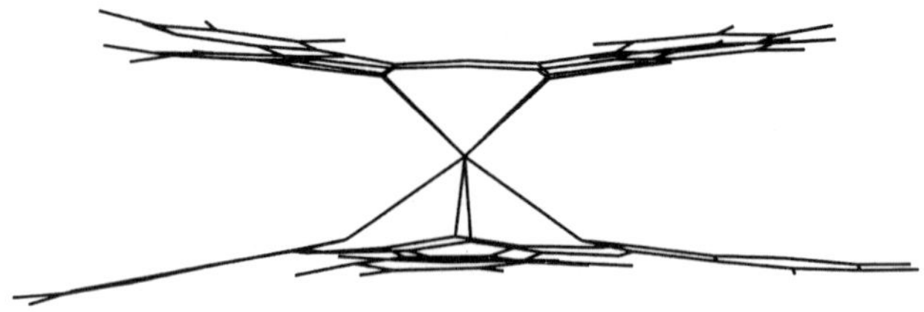

Appendix 18-35. Pc_2In, ZAKZOA[224]

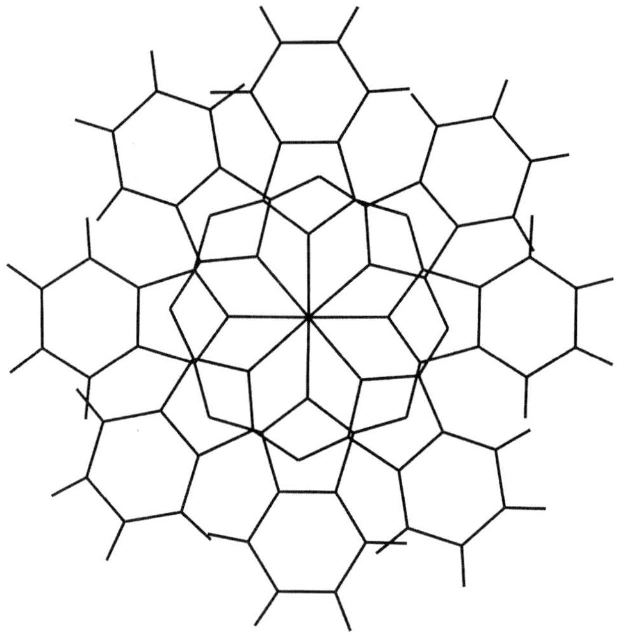

Appendix 18-36. Pc_2In, ZAKZOA[224]

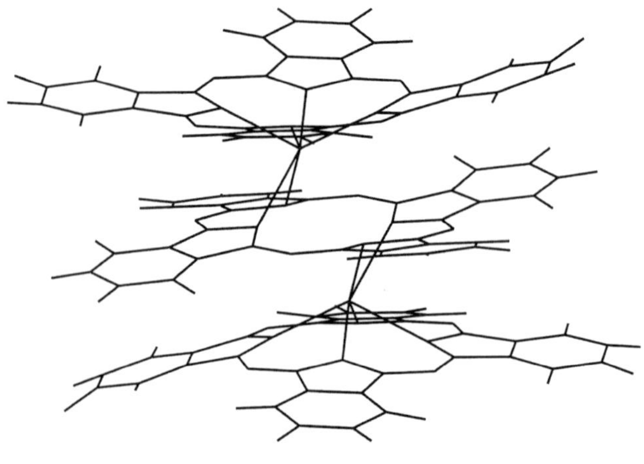

Appendix 18-37. Pc_3In_2, PIFDEN[53]

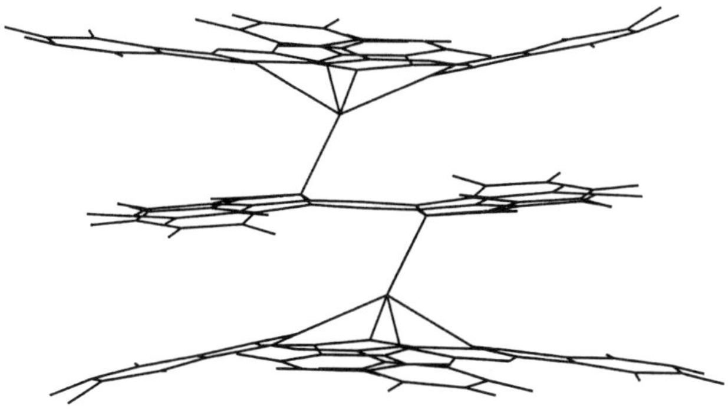

Appendix 18-38. Pc_3In_2, PIFDEN[53]

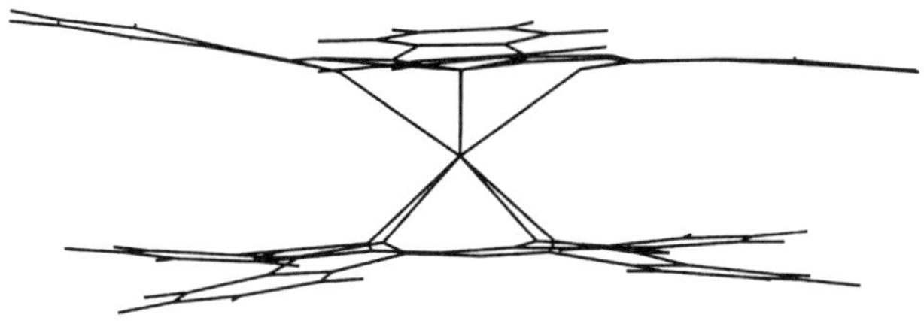

Appendix 18-39. $[Pc_2In][I][PNP]_2[2Et_2O]$, QOCLUP[198]

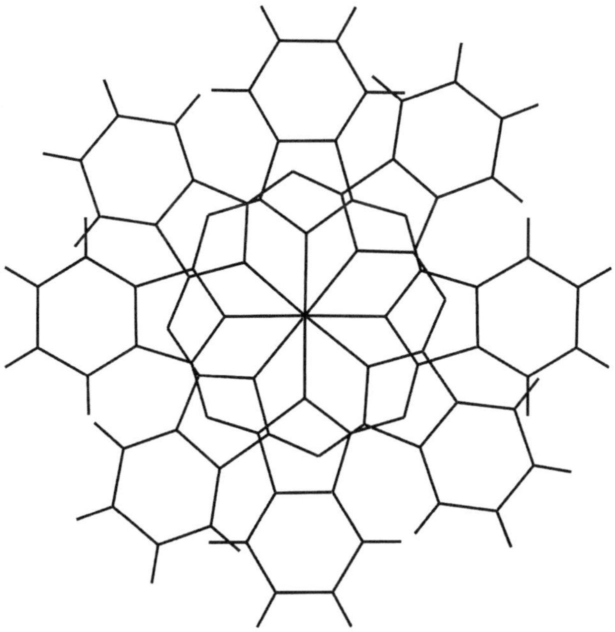

Appendix 18-40. $[Pc_2In][I][PNP]_2[2Et_2O]$, QOCLUP[198]

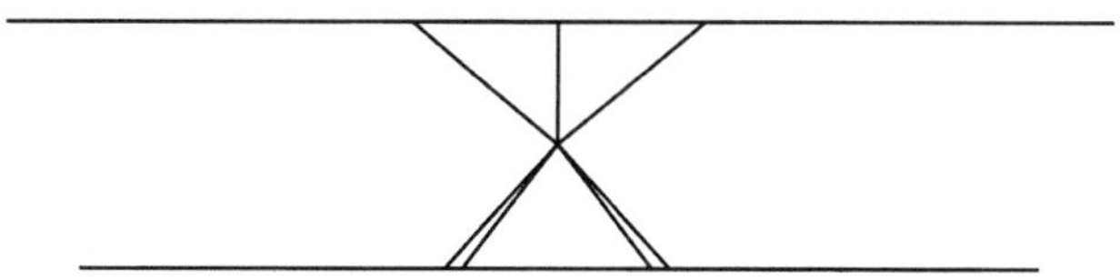

Appendix 18-41. $[Pc_2In][I_2]$, NOFMAW[409]

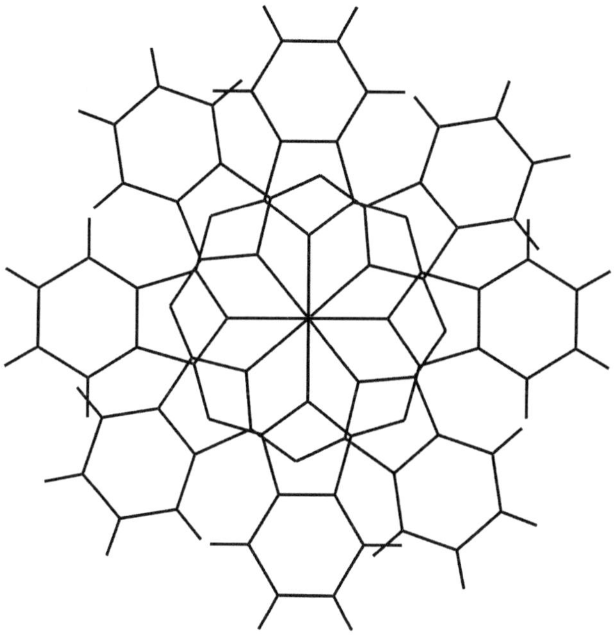

Appendix 18-42. $[Pc_2In][I_2]$, NOFMAW[409]

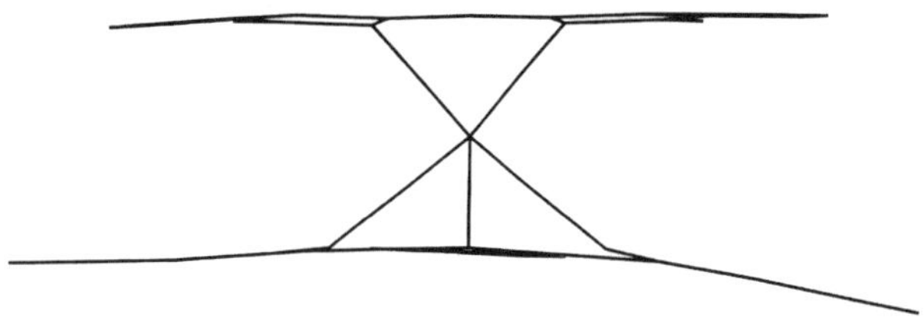

Appendix 18-43. $[Pc_2La][CH_2Cl_2]$, ZEHTUB[226]

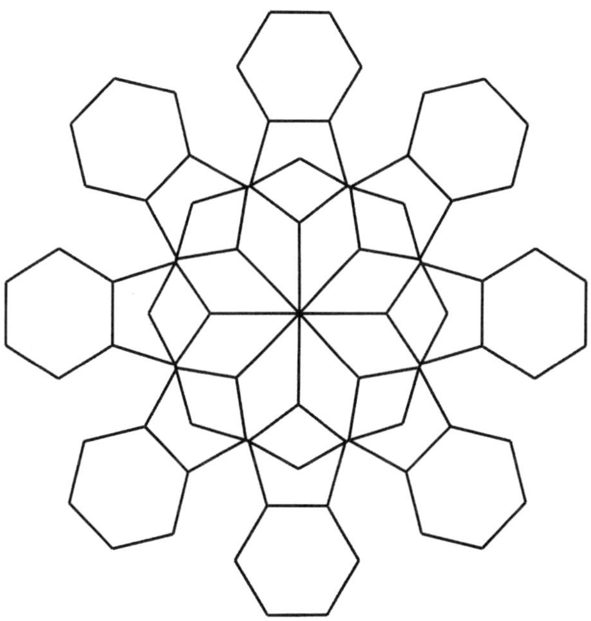

Appendix 18-44. $[Pc_2La][CH_2Cl_2]$, ZEHTUB[226]

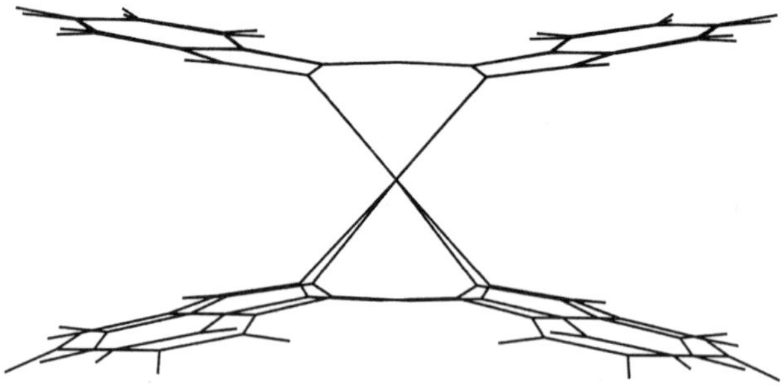

Appendix 18-45. $[Pc_2La][NBu_4]$, QOCJOH[198]

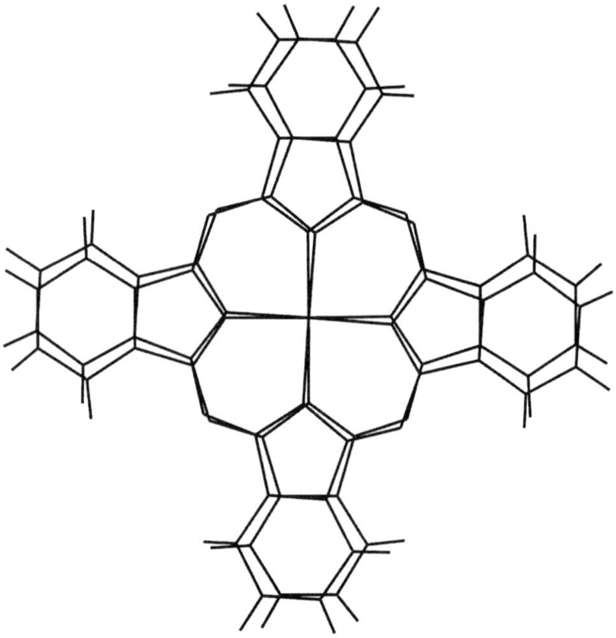

Appendix 18-46. $[Pc_2La][NBu_4]$, QOCJOH[198]

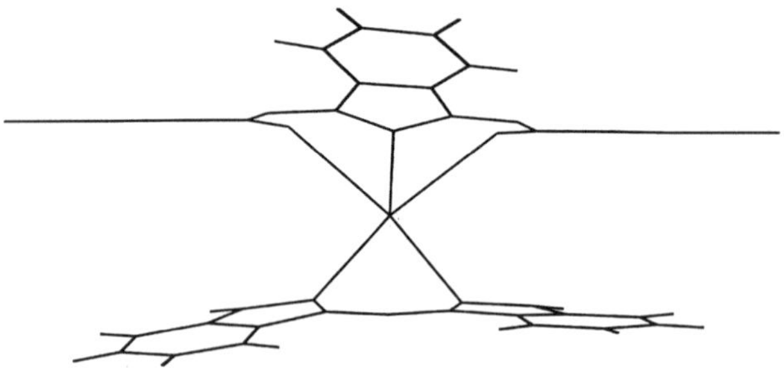

Appendix 18-47. $[Pc_2La][NPe_4]$, QOCJUN[198]

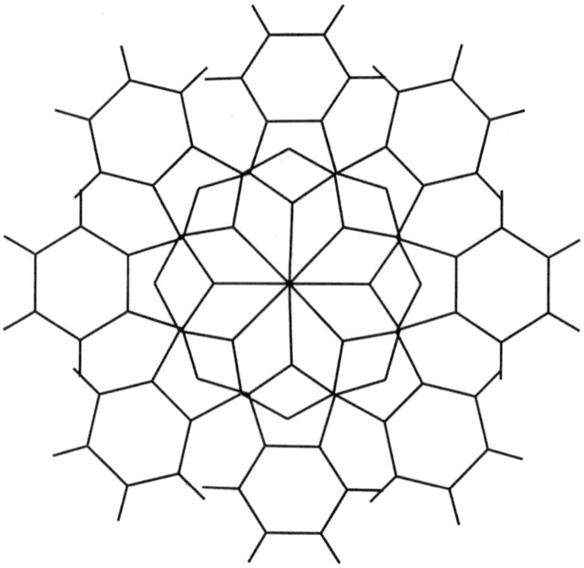

Appendix 18-48. $[Pc_2La][NPe_4]$, QOCJUN[198]

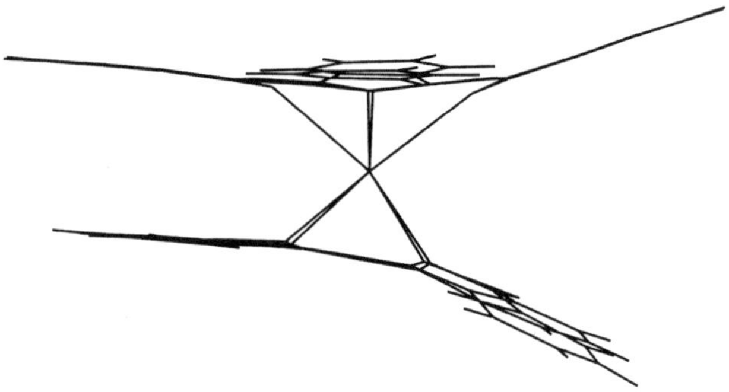

Appendix 18-49. $[Pc_2La][PNP]$, YEGTIN[220]

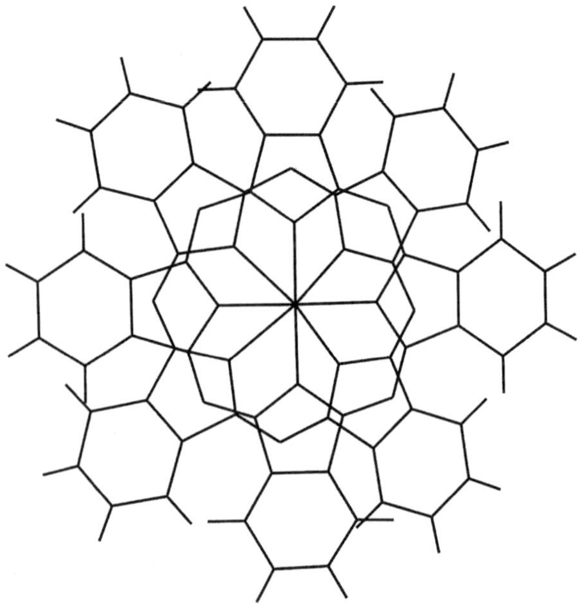

Appendix 18-50. $[Pc_2La][PNP]$, YEGTIN[220]

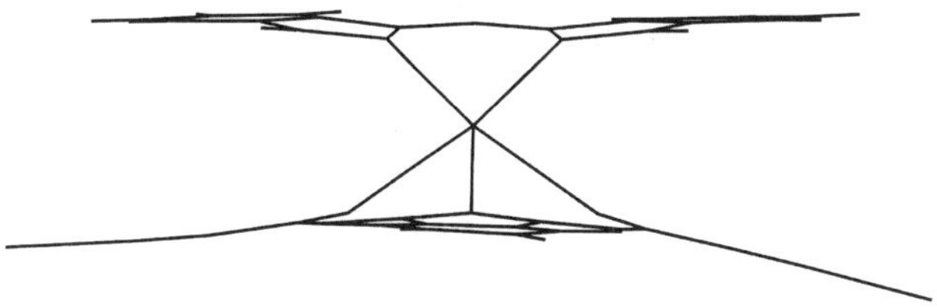

Appendix 18-51. $[Pc_2Lu][CH_2Cl_2]$, DICBUM[159]

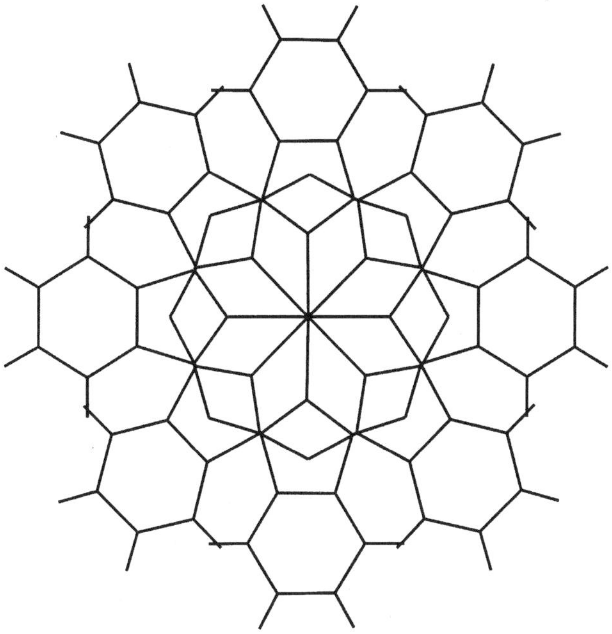

Appendix 18-52. $[Pc_2Lu][CH_2Cl_2]$, DICBUM[159]

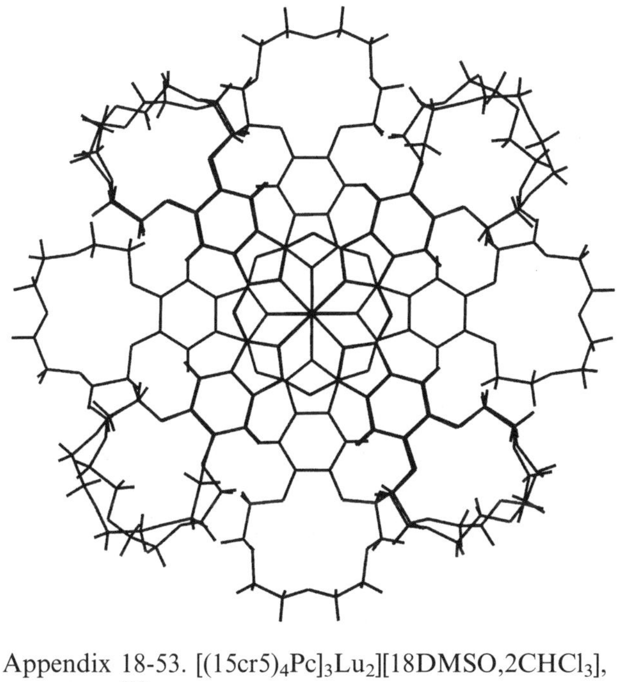

Appendix 18-53. $[(15cr5)_4Pc]_3Lu_2][18DMSO,2CHCl_3]$, LORKIM[195]

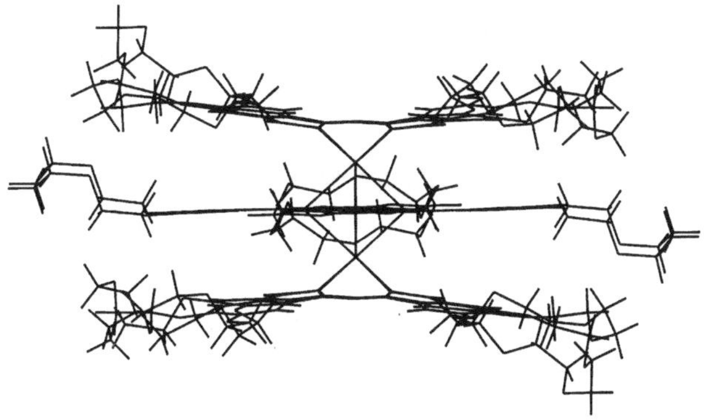

Appendix 18-54. $[(15cr5)_4Pc]_3Lu_2][18DMSO,2CHCl_3]$, LORKIM[195]

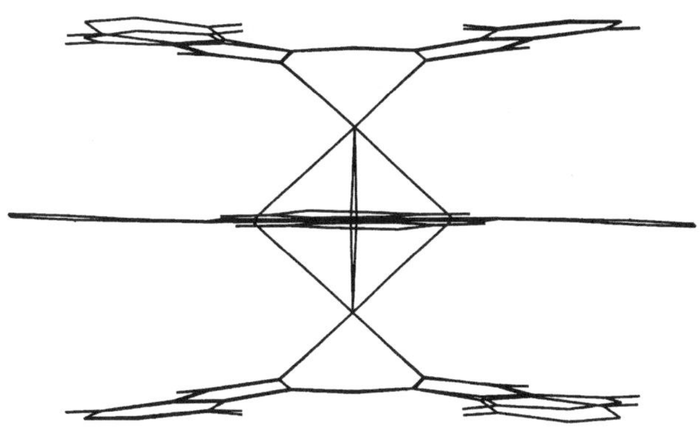

Appendix 18-55. Macrocyclic skeleton of $[(15cr5)_4Pc]_3Lu_2][18DMSO,2CHCl_3]$, LORKIM[195]

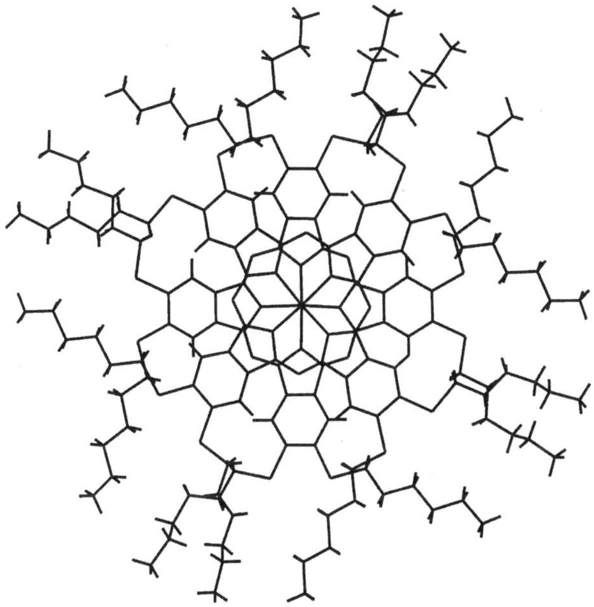

Appendix 18-56. $[(H_{13}C_6S)_8Pc]_2Lu$, VODDIB[247]

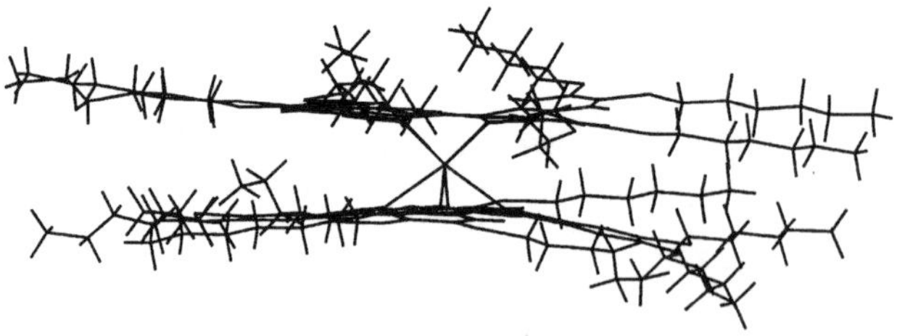

Appendix 18-57. $[(H_{13}C_6S)_8Pc]_2Lu$, VODDIB[247]

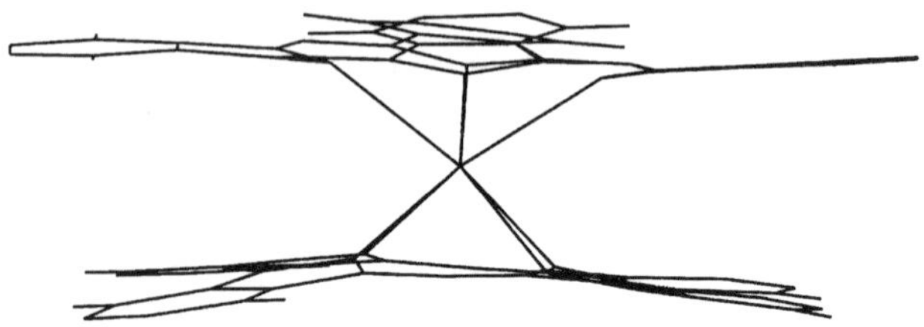

Appendix 18-58. Macrocyclic skeleton of $[(H_{13}C_6S)_8Pc]_2Lu$, VODDIB[247]

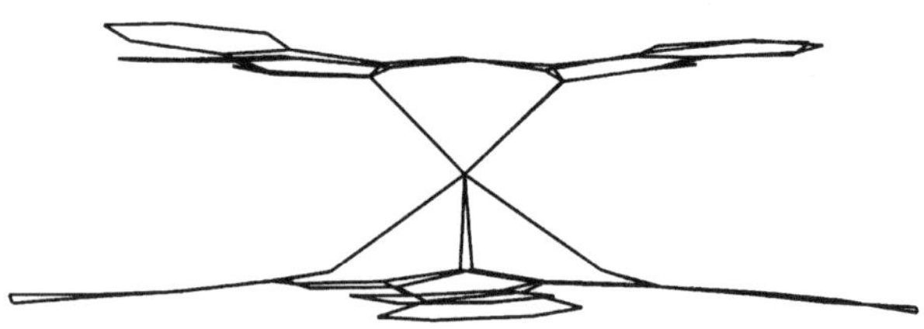

Appendix 18-59. γ-Pc_2Lu, DULZAL[208]

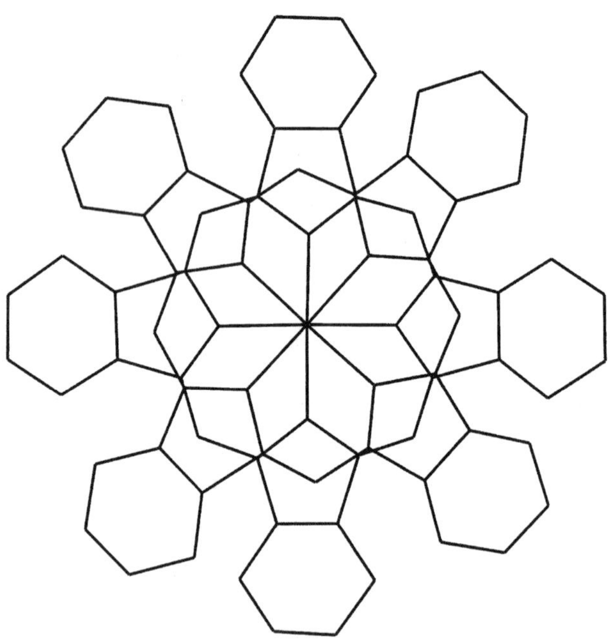

Appendix 18-60. γ-Pc_2Lu, DULZAL[208]

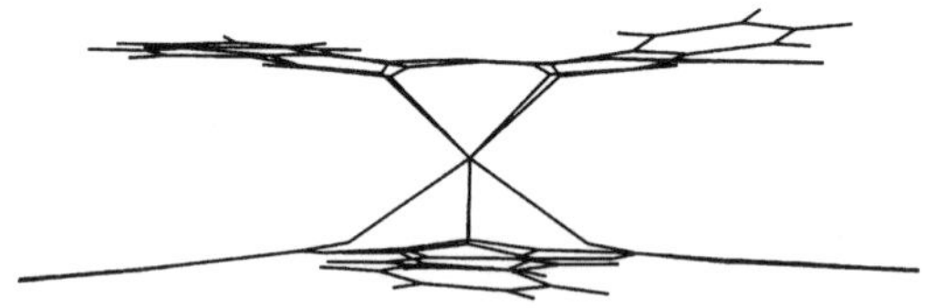

Appendix 18-61. Pc_2LuH, GAWBAH[212]

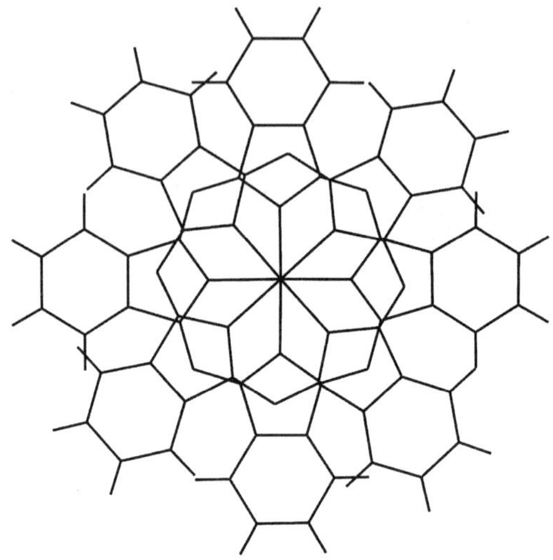

Appendix 18-62. Pc_2LuH, GAWBAH[212]

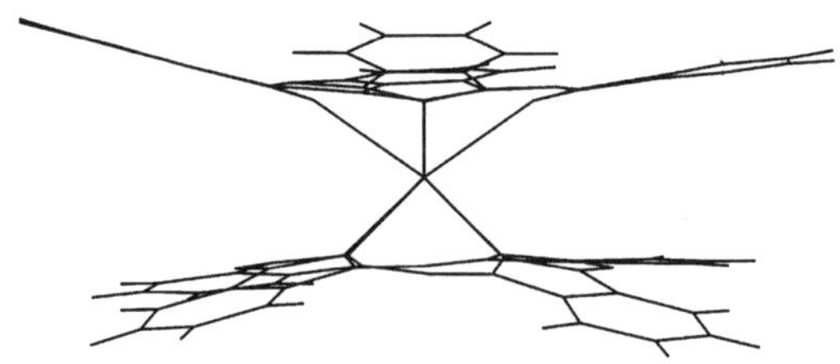

Appendix 18-63. $[Pc_2Lu][NBu_4][2DMF]$, GAVZUY[212]

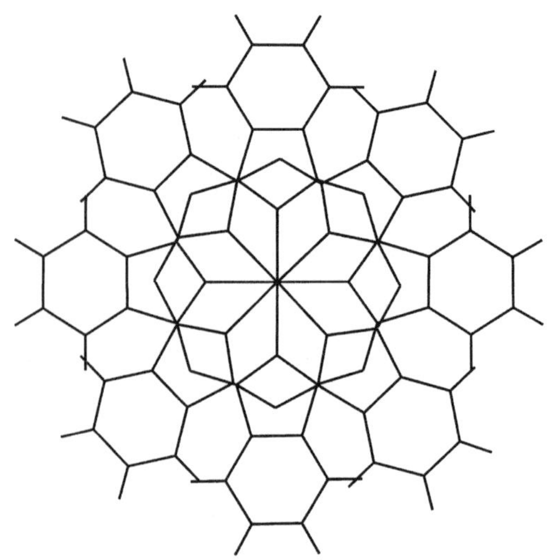

Appendix 18-64. $[Pc_2Lu][NBu_4][2DMF]$, GAVZUY[212]

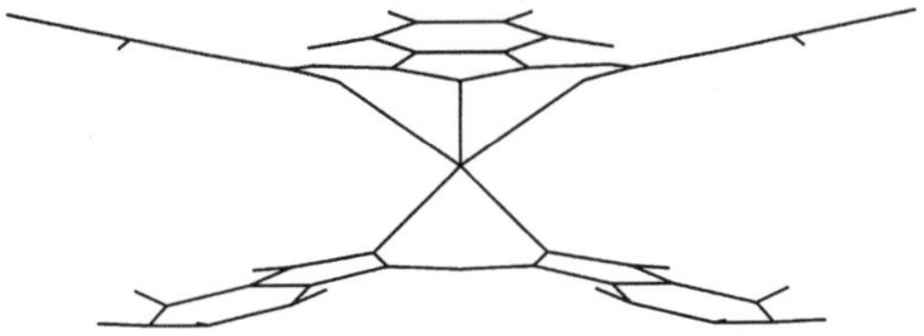

Appendix 18-65. $[Pc_2Lu][NBu_4][H_2O]$, NAKYIH[197]

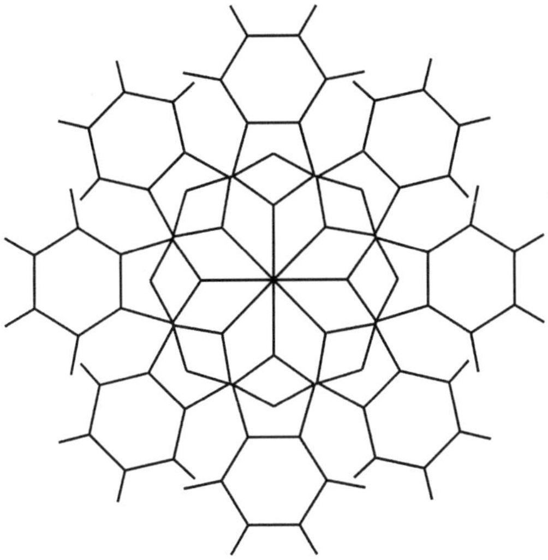

Appendix 18-66. $[Pc_2Lu][NBu_4][H_2O]$, NAKYIH[197]

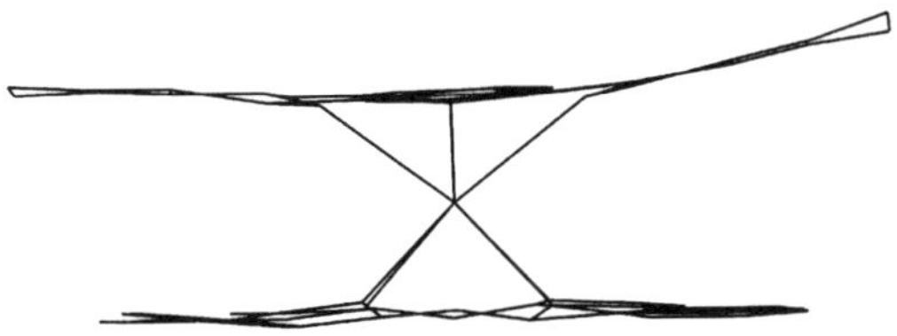

Appendix 18-67. $[Pc_2Nd][CH_2Cl_2]$, PTCNND[205,206] (App. 38)

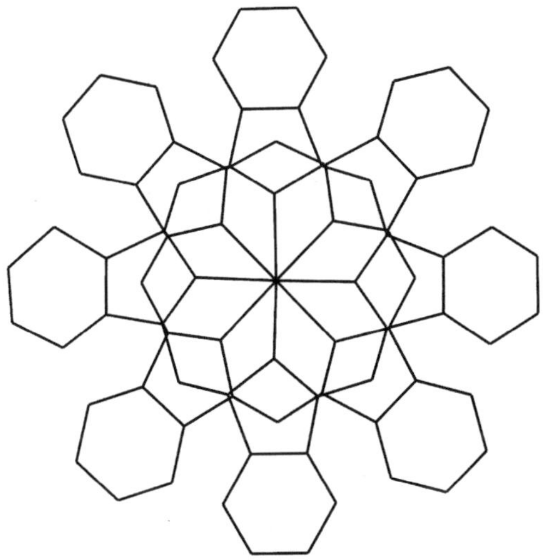

Appendix 18-68. $[Pc_2Nd][CH_2Cl_2]$, PTCNND[205,206] (App. 38)

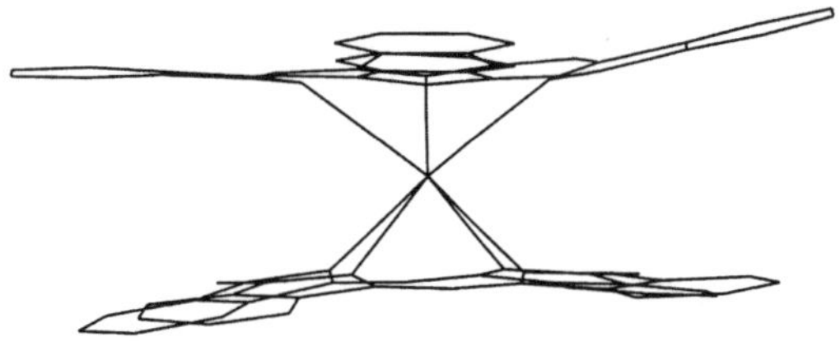

Appendix 18-69. β-Pc_2Nd, CIZGIB02[196]

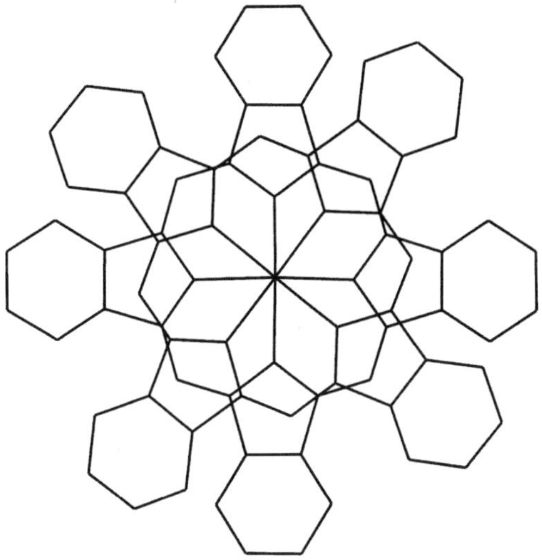

Appendix 18-70. β-Pc_2Nd, CIZGIB02[196]

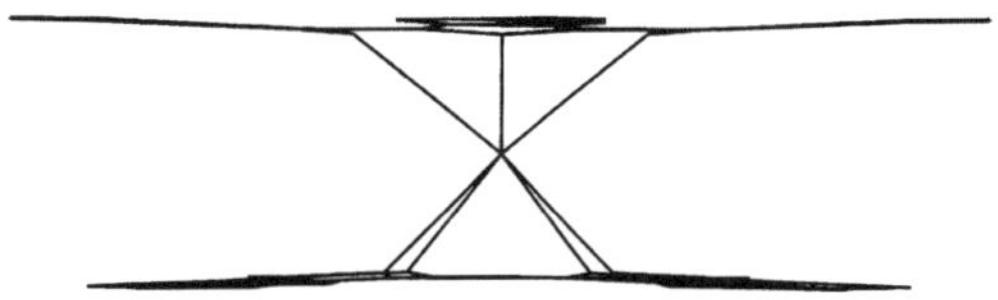

Appendix 18-71. Pc_2Nd, CIZGIB08[222]

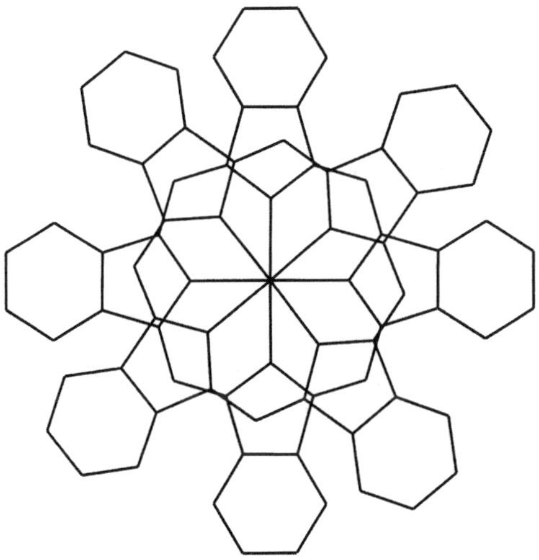

Appendix 18-72. Pc_2Nd, CIZGIB08[222]

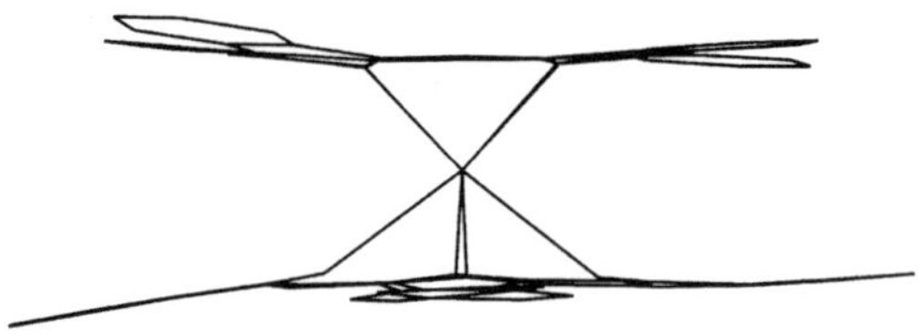

Appendix 18-73. Pc_2Nd (ox1), CIZGIB06[210] (App. 39)

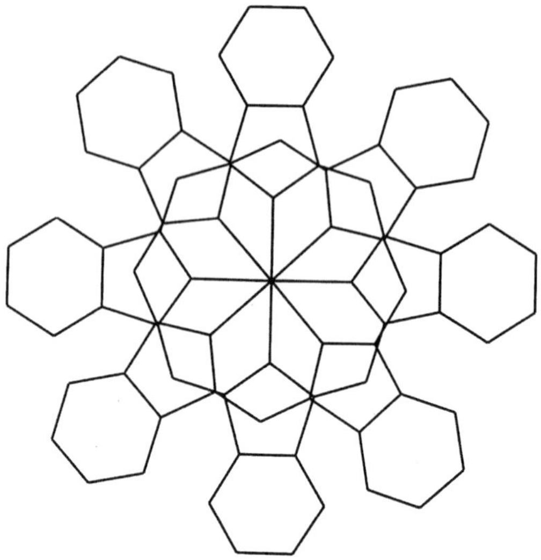

Appendix 18-74. Pc_2Nd (ox1), CIZGIB06[210] (App. 39)

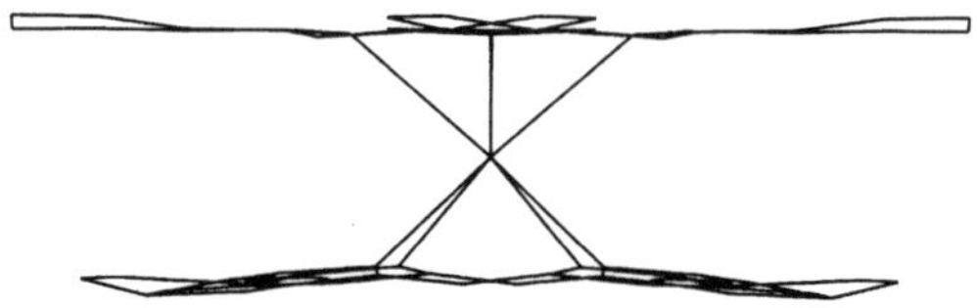

Appendix 18-75. α-Pc_2Nd, CIZGIB07[209]

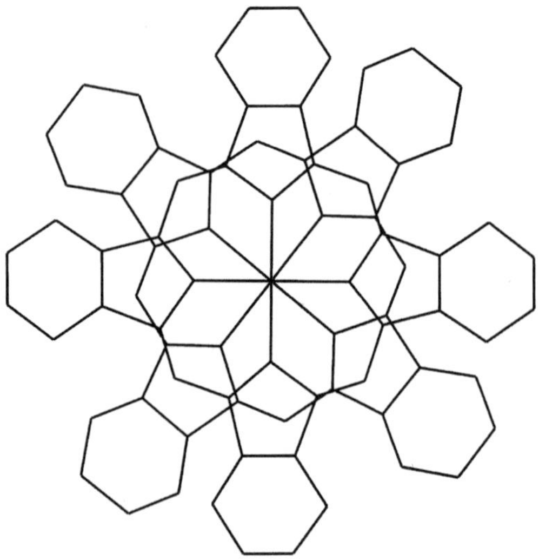

Appendix 18-76. α-Pc_2Nd, CIZGIB07[209]

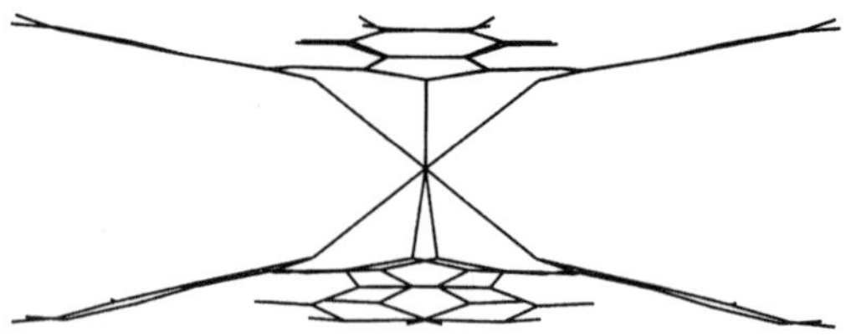

Appendix 18-77. $[Pc_2Nd][NBu_4]$, NAKXUS[197]

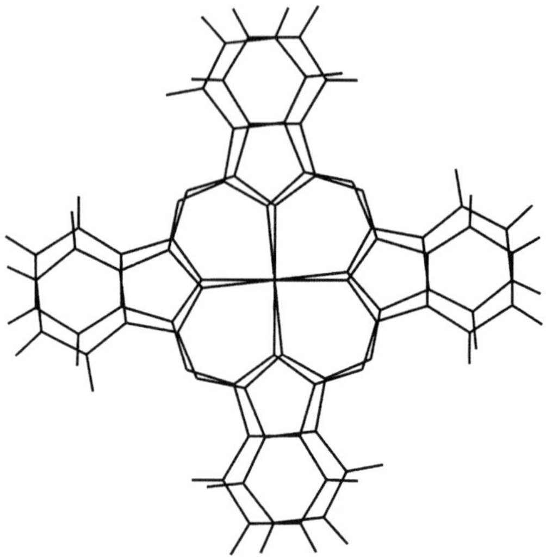

Appendix 18-78. $[Pc_2Nd][NBu_4]$, NAKXUS[197]

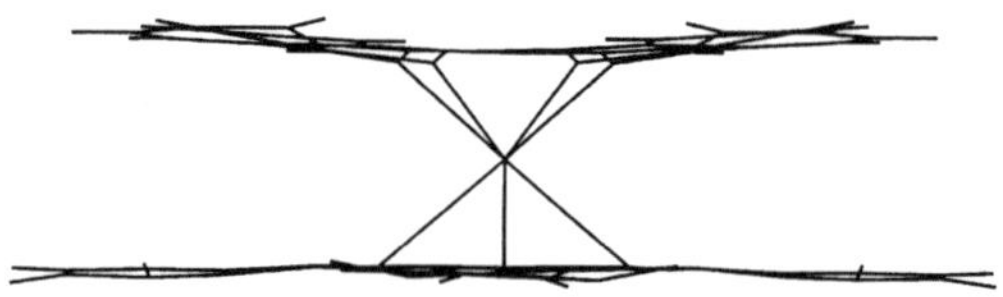

Appendix 18-79. $[Pc_2Nd][O]$, SAFPUK[213] (App. 40)

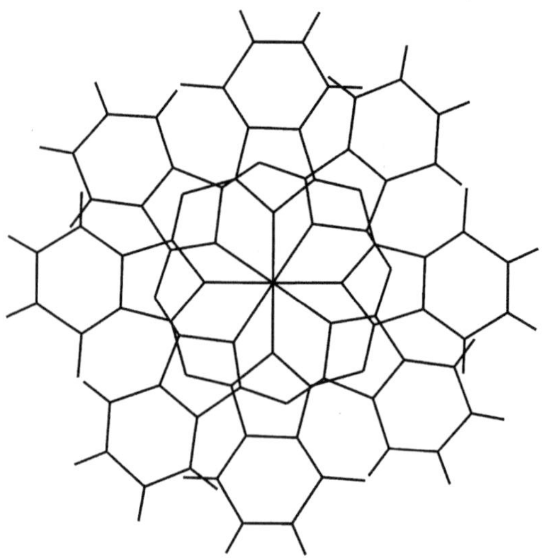

Appendix 18-80. $[Pc_2Nd][O]$, SAFPUK[213] (App. 40)

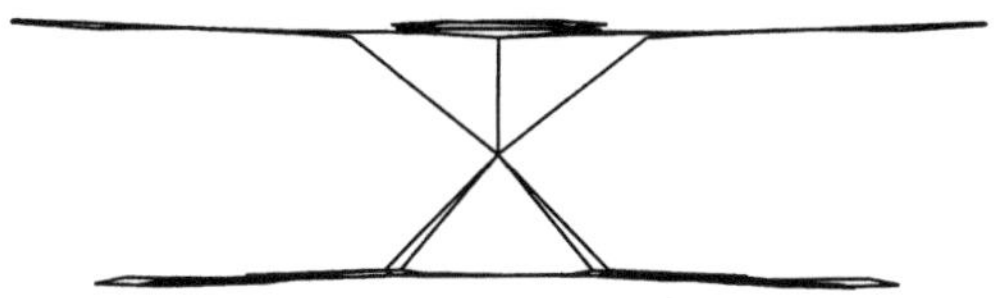

Appendix 18-81. Pc_2Pr, KOBRUO02[222]

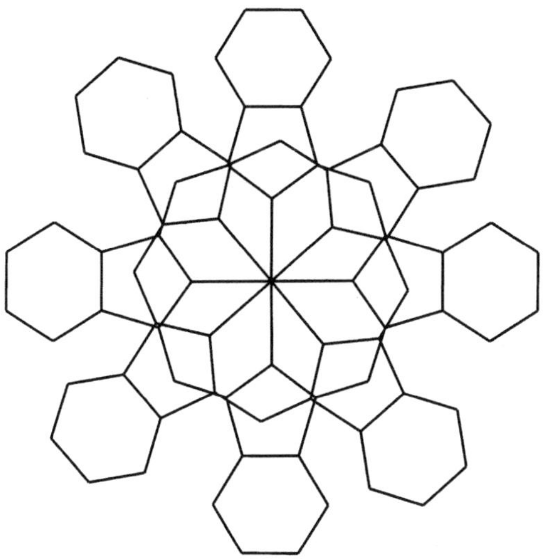

Appendix 18-82. Pc_2Pr, KOBRUO02[222]

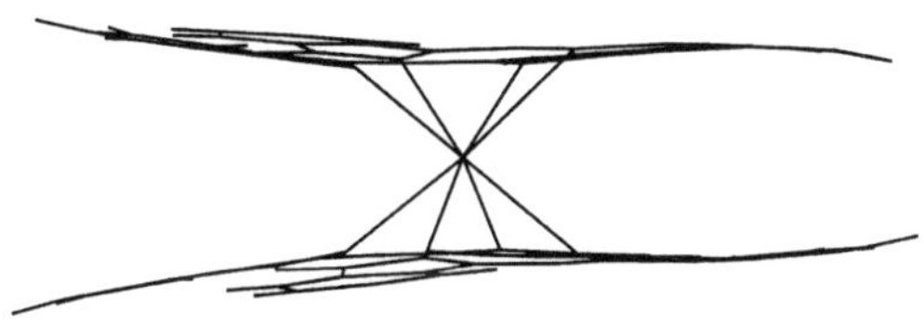

Appendix 18-83. β1-Pc_2Pr, KOBRUO[211]

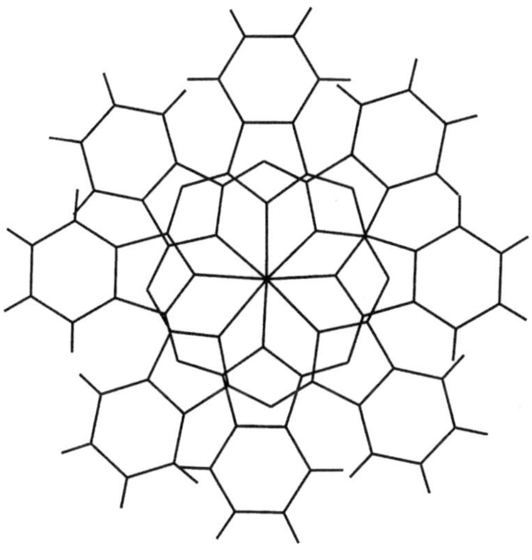

Appendix 18-84. β1-Pc_2Pr, KOBRUO[211]

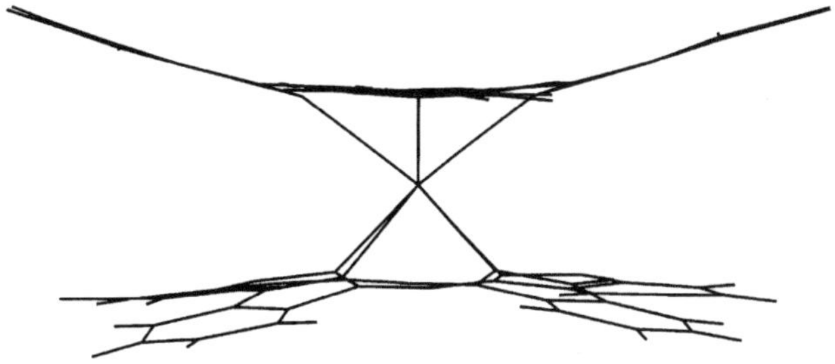

Appendix 18-85. $[Pc_2Pr][NBu_4][0.75MeOH]$, QOCKEY[198]

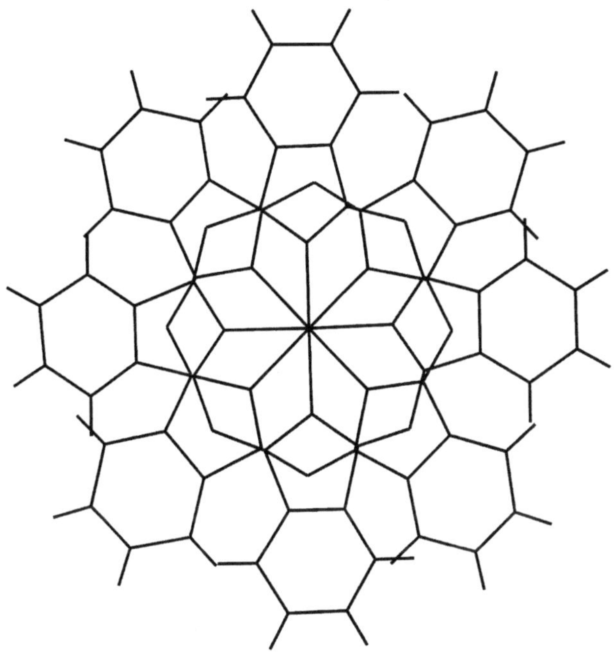

Appendix 18-86. $[Pc_2Pr][NBu_4][0.75MeOH]$, QOCKEY[198]

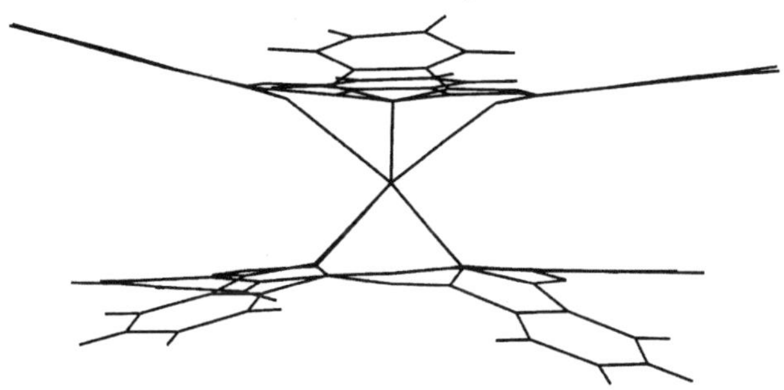

Appendix 18-87. $[Pc_2Pr][NBu_4][2py]$, QOCLAV[198]

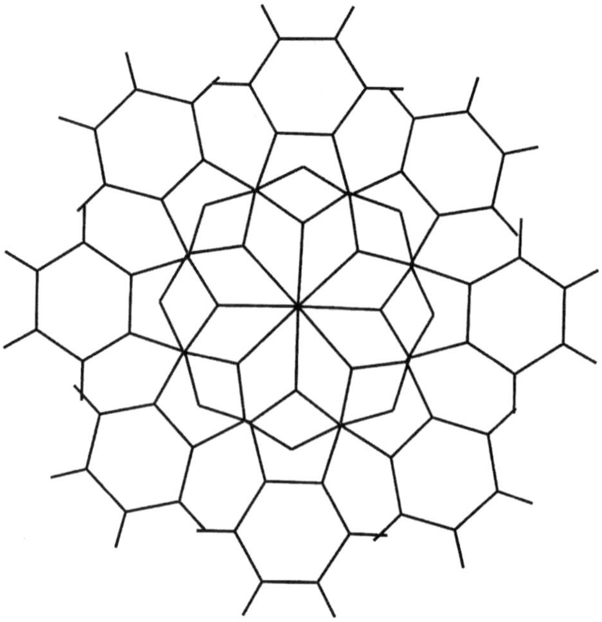

Appendix 18-88. $[Pc_2Pr][NBu_4][2py]$, QOCLAV[198]

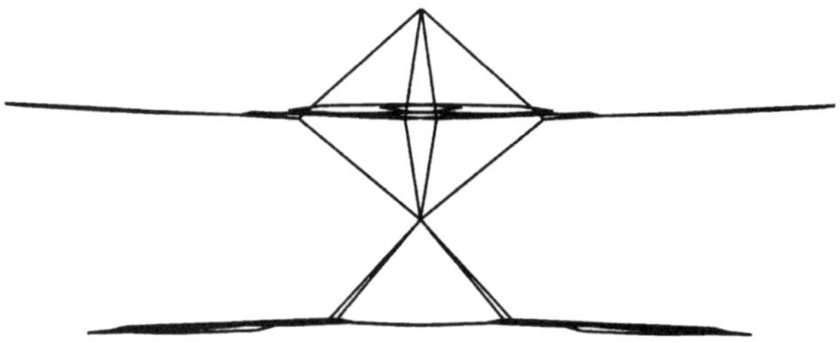

Appendix 18-89. α-$[Pc_2Pr][Br_{1.5}]$, WAYSIY[218]

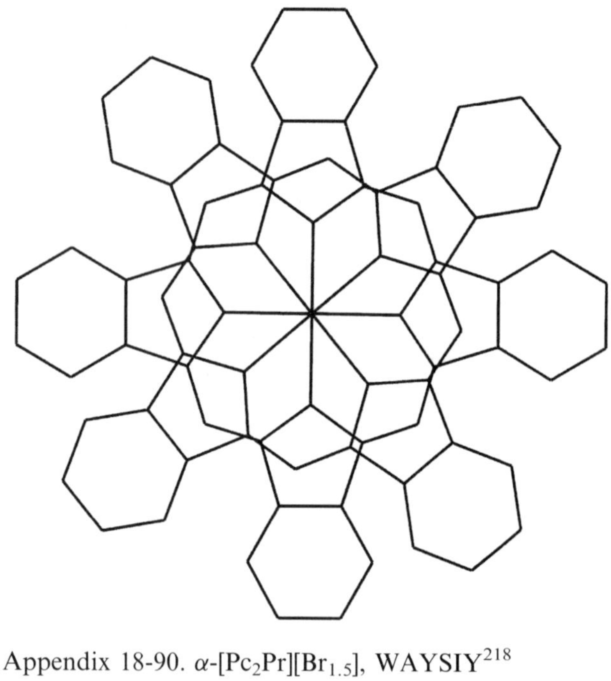

Appendix 18-90. α-$[Pc_2Pr][Br_{1.5}]$, WAYSIY[218]

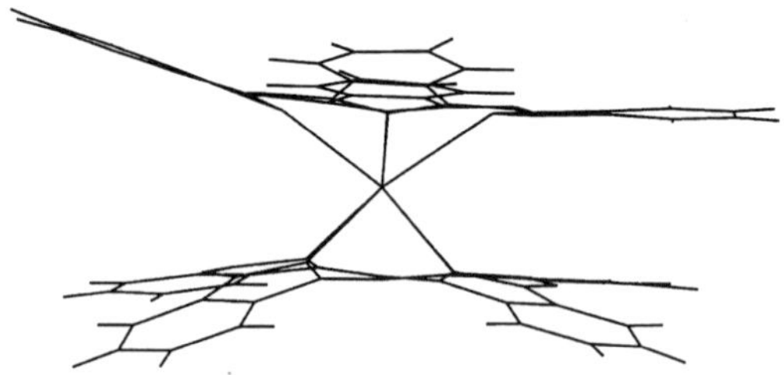

Appendix 18-91. $[Pc_2Sb][NBu_4][2THF]$, QOCLEZ[198]

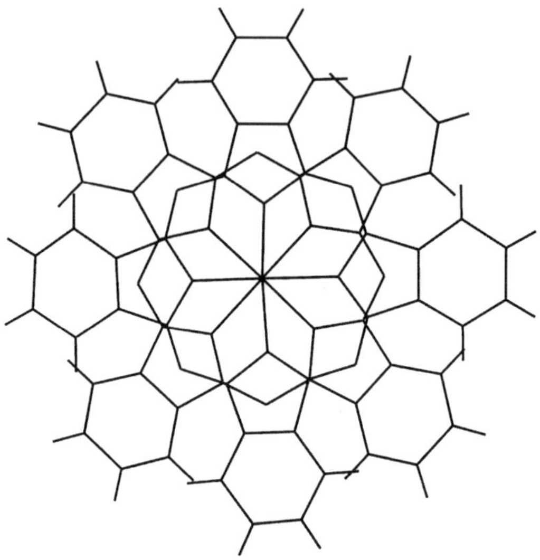

Appendix 18-92. $[Pc_2Sb][NBu_4][2THF]$, QOCLEZ[198]

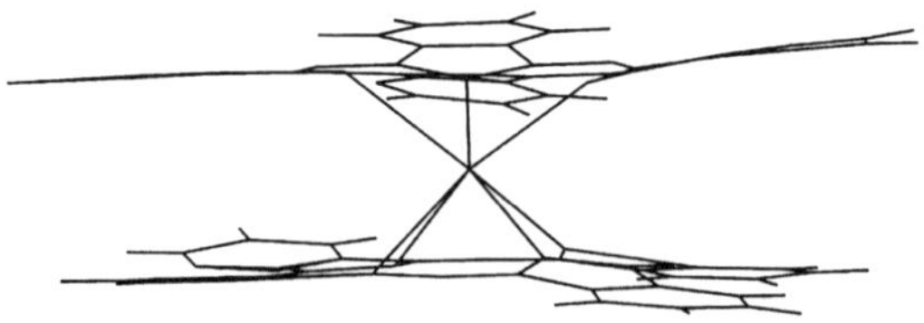

Appendix 18-93. $[Pc_2Sb][Br][PNP]_2[2Et_2O]$, QOCLID[198]

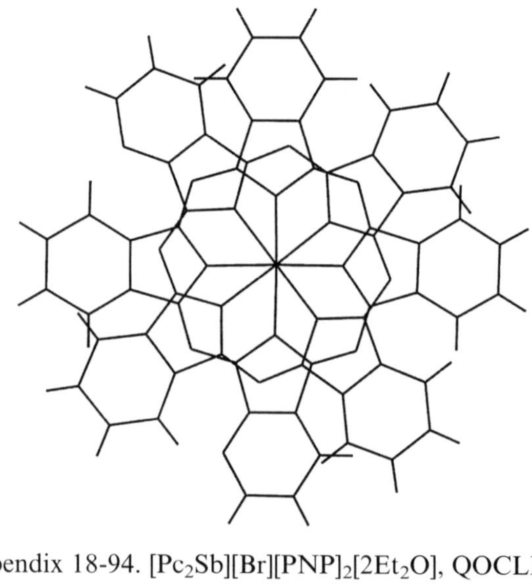

Appendix 18-94. $[Pc_2Sb][Br][PNP]_2[2Et_2O]$, QOCLID[198]

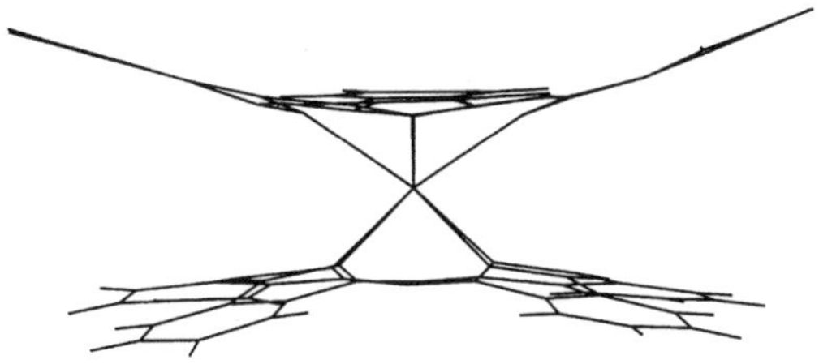

Appendix 18-95. $[Pc_2Sc][NBu_4]$, QOCKUO[198]

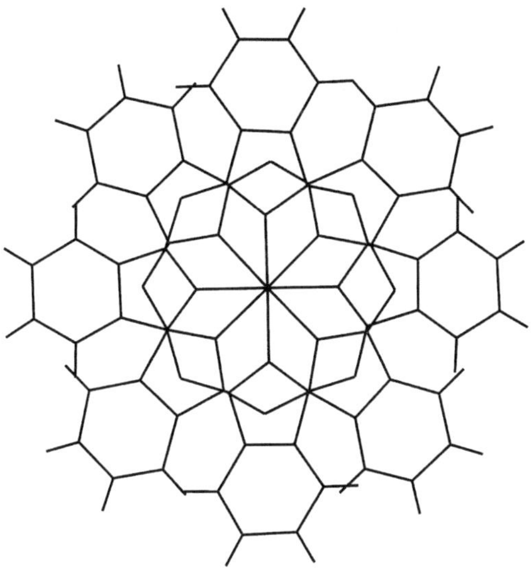

Appendix 18-96. $[Pc_2Sc][NBu_4]$, QOCKUO[198]

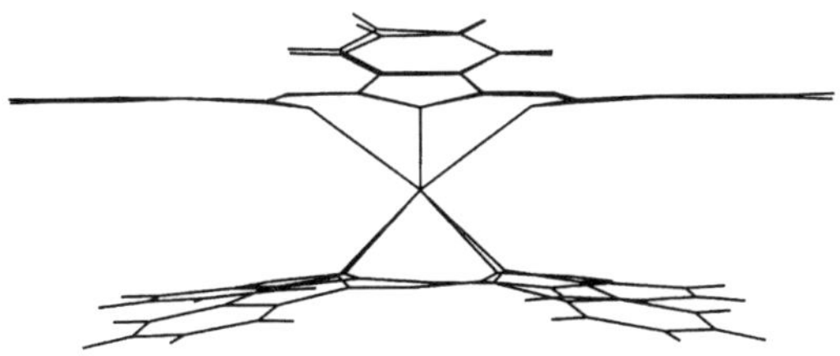

Appendix 18-97. $[Pc_2Sm][NBu_4][MeOH]$, QOCKIC[198]

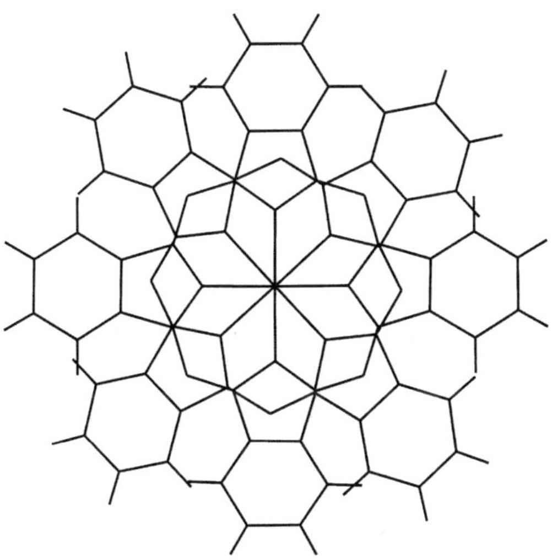

Appendix 18-98. $[Pc_2Sm][NBu_4][MeOH]$, QOCKIC[198]

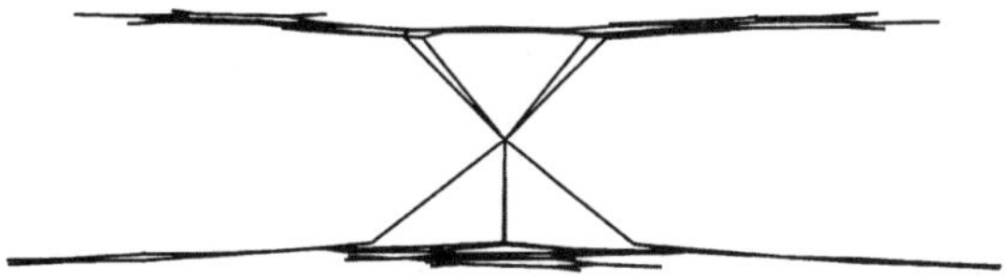

Appendix 18-99. α-$[Pc_2Sm][Br_{1.45}]$, ZOMWUT[227]

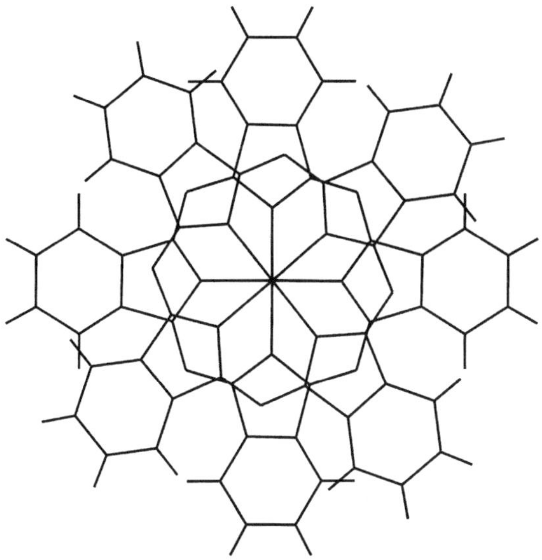

Appendix 18-100. α-$[Pc_2Sm][Br_{1.45}]$, ZOMWUT[227]

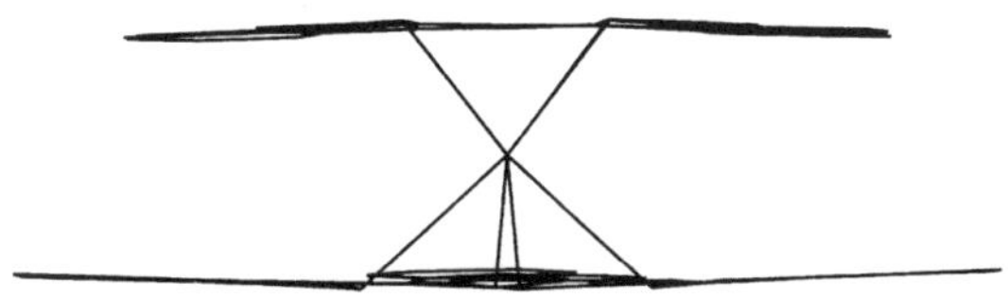

Appendix 18-101. $[Pc_2Sm][ClO_4]_{0.63}$, ZOMXAA[227]

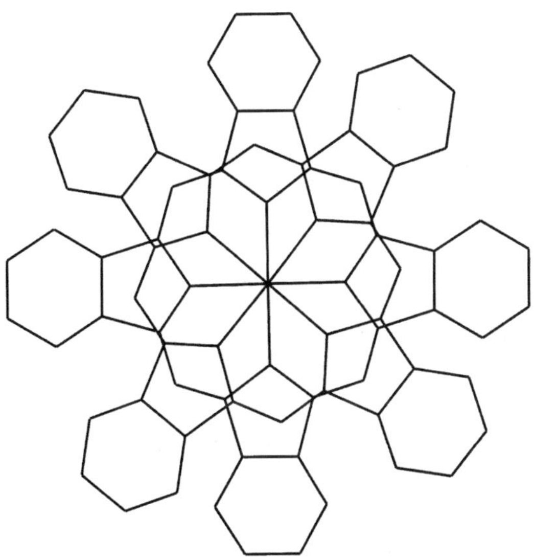

Appendix 18-102. $[Pc_2Sm][ClO_4]_{0.63}$, ZOMXAA[227]

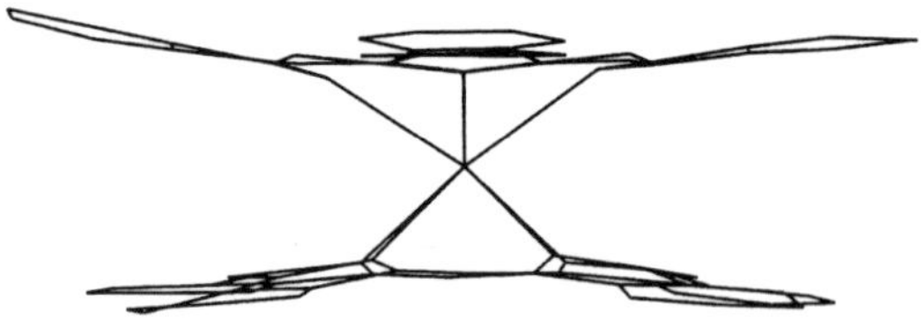

Appendix 18-103. Pc_2Sn, SNPTCY[201]

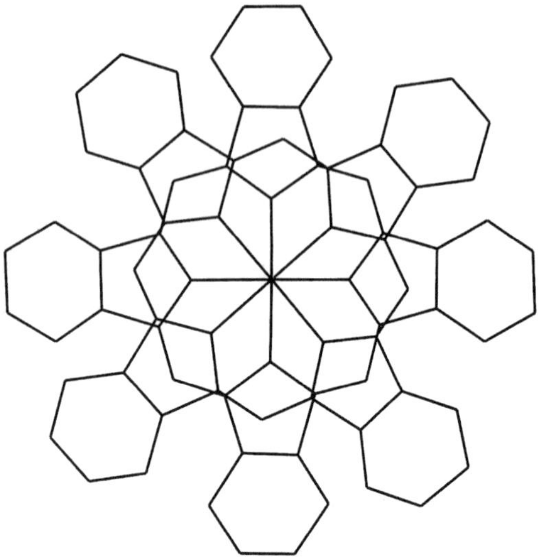

Appendix 18-104. Pc_2Sn, SNPTCY[201]

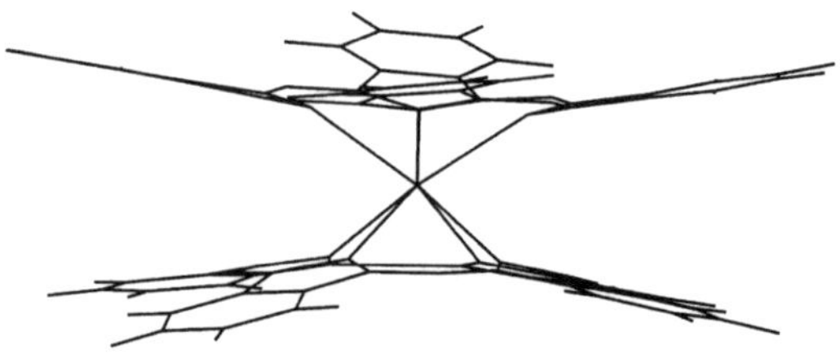

Appendix 18-105. Pc_2Sn, SNPTCY01[223]

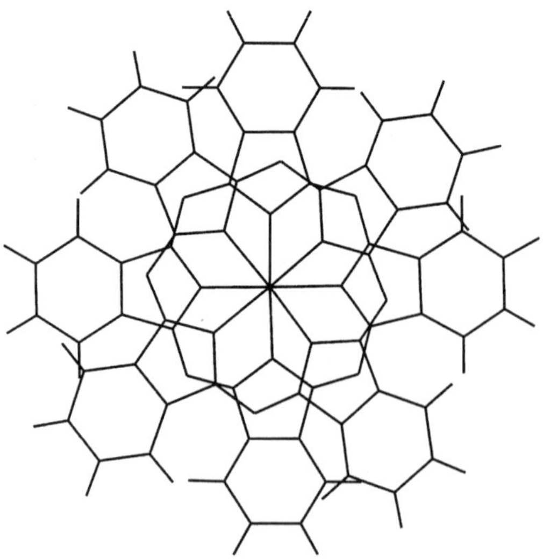

Appendix 18-106. Pc_2Sn, SNPTCY01[223]

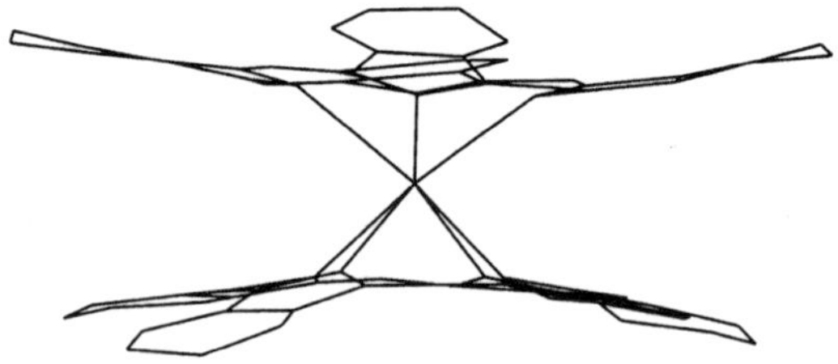

Appendix 18-107. Pc_2Th, DPCYTH[203]

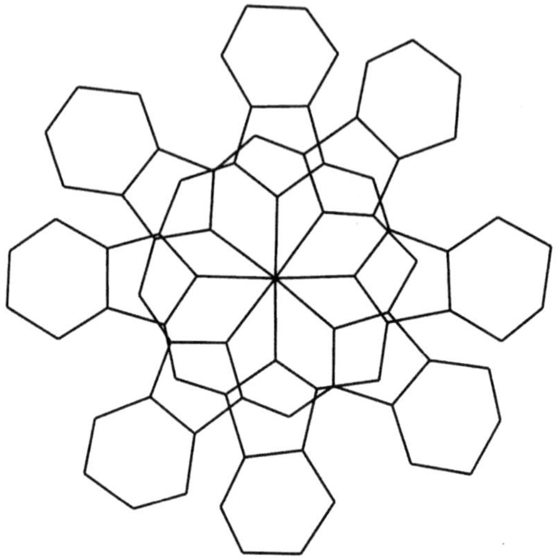

Appendix 18-108. Pc_2Th, DPCYTH[203]

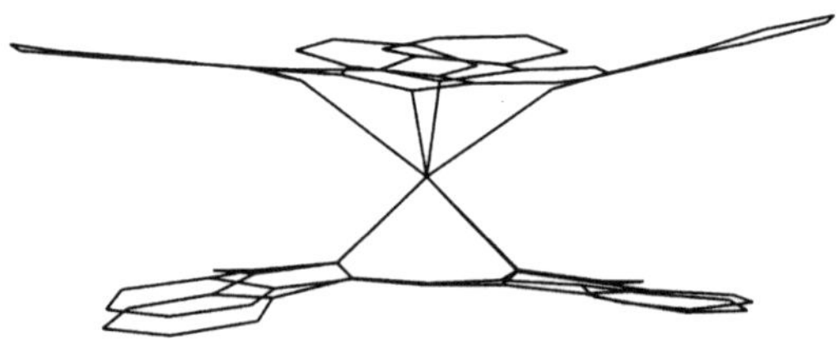

Appendix 18-109. Pc_2Th, DPCYTH01[204]

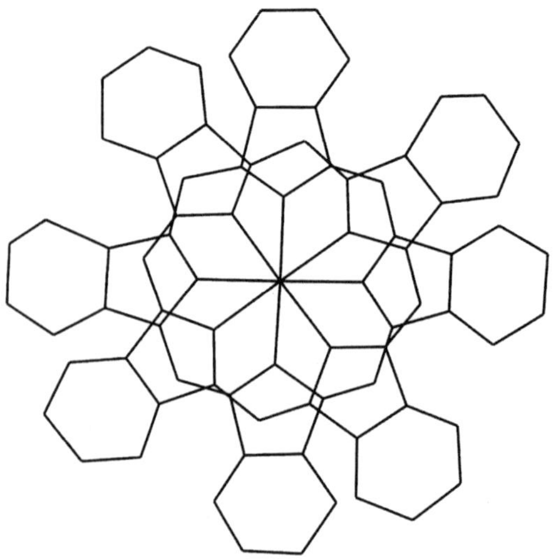

Appendix 18-110. Pc_2Th, DPCYTH01[204]

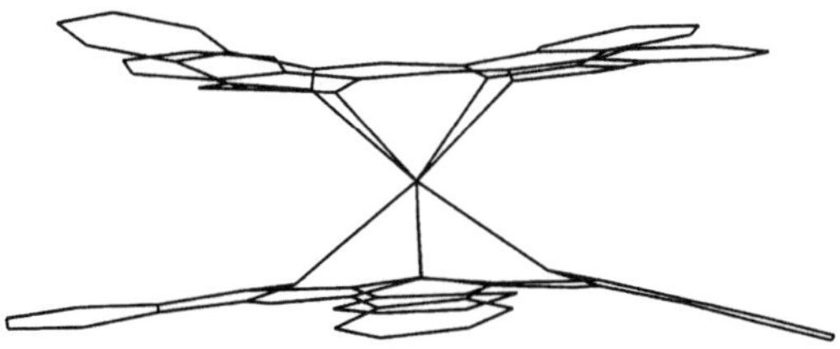

Appendix 18-111. Pc_2Th, DPCYTH03[207]

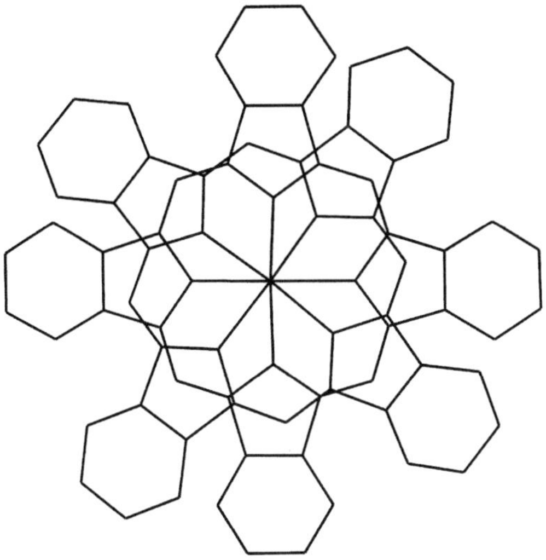

Appendix 18-112. Pc_2Th, DPCYTH03[207]

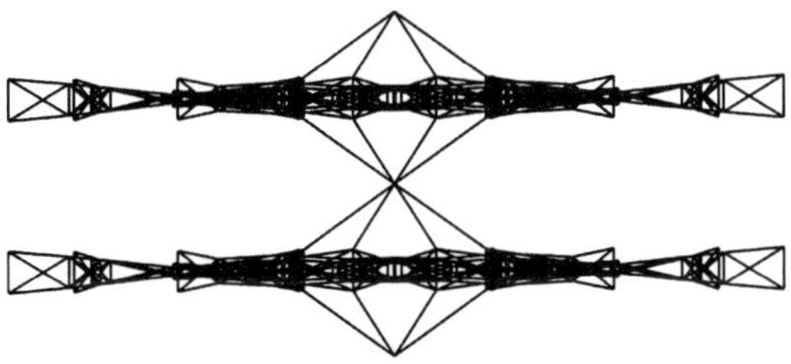

Appendix 18-113. $[Pc_2Ti][I_3]_{0.66}$, PIBJIT[219]

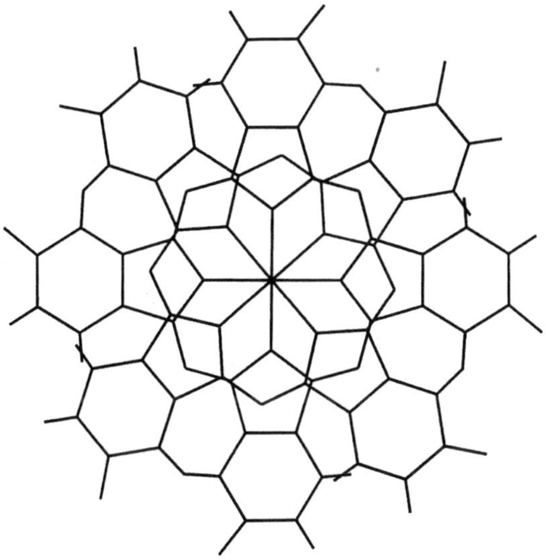

Appendix 18-114. $[Pc_2Ti][I_3]_{0.66}$, PIBJIT[219]

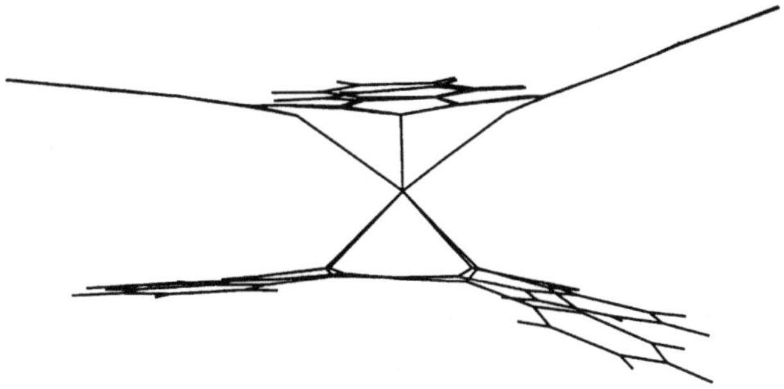

Appendix 18-115. $[Pc_2Tm][PNP][xH_2O]$, YEGTUZ[220]

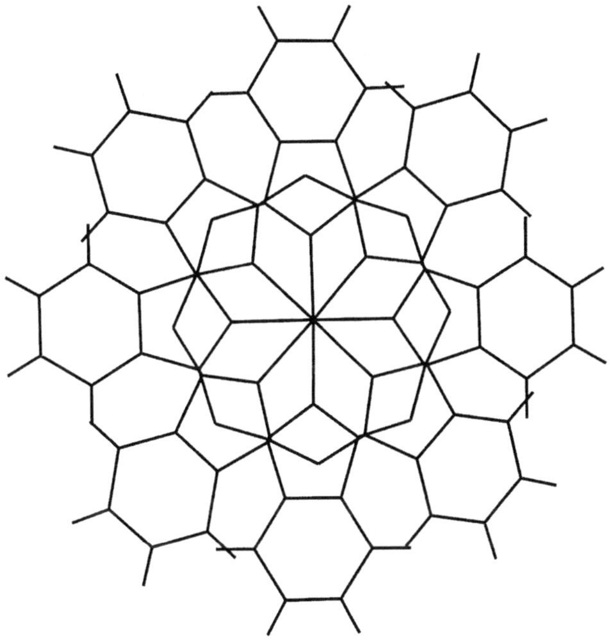

Appendix 18-116. $[Pc_2Tm][PNP][xH_2O]$, YEGTUZ[220]

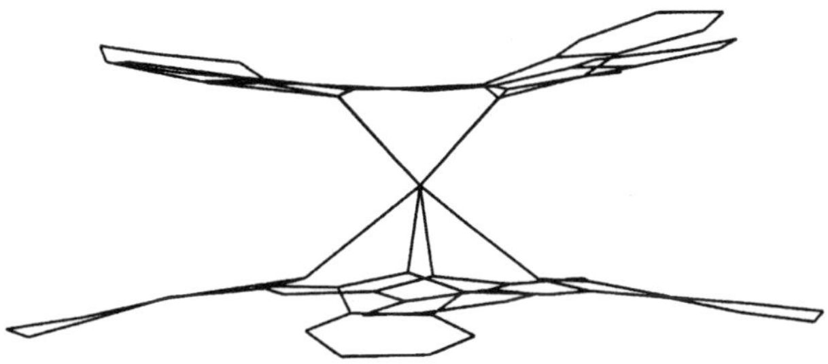

Appendix 18-117. Pc_2U, PHALCU01[202,203]

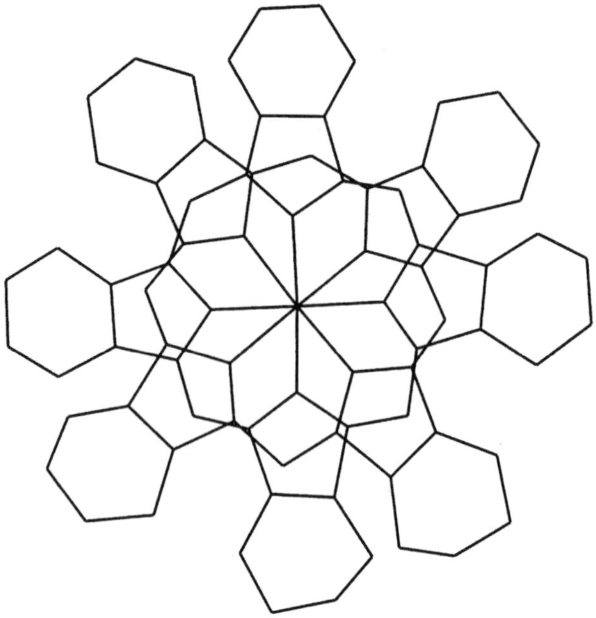

Appendix 18-118. Pc_2U, PHALCU01[202,203]

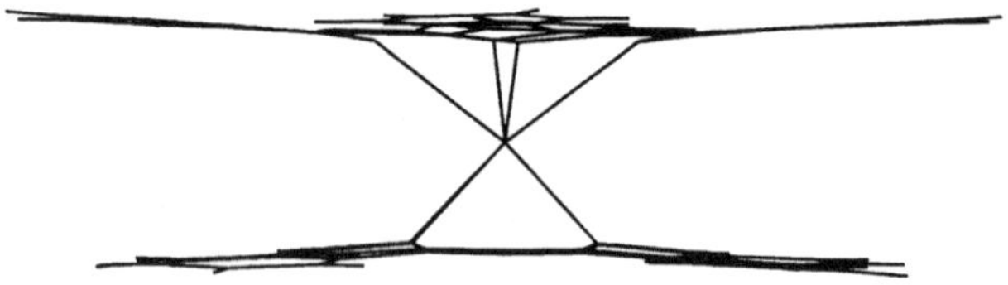

Appendix 18-119. $[Pc_2U][I_3]_{2/3}$, MEGNOB[235]

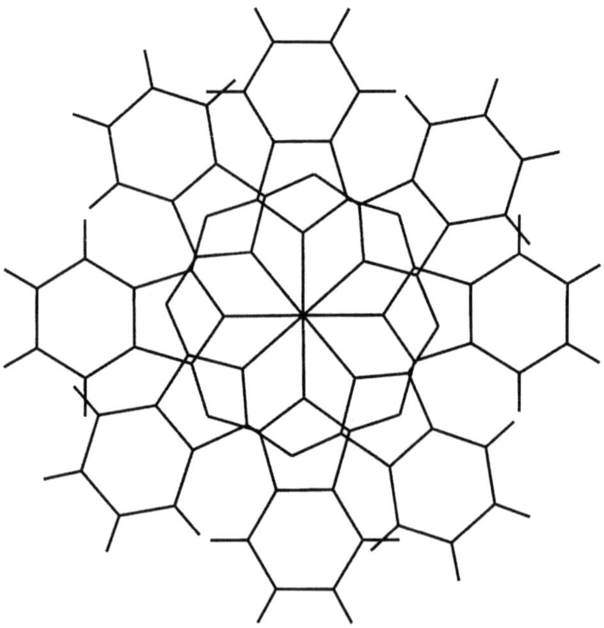

Appendix 18-120. $[Pc_2U][I_3]_{2/3}$, MEGNOB[235]

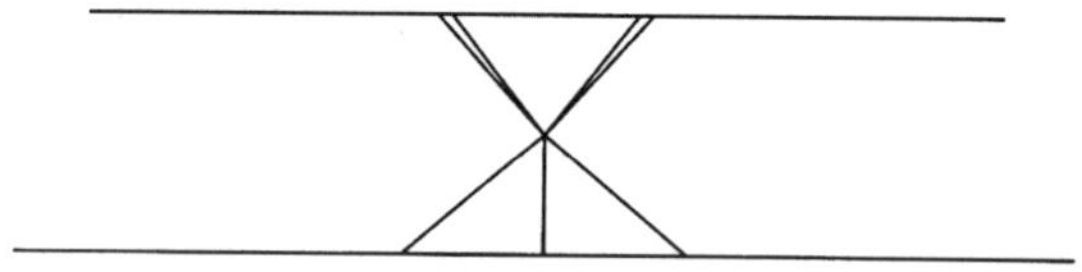

Appendix 18-121. $[Pc_2U][I_{5/3}]$, HOTWOC[234]

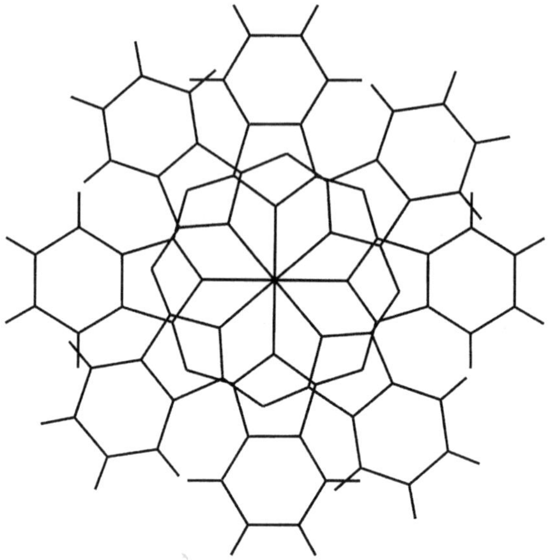

Appendix 18-122. $[Pc_2U][I_{5/3}]$, HOTWOC[234]

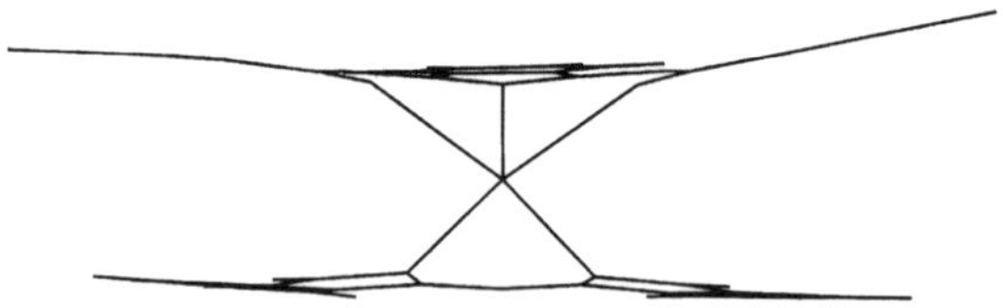

Appendix 18-123. $[Pc_2Y][CH_2Cl_2]$, JUVJIT[216]

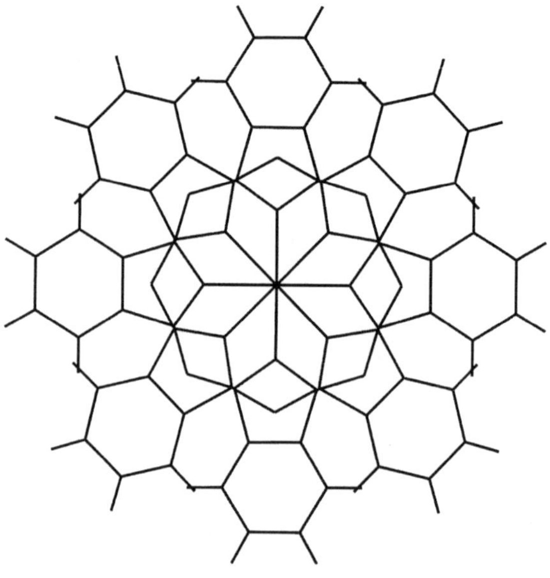

Appendix 18-124. $[Pc_2Y][CH_2Cl_2]$, JUVJIT[216]

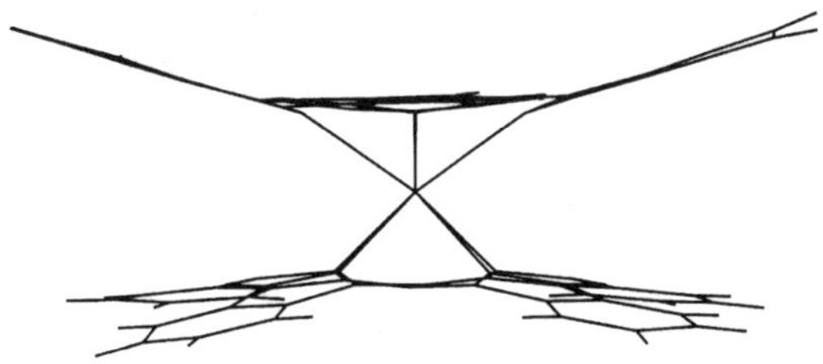

Appendix 18-125. $[Pc_2Y][NBu_4]$, QOCKOI[198]

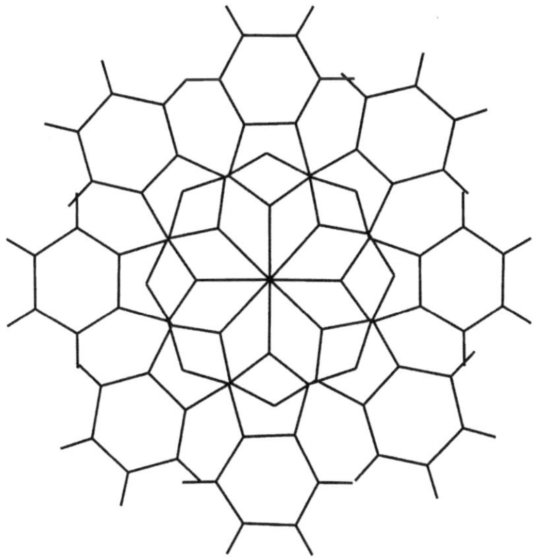

Appendix 18-126. $[Pc_2Y][NBu_4]$, QOCKOI[198]

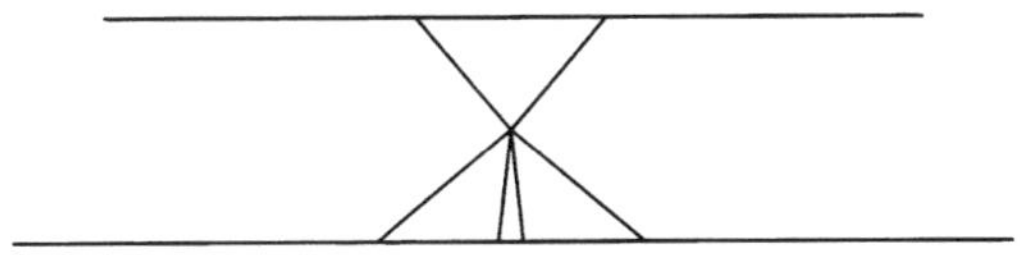

Appendix 18-127. $[Pc_2Yb][I_2]$, BECSUX[233]

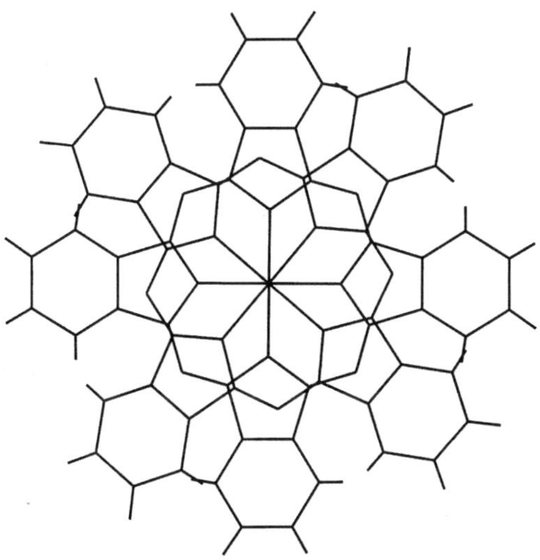

Appendix 18-128. $[Pc_2Yb][I_2]$, BECSUX[233]

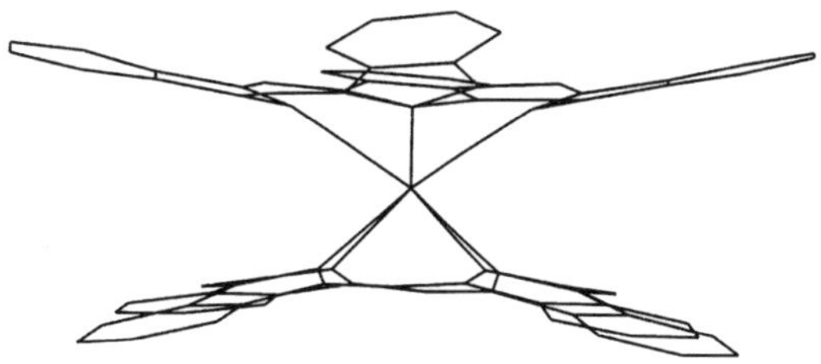

Appendix 18-129. Pc_2Zr, SOTGAJ[215]

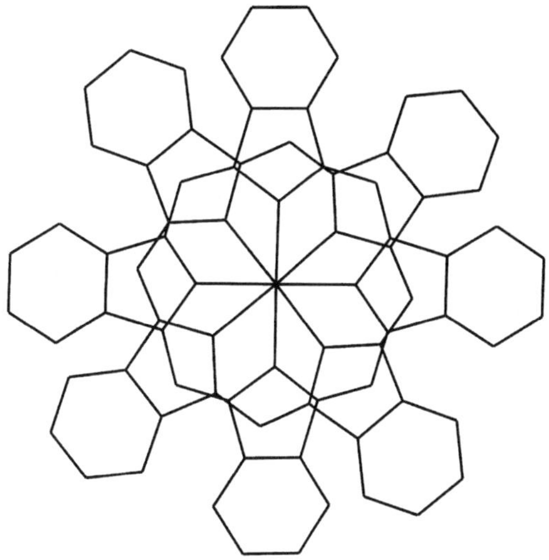

Appendix 18-130. Pc_2Zr, SOTGAJ[215]

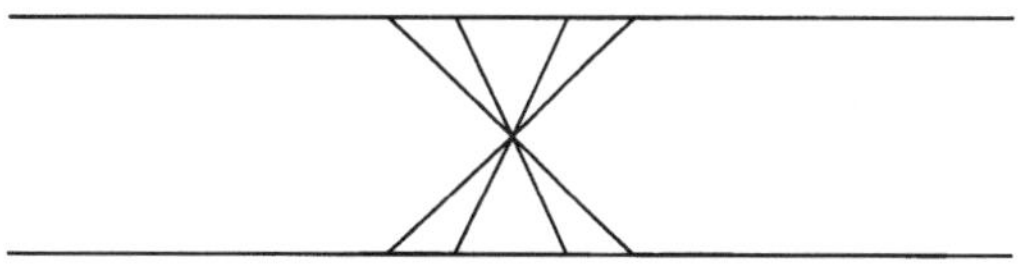

Appendix 18-131. $[Pc_2Zr][I_2]$, NOFMEA[409]

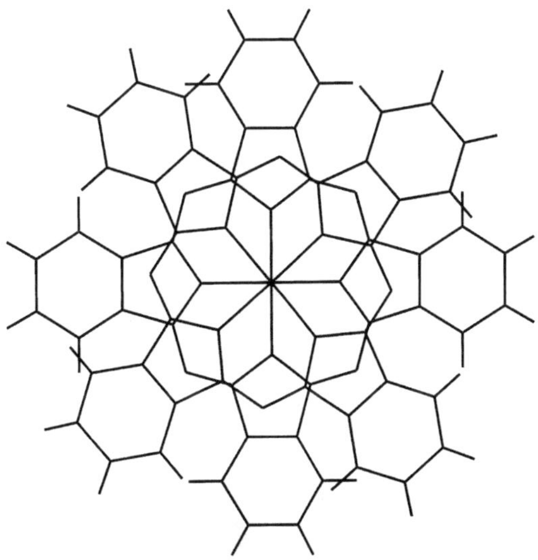

Appendix 18-132. $[Pc_2Zr][I_2]$, NOFMEA[409]

Appendix 19. Molecular Structures of Naphthalocyanine Complexes

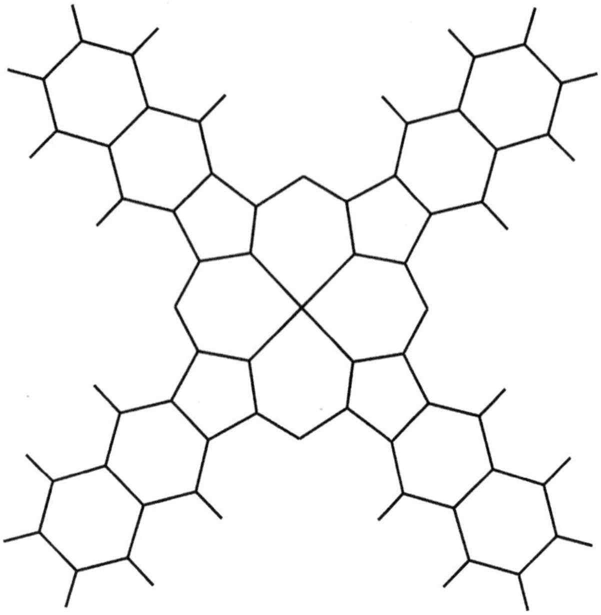

Appendix 19-1. 2,3-NcCu, RADPIV[104]

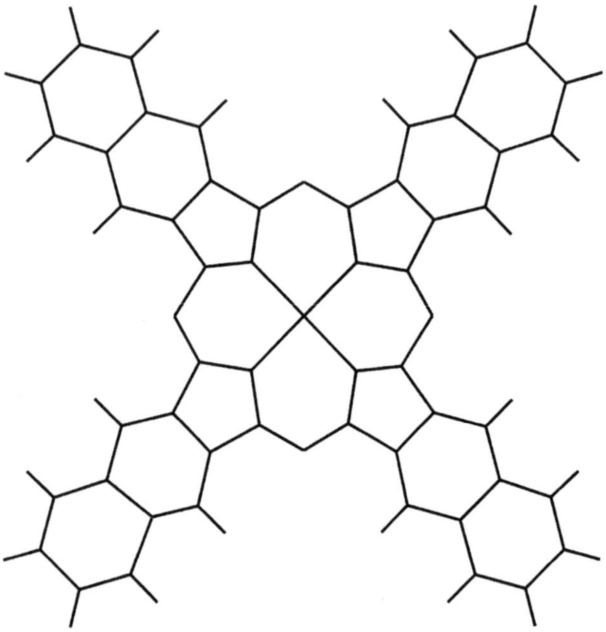

Appendix 19-2. 2,3-NcNi, RADPER[104]

Appendix 19-3. 2,3-NcNi, RADPER[104]

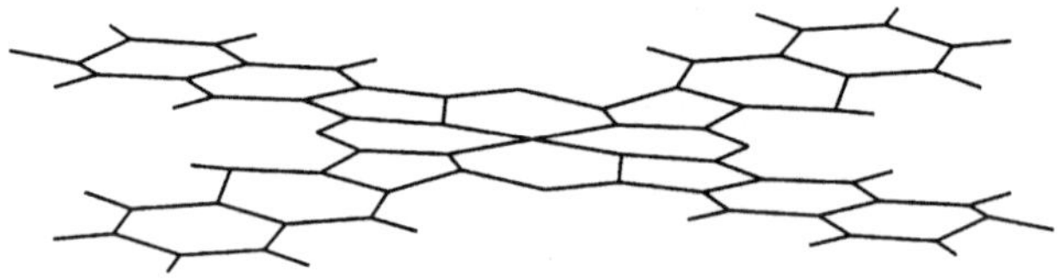

Appendix 19-4. NcZn, RADPOB[104]

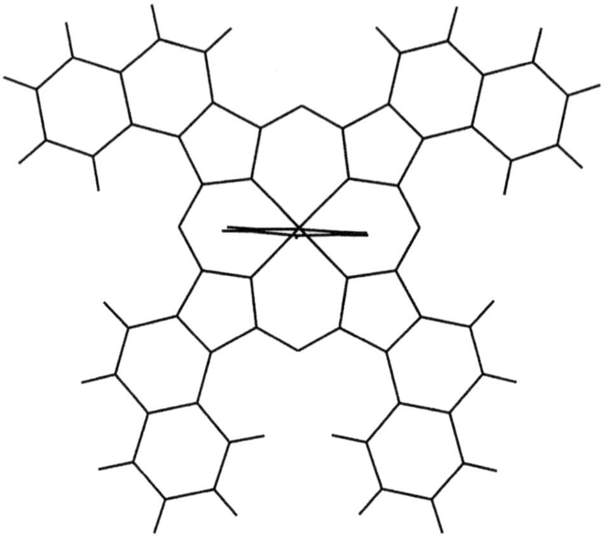

Appendix 19-5. [C_{2v}-1,2-NcCo-py][2py], #2001-03[135]

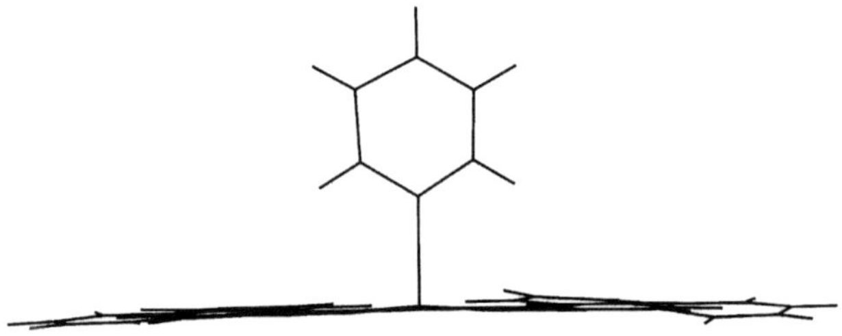

Appendix 19-6. [C_{2v}-1,2-NcCo-py][2py], #2001-03[135]

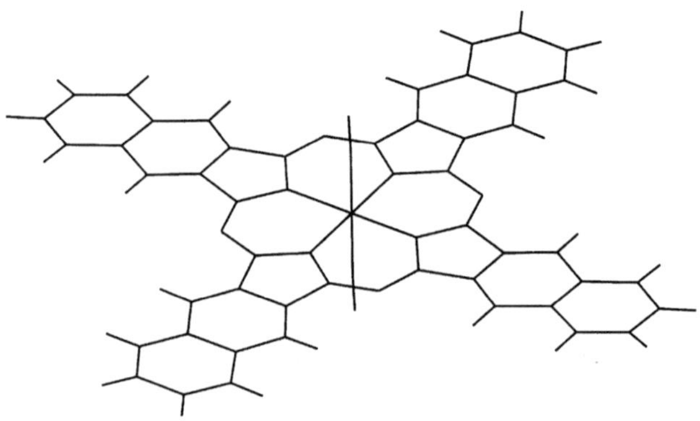

Appendix 19-7. [2,3-NcCo$(CN)_2$][2DCB], WOHDUH[251]

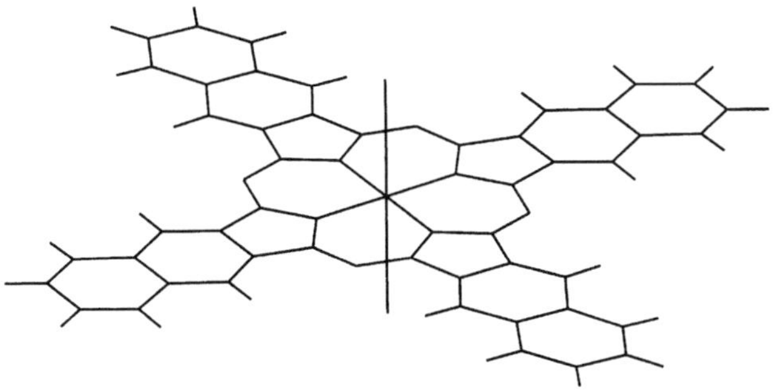

Appendix 19-8. [2,3-NcCo(CN)$_2$][CH$_3$CN], WOWDOB[251]

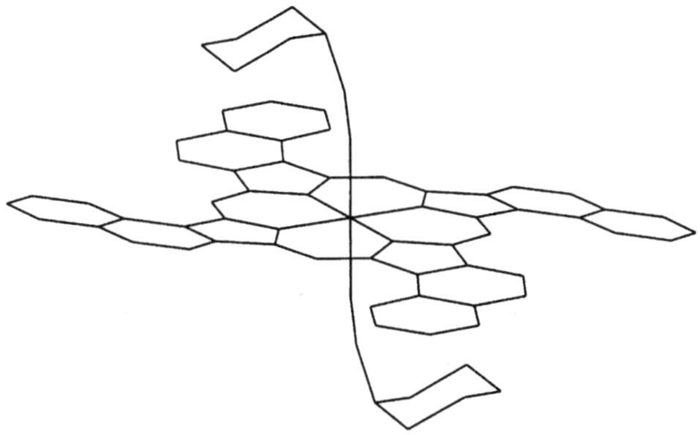

Appendix 19-9. D_{4h}-1,2-NcFe(NC–C$_6$H$_{11}$)$_2$, VACXOM[249]

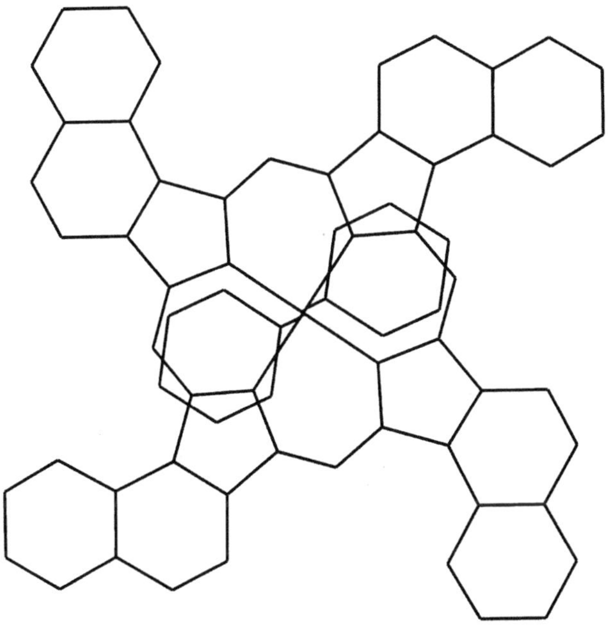

Appendix 19-10. D_{4h}-1,2-NcFe(NC–C$_6$H$_{11}$)$_2$, VACXOM[249]

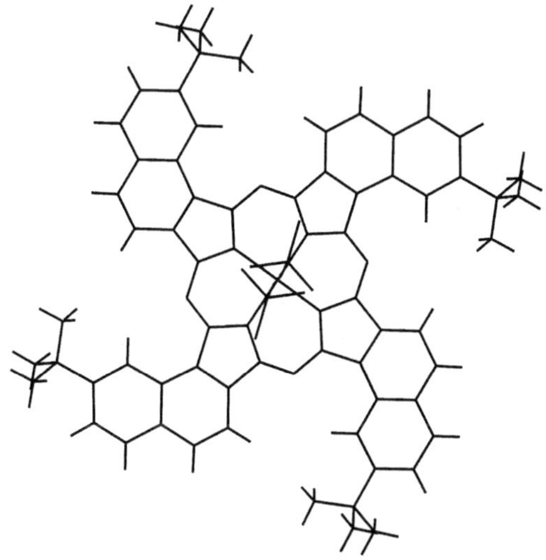

Appendix 19-11. Mol. 1 of tBu$_4$-D_{4h}-1,2-NcFe(NC–tBu)$_2$, VIWMOD[250]

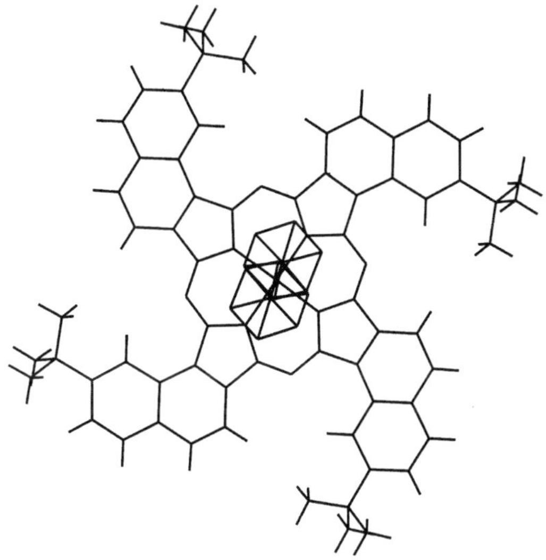

Appendix 19-12. Mol. 2 of tBu$_4$-D_{4h}-1,2-NcFe(NC–tBu)$_2$, VIWMOD[250]

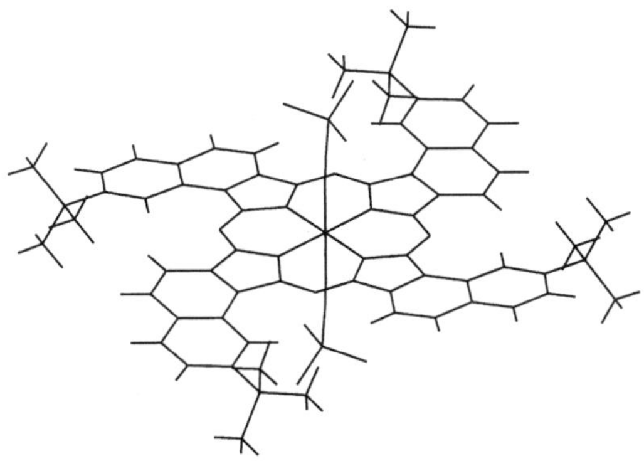

Appendix 19-13. Mol. 1 of tBu$_4$-D_{4h}-1,2-NcFe(NC–tBu)$_2$, VIWMOD[250]

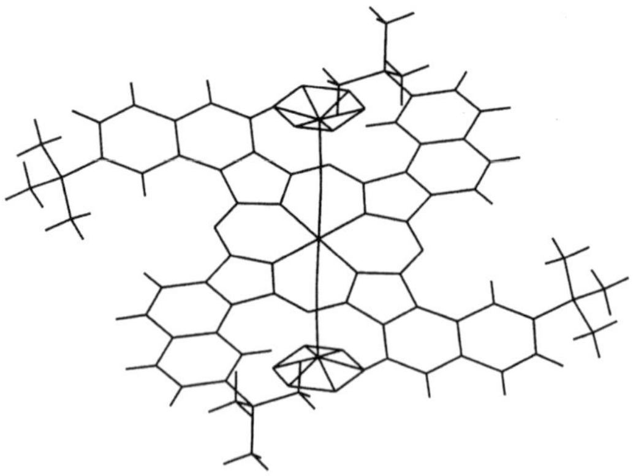

Appendix 19-14. Mol. 2 of tBu_4-D_{4h}-1,2-NcFe(NC–tBu)$_2$, VIWMOD[250]

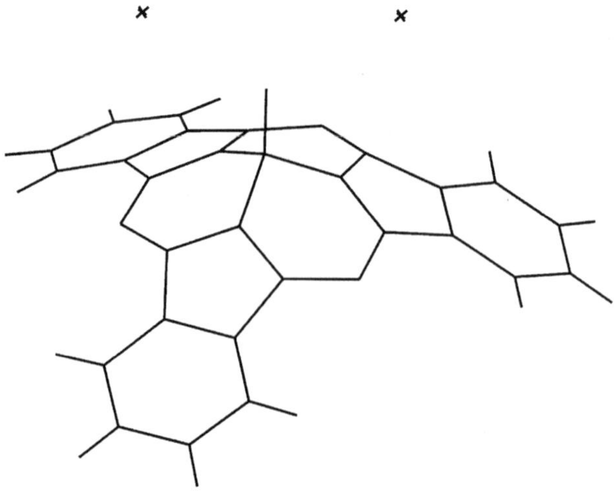

Appendix 20-3. [SubpcB-OH][$2H_2O$], KAJGOR[141]

Appendix 20. Molecular Structures of Sub- and Superphthalocyanine Complexes

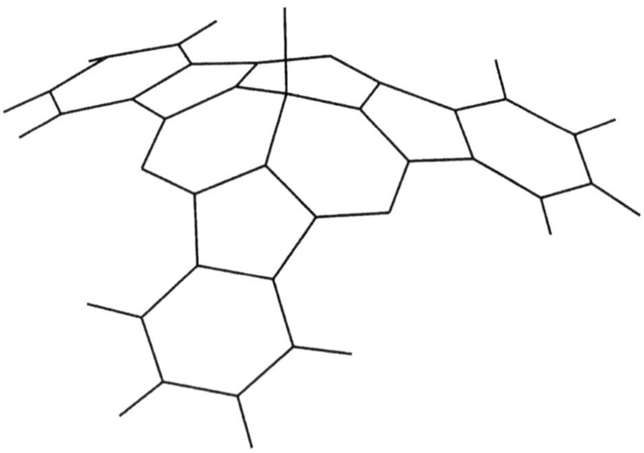

Appendix 20-1. SubpcB-Cl, PHTLBC[138]

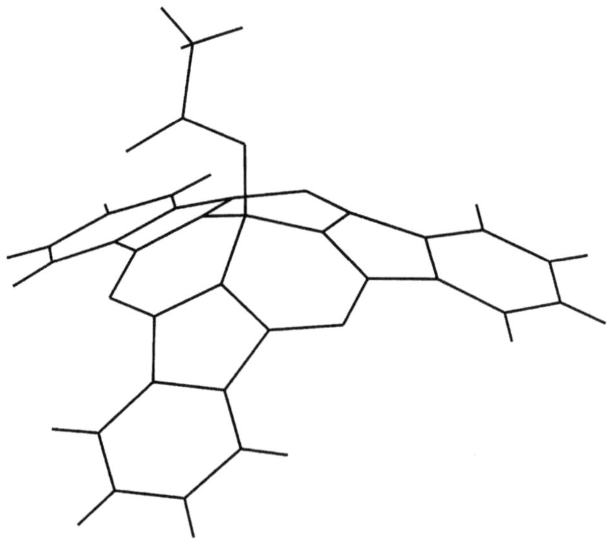

Appendix 20-4. [SubpcB-OAc][$0.5H_2O$,EtOH], KAJQAN[141]

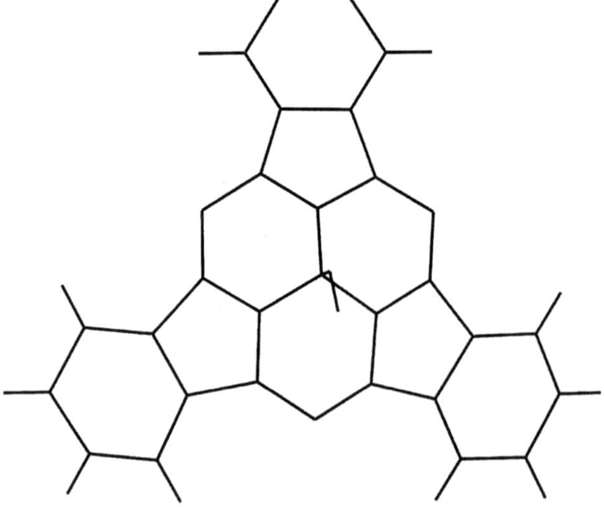

Appendix 20-2. SubpcB-OH, KAJGIL[141]

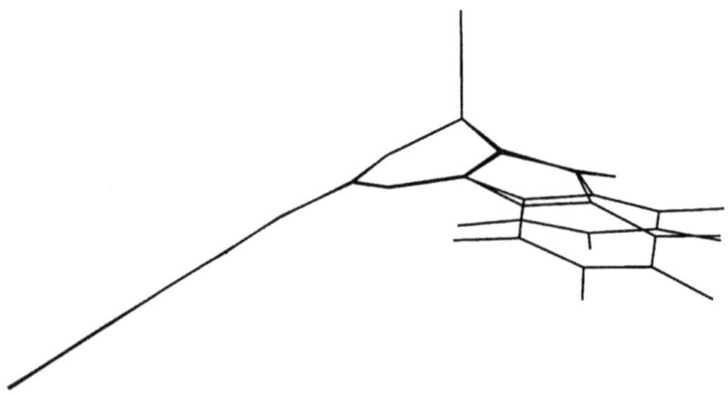

Appendix 20-5. MonobenzosubpcB-Cl, XAQGEB[264]

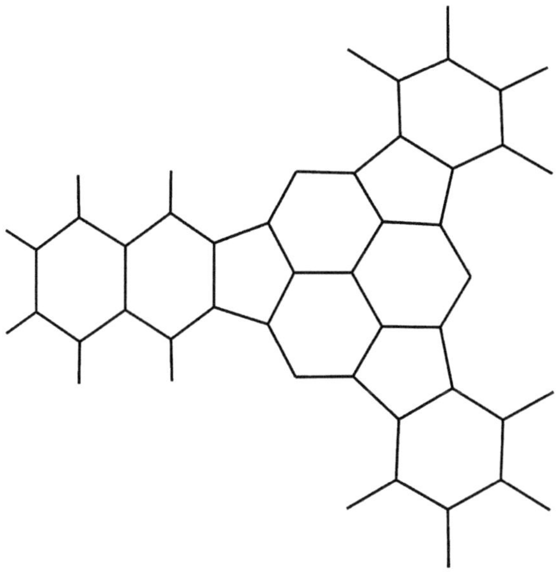

Appendix 20-6. MonobenzosubpcB-Cl, XAQGEB[264]

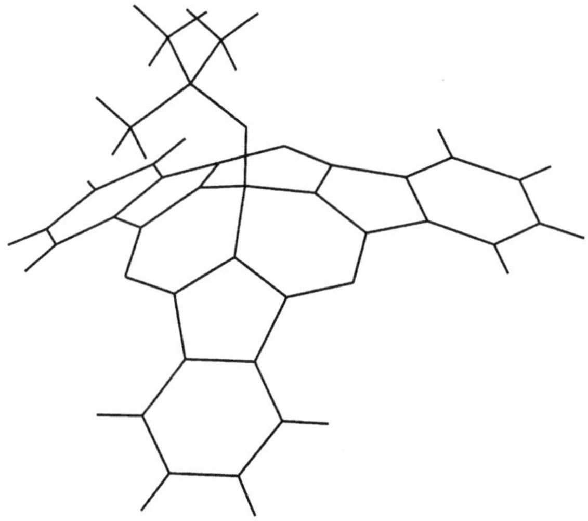

Appendix 20-7. SubpcB-OtBu, KAJPEQ[141]

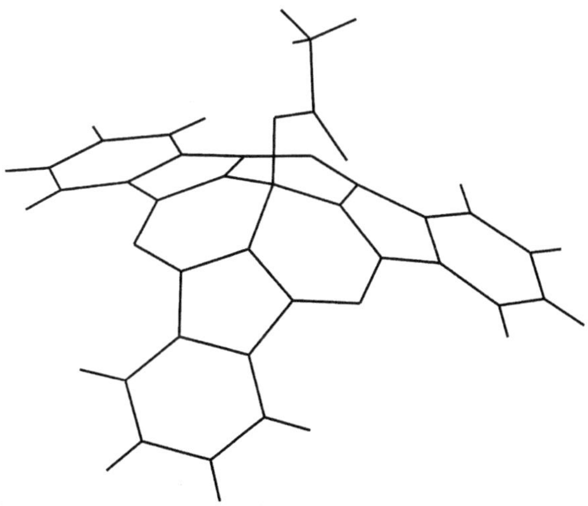

Appendix 20-8. [SubpcB-OAc][H_2O,py], KAJQER[141]

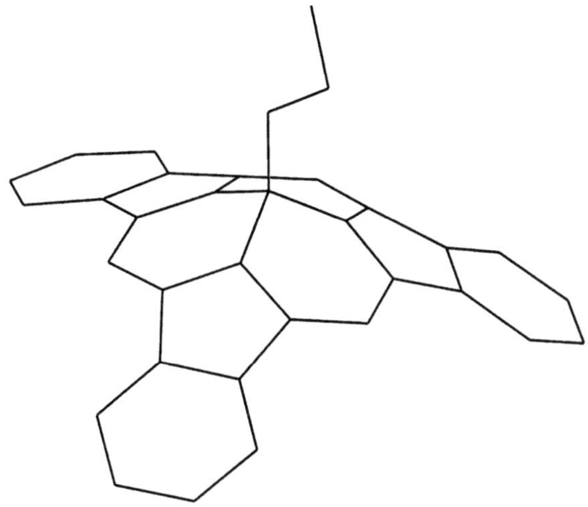

Appendix 20-9. SubpcB-OEt, RUGJOS[140]

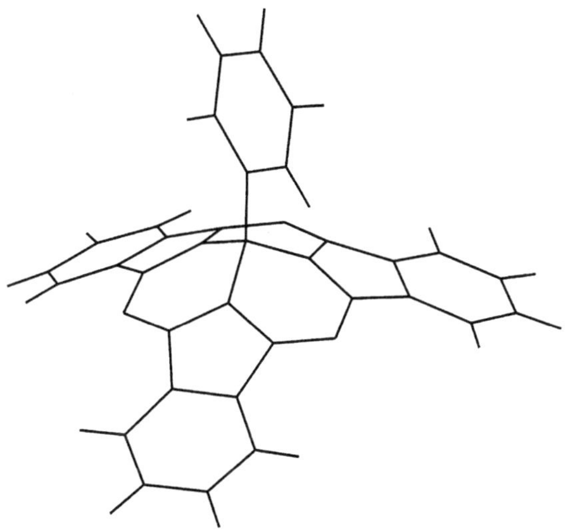

Appendix 20-10. SubpcB-ϕ, YOKMUG01[141]

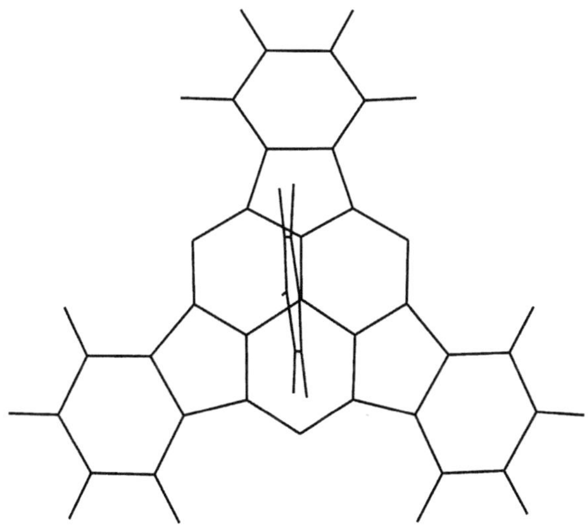

Appendix 20-11. SubpcB-ϕ, #1998-07[137]

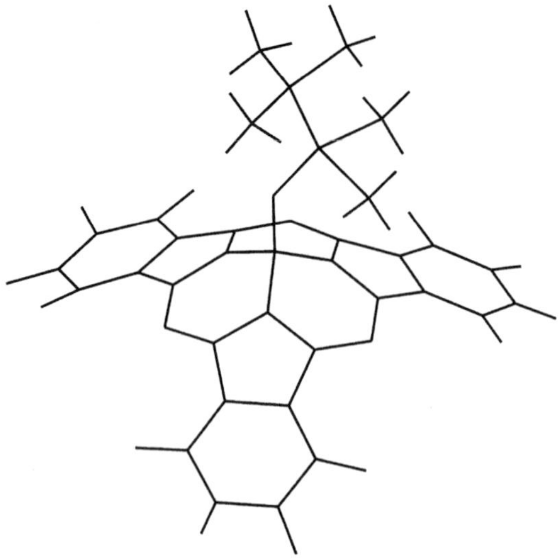

Appendix 20-12. SubpcB-OSi(tBu)Me_2, #1998-06[137]

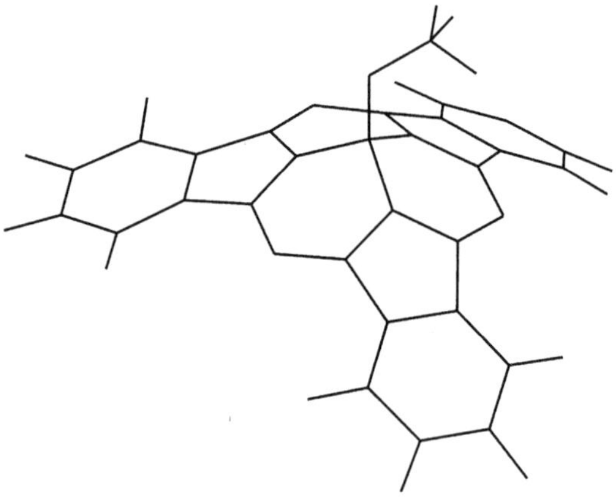

Appendix 20-13. SubpcB-OMe, #1998-04[137]

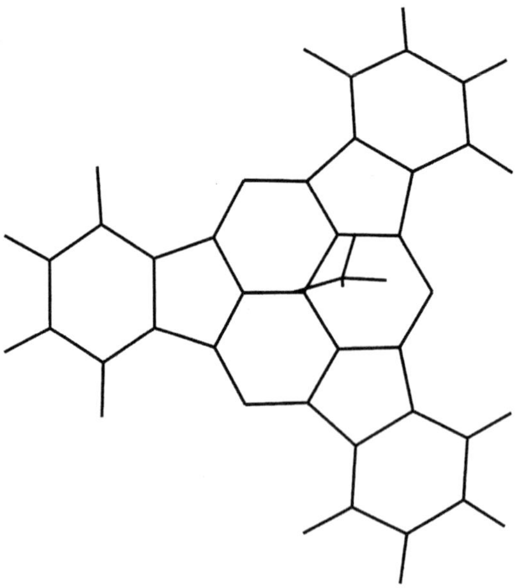

Appendix 20-14. SubpcB-OMe, #1998-04[137]

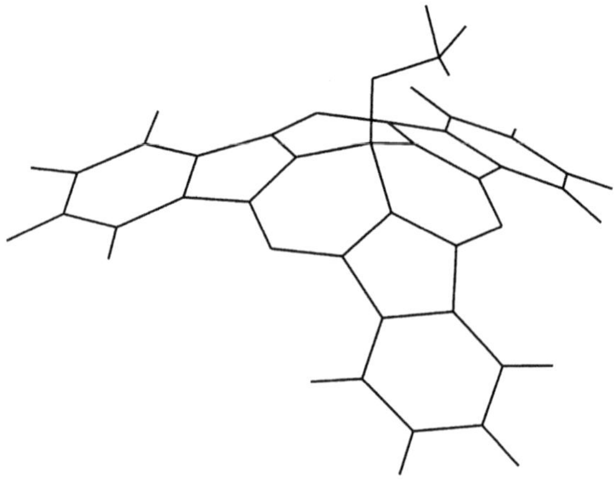

Appendix 20-15. SubpcB-OMe, #1998-05[137]

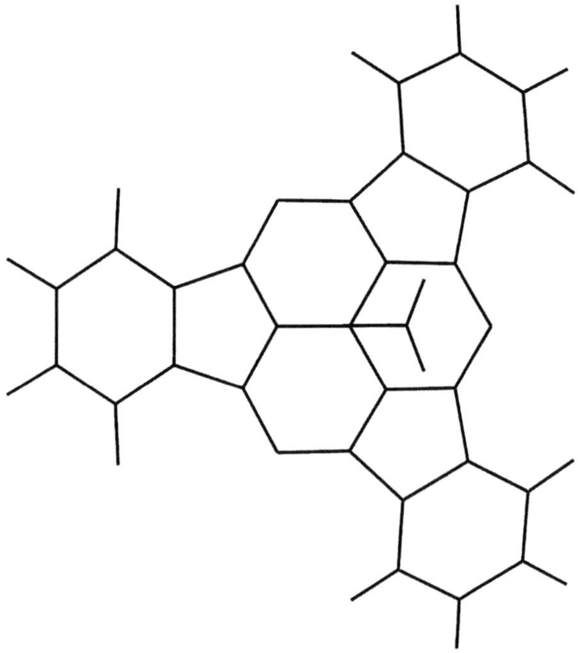

Appendix 20-16. SubpcB-OMe, #1998-05[137]

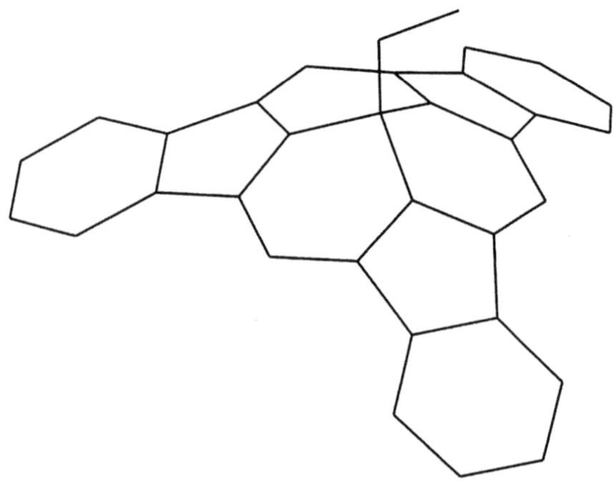

Appendix 20-17. SubpcB-OMe, #1992-01[262,263] (App. 41)

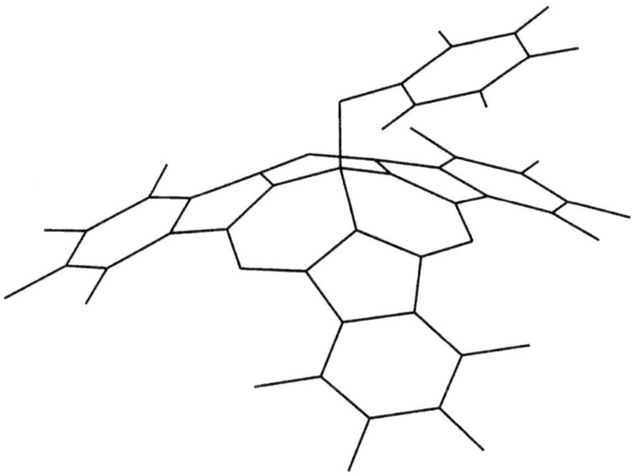

Appendix 20-18. F_{12}SubpcB-Oϕ, #2002-04[265]

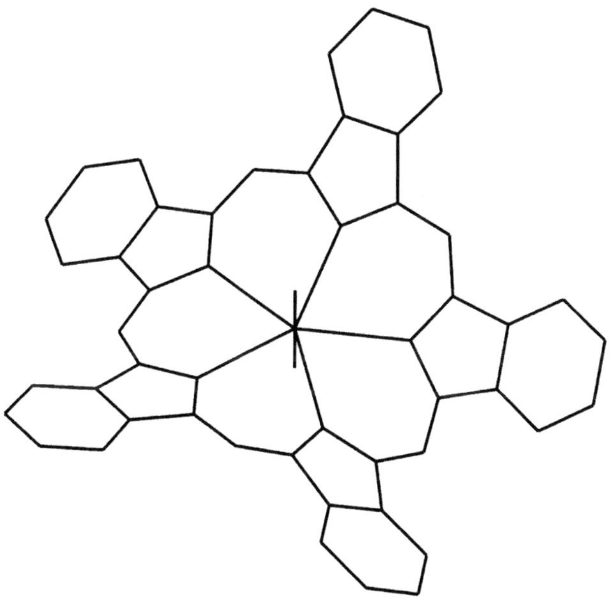

Appendix 20-19. SuperpcUO_2, CIMINU10[260]

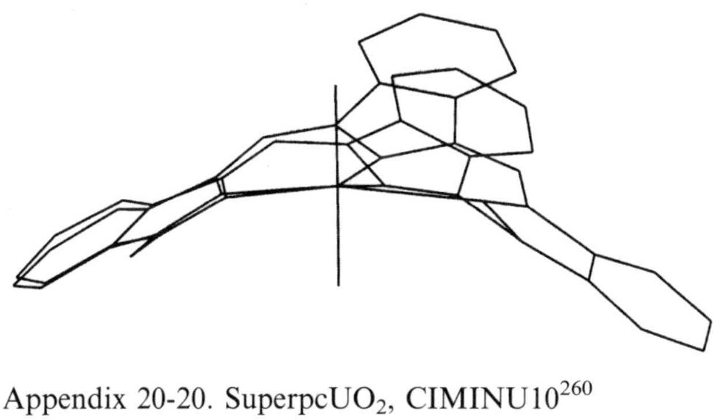

Appendix 20-20. SuperpcUO_2, CIMINU10[260]

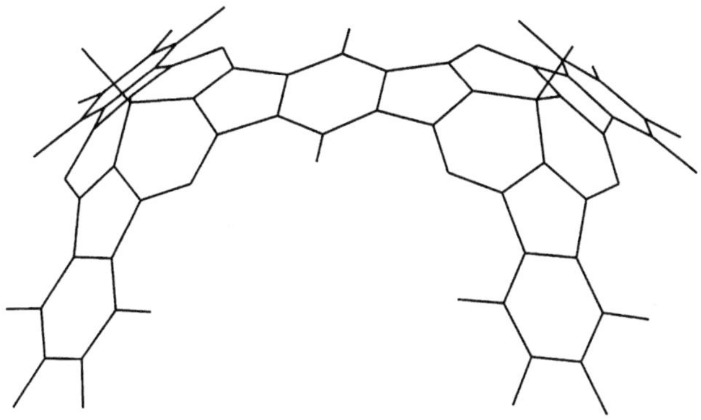

Appendix 20-21. [Cis-binuclear-$F_{16}H_2$(SubpcBCl)$_2$], #2002-11[261]

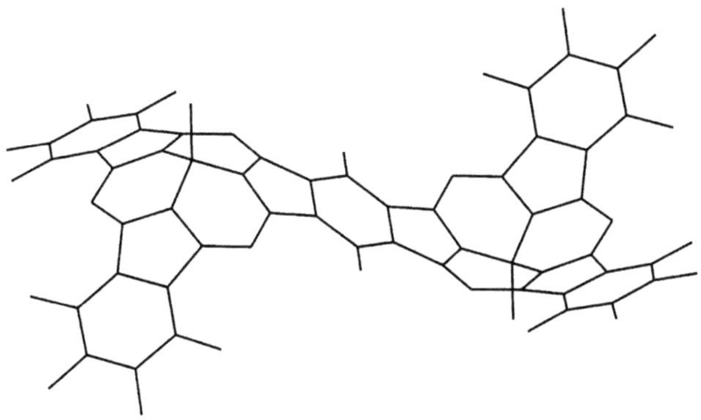

Appendix 20-22. [Trans-binuclear-$F_{16}H_2$(SubpcBCl)$_2$], #2002-12[261]

Appendix 21. Molecular Structures of Phthalocyanine-Related Isoindole Complexes

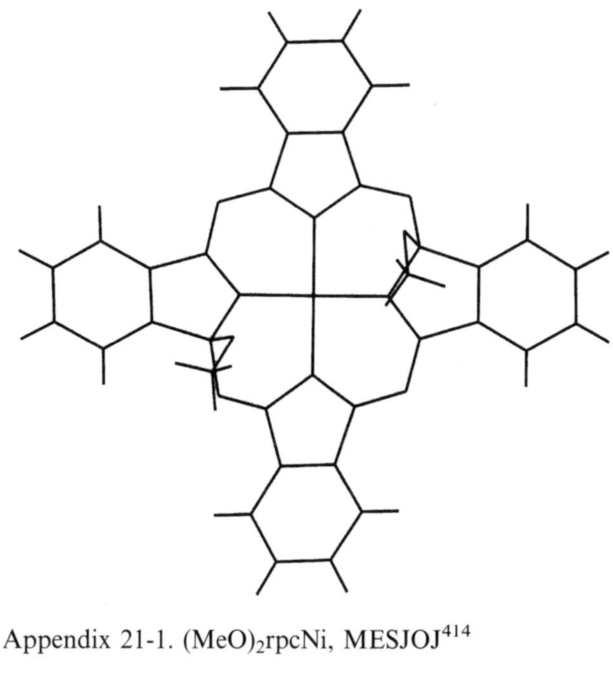

Appendix 21-1. (MeO)$_2$rpcNi, MESJOJ[414]

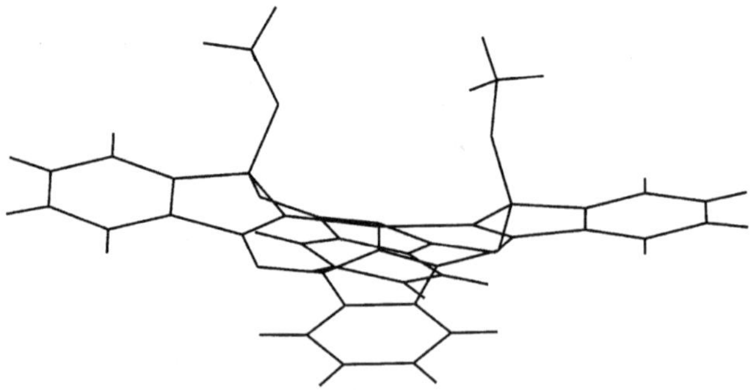

Appendix 21-2. $(MeO)_2$rpcNi, MESJOJ[414]

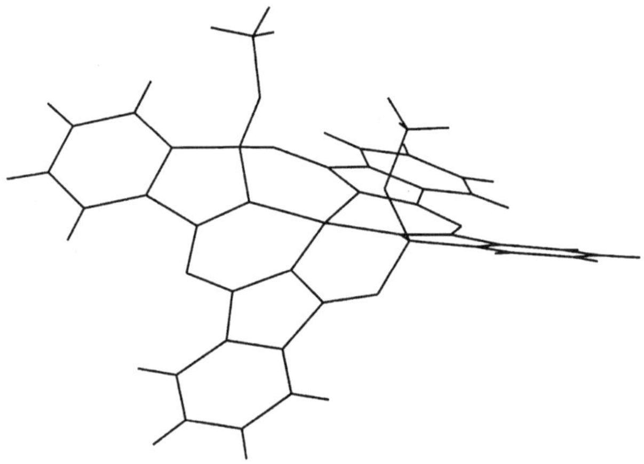

Appendix 21-3. $(MeO)_2$rpcNi, MESJOJ[414]

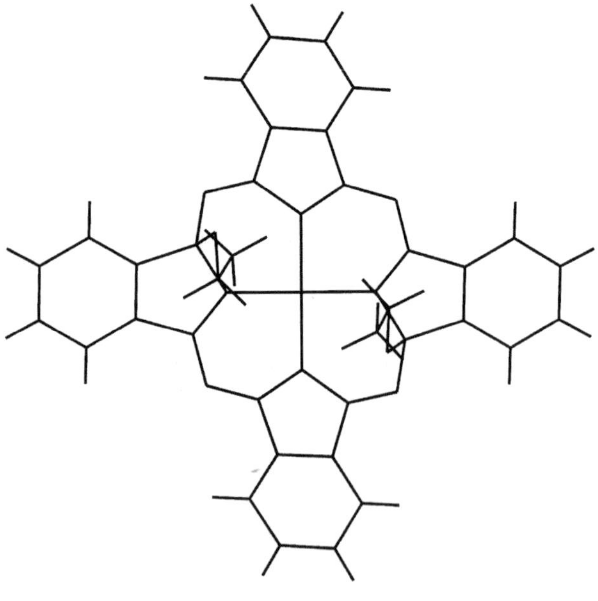

Appendix 21-4. $(EtO)_2$rpcNi, MESJUP[414]

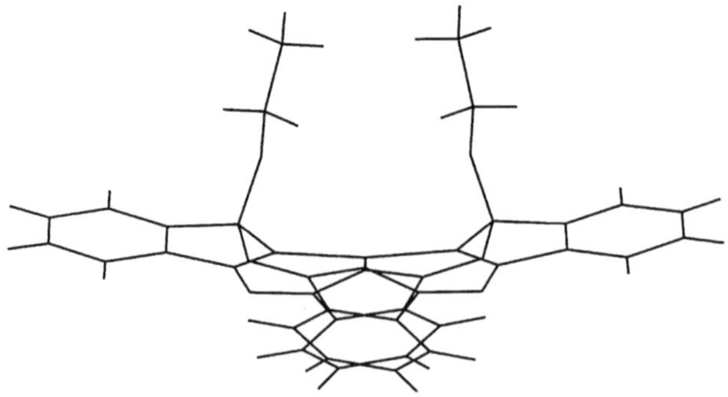

Appendix 21-5. $(EtO)_2$rpcNi, MESJUP[414]

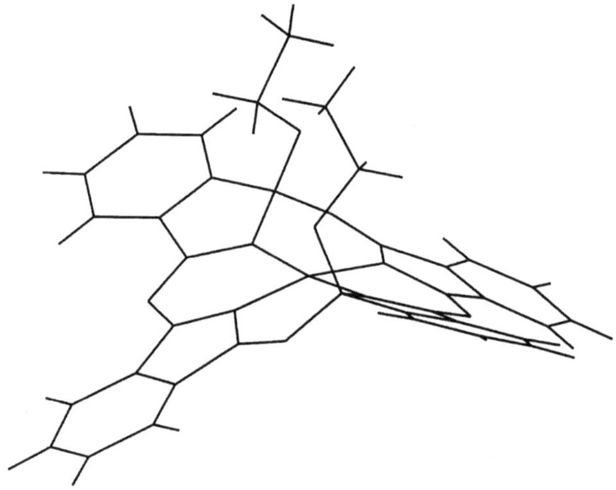

Appendix 21-6. $(EtO)_2$rpcNi, MESJUP[414]

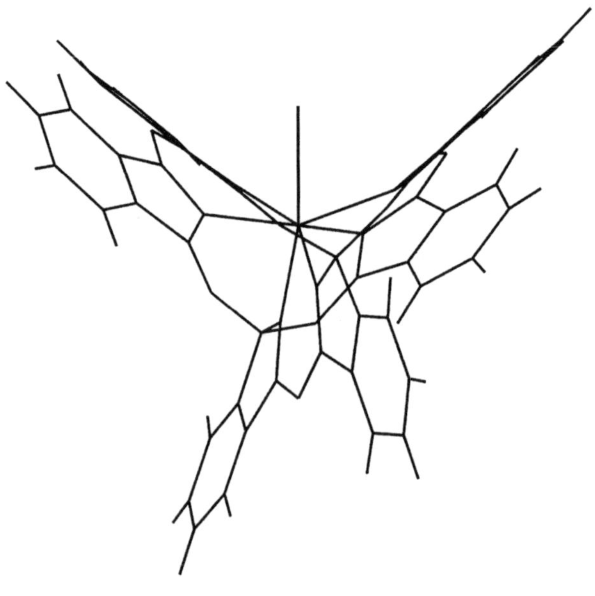

Appendix 21-7. PccNbCl, KEYPAF[254]

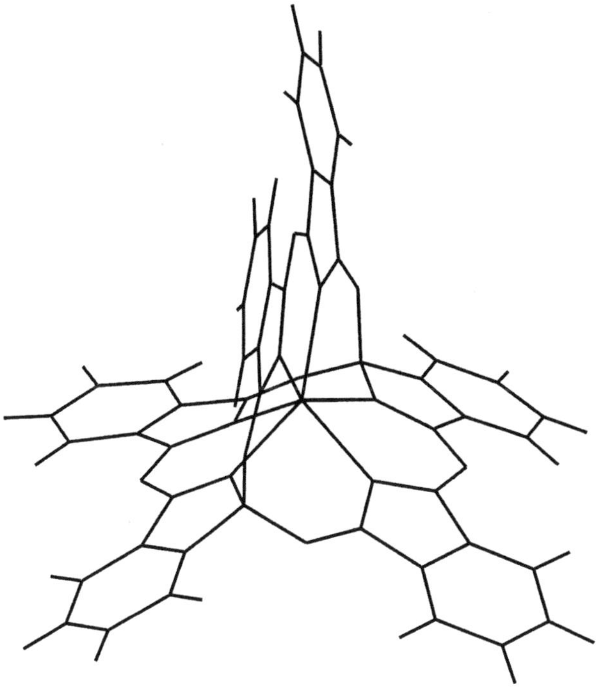

Appendix 21-8. PccGd, ZENGUU[256]

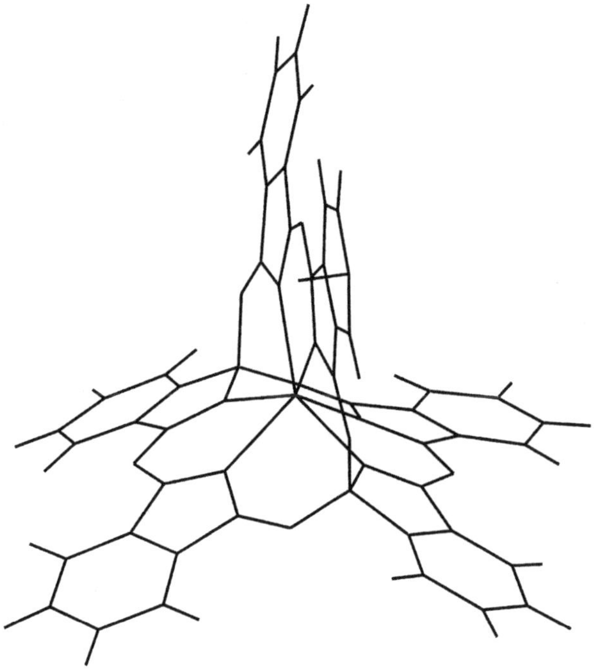

Appendix 21-10. PccIn, LIFJOZ[255]

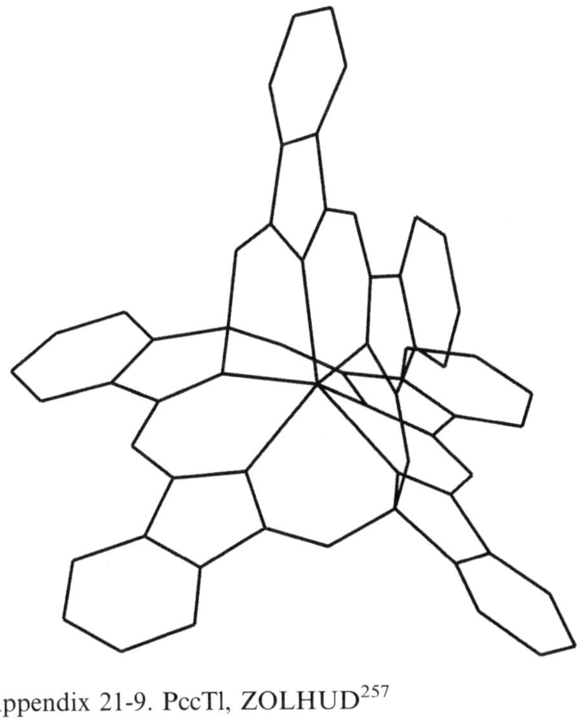

Appendix 21-9. PccTl, ZOLHUD[257]

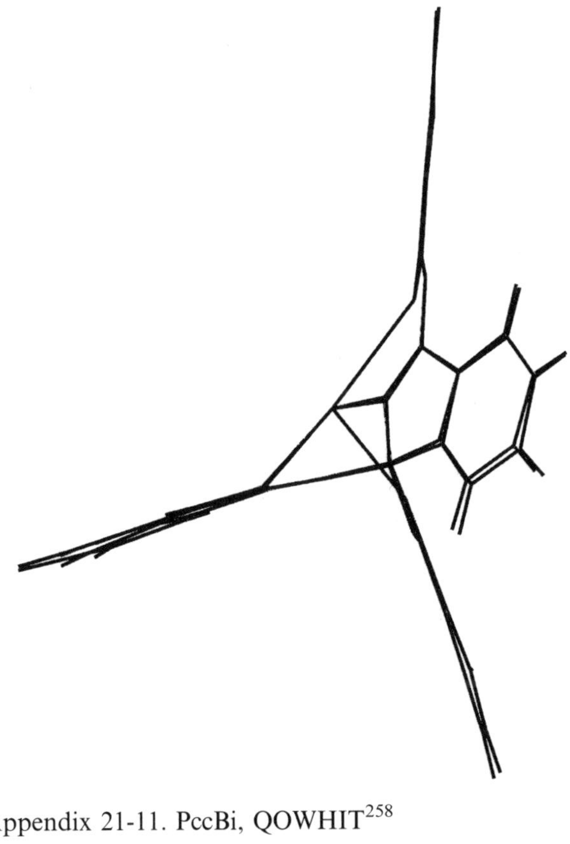

Appendix 21-11. PccBi, QOWHIT[258]

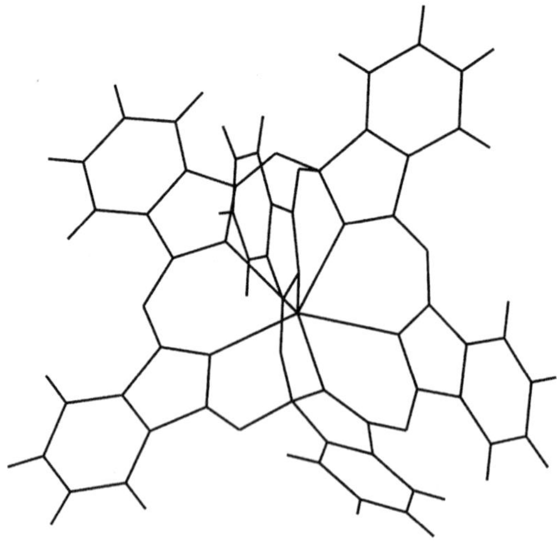

Appendix 21-12. PccBi, QOWHIT[258]

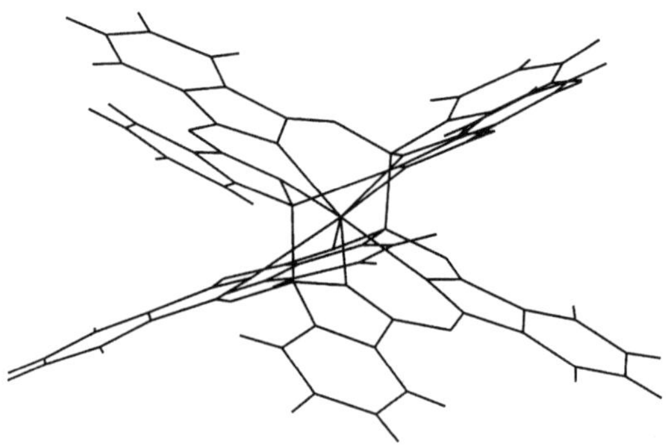

Appendix 21-13. $[Pc_2Nb][I_3][0.5I_2,3.5(1\text{-}ClN)]$, NOCYEJ[231]

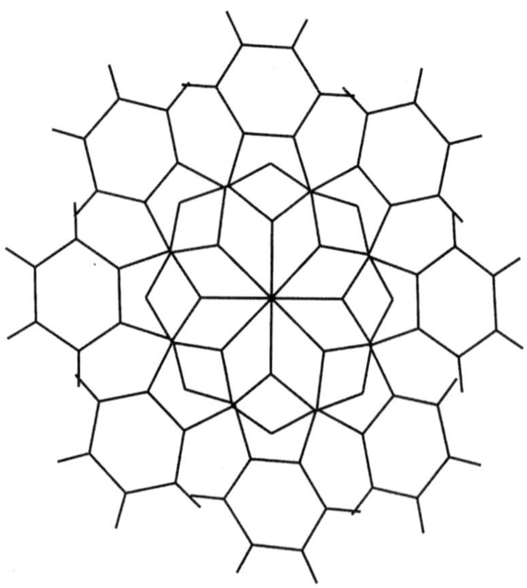

Appendix 21-14. $[Pc_2Nb][I_3][0.5I_2,3.5(1\text{-}ClN)]$, NOCYEJ[231]

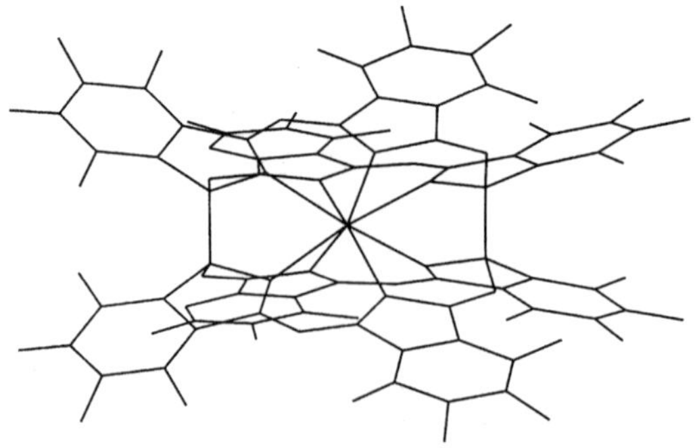

Appendix 21-15. $[Pc_2Ti][1\text{-}ClN]$, SEWRER[214]

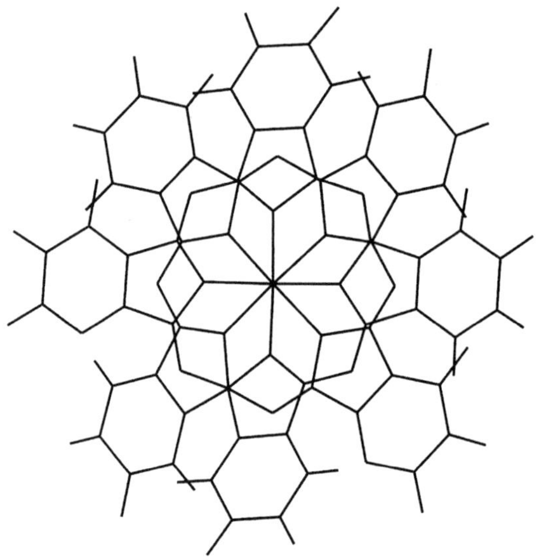

Appendix 21-16. $[Pc_2Ti][1\text{-}ClN]$, SEWRER[214]

Appendix 22. Molecular Structures of Porphyrazine Complexes

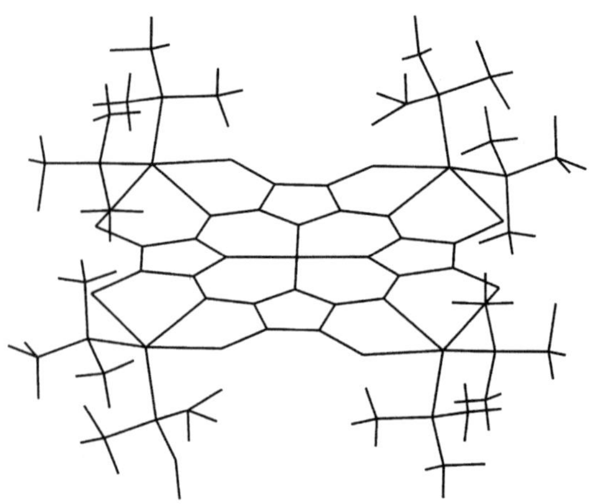

Appendix 22-1. $[(tBu_2SnS_2)_4PazNi][C_7H_8][1.25CH_2Cl_2]$, JEWRAE[415]

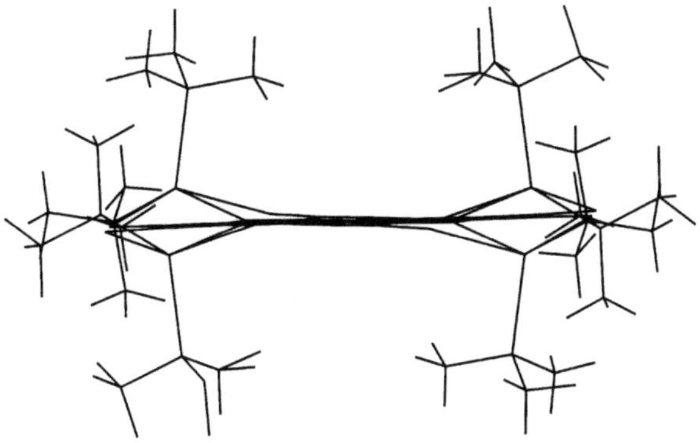

Appendix 22-2. [(tBu$_2$SnS$_2$)$_4$PazNi][C$_7$H$_8$][1.25CH$_2$Cl$_2$], JEWRAE[415]

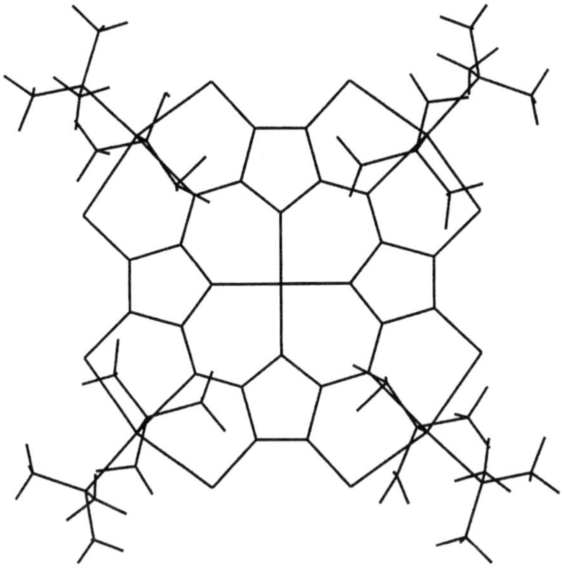

Appendix 22-3. [(tBu$_2$SnS$_2$)$_4$PazNi][C$_7$H$_8$][1.25CH$_2$Cl$_2$], JEWRAE10[416]

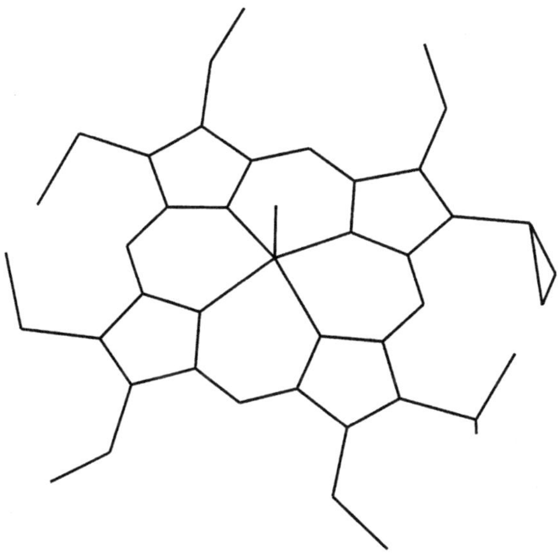

Appendix 22-4. [(MeS)$_8$PazMg(OH$_2$)], VUMZOS[416]

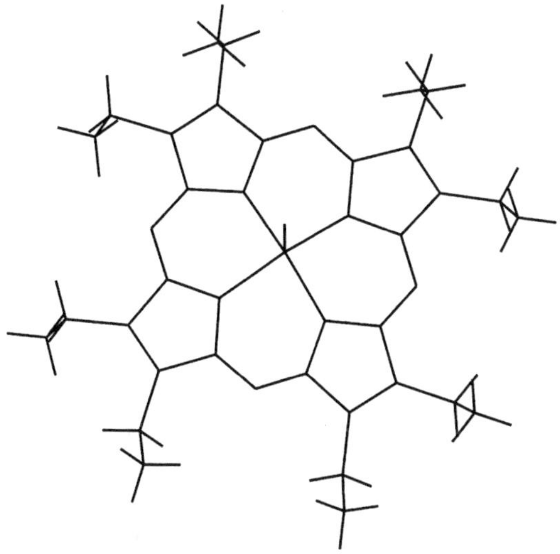

Appendix 22-5. [PazFeCl], PAGYAX[417]

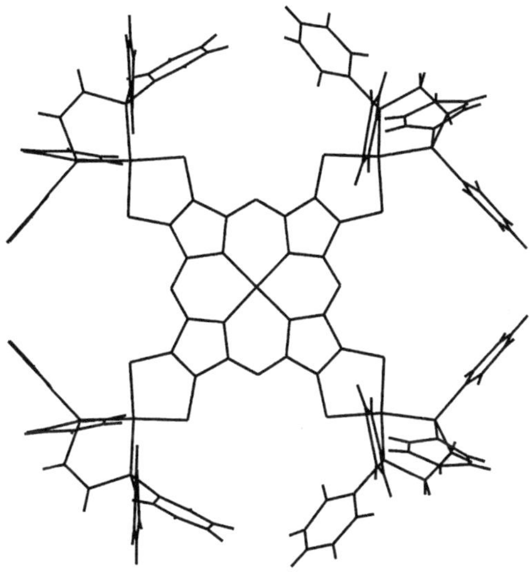

Appendix 22-6. [(dppeNiS$_2$)$_4$PazNi][4DMF], HATDEL[418]

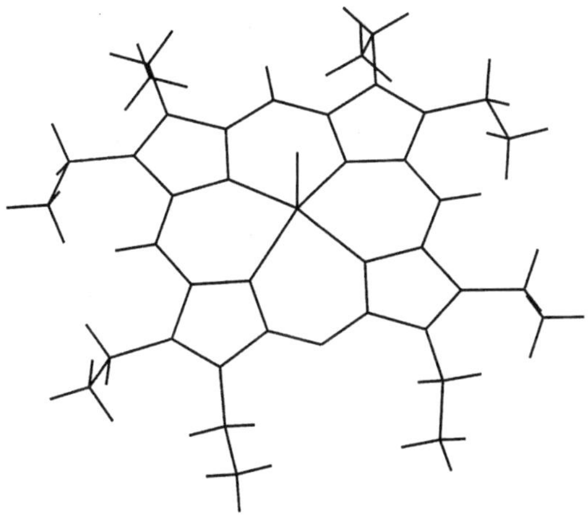

Appendix 22-7. [(Et)$_8$N$_1$PorFeCl], JUXNOF[419]

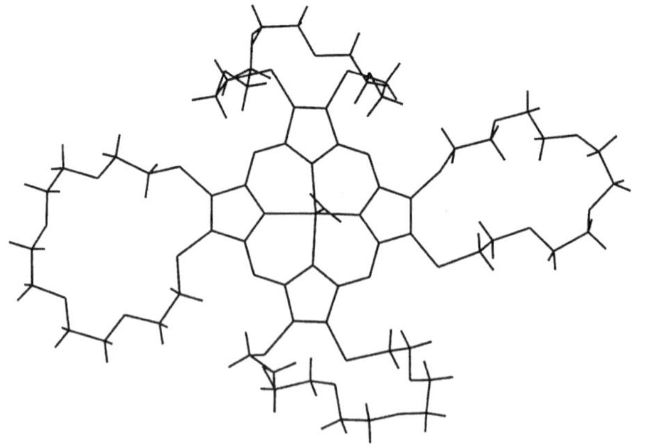

Appendix 22-8. [(cr18S2)$_4$PazMg(OH$_2$)], WETGAD,[420] WETGAD10[421]

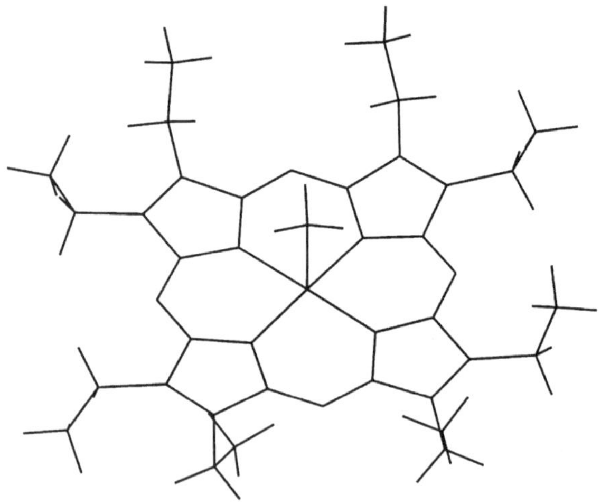

Appendix 22-9. Et$_8$PazRhMe, PITROZ[422]

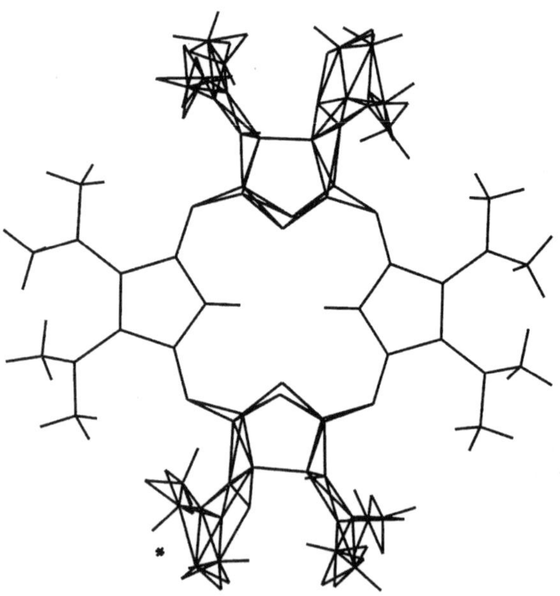

Appendix 22-10. [(Me$_2$N)$_8$-2secoPazH$_2$-2,3-dione], POGKEB,[423] POGKEB01[424]

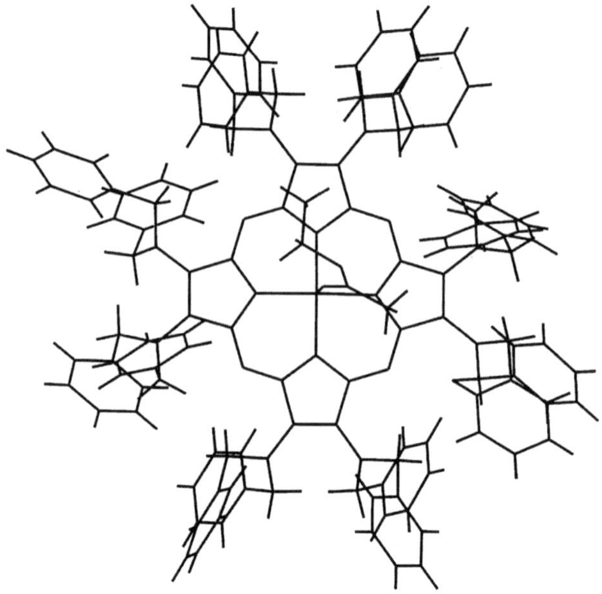

Appendix 22-11. [(Bn$_2$N)$_8$PazMg(O-C(OEt)CH$_3$), POFNON[425]

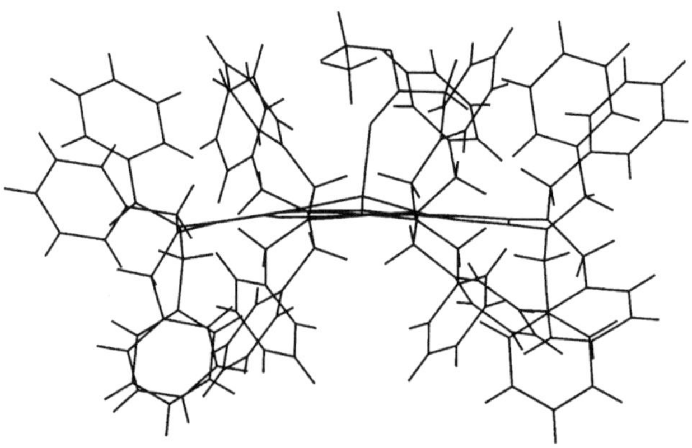

Appendix 22-12. [(Bn$_2$N)$_8$PazMg(O-C(OEt)CH$_3$), POFNON[425]

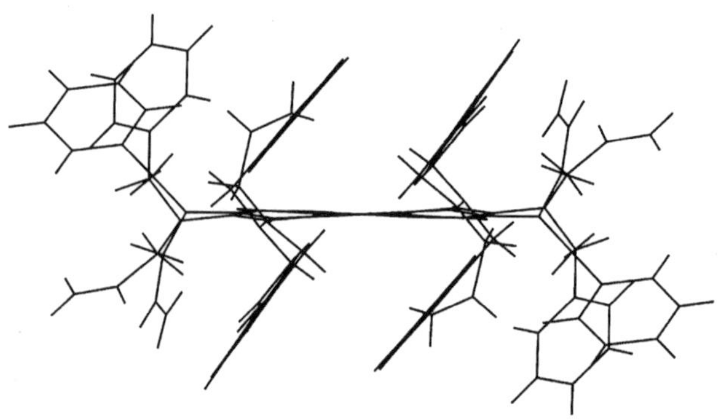

Appendix 22-13. [{(allyl)(Bn)N}$_8$PazNi][ϕCl], POFNUT[425]

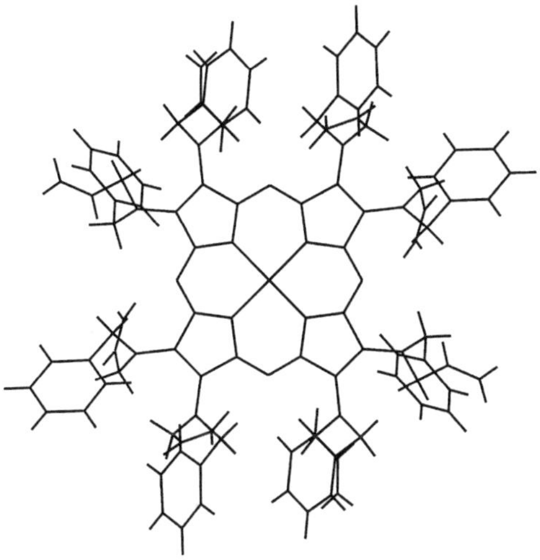

Appendix 22-14. [{(allyl)(Bn)N}$_8$PazNi][ϕCl], POFNUT[425]

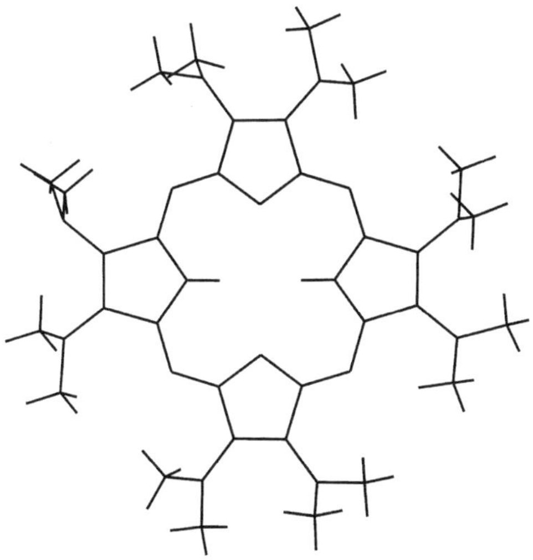

Appendix 22-15. [(Me$_2$N)$_8$PazH$_2$]$_2$[C$_{60}$], ZAPNEJ[426]

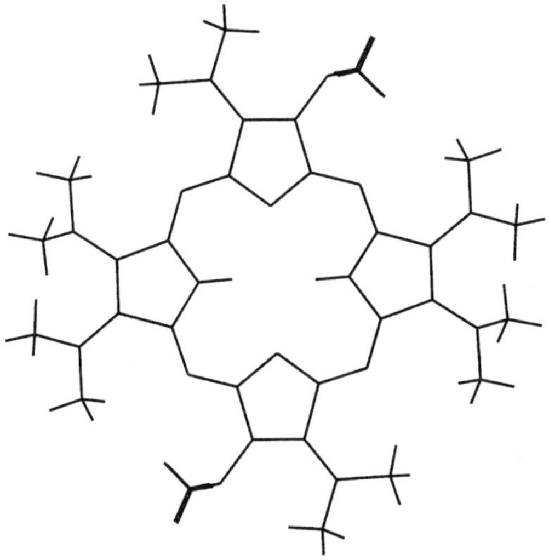

Appendix 22-16. [(Me$_2$N)$_8$PazH$_2$][TCNQ], ZAPNAF[426]

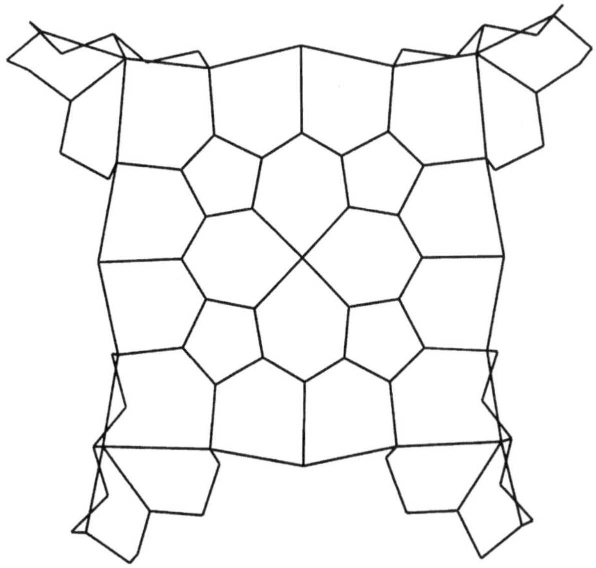

Appendix 22-17. [(Ag-contng-crownS$_2$)$_4$Ag$_4$PazNi][8BF$_4$], SURDIS[427]

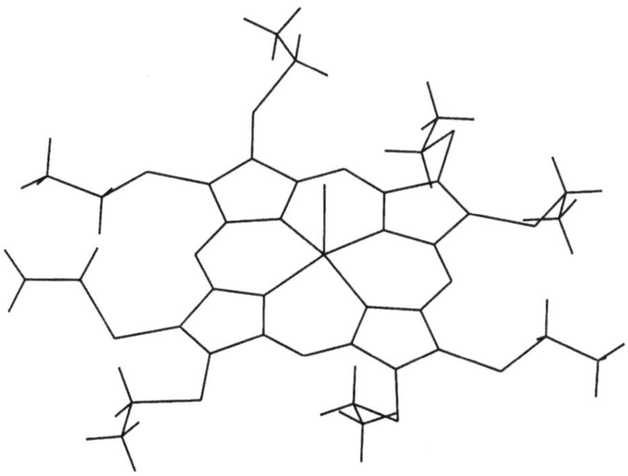

Appendix 22-18. [(EtS)$_8$PazMnCl], TISDOO[428]

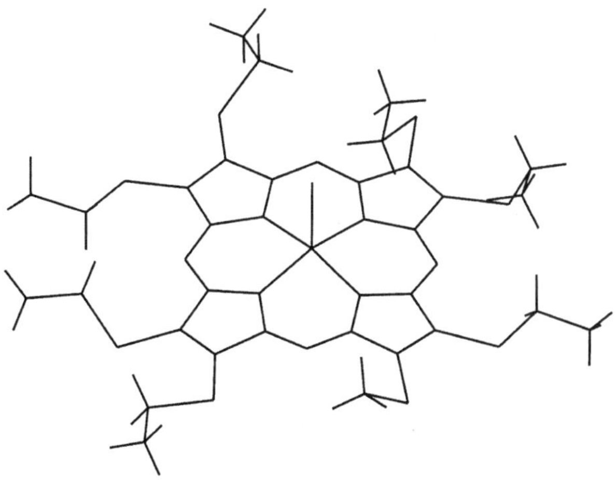

Appendix 22-19. [(EtS)$_8$PazFeCl], TISDUU[428]

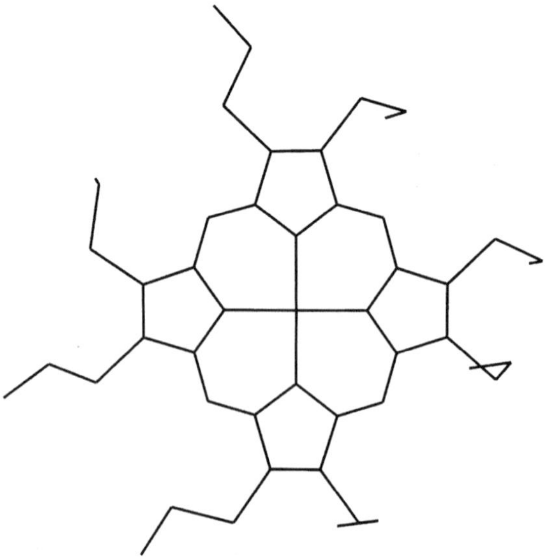

Appendix 22-20. [$(EtS)_8$PazMn], TOGPEK[429]

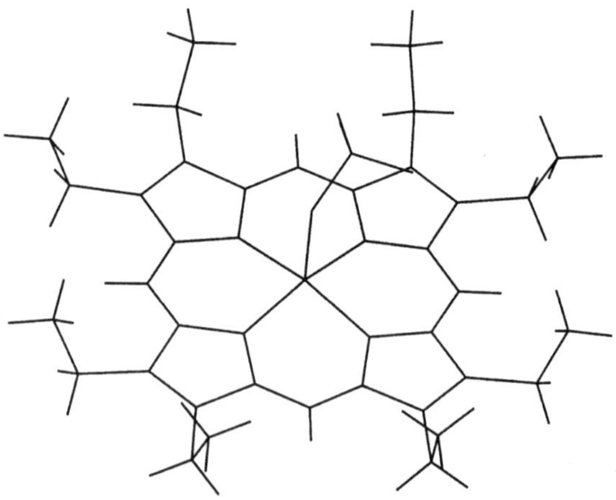

Appendix 22-21. [$(Et)_8$PazFe($OClO_3$)], ZUQDOE[431]

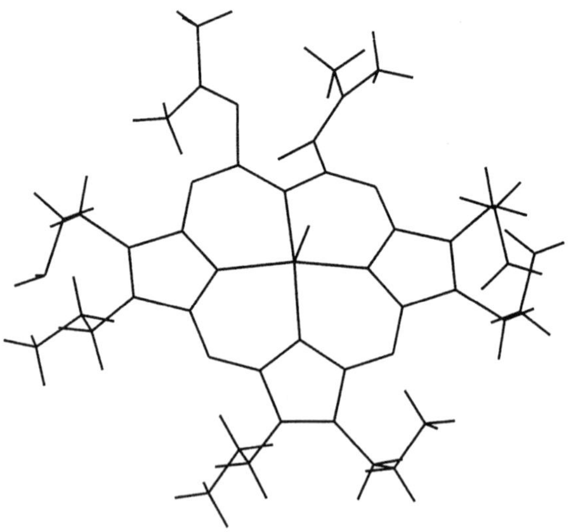

Appendix 22-22. [$(Me_2N)_8$-2secoPazZn-2,3-dione], POWLUI[424]

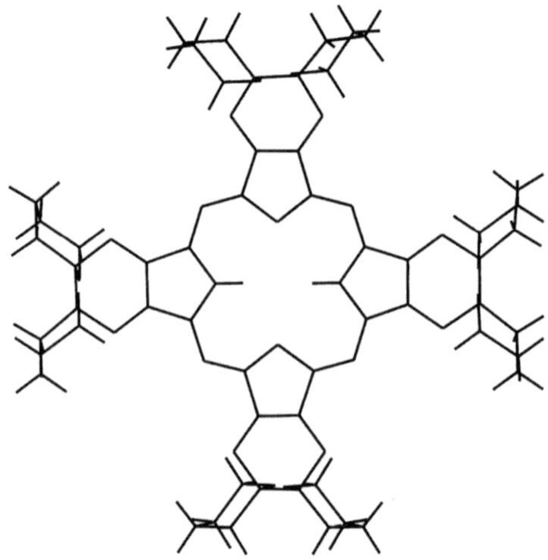

Appendix 22-23. [$(RO_2)_4PazH_2$], ROPLUD[432]

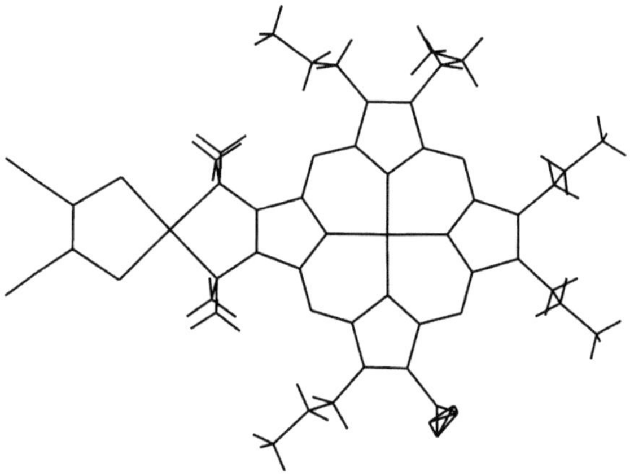

Appendix 22-24. [$(Pr)_6$((mnt)PdNMe$)_2$PazNi], HOGGAL[434]

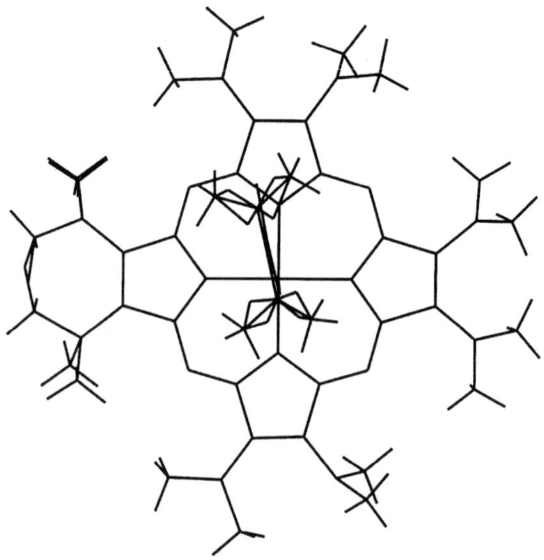

Appendix 22-25. [$(Me_2N)_8$PazMn(Et_2dtc)], NUGSUD[435]

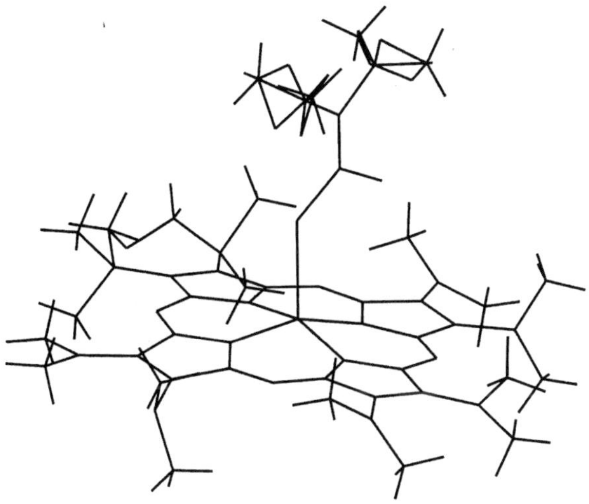

Appendix 22-26. $[(Me_2N)_8PazMn(Et_2dtc)]$, NUGSUD[435]

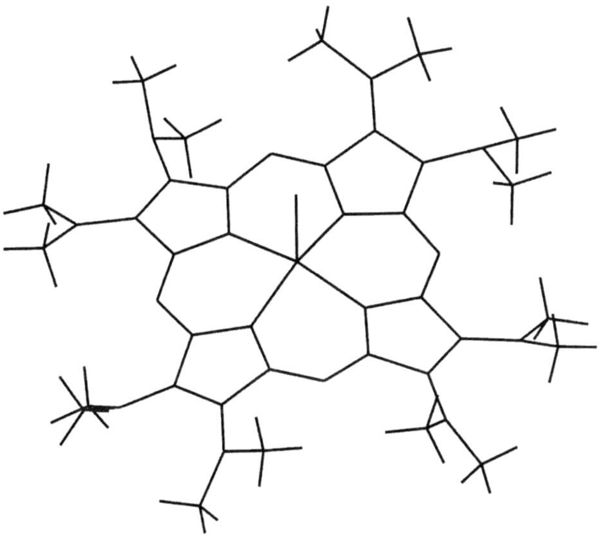

Appendix 22-27. $[(Me_2N)_8PazMnCl]$, NUGTAK[435]

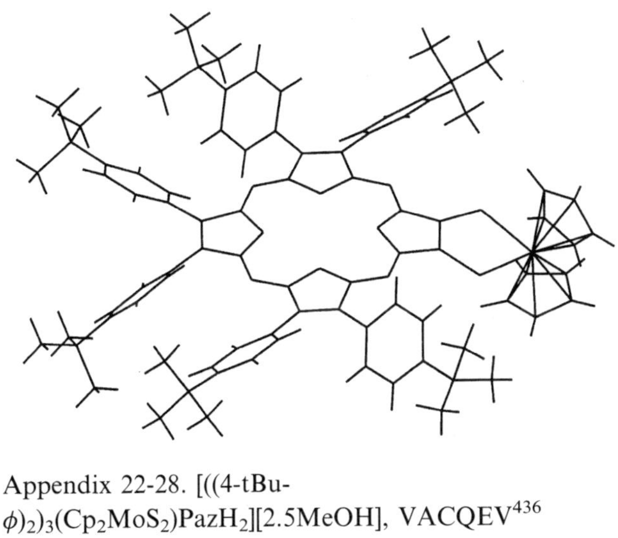

Appendix 22-28. $[((4\text{-tBu-}\phi)_2)_3(Cp_2MoS_2)PazH_2][2.5MeOH]$, VACQEV[436]

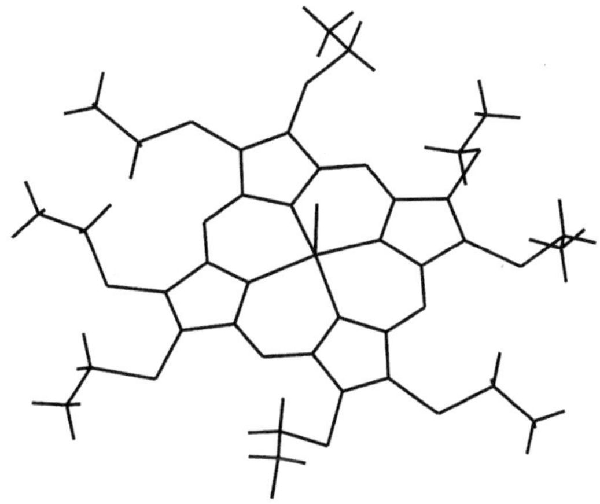

Appendix 22-29. $[(EtS)_8PazMn(SH)]$, PUNTEX[437]

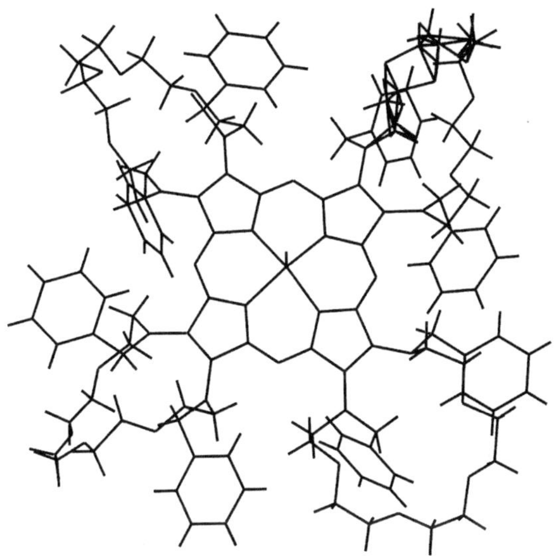

Appendix 22-30. $[(cr18N_2Bn_2)_4PazMg][0.75EtOAc]$, HEXJAV[438]

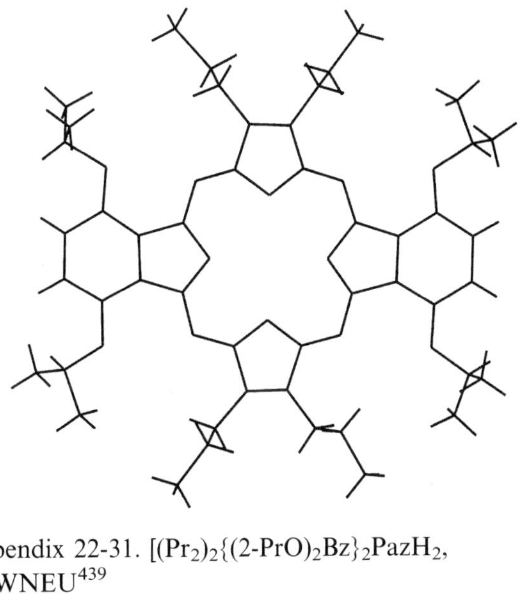

Appendix 22-31. $[(Pr_2)_2\{(2\text{-PrO})_2Bz\}_2PazH_2$, POWNEU[439]

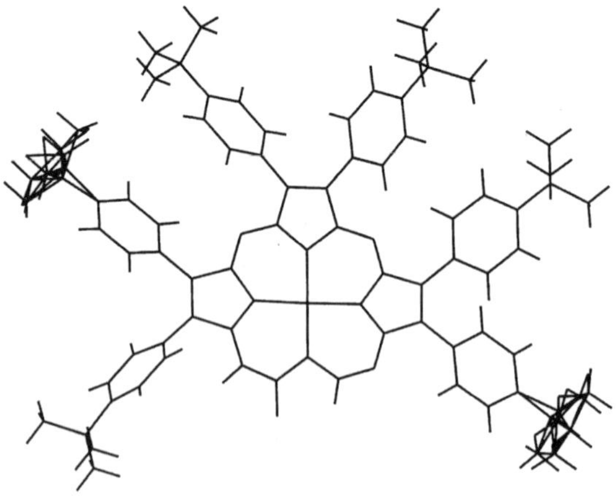

Appendix 22-32. (4-tBu-ϕ-)$_6$Seco′PazNi, JUTQIY[440]

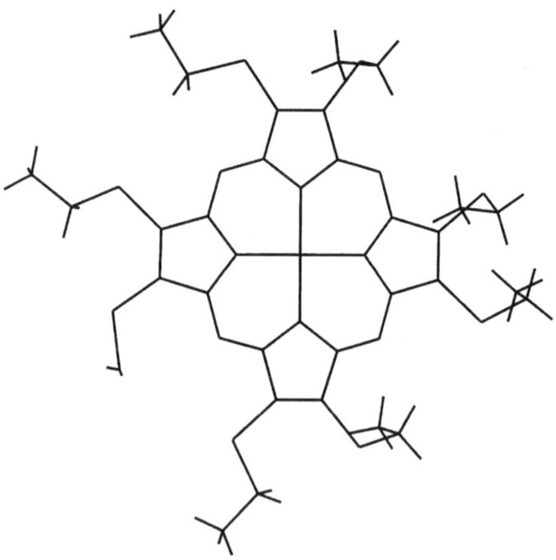

Appendix 22-33. [(EtS)$_8$PazCo], HOSYUJ[441]

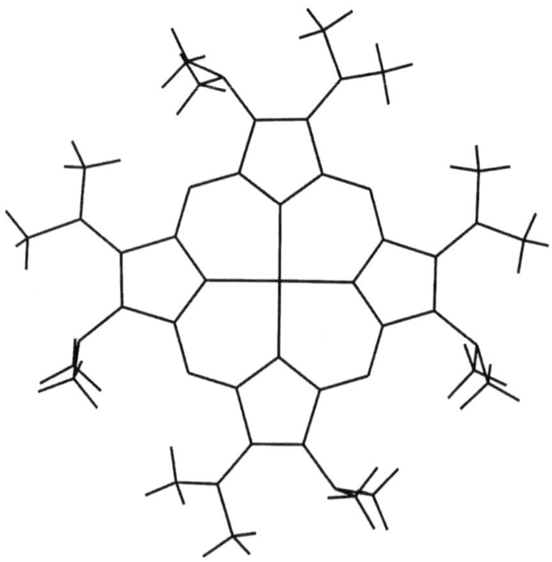

Appendix 22-34. [(Me$_2$N)$_8$PazCu]$_2$[C$_{60}$]173K, FOPYEO01[442]

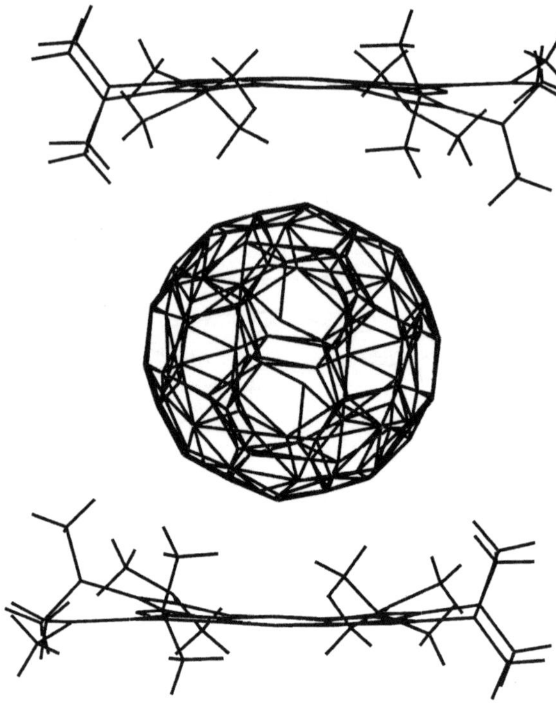

Appendix 22-35. [(Me$_2$N)$_8$PazCu]$_2$[C$_{60}$]293K, FOPYEO[442]

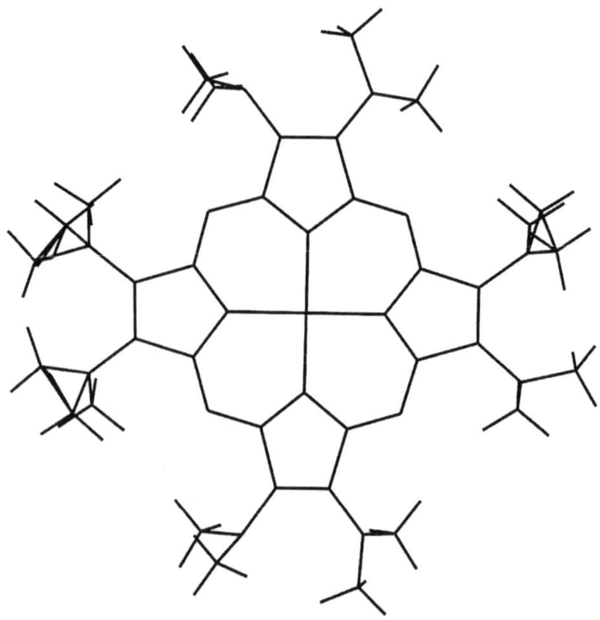

Appendix 22-36. [(Me$_2$N)$_8$PazNi][C$_{60}$]123K, FOPZEP01[442]

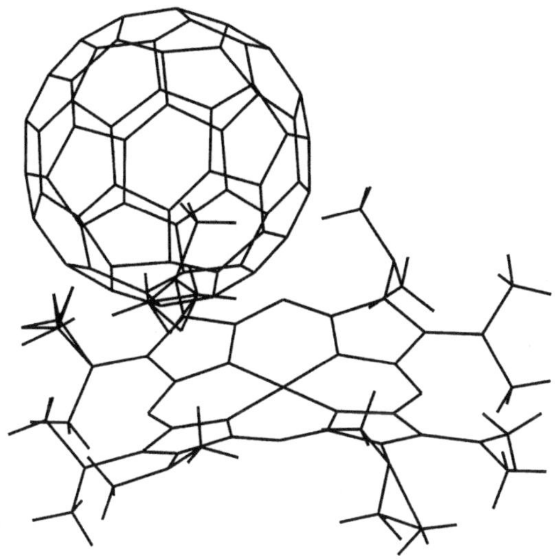

Appendix 22-37. $[(Me_2N)_8PazNi][C_{60}]293K$, FOPZEP[442]

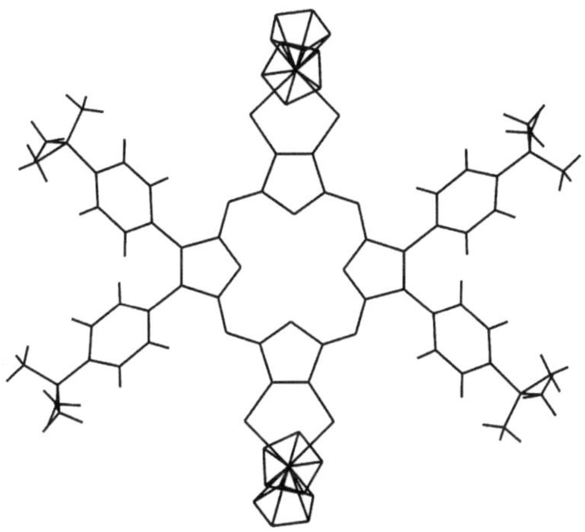

Appendix 22-38. $[((4\text{-tBu-}\phi)_2)_2(Cp_2VS_2)_2PazH_2]$, #2001-04[444]

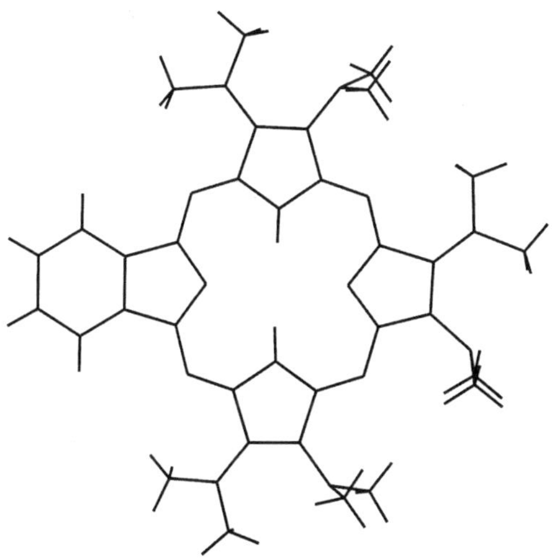

Appendix 22-39. $[((Me_2N)_2)_3(benzo)PazH_2]$, QOTPIY[443]

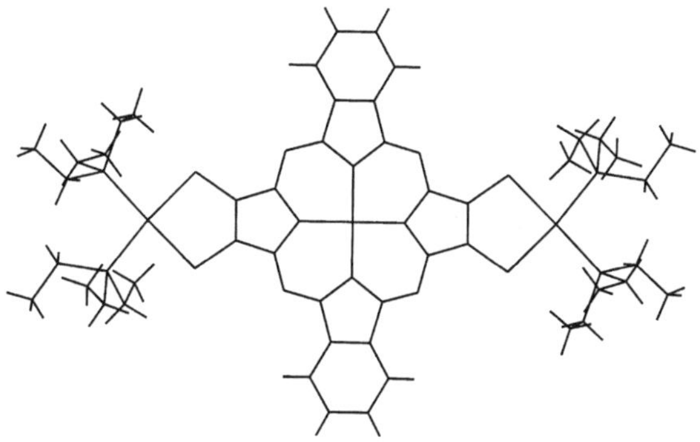

Appendix 22-40. $[tr((Et_3P)_2PtS_2)_2PazNi][C_6H_4Cl_2]_2$, TOZCOA[430]

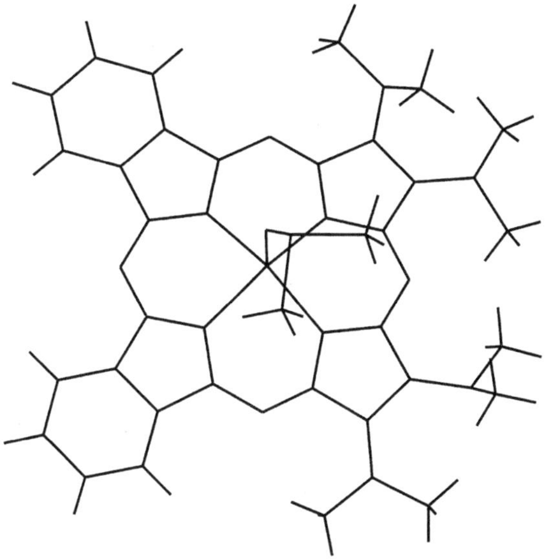

Appendix 22-41. [*cis*$((Me_2N)_2)_2(benzo)_2PazMg$], QOTPOE[443]

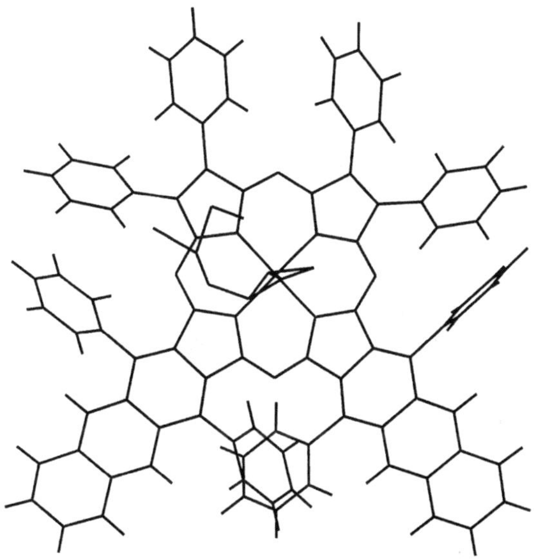

Appendix 22-42. $[Ac_2pazCo(OH_2)(OC_6H_{13})][acetone]$, ECOSUK[239]

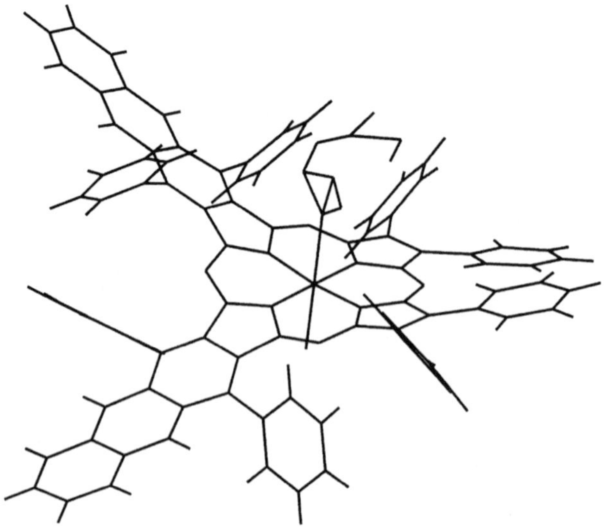

Appendix 22-43. $[Ac_2pazCo(OH_2)(OC_6H_{13})][acetone]$, ECOSUK[239]

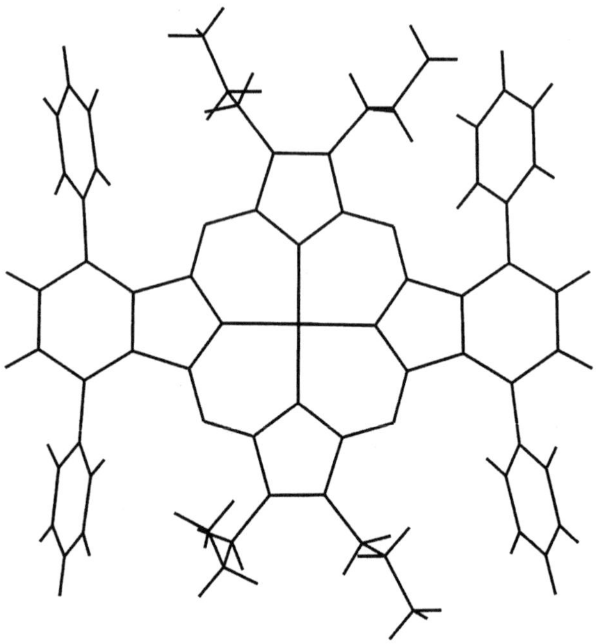

Appendix 22-44. $[(\phi_2benzo)_2Pr_2PazCo]$, #2002-05[445]

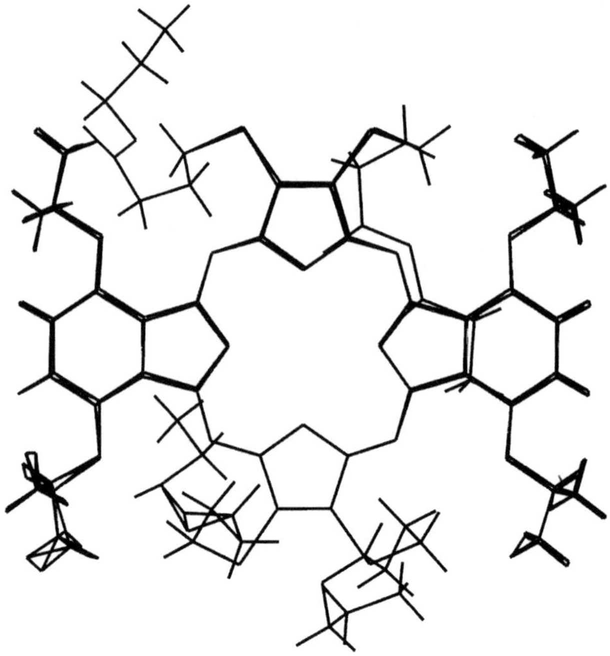

Appendix 22-45. $[((PrO(CO)(CH_2)_3)_2)_2((iPrO)_2benzo)_2PazH_2]$, UDIHEU[446]

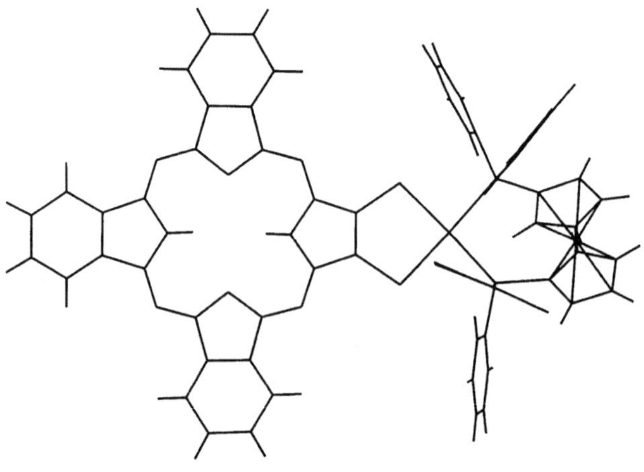

Appendix 22-46. $[(dppf)PdS_2NorpcH_2][CH_2Cl_2]$, WETJUA[43], WETJUA10[44]

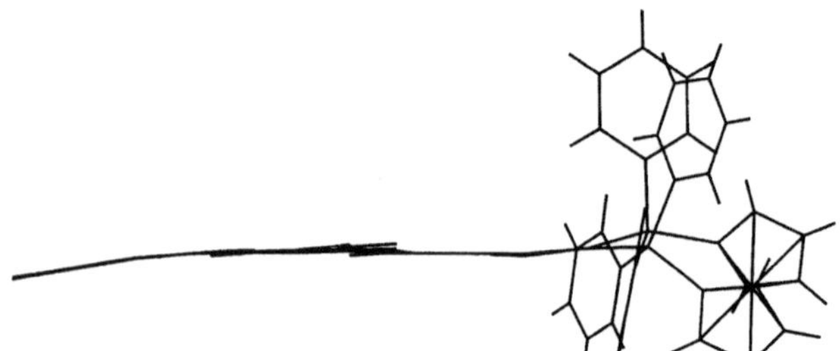

Appendix 22-47. $[(dppf)PdS_2NorpcH_2][CH_2Cl_2]$, WETJUA[43], WETJUA10[44]

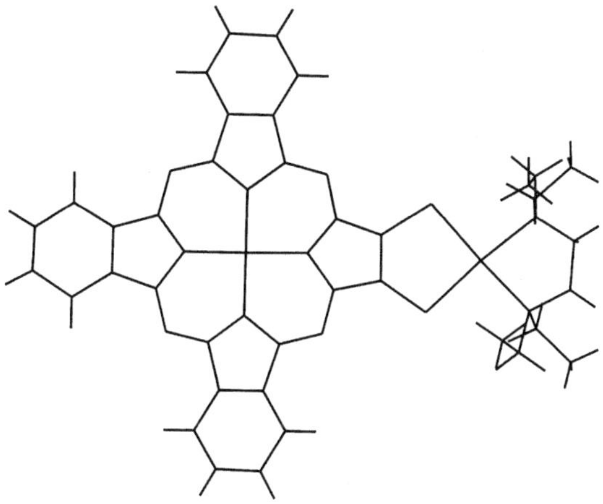

Appendix 22-48. [(teeda)PtS_2NorpcNi][0.5$CHCl_3$], TOZCIU[44]

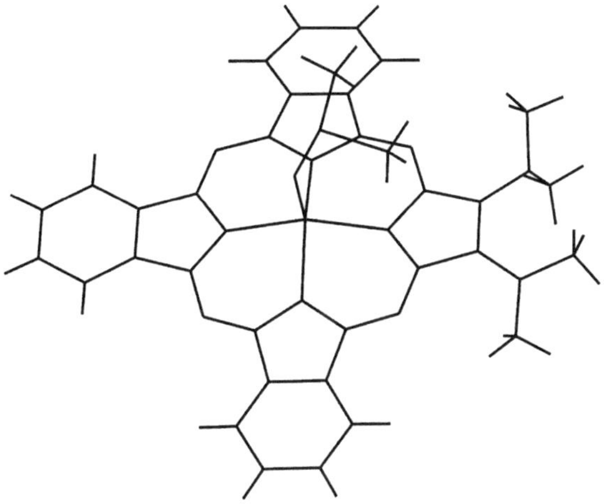

Appendix 22-49. [$(Me_2N)_2$$(benzo)_3$PazMg], QOTPUK[443]

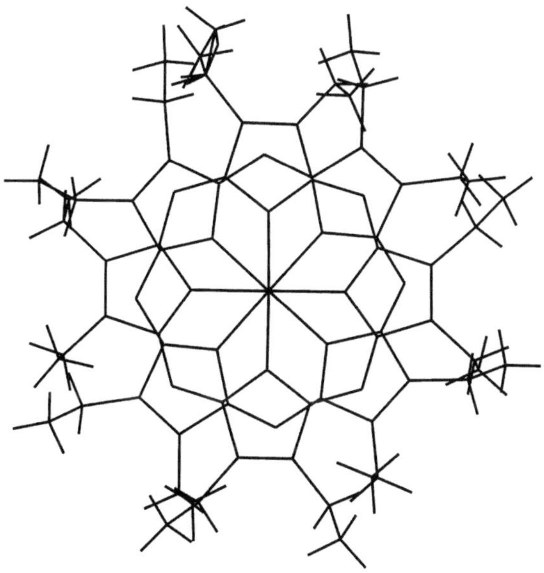

Appendix 22-50. [$((Et)_8Paz)_2$Zr], NULSUI[433]

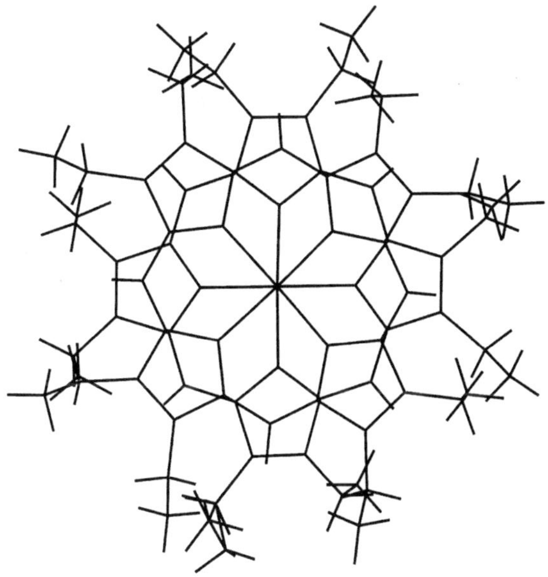

Appendix 22-51. [$((Et)_8$Paz)(OEP)Zr], NULSOC[433]

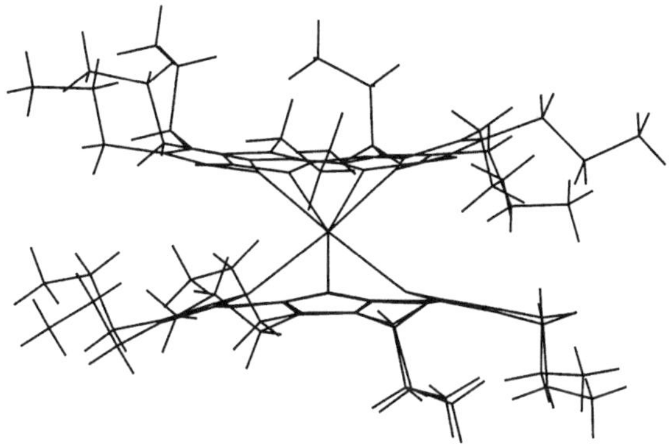

Appendix 22-52. [$((Pr)_8Paz)_2$Ce], ICISIW[447]

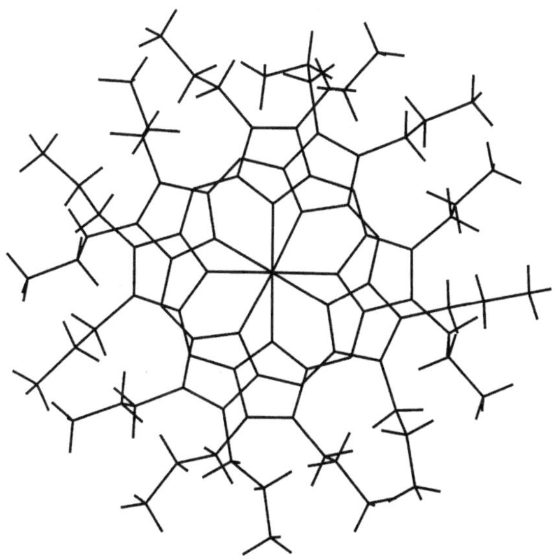

Appendix 22-53. [$((Pr)_8Paz)_2$Ce], ICISIW[447]

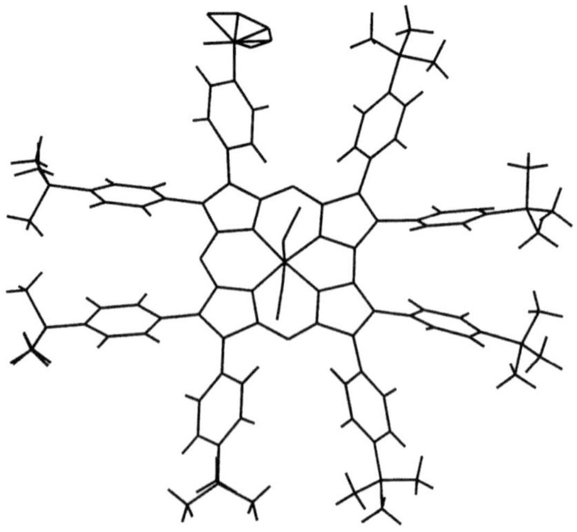

Appendix 22-54. [(4-tBu-ϕ)$_8$CzP(OMe)$_2$][$_2$CH$_2$Cl$_2$,0.75H$_2$O], BUDQOC[448]

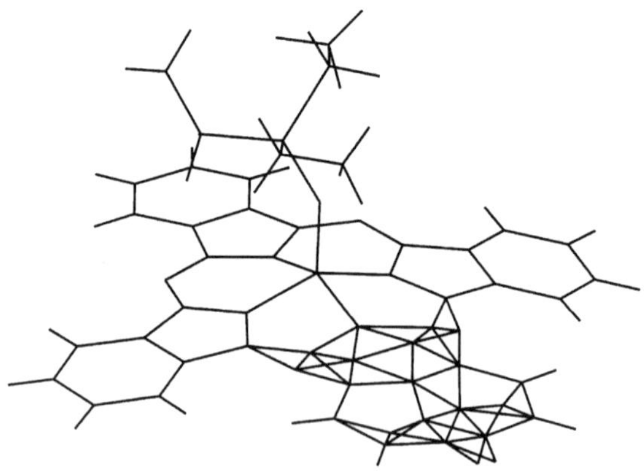

Appendix 22-55. [TBCSiOSi(Et)$_3$], #2002-02[450]

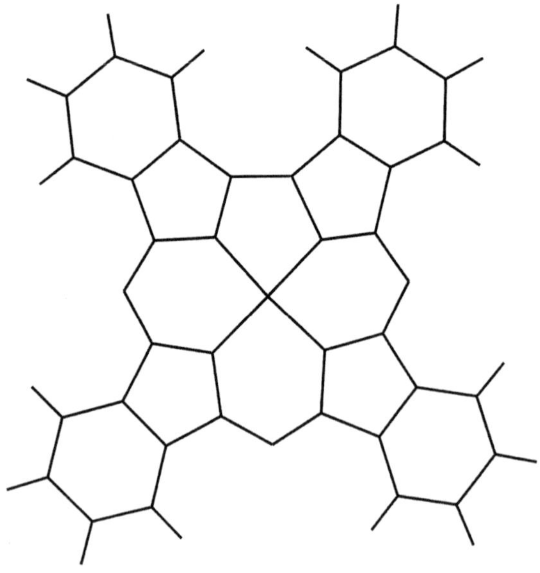

Appendix 22-56. [TBCSiOSi(Et)$_3$], #2002-02[450]

Appendix 23. Molecular Structures of Phthalocyanine–Porphyrin Complexes

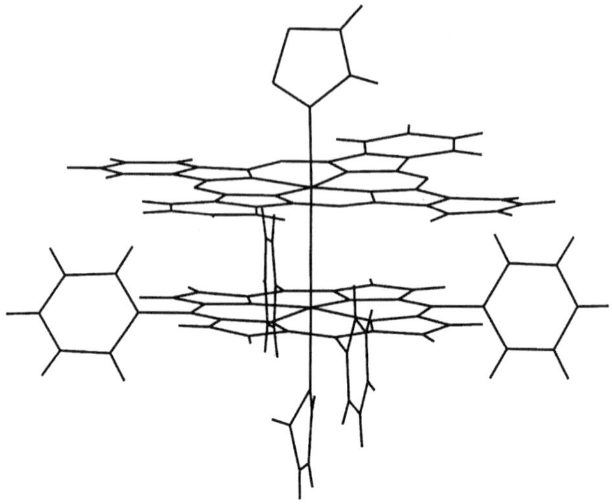

Appendix 23-1. THF-(Pc)Fe-C-Fe(tpp)-THF, PAZRAJ[353]

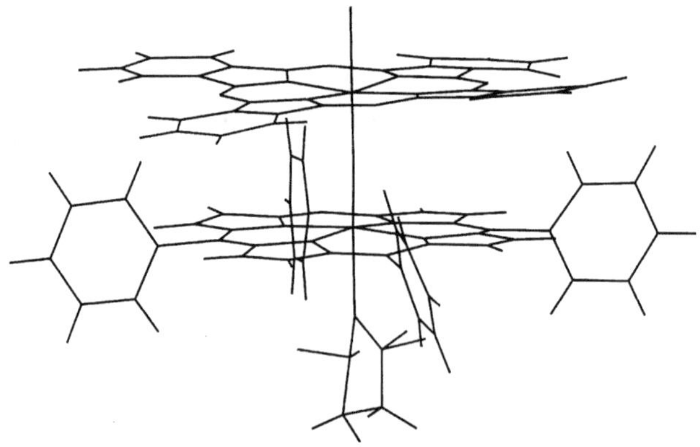

Appendix 23-2. [H$_2$O-(Pc)Fe-N-Fe(tpp)-OH$_2$][I$_5$][2THF], YUGKOA[361]

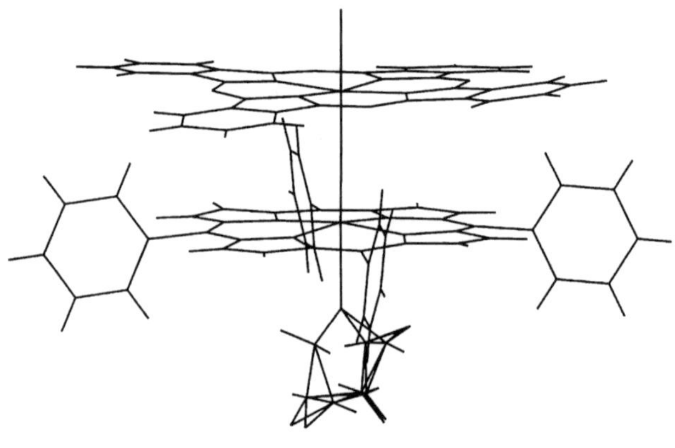

Appendix 23-3. [THF-(Pc)Fe-N-Mn(TPP)-THF][I$_5$][2THF], #2001-01[362]

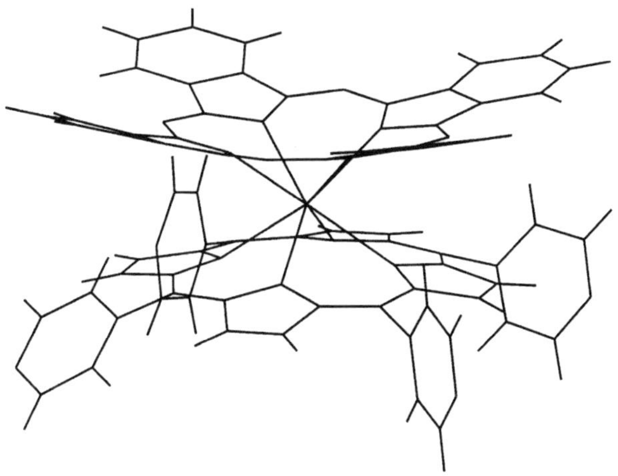

Appendix 23-4.
[(Pc)Ce(T(pMeOP)P)][2CHCl$_3$,pentane], SACNOZ[405]

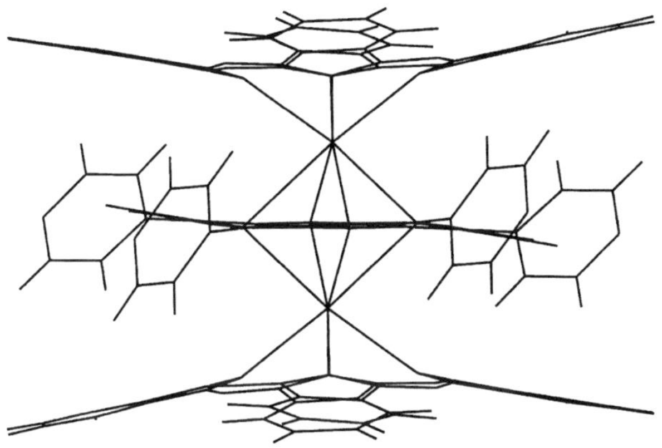

Appendix 23-5. [(Pc)Ce(T(pMeOP)P)Ce(Pc)][2tol], JUDPAZ[406]

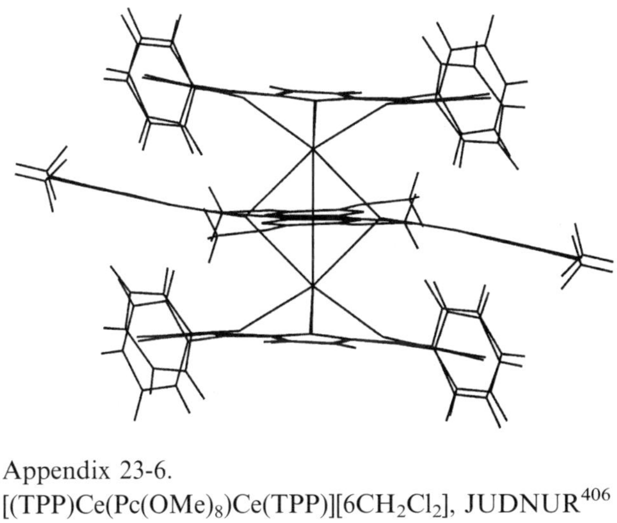

Appendix 23-6.
[(TPP)Ce(Pc(OMe)$_8$)Ce(TPP)][6CH$_2$Cl$_2$], JUDNUR[406]

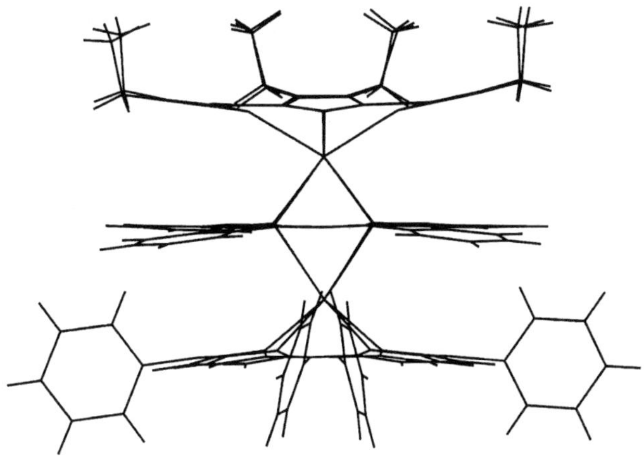

Appendix 23-7. [(TPP)Ce(Pc)Gd(OEP)][2ClB], TANMUQ[407]

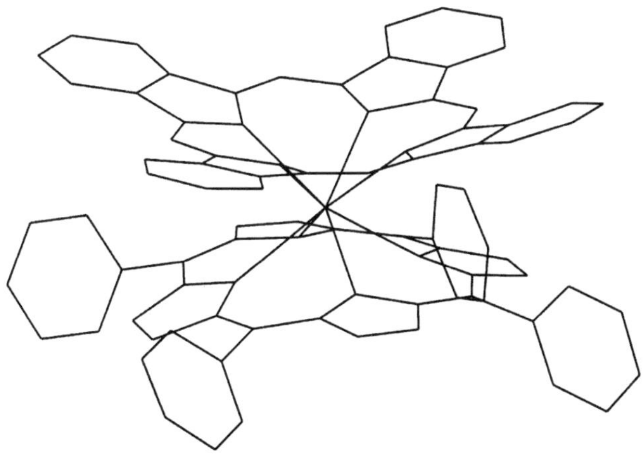

Appendix 23-8. [PcCeTPP][SbCl$_6$], SACNUF[405]

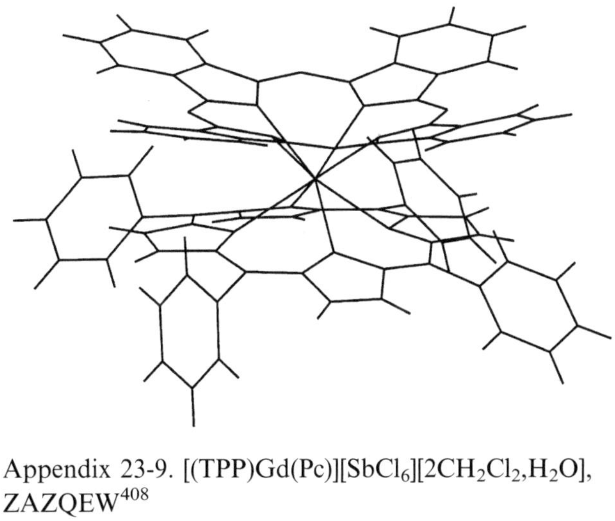

Appendix 23-9. [(TPP)Gd(Pc)][SbCl$_6$][2CH$_2$Cl$_2$,H$_2$O], ZAZQEW[408]

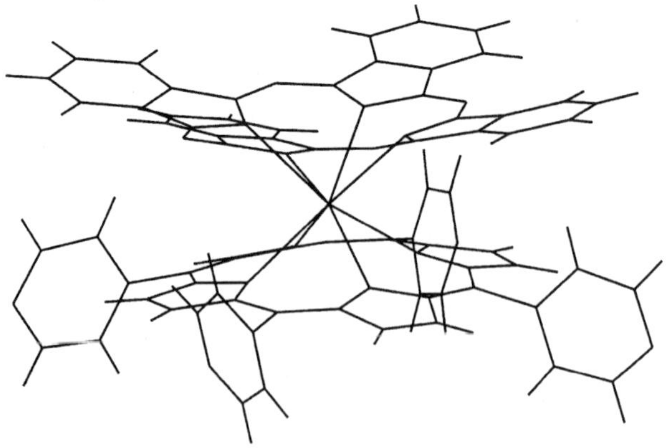

Appendix 23-10. [PcLaTPyP][OH][$CHCl_3$,MeOH], TUPWEG[410]

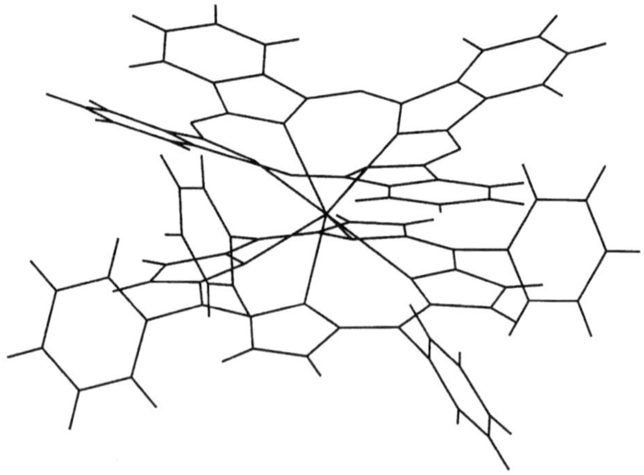

Appendix 23-11. [TPPLaPc][$2CH_2Cl_2$], ZAZQAS[408]

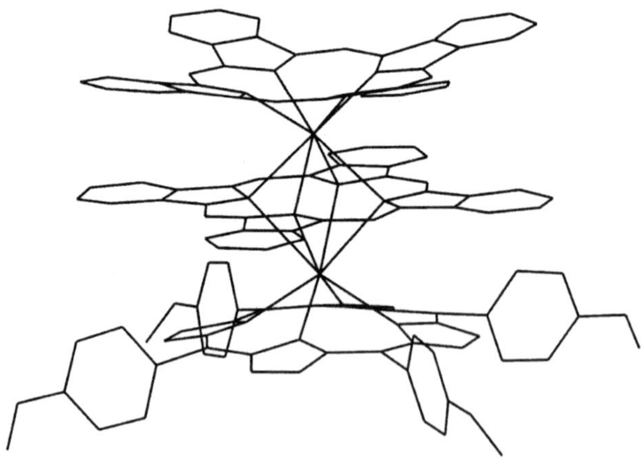

Appendix 23-12. [(Pc)Nd(Pc)Nd(T(4-OMe)PP)] [1,1,2-trichloroethane], DULBIV[412]

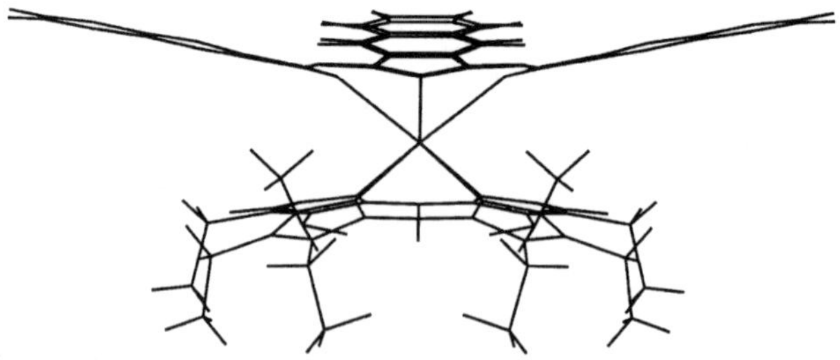

Appendix 23-13. [2,3-NcEuOEP][$4CHCl_3$], OCOBIR[252]

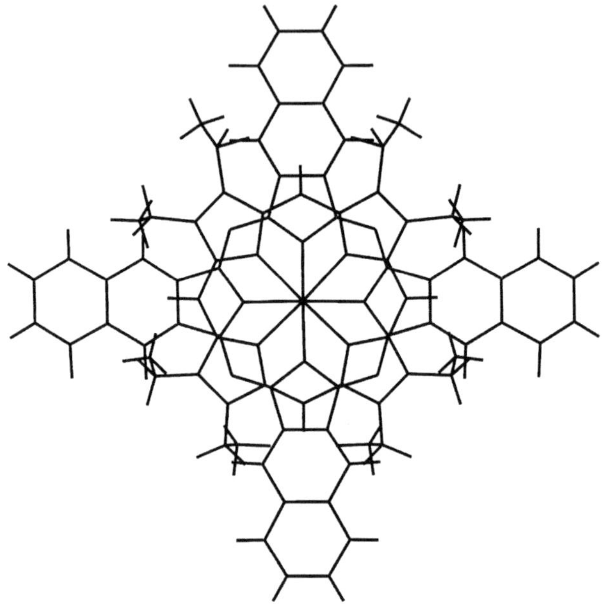

Appendix 23-14. [2,3-NcEuOEP][$4CHCl_3$], OCOBIR[252]

Appendix 24. The Atomic Coordinates of β-PcH_2, #1961-01

Space group $P2_1/a$
$a = 19.85$ Å, $c = 14.8$ Å, $\beta = 122.25°$

N(1)	0.163	0.256
N(2)	0.098	0.074
N(3)	0.081	−0.074
N(4)	−0.033	−0.132
C(1)	0.199	0.103
C(2)	0.222	0.199
C(3)	0.159	0.179
C(4)	0.119	0.026
C(5)	0.008	−0.141
C(6)	−0.102	−0.228
C(7)	−0.031	−0.257
C(8)	−0.102	−0.309
C(9)	0.344	0.285
C(10)	0.320	0.189
C(11)	0.247	0.099
C(12)	0.298	0.294
C(13)	−0.139	−0.472
C(14)	−0.069	−0.420
C(15)	−0.014	−0.311
C(16)	−0.154	−0.420

Reproduced with permission from the doctoral thesis of Everly B. Fleischer.[34,39]

Appendix 25. The Atomic Coordinates of β-PcH_2, #1970-02

Space group $P2_1/a$
$a = 19.847,4$ Å, $b = 4.727,1$ Å, $c = 14.792,3$ Å, $\beta = 121.97,3°$

N(11)	0.0793,1	0.5010,5	−0.0734,2
N(13)	−0.0347,1	0.1929,5	−0.1347,2
N(7)	0.1620,1	0.0275,5	0.2548,2
N(9)	0.0993,1	0.2163,5	0.0758,2
C(12)	0.0092,2	0.3966,6	−0.1469,2
C(18)	−0.0325,2	0.4879,6	−0.2571,2
C(17)	−0.0130,2	0.6849,7	−0.3092,2
C(16)	−0.0663,2	0.7221,7	−0.4175,2
C(15)	−0.1369,2	0.5692,7	−0.4716,3
C(20)	−0.1563,2	0.3740,7	−0.4191,2
C(19)	−0.1028,2	0.3355,6	−0.3109,2
C(14)	−0.1032,2	0.1506,6	−0.2322,?
C(8)	0.1599,2	0.1963,6	0.1814,2
C(5)	0.2225,2	0.3917,6	0.2014,2
C(6)	0.2958,2	0.4518,7	0.2916,2
C(1)	0.3429,2	0.6513,8	0.2824,3
C(2)	0.3176,2	0.7879,7	0.1862,3
C(3)	0.2446,2	0.7294,7	0.0962,3
C(4)	0.1977,2	0.5284,6	0.1053,2
C(10)	0.1200,2	0.4163,6	0.0272,2
H(27)	0.0377,16	0.8007,63	−0.2715,20
H(26)	−0.0513,18	0.8555,71	−0.4575,23
H(25)	−0.1737,17	0.5928,66	−0.5506,23
H(28)	−0.2037,17	0.2715,65	−0.4549,22
H(24)	0.3135,16	0.3628,64	0.3576,22
H(21)	0.3968,17	0.6889,66	0.3469,22
H(22)	0.3511,17	0.9226,68	0.1824,22
H(23)	0.2265,15	0.8193,60	0.0300,2?

Reproduced with permission from the doctoral thesis of Edward C. Bissell.[40]

Appendix 26. The Atomic Coordinates of β-PcH_2 (PHTHCY,[38] #1971-03[31])

Space group $P2_1/a$
a= 19.847,4 Å, b= 4.727,1 Å, c= 14.792,3 Å, β= 121.97,3°

N(1)	−0.0792,?	−0.5012,?	0.0731,?
N(2)	0.0349,?	−0.1925,?	0.1348,?
N(3)	0.1617,?	0.0279,?	0.2545,?
N(4)	0.1001,?	0.2183,?	0.0762,?
C(1)	−0.0087,2	−0.3965,7	0.1463,2
C(2)	0.0321,2	−0.4889,7	0.2571,3
C(3)	0.0128,2	−0.6856,8	0.3083,3
C(4)	0.0664,2	−0.7234,9	0.4177,3
C(5)	0.1379,2	−0.5682,9	0.4727,3
C(6)	0.1564,3	−0.3711,8	0.4197,3
C(7)	0.1029,2	−0.3346,7	0.3109,3
C(8)	0.1035,2	−0.1498,7	0.2322,2
C(9)	0.1597,2	0.1950,7	0.1817,2
C(10)	0.2226,2	0.3913,7	0.2015,2
C(11)	0.2963,2	0.4530,8	0.2918,3
C(12)	0.3429,3	0.6510,9	0.2828,3
C(13)	0.3183,3	0.7909,9	0.1860,3
C(14)	0.2443,2	0.7310,8	0.0952,3
C(15)	0.1975,2	0.5289,7	0.1044,2
C(16)	0.1196,2	0.4171,7	0.0275,2
H(1)	−0.0426	−0.8090	0.2645
H(2)	0.0540	−0.8783	0.4613
H(3)	0.1782	−0.6047	0.5571
H(4)	0.2120	−0.2515	0.4624
H(5)	0.3157	0.3504	0.3678
H(6)	0.4021	0.7026	0.3509
H(7)	0.3559	0.9452	0.1813
H(8)	0.2240	0.8385	0.0213
H(9)	0.0188	−0.0966	0.0712
H(10)	0.0504	0.1046	0.0438

Reproduced with permission from the doctoral thesis of Sax A. Mason.[31]

Appendix 27. The Atomic Coordinates of PcSn (SNPHCY,[142] #1971-02[31])

Space group P-1
a= 12.060,1 Å, b= 12.618,1 Å, c= 8.675,1 Å, α=95.89,1°, β= 95.08,1°, γ=68.17,1°

Sn	0.24818,2	0.00579,2	0.20174,3
N(1)	0.3550,3	−0.2976,3	0.2341,4
N(2)	0.2259,3	−0.1497,3	0.0709,4
N(3)	0.0837,3	−0.0636,3	−0.1381,4
N(4)	0.1958,3	0.0569,3	−0.0427,4
N(5)	0.2724,3	0.2110,3	−0.0356,4
N(6)	0.3911,3	0.0688,3	0.1425,4
N(7)	0.5527,3	−0.0302,3	0.3247,4
N(8)	0.4230,3	−0.1376,3	0.2583,4
C(1)	0.2639,4	−0.2542,3	0.1304,5
C(2)	0.1976,4	−0.3213,3	0.0536,5
C(3)	0.2050,4	−0.4322,4	0.0687,5
C(4)	0.1315,5	−0.4736,4	−0.0312,6
C(5)	0.0523,5	−0.4054,4	−0.1413,6
C(6)	0.0446,4	−0.2947,4	−0.1558,5
C(7)	0.1188,4	−0.2532,3	−0.0566,5
C(8)	0.1389,3	−0.1460,3	−0.0446,5

(*Continued*)

Appendix 27. Continued

C(9)	0.1117,4	0.0288,3	−0.1383,4
C(10)	0.0563,4	0.1155,3	−0.2487,5
C(11)	−0.0319,4	0.1263,4	−0.3676,5
C(12)	−0.0666,5	0.2232,5	−0.4500,6
C(13)	−0.0140,5	0.3047,4	−0.4164,6
C(14)	0.0745,4	0.2936,4	−0.2998,5
C(15)	0.1093,4	0.1971,3	−0.2156,5
C(16)	0.1988,4	0.1568,3	−0.0892,5
C(17)	0.3641,4	0.1663,3	0.0661,5
C(18)	0.4514,4	0.2193,4	0.1139,5
C(19)	0.4619,4	0.3183,4	0.0718,6
C(20)	0.5532,4	0.3490,4	0.1484,6
C(21)	0.6327,4	0.2810,4	0.2578,6
C(22)	0.6245,4	0.1799,4	0.2936,6
C(23)	0.5316,4	0.1506,4	0.2217,5
C(24)	0.4936,3	0.0548,4	0.2357,5
C(25)	0.5216,3	−0.1208,4	0.3287,5
C(26)	0.5943,4	−0.2211,4	0.4103,5
C(27)	0.7000,4	−0.2463,5	0.5016,5
C(28)	0.7457,5	−0.3524,5	0.5636,6
C(29)	0.6897,5	−0.4321,5	0.5313,6
C(30)	0.5846,4	−0.4068,4	0.4388,5
C(31)	0.5375,4	−0.2997,4	0.3808,5
C(32)	0.4293,4	−0.2446,4	0.2851,4
H(1)	0.259,5	−0.480,4	0.142,6
H(2)	0.137,5	−0.551,5	−0.023,7
H(3)	−0.002,7	−0.436,6	−0.201,9
H(4)	−0.003,5	−0.252,4	−0.234,6
H(5)	−0.059,4	0.067,4	−0.391,5
H(6)	−0.136,7	0.237,7	−0.544,10
H(7)	−0.039,6	0.375,5	−0.474,7
H(8)	0.113,5	0.348,4	−0.290,6
H(9)	0.408,4	0.363,4	−0.004,6
H(10)	0.555,10	0.425,10	0.101,13
H(11)	0.688,5	0.302,5	0.309,6
H(12)	0.674,6	0.140,5	0.371,7
H(13)	0.748,6	−0.197,6	0.520,8
H(14)	0.819,6	−0.374,5	0.624,7
H(15)	0.720,6	−0.519,6	0.566,8
H(16)	0.541,5	−0.459,4	0.411,6

Reproduced with permission from the doctoral thesis of Sax A. Mason[31]

Appendix 28. The Atomic Coordinates of PcAlCl, #1995-01

Space group P-1
$a = 13.746,2$ Å, $b = 14.059,2$ Å, $c = 13.746,2$ Å, $\alpha = 98.37,1^\circ$, $\beta = 90.01,1^\circ$, $\gamma = 108.620,9^\circ$

Cl(1)	0.4874,3	0.3258,3	0.0955,3
Cl(2)	1.0623,3	0.8249,3	0.4268,3
Al(1)	0.4300,4	0.1596,4	0.0703,4
Al(2)	1.0068,4	0.6601,4	0.4015,3
N(1)	0.4817,9	0.1306,8	−0.0574,9
N(2)	0.299,1	0.131,1	0.004,1
N(3)	0.3585,10	0.1312,10	0.1908,9
N(4)	0.543,1	0.127,1	0.127,1
N(5)	0.650,1	0.121,1	−0.012,1
N(6)	0.343,1	0.135,1	−0.168,1
N(7)	0.187,1	0.124,1	0.146,1
N(8)	0.499,1	0.129,1	0.300,1

(*Continued*)

Appendix 28. Continued

N(1)	1.0580,1	0.6293,10	0.2724,9
N(10)	0.870,1	0.632,1	0.337,1
N(11)	0.936,1	0.6305,9	0.5225,9
N(12)	1.121,1	0.6292,10	0.4587,10
N(13)	1.227,1	0.629,1	0.317,1
N(14)	0.9155,10	0.6227,10	0.1662,10
N(15)	0.765,1	0.633,1	0.474,1
N(16)	1.072,1	0.622,1	0.626,1
C(1)	0.436,2	0.128,1	−0.150,1
C(2)	0.276,1	0.130,1	−0.094,1
C(3)	0.170,1	0.129,1	−0.111,1
C(4)	0.118,1	0.126,1	−0.197,1
C(5)	0.018,2	0.125,1	−0.188,1
C(6)	−0.029,1	0.122,1	−0.097,1
C(7)	0.027,1	0.125,1	−0.011,1
C(8)	0.130,1	0.126,1	−0.021,1
C(9)	0.208,1	0.130,1	0.050,1
C(10)	0.257,1	0.122,1	0.209,1
C(11)	0.239,1	0.120,1	0.313,1
C(12)	0.152,1	0.111,1	0.361,1
C(13)	0.157,2	0.108,2	0.462,2
C(14)	0.250,2	0.110,2	0.508,1
C(15)	0.340,1	0.120,1	0.457,1
C(16)	0.332,1	0.125,1	0.358,1
C(17)	0.404,2	0.131,1	0.282,1
C(18)	0.563,1	0.128,1	0.228,1
C(19)	0.667,1	0.125,1	0.243,1
C(20)	0.719,2	0.122,1	0.329,1
C(21)	0.821,2	0.117,1	0.323,2
C(22)	0.864,1	0.118,1	0.226,1
C(23)	0.808,1	0.122,1	0.145,1
C(24)	0.705,1	0.123,1	0.149,1
C(25)	0.630,1	0.124,1	0.081,1
C(26)	0.582,1	0.124,1	−0.077,1
C(27)	0.597,1	0.124,1	−0.185,1
C(28)	0.679,1	0.119,1	−0.239,1
C(29)	0.670,2	0.119,1	−0.335,2
C(30)	0.578,2	0.124,2	−0.379,1
C(31)	0.497,1	0.128,1	−0.326,1
C(32)	0.504,1	0.127,1	−0.225,1
C(33)	1.013,1	0.630,1	0.184,1
C(34)	0.853,1	0.625,1	0.235,1
C(35)	0.749,1	0.620,1	0.217,1
C(36)	0.690,1	0.612,1	0.132,1
C(37)	0.587,2	0.606,2	0.140,2
C(38)	0.548,2	0.614,2	0.231,2
C(39)	0.602,1	0.623,2	0.318,1
C(40)	0.702,1	0.628,1	0.307,1
C(41)	0.779,1	0.631,1	0.379,1
C(42)	0.838,2	0.635,1	0.539,1
C(43)	0.816,1	0.626,1	0.640,1
C(44)	0.732,1	0.626,1	0.703,2
C(45)	0.741,2	0.620,1	0.798,1
C(46)	0.833,1	0.622,1	0.842,1
C(47)	0.916,1	0.623,1	0.789,1
C(48)	0.906,1	0.624,1	0.687,1
C(49)	0.978,1	0.625,1	0.610,1
C(50)	1.137,1	0.624,1	0.555,1
C(51)	1.240,1	0.621,1	0.576,1
C(52)	1.296,1	0.620,1	0.658,1
C(53)	1.395,2	0.621,2	0.645,2

(*Continued*)

Appendix 28. Continued

C(54)	1.440,1	0.623,1	0.560,2
C(55)	1.387,1	0.623,1	0.472,2
C(56)	1.288,1	0.623,1	0.486,1
C(57)	1.214,1	0.628,1	0.412,1
C(58)	1.155,1	0.628,1	0.253,1
C(59)	1.174,1	0.628,1	0.153,1
C(60)	1.258,1	0.626,1	0.096,1
C(61)	1.246,2	0.625,2	−0.009,2
C(62)	1.159,1	0.626,1	−0.049,1
C(63)	1.071,2	0.627,1	0.005,1
C(64)	1.084,1	0.627,1	0.105,1
H(1)	0.1495	0.1257	−0.2587
H(2)	−0.0186	0.1262	−0.2459
H(3)	−0.0991	0.1164	−0.0960
H(4)	−0.0019	0.1284	0.0520
H(5)	0.0878	0.1056	0.3291
H(6)	0.0998	0.1007	0.5006
H(7)	0.2535	0.1051	0.5760
H(8)	0.4035	0.1231	0.4874
H(9)	0.6867	0.1222	0.3904
H(10)	0.8581	0.1120	0.3793
H(11)	0.9316	0.1167	0.2186
H(12)	0.8400	0.1244	0.0837
H(13)	0.7398	0.1157	−0.2103
H(14)	0.7235	0.1165	−0.3761
H(15)	0.5718	0.1225	−0.4489
H(16)	0.4356	0.1307	−0.3558
H(17)	0.7208	0.6112	0.0694
H(18)	0.5438	0.5955	0.0831
H(19)	0.4779	0.6137	0.2370
H(20)	0.5721	0.6243	0.3791
H(21)	0.6699	0.6288	0.6749
H(22)	0.6817	0.6155	0.8358
H(23)	0.8368	0.6230	0.9122
H(24)	0.9784	0.6242	0.8199
H(25)	1.2683	0.6185	0.7213
H(26)	1.4343	0.6194	0.7000
H(27)	1.5105	0.6225	0.5582
H(28)	1.4182	0.6257	0.4101
H(29)	1.3197	0.6243	0.1254
H(30)	1.3027	0.6235	−0.0493
H(31)	1.1554	0.6277	−0.1183
H(32)	1.0077	0.6230	−0.0271

Reproduced with permission from the master thesis of Masahiro Hasegawa,[155,156] personal communication N. Sato.

Appendix 29. The Atomic Coordinates of α-PcGe(OH)$_2$, #1985-01

Space group P2$_1$/c
a= 6.721,3 Å, b= 11.734,5 Å, c= 16.530,9 Å, β= 105.97,5°

Ge	0	0	0
O	−0.0649,9	0.0937,5	0.0771,4
N(1)	−0.093,1	0.2137,6	−0.1423,4
N(2)	0.492,1	0.0918,6	0.0775,4
N(3)	0.168,1	0.1259,6	−0.0274,4
N(4)	0.243,1	−0.0508,6	0.0902,4
C(1)	0.368,1	0.1483,7	0.0134,5
C(2)	0.432,1	0.2514,7	−0.0228,5
C(3)	0.616,1	0.3099,8	−0.0059,5

(*Continued*)

Appendix 29. Continued

C(4)	0.624,1	0.4043,8	−0.0540,6
C(5)	0.455,1	0.4389,8	−0.1179,6
C(6)	0.269,1	0.3806,8	−0.1349,5
C(7)	0.260,1	0.2854,8	−0.0867,5
C(8)	0.095,1	0.2065,7	−0.0882,5
C(9)	0.243,1	−0.1411,8	0.1433,5
C(10)	0.446,1	−0.1474,7	0.2039,5
C(11)	0.529,1	−0.2204,8	0.2693,6
C(12)	0.731,1	−0.2042,9	0.3137,6
C(13)	0.849,1	−0.117,1	0.2955,6
C(14)	0.767,1	−0.0433,9	0.2306,5
C(15)	0.566,1	−0.0581,7	0.1844,5
C(16)	0.432,1	0.001,1	0.1127,4

Reproduced with permission from the doctoral thesis of Thomas Krämer,[120] personal communication J. Strähle.

Appendix 30. The Atomic Coordinates of β-PcGe(OH)$_2$, #1985-02

Space group P2$_1$/n
a = 6.812,1 Å, b = 14.541,2 Å, c = 12.842,2 Å, β = 94.38,1°

Ge	0.5	0.5	0.5
O	0.6297,7	0.4163,3	0.5862,3
N(1)	0.3095,7	0.4020,4	0.4545,4
N(2)	0.3345,7	0.5304,3	0.6146,4
N(3)	0.4693,8	0.3458,4	0.3046,4
N(4)	0.0766,7	0.4175,4	0.5860,4
C(1)	0.3309,10	0.3451,5	0.3719,4
C(2)	0.1702,9	0.2805,4	0.3653,5
C(3)	0.1234,10	0.2090,5	0.2969,5
C(4)	−0.0436,12	0.1583,5	0.3107,6
C(5)	−0.1583,11	0.1781,5	0.3935,6
C(6)	−0.1162,10	0.2488,5	0.4616,5
C(7)	0.0526,10	0.3004,4	0.4470,5
C(8)	0.1435,9	0.3792,4	0.5022,5
C(9)	0.1615,8	0.4866,4	0.6373,4
C(10)	0.0950,9	0.5274,4	0.7323,4
C(11)	−0.0690,9	0.5097,5	0.7882,5
C(12)	−0.0861,11	0.5614,6	0.8765,5
C(13)	0.0483,12	0.6291,6	0.9085,5
C(14)	0.2113,11	0.6473,6	0.8554,5
C(15)	0.2303,9	0.5956,5	0.7644,4
C(16)	0.3795,9	0.5970,5	0.6891,4

Reproduced with permission from the doctoral thesis of Thomas Krämer,[120] personal communication J. Strähle.

Appendix 31. The Atomic Coordinates of [PcMn(NCMe)$_2$][C$_4$(CN)$_6$], #1996-02

Space group P-1
a = 8.302,20 Å, b = 8.787,2 Å, c = 14.446,5 Å, α = 82.52,2°, β = 73.21,3°, γ = 74.31,2°

Mn(1)	0.5000	0.5000	0.5000,
N(3)	−0.1497,8	0.2686,7	1.1467,4
N(4)	0.3025,9	0.1002,9	1.0855,4
N(5)	0.4924,11	0.4475,9	0.8759,7
N(6)	0.5090,6	0.6597,5	0.6163,3
N(11)	0.6971,6	0.3489,4	0.5314,2
N(12)	0.5468,7	0.2061,5	0.6716,3

(*Continued*)

Appendix 31. Continued

N(13)	0.3423,6	0.4076,4	0.6063,3
N(14)	0.0660,6	0.5747,5	0.5917,3
C(1)	0.0357,26	0.4271,12	1.0231,7
C(1')	−0.0839,26	0.5089,30	1.0226,15
C(2)	0.2277,10	0.3659,9	0.9943,5
C(3)	−0.0876,10	0.3524,7	1.0928,4
C(4)	0.2676,9	0.2181,10	1.0448,5
C(5)	0.3726,13	0.4149,8	0.9290,7
C(6)	0.5315,7	0.7078,7	0.6781,4
C(7)	0.5671,8	0.7773,7	0.7531,4
C(11)	0.8729,7	0.3352,6	0.4852,3
C(12)	0.6919,8	0.2345,6	0.6078,3
C(13)	0.3888,8	0.2879,6	0.6699,3
C(14)	0.1621,8	0.4613,6	0.6341,3
C(21)	0.9755,8	0.2130,6	0.5321,3
C(22)	1.1541,8	0.1524,6	0.5142,3
C(23)	1.2151,7	0.0320,6	0.5745,4
C(24)	1.1044,9	−0.0323,6	0.6517,4
C(25)	0.9256,8	0.0271,6	0.6707,3
C(26)	0.8646,8	0.1477,6	0.6095,3
C(31)	0.2351,8	0.2616,6	0.7428,3
C(32)	0.2149,8	0.1523,6	0.8224,3
C(33)	0.0453,9	0.1544,6	0.8750,3
C(34)	−0.0927,8	0.2569,7	0.8504,4
C(35)	−0.0747,7	0.3671,6	0.7728,3
C(36)	0.0947,7	0.3664,6	0.7192,3
H(7)	0.4889	0.7666	0.8138
H(7')	0.5630	0.8870	0.7379
H(7'')	0.6820	0.7270	0.7587
H(22)	1.2342	0.1927	0.4613
H(23)	1.3385	−0.0094	0.5639
H(24)	1.1550	−0.1189	0.6890
H(25)	0.8545	−0.0182	0.7255
H(32)	0.3142	0.0791	0.8384
H(33)	0.0260	0.0810	0.9291
H(34)	−0.2064	0.2508	0.8887
H(35)	−0.1711	0.4433	0.7578

Personal communication Joel S. Miller, reproduced with permission.

Appendix 32. The Atomic Coordinates of PcMoOCl, #1990-03

Space group $P2_1/n$
$a = 10.504,3$ Å, $b = 13.974,2$ Å, $c = 8.989,3$ Å, $\beta = 95.25,2°$

Mo	0.45254,9	0.49608,7	0.5198,1
Cl	0.6767,3	0.5226,2	0.4306,3
O	0.2967,7	0.4764,5	0.5781,8
N(1)	0.5776,6	0.4414,4	0.6911,6
N(2)	0.5112,5	0.6304,4	0.5961,6
N(3)	0.6081,5	0.5844,4	0.8424,5
N(4)	0.4221,5	0.7258,4	0.3857,5
C(1)	0.6006,6	0.3450,5	0.7138,7
C(2)	0.6534,5	0.3309,4	0.8662,6
C(3)	0.6918,6	0.2505,5	0.9468,7
C(4)	0.7397,6	0.2623,5	1.0950,7
C(5)	0.7486,7	0.3532,6	1.1582,7
C(6)	0.7101,6	0.4352,5	1.0799,6
C(7)	0.6640,5	0.4217,5	0.9324,6
C(8)	0.6158,6	0.4900,5	0.8191,6

(*Continued*)

Appendix 32. Continued

C(9)	0.5600,6	0.6476,4	0.7399,7
C(10)	0.5547,5	0.7501,4	0.7644,6
C(11)	0.5941,6	0.8053,5	0.8893,7
C(12)	0.5714,6	0.9026,5	0.8759,8
C(13)	0.5143,7	0.9443,5	0.7473,9
C(14)	0.4745,6	0.8887,5	0.6214,8
C(15)	0.4953,5	0.7911,4	0.6343,6
C(16)	0.4719,6	0.7140,4	0.5274,7
H(3)	0.682	0.186	0.902
II(4)	0.766	0.207	1.151
H(5)	0.782	0.358	1.261
H(6)	0.717	0.498	1.123
H(11)	0.636	0.777	0.979
H(12)	0.596	0.945	0.959
H(13)	0.505	1.012	0.743
H(14)	0.436	0.917	0.530

Reproduced with permission from the doctoral thesis of Franz Gingl,[124] personal communication J. Strähle and F. Gingl.

Appendix 33. The Atomic Coordinates of [PcSiOSiEt$_3$,Me][0.5C$_6$H$_{12}$], #1971-01

Space group P-1
a= 12.050,6 Å, b= 14.892,6 Å, c= 12.003,5 Å, α= 117.17,2°,
β= 95.12,2°, γ= 94.33,2°

Si(1)	0.1207,1	0.1605,1	0.4026,1
Si(2)	0.2084,2	0.3916,1	0.4621,2
O(1)	0.1655,3	0.2826,2	0.4475,3
N(1A)	0.0665,3	0.2038,3	0.5637,4
N(2B)	0.2647,3	0.1492,3	0.4749,4
N(3C)	0.1750,3	0.1221,3	0.2426,4
N(4D)	−0.0227,3	0.1762,3	0.3307,4
N(5AB)	0.2326,4	0.2030,4	0.6915,4
N(6BC)	0.3676,3	0.1011,3	0.2958,4
N(7CD)	0.0127,3	0.1335,3	0.1177,4
N(8DA)	−0.1232,4	0.2333,3	0.5122,4
C(1A)	−0.0378,5	0.2324,4	0.5901,5
C(2A)	−0.0438,5	0.2671,4	0.7241,5
C(3A)	−0.1303,5	0.3054,5	0.7965,6
C(4A)	−0.1053,6	0.3330,6	0.9239,6
C(5A)	−0.0032,6	0.3240,6	0.9765,6
C(6A)	0.0829,6	0.2868,5	0.9049,5
C(7A)	0.0590,5	0.2588,4	0.7763,5
C(8A)	0.1273,5	0.2182,4	0.6737,5
C(9B)	0.2956,5	0.1713,4	0.5992,5
C(10B)	0.4121,5	0.1579,4	0.6176,5
C(11B)	0.4822,5	0.1703,5	0.7250,6
C(12B)	0.5918,5	0.1511,5	0.7092,6
C(13B)	0.6323,5	0.1222,5	0.5943,6
C(14B)	0.5628,5	0.1098,5	0.4871,6
C(15B)	0.4324,5	0.1279,4	0.5025,5
C(16B)	0.3579,4	0.1233,4	0.4144,5
C(17C)	0.2825,4	0.1025,4	0.2182,5
C(18C)	0.2937,4	0.0842,4	0.0908,4
C(19C)	0.3855,5	0.0648,4	0.0245,5
C(20C)	0.3665,5	0.0485,4	−0.0993,5
C(21C)	0.2624,5	0.0545,4	−0.1533,5
C(22C)	0.1718,5	0.0763,4	−0.0864,5
C(23C)	0.1902,4	0.0910,4	0.0378,5
C(24C)	0.1164,4	0.1164,4	0.1348,4

(*Continued*)

Appendix 33. Continued

C(25D)	−0.0504,4	0.1630,4	0.2101,5
C(26D)	−0.1631,4	0.1873,4	0.1943,5
C(27D)	−0.2274,5	0.1870,4	0.0916,5
C(28D)	−0.3332,5	0.2156,5	0.1090,6
C(29D)	−0.3738,5	0.2452,5	0.2254,6
C(30D)	−0.3092,4	0.2458,4	0.3277,5
C(31D)	−0.2027,4	0.2157,4	0.3089,5
C(32D)	−0.1134,4	0.2089,4	0.3935,5
C(Si)	0.0694,5	0.0204,4	0.3574,5
C(1*)	0.1382,12	0.4900,6	0.5926,8
C(1**)	0.1499,9	0.4927,7	0.7154,9
C(2*)	0.3722,10	0.4004,11	0.4713,17
C(2**)	0.4382,10	0.3898,10	0.5473,20
C(3*)	0.1575,7	0.3940,5	0.3094,7
C(3**)	0.1926,9	0.5000,6	0.3153,9
C(1N)	0.3770,22	0.5010,20	0.0304,24
C(2N)	0.4055,23	0.3664,21	0.0706,26
C(3N)	0.5673,13	0.4371,10	0.0213,13
C(4N)	0.4418,17	0.4276,14	0.0336,17
H(3A)	−0.206	0.312	0.7590
H(4A)	−0.167	0.361	0.9790
H(5A)	0.012	0.346	1.0700
H(6A)	0.159	0.281	0.9450
H(3B)	0.454	0.192	0.8100
H(4B)	0.643	0.161	0.7870
H(5B)	0.714	0.108	0.5860
H(6B)	0.591	0.088	0.4030
H(3C)	0.463	0.061	0.0640
H(4C)	0.433	0.034	−0.1500
H(5C)	0.251	0.044	−0.2440
H(6C)	0.095	0.080	−0.1260
H(3D)	−0.199	0.166	0.0070
H(4D)	−0.381	0.217	0.0350
H(5D)	−0.452	0.267	0.2380
H(6D)	−0.338	0.267	0.4120

Reproduced with permission from the doctoral thesis of Anthony P. D'Addario.[270]

Appendix 34. The Atomic Coordinates of $(Me_3SiO)_2MeSiO\text{-}((Pc)Si\text{-}O)_3\text{-}SiMe(OSiMe_3)_2$, #1970-01

Space group P-1
a= 14.642,6 Å, b= 15.831,7 Å, c= 14.318,9 Å, α= 90.986,29°,
β= 94.866,29°, γ= 128.049,30°

Si(1)	0	0	0
Si(2)	−0.1120,1	0.0583,1	0.1575,1
Si(3)	−0.2648,1	0.0706,1	0.2967,1
Si(4)	−0.1506,2	0.3048,2	0.3603,2
Si(5)	−0.3976,2	−0.1026,2	0.4347,1
O(1)	−0.0557,3	0.0296,2	0.0791,2
O(2)	−0.1665,3	0.0918,3	0.2381,2
O(3)	−0.2114,4	0.1781,4	0.3637,3
O(4)	−0.3027,4	−0.0238,4	0.3649,3
N(1)	0.0400,3	0.1874,3	0.2075,3
N(2)	−0.0968,3	−0.0211,3	0.2481,3
N(3)	−0.2671,3	−0.0692,3	0.1124,3
N(4)	−0.1304,3	0.1392,3	0.0705,3
N(5)	0.0956,4	0.1180,4	0.3292,3
N(6)	−0.2844,4	−0.1981,4	0.2129,3
N(7)	−0.3231,4	0.0001,4	−0.0095,3

(*Continued*)

Appendix 34. Continued

N(8)	0.0520,4	0.3187,3	0.1134,3
N(9)	0.1526,3	0.1011,3	0.0694,3
N(10)	−0.0258,4	−0.1128,3	0.0726,3
N(11)	0.1554,4	−0.0135,4	0.1771,3
N(12)	−0.2193,4	−0.2642,4	0.0030,3
C(1)	0.0910,4	0.2896,4	0.1815,3
C(2)	0.1992,4	0.3663,3	0.2418,3
C(3)	0.2781,4	0.4776,4	0.2476,3
C(4)	0.3720,5	0.5270,5	0.3164,4
C(5)	0.3873,5	0.4685,5	0.3770,4
C(6)	0.3089,4	0.3575,4	0.3726,3
C(7)	0.2129,4	0.3079,3	0.3028,3
C(8)	0.1123,4	0.1950,4	0.2816,3
C(9)	−0.0016,4	0.0172,4	0.3145,3
C(10)	−0.0239,4	−0.0670,4	0.3692,3
C(11)	0.0419,4	−0.0694,4	0.4427,4
C(12)	−0.0071,5	−0.1649,5	0.4821,4
C(13)	−0.1194,5	−0.2569,4	0.4499,4
C(14)	−0.1853,4	−0.2566,4	0.3762,4
C(15)	−0.1354,4	−0.1593,4	0.3363,3
C(16)	−0.1792,4	−0.1286,4	0.2608,4
C(17)	−0.3235,4	−0.1694,4	0.1444,4
C(18)	−0.4377,4	−0.2433,4	0.0914,3
C(19)	−0.5265,4	−0.3523,4	0.0963,3
C(20)	−0.6284,5	−0.3999,5	0.0359,4
C(21)	−0.6411,5	−0.3421,5	−0.0272,4
C(22)	−0.5524,4	−0.2336,4	−0.0333,3
C(23)	−0.4498,4	−0.1858,4	0.0274,3
C(24)	−0.3415,4	−0.0759,4	0.0408,4
C(25)	−0.2250,5	0.1012,4	0.0045,4
C(26)	−0.2051,4	0.1855,4	−0.0478,3
C(27)	−0.2715,4	0.1891,4	−0.1197,4
C(28)	−0.2259,5	0.2869,5	−0.1538,4
C(29)	−0.1149,5	0.3800,4	−0.1193,4
C(30)	−0.0480,4	0.3764,4	−0.0490,3
C(31)	−0.0946,4	0.2786,3	−0.0130,3
C(32)	−0.0503,4	0.2476,4	0.0608,3
C(33)	0.2307,4	0.2100,4	0.0585,4
C(34)	0.3328,5	0.2587,5	0.1273,4
C(35)	0.4336,5	0.3657,5	0.1465,4
C(36)	0.5130,6	0.3869,6	0.2197,5
C(37)	0.4961,6	0.3055,6	0.2717,5
C(38)	0.3956,6	0.1992,6	0.2532,5
C(39)	0.3147,5	0.1789,5	0.1799,4
C(40)	0.2013,4	0.0798,5	0.1422,4
C(41)	0.0502,4	−0.1024,4	0.1438,3
C(42)	−0.0017,5	−0.2043,5	0.1824,4
C(43)	0.0386,5	−0.2361,5	0.2535,4
C(44)	−0.0361,6	−0.3421,6	0.2741,5
C(45)	−0.1474,6	−0.4150,6	0.2250,5
C(46)	−0.1882,5	−0.3837,5	0.1530,4
C(47)	−0.1122,5	−0.2767,5	0.1327,4
C(48)	−0.1254,5	−0.2187,4	0.0634,4
C(49)	−0.3963,7	0.0326,7	0.2184,6
C(50)	−0.1679,10	0.3322,10	0.2328,9
C(51)	−0.2263,12	0.3393,11	0.4291,9
C(52)	0.0072,11	0.3862,11	0.3983,9
C(53)	−0.3771,9	−0.0181,8	0.5407,7
C(54)	−0.5491,8	−0.1801,8	0.3709,7
C(55)	−0.3713,7	−0.1994,6	0.4703,5

Reproduced with permission from the doctoral thesis of David R. Swift.[269]

Appendix 35. The Atomic Coordinates of $PcSnCl_2$, #1972-01, CPHCYT

Space group C2/c
a= 21.104,2 Å, b= 11.060,1 Å, c= 11.392,1 Å, β= 96.04,1°

Sn	0	0	0
Cl(1)	0.05506,4	0.08398,11	0.18169,9
N(1)	−0.14142,12	0.12288,27	0.05644,26
N(2)	−0.04383,12	0.16389,28	−0.03160,27
N(3)	0.04111,12	0.26505,27	−0.12247,25
N(4)	0.07539,12	0.05712,26	−0.08552,25
C(1)	−0.10156,15	0.19498,34	0.00614,31
C(2)	−0.11171,15	0.32196,33	−0.02158,30
C(3)	−0.16094,17	0.39980,37	−0.00247,33
C(4)	−0.15685,19	0.51801,38	−0.04130,34
C(5)	−0.10522,18	0.55687,36	−0.09644,34
C(6)	−0.05536,16	0.48092,34	−0.11422,31
C(7)	−0.05898,15	0.36282,33	−0.07652,30
C(8)	−0.01582,15	0.26136,34	−0.08120,31
C(9)	0.08275,15	0.17468,33	−0.12168,30
C(10)	0.14747,15	0.18482,33	−0.15547,29
C(11)	0.17942,17	0.28138,35	−0.19952,33
C(12)	0.24275,17	0.26237,39	−0.21731,35
C(13)	0.27291,16	0.15241,39	−0.19224,34
C(14)	0.24106,15	0.05592,35	−0.14873,31
C(15)	0.17707,15	0.07253,34	−0.13252,29
C(16)	0.13092,15	0.00814,35	−0.08749,29
H(1)	−0.196,1	0.375,3	0.041,3
H(2)	−0.193,2	0.569,4	−0.034,3
H(3)	−0.102,2	0.631,3	−0.118,3
H(4)	−0.021,1	0.511,3	−0.154,3
H(5)	0.159,1	0.360,3	−0.216,3
H(6)	0.267,1	0.328,3	−0.247,3
H(7)	0.317,2	0.136,4	−0.205,3
H(8)	0.262,2	0.016,3	−0.132,3

Reproduced from the doctoral thesis of Ruth S. Osborn.[116,117]

Appendix 36. The Atomic Coordinates of PcWOCl, #1990-02

Space group $P2_1/n$
a= 10.504,4 Å, b= 13. 910,2 Å, c= 8. 944,3 Å, β= 95.32,3°

W	0.45293,5	0.49515,4	0.52016,6
Cl	0.6757,3	0.5236,3	0.4276,4
O	0.300,1	0.4851,9	0.572,1
N(1)	0.5804,7	0.4407,5	0.6908,7
N(2)	0.5128,7	0.6305,5	0.5966,8
N(3)	0.6086,7	0.5851,5	0.8440,7
N(4)	0.4219,6	0.7266,5	0.3870,7
C(1)	0.6007,8	0.3450,6	0.7127,8
C(2)	0.6536,7	0.3312,6	0.8681,8
C(3)	0.6921,8	0.2481,6	0.9474,9
C(4)	0.7373,9	0.2609,7	1.0967,9
C(5)	0.7511,9	0.3520,7	1.163,1
C(6)	0.7116,7	0.4337,6	1.0814,9
C(7)	0.6647,6	0.4225,6	0.9335,8
C(8)	0.6161,8	0.4906,6	0.8206,9
C(9)	0.5613,8	0.6487,6	0.7428,8
C(10)	0.5556,7	0.7510,5	0.7670,9
C(11)	0.5935,7	0.8066,6	0.8908,9
C(12)	0.5697,9	0.9060,7	0.877,1

(*Continued*)

Appendix 36. Continued

C(13)	0.5146,9	0.9451,7	0.746,1
C(14)	0.4738,8	0.8908,6	0.623,1
C(15)	0.4958,7	0.7917,6	0.6340,9
C(16)	0.4717,8	0.7149,5	0.5270,9
H(3)	0.686	0.186	0.901
H(4)	0.762	0.205	1.157
H(5)	0.788	0.359	1.266
H(6)	0.718	0.497	1.127
H(11)	0.636	0.779	0.982
H(12)	0.591	0.948	0.961
H(13)	0.506	1.014	0.738
H(14)	0.431	0.919	0.533

Reproduced with permission from the doctoral thesis of Franz Gingl,[124] personal communication J. Strähle and F. Gingl.

Appendix 37. The Atomic Coordinates of $PcNbCl_3$, #1990-01

Space group P-1
$a = 8.360{,}1$ Å, $b = 12.090{,}2$ Å, $c = 13.655{,}1$ Å, $\alpha = 79.85{,}1°$,
$\beta = 93.48{,}1°$, $\gamma = 101.06{,}?$ °

Nb	0.53260,5	0.30706,3	0.26338,3
Cl(1)	0.5114,2	0.1393,1	0.3921,1
Cl(2)	0.7662,2	0.3817,1	0.3554,1
Cl(3)	0.7316,2	0.1996,1	0.2149,1
N(1)	0.2993,5	0.2968,3	0.3306,3
N(2)	0.5144,5	0.4849,3	0.2624,3
N(3)	0.5844,5	0.3871,3	0.1098,3
N(4)	0.3662,5	0.1965,3	0.1771,3
N(5)	0.3524,5	0.4666,3	0.4082,3
N(6)	0.6942,5	0.5829,3	0.1282,3
N(7)	0.5082,5	0.2253,3	0.0229,3
N(8)	0.1490,5	0.1161,3	0.2932,3
C(1)	0.1772,6	0.1998,4	0.3447,4
C(2)	0.0795,6	0.2034,4	0.4279,4
C(3)	−0.0588,6	0.1286,5	0.4695,4
C(4)	−0.1220,7	0.1569,5	0.5507,4
C(5)	−0.0546,7	0.2567,5	0.5879,4
C(6)	0.0778,7	0.3333,5	0.5448,4
C(7)	0.1436,6	0.3027,4	0.4644,4
C(8)	0.2776,6	0.3621,4	0.4005,4
C(9)	0.4552,6	0.5252,4	0.3394,4
C(10)	0.5147,6	0.6462,4	0.3344,4
C(11)	0.4845,7	0.7240,5	0.3915,4
C(12)	0.5547,8	0.8371,5	0.3616,5
C(13)	0.6480,8	0.8715,5	0.2753,5
C(14)	0.6762,7	0.7951,4	0.2194,4
C(15)	0.6091,6	0.6803,4	0.2502,4
C(16)	0.6115,6	0.5795,4	0.2076,4
C(17)	0.6801,6	0.4938,4	0.0835,4
C(18)	0.7570,6	0.4988,4	−0.0103,4
C(19)	0.8660,6	0.5866,5	−0.0665,4
C(20)	0.9194,6	0.5629,5	−0.1530,4
C(21)	0.8666,7	0.4573,5	−0.1835,4
C(22)	0.7555,6	0.3708,5	−0.1295,4
C(23)	0.7033,6	0.3952,4	−0.0407,4
C(24)	0.5916,6	0.3259,4	0.0334,3
C(25)	0.4021,6	0.1663,4	0.0887,4
C(26)	0.2966,6	0.0598,4	0.0743,4
C(27)	0.2853,6	−0.0068,5	−0.0009,4

(*Continued*)

Appendix 37. Continued

C(28)	0.1656,7	−0.1046,5	0.0084,4
C(29)	0.0616,7	−0.1351,5	0.0890,4
C(30)	0.0726,7	−0.0694,5	0.1630,4
C(31)	0.1928,6	0.0292,4	0.1544,4
C(32)	0.2346,6	0.1158,4	0.2160,4

Reproduced with permission from the doctoral thesis of Franz Gingl,[124] personal communication J. Strähle and F. Gingl.

Appendix 38. The Atomic Coordinates of $[Pc_2Nd][CH_2Cl_2]$, PTCNND[205,206]

Space group $P2_12_12_1$ (Pnma)
$a = 8.030,4$ Å, $b = 22.925,7$ Å, $c = 28.315,7$ Å

Nd	−0.1155,1	1.0002,1	0.3854,0
N(1)	0.0402,15	0.9185,5	0.3567,4
N(2)	0.2223,13	0.8954,4	0.4229,4
N(3)	0.1743,12	1.0057,8	0.4131,3
N(4)	0.2062,19	1.1038,6	0.4236,5
N(5)	0.0305,13	1.0917,4	0.3563,4
N(6)	−0.1171,14	1.1040,5	0.2839,4
N(7)	−0.1032,11	0.9979,9	0.2986,3
N(8)	−0.1107,17	0.8943,6	0.2833,5
N(9)	−0.3693,13	0.9341,4	0.3754,4
N(10)	−0.5177,12	0.9996,11	0.3199,3
N(11)	−0.3639,16	1.0610,6	0.3747,5
N(12)	−0.2892,12	1.1461,4	0.4194,3
N(13)	−0.1743,14	1.0599,5	0.4558,4
N(14)	−0.0300,12	1.0004,11	0.5138,3
N(15)	−0.1574,14	0.9380,5	0.4547,4
N(16)	−0.2690,18	0.8506,6	0.4174,5
Cl(1)	0.1913,?	−0.0629,?	0.1619,?
Cl(2)	0.1924,?	0.0633,?	0.1638,?
C(1)	0.2419,?	−0.0011,?	0.1963,?
C(2)	−0.0274,17	0.8778,6	0.3241,5
C(3)	−0.0071,14	0.8187,3	0.3319,4
C(4)	−0.0454,14	0.7660,3	0.3099,4
C(5)	0.0086,14	0.7135,3	0.3295,4
C(6)	0.1010,14	0.7137,3	0.3712,4
C(7)	0.1394,14	0.7664,3	0.3933,4
C(8)	0.0854,14	0.8189,3	0.3736,4
C(9)	0.1341,28	0.8815,10	0.3875,7
C(10)	0.2551,15	0.9576,5	0.4330,4
C(11)	0.3566,20	0.9688,7	0.4725,6
C(12)	0.4403,26	0.9344,10	0.5022,8
C(13)	0.5228,22	0.9622,7	0.5383,6
C(14)	0.5413,22	1.0198,7	0.5381,6
C(15)	0.4536,20	1.0616,7	0.5071,6
C(16)	0.3607,22	1.0277,8	0.4707,6
C(17)	0.2331,19	1.0552,6	0.4343,5
C(18)	0.1153,12	1.1219,4	0.3866,4
C(19)	0.1011,12	1.1817,3	0.3751,3
C(20)	0.1454,12	1.2342,3	0.3966,3
C(21)	0.0924,12	1.2870,3	0.3773,3
C(22)	−0.0050,12	1.2873,3	0.3365,3
C(23)	−0.0494,12	1.2347,3	0.3150,3
C(24)	0.0036,12	1.1819,3	0.3343,3
C(25)	−0.0446,12	1.1205,7	0.3205,6
C(26)	−0.1438,15	1.0462,5	0.2724,4
C(27)	−0.2089,19	1.0220,6	0.2260,5
C(28)	−0.2680,20	1.0527,7	0.1863,6

(Continued)

Appendix 38. Continued

C(29)	−0.3094,21	1.0217,6	0.1466,6
C(30)	−0.3365,18	0.9615,6	0.1519,5
C(31)	−0.2669,18	0.9298,7	0.1895,5
C(32)	−0.2126,16	0.9632,5	0.2273,4
C(33)	−0.1353,22	0.9486,8	0.2717,6
C(34)	−0.3645,24	0.8797,8	0.3891,6
C(35)	−0.4507,12	0.8486,4	0.3497,3
C(36)	−0.4725,12	0.7898,4	0.3388,3
C(37)	−0.5689,12	0.7737,4	0.3000,3
C(38)	−0.6434,12	0.8164,4	0.2720,3
C(39)	−0.6216,12	0.8752,4	0.2829,3
C(40)	−0.5252,12	0.8914,4	0.3217,3
C(41)	−0.4689,14	0.9472,5	0.3387,4
C(42)	−0.4661,24	1.0498,8	0.3378,6
C(43)	−0.5283,11	1.1041,3	0.3188,3
C(44)	−0.6272,11	1.1175,3	0.2798,3
C(45)	−0.6645,11	1.1756,3	0.2694,3
C(46)	−0.6028,11	1.2201,3	0.2980,3
C(47)	−0.5039,11	1.2067,3	0.3370,3
C(48)	−0.4666,11	1.1487,3	0.3474,3
C(49)	−0.3580,12	1.1205,4	0.3811,4
C(50)	−0.1839,19	1.1182,6	0.4514,5
C(51)	−0.1035,11	1.1477,4	0.4921,3
C(52)	−0.0735,11	1.2058,4	0.5041,3
C(53)	0.0254,11	1.2191,4	0.5432,3
C(54)	0.0942,11	1.1743,4	0.5702,3
C(55)	0.0642,11	1.1162,4	0.5582,3
C(56)	−0.0347,11	1.1029,4	0.5192,3
C(57)	−0.0786,23	1.0482,8	0.4955,7
C(58)	−0.0758,13	0.9469,5	0.4954,4
C(59)	−0.0371,12	0.8903,3	0.5153,3
C(60)	0.0708,12	0.8729,3	0.5512,3
C(61)	0.0959,12	0.8137,3	0.5601,3
C(62)	0.0131,12	0.7719,3	0.5331,3
C(63)	−0.0948,12	0.7893,3	0.4972,3
C(64)	−0.1200,12	0.8485,3	0.4883,3
C(65)	−0.1980,15	0.8783,5	0.4509,4

Personal communication Kuninobu Kasuga, reproduced with permission.

Appendix 39. The Atomic Coordinates of $Pc_2Nd(ox1)$, CIZGIB06[210]

Space group C2/c
$a = 28.29{,}1$ Å, $b = 14.187{,}5$ Å, $c = 13.325{,}5$ Å, $\beta = 115.35{,}4°$

Nd(1)	0.000	0.09248(2)	0.250
N(5)	−0.0543(1)	−0.1290(2)	0.0909(3)
N(1)	0.0240(1)	−0.0335(2)	0.1525(3)
C(1)	0.0026(1)	−0.1171(3)	0.1280(3)
C(2)	0.0342(1)	−0.1954(3)	0.1508(3)
C(3)	0.0284(1)	−0.2920(3)	0.1523(3)
C(4)	0.0733(1)	−0.3462(3)	0.1892(3)
C(5)	0.1223(1)	−0.3058(3)	0.2219(3)
C(6)	0.1281(1)	−0.2092(3)	0.3196(3)
C(7)	0.0834(1)	−0.1549(3)	0.1847(3)
C(8)	0.0756(1)	−0.0530(3)	0.1825(3)
N(6)	0.1150(1)	0.0084(2)	0.2097(3)
N(2)	0.0630(1)	0.1502(2)	0.1815(3)
C(9)	0.1084(1)	0.1014(3)	0.2064(3)
C(10)	0.1512(1)	0.1684(3)	0.2283(3)

(*Continued*)

Appendix 39. Continued

C(11)	0.2046(1)	0.1560(3)	0.2641(3)
C(12)	0.2344(1)	0.2363(3)	0.2776(3)
C(13)	0.2121(1)	0.3247(3)	0.2542(3)
C(14)	0.1591(1)	0.3378(3)	0.2203(3)
C(15)	0.1289(1)	0.2577(3)	0.2086(3)
C(16)	0.0733(1)	0.2438(3)	0.1789(3)
N(7)	0.0394(1)	0.3156(2)	0.1524(3)
N(3)	−0.0378(1)	0.2215(2)	0.1154(3)
C(17)	−0.0118(1)	0.3041(3)	0.1210(3)
C(18)	−0.0492(1)	0.3819(3)	0.0821(1)
C(19)	−0.0433(1)	0.4771(3)	0.0712(3)
C(20)	−0.0.880(1)	0.5317(3)	0.0270(3)
C(21)	−0.1371(1)	0.4922(3)	−0.0042(3)
C(22)	−0.1432(1)	0.3966(3)	0.0071(3)
C(23)	−0.0985(1)	0.3419(3)	0.0503(3)
C(24)	−0.0903(1)	0.2404(3)	0.0701(3)
N(8)	−0.1294(1)	0.1791(2)	0.0421(3)
N(4)	−0.0766(1)	0.0369(2)	0.0853(3)
C(25)	−0.1225(1)	0.0860(3)	0.0485(3)
C(26)	−0.1663(1)	0.0199(3)	0.0117(3)
C(27)	−0.2194(1)	0.0328(3)	−0.0267(3)
C(28)	−0.2507(1)	−0.0466(3)	−0.0538(3)
C(29)	−0.2295(1)	−0.1360(3)	−0.0413(3)
C(30)	−0.1758(1)	−0.1514(3)	−0.0010(3)
C(31)	−0.1448(1)	−0.0702(3)	0.0252(3)
C(32)	−0.0878(1)	−0.0572(3)	0.0698(3)

Personal communication Mireille Mossoyan-Deneux, reproduced with permission.

Appendix 40. The Atomic Coordinates of $[Pc_2Nd][O]$, SAFPUK[213]

Space group P4nc
$a = 19.514{,}5$ Å, $b = 19.514{,}5$ Å, $c = 6.511{,}3$ Å

Nd	0.000	0.000	0.000
O	0.500	0.500	0.000
N(1)	0.0267(9)	0.0965(9)	0.217(3)
N(5)	−0.153(1)	0.0730(9)	0.241(3)
C(1)	−0.103(1)	0.112(1)	0.242(3)
C(2)	−0.107(1)	0.187(1)	0.255(4)
C(3)	−0.166(1)	0.232(1)	0.291(4)
C(4)	−0.148(1)	0.299(1)	0.291(5)
C(5)	−0.081(1)	0.328(1)	0.280(4)
C(6)	−0.028(1)	0.283(1)	0.268(4)
C(7)	−0.042(1)	0.212(1)	0.256(5)
C(8)	0.003(9)	0.1531(9)	0.245(3)
N'(1)	0.093(1)	−0.031(1)	−0.247(4)
N'(5)	0.162(1)	0.076(1)	−0.247(4)
C'(1)	0.1555(1)	0.010(1)	−0.244(4)
C'(2)	0.215(1)	−0.033(1)	−0.255(4)
C'(3)	0.285(2)	−0.013(2)	−0.253(5)
C'(4)	0.332(2)	−0.068(2)	−0.256(7)
C'(5)	0.310(2)	−0.135(2)	−0.265(5)
C'(6)	0.235(2)	−0.155(2)	−0.270(7)
C'(7)	0.194(1)	−0.103(1)	−0.264(5)
C'(8)	0.120(1)	−0.097(1)	−0.245(4)
H(1)	−0.2114	0.2164	0.3116
H(2)	−0.1852	0.3305	0.2995
H(3)	−0.0738	0.3761	0.2803
H(4)	0.0184	0.2986	0.2681

(*Continued*)

Appendix 40. Continued

H(5)	0.2989	0.0332	−0.2495
H(6)	0.3793	−0.0589	−0.2515
H(7)	0.3452	−0.1675	−0.2837
H(8)	0.2200	−0.2142	−0.2362

Personal communication Mireille Mossoyan-Deneux, reproduced with permission.

Appendix 41. The Atomic Coordinates of SubpcB-OMe, #1992-01

Space group Pnma
$a = 12.339,2$ Å, $b = 15.290,4$ Å, $c = 10.567,3$ Å

B1	0.2448,4	0.2500,0	0.5646,5
N1	0.1926,2	0.1732,2	0.5009,2
N2	0.3579,3	0.2500,0	0.5127,3
N3	0.3595,2	0.0985,2	0.4629,2
N4	0.0365,3	0.2500,0	0.4302,4
C1	0.0909,2	0.1752,2	0.4493,3
C2	0.0731,3	0.0885,2	0.3976,3
C3	−0.0151,3	0.0524,2	0.3339,3
C4	−0.0043,3	−0.0315,2	0.2854,4
C5	0.0926,3	−0.0785,2	0.2978,3
C6	0.1821,3	−0.0421,2	0.3582,3
C7	0.1723,3	0.0418,2	0.4082,3
C8	0.2500,3	0.1001,2	0.4665,3
C9	0.4103,3	0.1752,2	0.4789,3
C10	0.5186,3	0.2035,2	0.4408,3
C11	0.6109,3	0.1560,3	0.4048,4
C12	0.7015,3	0.2045,3	0.3704,4
O1	0.2476,3	0.2500,0	0.6889,3
C13	0.1478,5	0.2500,0	0.7649,5

REFERENCES

1. Robertson, J. M. *J. Chem. Soc.* **1936**, 1195.
2. Hiller, W.; Strähle, J.; Kobel, W.; Hanack, M. *Z. Kristallogr.* **1982**, *159*, 173.
3. Okada, O.; Oka, K.; Iijima, M. *Jpn. J. Appl. Phys.* **1993**, *32*, 3556.
4. Gregory, P. *J. Porph. Phthalocyan.* **1998**, *3*, 468.
5. Barton, D. H. R.; Rydon, H. N.; Elvidge, J. A. *Biographical Memoirs of Fellows of the Royal Society* **1968**, *14*, 309.
6. Linstead, R. P. *J. Chem. Soc.* **1934**, 1016.
7. Byrne, G. T.; Linstead, R. P.; Lowe, A. R. *J. Chem. Soc.* **1934**, 1017.
8. Linstead, R. P.; Lowe, A. R. *J. Chem. Soc.* **1934**, 1022.
9. Dent, C. E.; Linstead, R. P. *J. Chem. Soc.* **1934**, 1027.
10. Linstead, R. P.; Lowe, A. R. *J. Chem. Soc.* **1934**, 1031.
11. Dent, C. E.; Linstead, R. P.; Lowe, A. R. *J. Chem. Soc.* **1934**, 1033.
12. de Diesbach, H.; von der Weid, E. *Helv. Chim. Acta* **1927**, *10*, 886.
13. Robertson, J. M.; Linstead, R. P.; Dent, C. E. *Nature* **1935**, 506.
14. Robertson, J. M.; Woodward, I. *J. Chem. Soc.* **1937**, 219.
15. Robertson, J. M.; Woodward, I. *J. Chem. Soc.* **1940**, 36.
16. Tanaka, M. In *High Performance Pigments.* Smith, H. M., Ed.; Wiley-VCH: Weinheim, **2002**; p. 263.
17. Erk, P.; Hengelsberg, H. In *The Porphyrin Handbook.* Kadish, K. M., Smith, K. M., Guilard, R., Eds.; Academic Press: Boston, **2003**; Chapter 119.
18. Kakuta, A. In *Infrared Absorbing Dyes.* Matsuoka, M., Ed.; Plenum Press: New York, **1990**; p. 155.
19. Law, K.-Y. *Chem. Rev.* **1993**, *93*, 449.
20. Borsenberger, P. M.; Weiss, D. S. *Organic Photoreceptors for Imaging Systems.* Marcel Dekker, Inc.: New York, **1993**.
21. Bouvet, M. In *The Porphyrin Handbook.* Kadish, K. M., Smith, K. M., Guilard, R., Eds.; Academic Press: Boston, **2003**; Chapter 118.
22. Erk, P. In *High Performance Pigments.* Smith, H. M., Ed.; Wiley-VCH: Weinheim, **2002**; p. 103.
23. Engel, M. K. *Kawamura Rikagaku Kenkyusho Hokoku* **1997**, *8(1996)*, 11; http://www.mkengel.de/phthalo/pcrev.htm.
24. Isoda, S.; Kobayashi, T. In *Phthalocyanines - Chemistry & Function.* Shirai, H., Kobayashi, N., Eds.; IPC Publisher: Tokyo, **1997**; 63.
25. Scheidt, W. R.; Lee, Y. J. In *Structure & Bonding*, **1987**; *64*, 1.
26. Scheidt, W. R. In *The Porphyrin Handbook.* Kadish, K. M., Smith, K. M., Guilard, R., Eds.; Academic Press: Boston, **2000**; Vol. 3, p. 49.
27. Williams, G. A.; Figgis, B. N.; Mason, R.; Mason, S. A.; Fielding, P. E. *J. Chem. Soc., Dalton Trans.* **1980**, 1688.
28. Williams, G. A.; Figgis, B. N.; Mason, R. *J. Chem. Soc., Dalton Trans.* **1981**, 734.
29. Matsumoto, S.; Matsuhama, K.; Mizuguchi, J. *Acta Crystallogr. C* **1999**, *55*, 131.
30. Mizuguchi, J.; Matsumoto, S. *J. Phys. Chem. A* **1999**, *103*, 614.
31. Mason, S. A. *Ph.D. thesis*, University of Melbourne, Australia **1971**.
32. Barrett, P. A.; Dent, C. E.; Linstead, R. P. *J. Chem. Soc.* **1936**, 1715.
33. Donohue, J. *J. Phys. Chem.* **1952**, *56*, 502.

34. Fleischer, E. B. *Ph.D. thesis*, Yale University, New Haven, USA, **1961**.
35. Mason, S. F. *J. Chem. Soc.* **1958**, 976.
36. Berezin, B. D. *Russ. J. Phys. Chem.* **1965**, *39*, 165.
37. Shurvell, H. F.; Pinzuti, L. *Can. J. Chem.* **1966**, *44*, 125.
38. Hoskins, B. F.; Mason, S. A.; White, J. C. B. *J. Chem. Soc., Chem. Commun.* **1969**, 554.
39. Fleischer, E. B. *Acc. Chem. Res.* **1970**, *3*, 105.
40. Bissell, E. C. *Ph.D. thesis*, Case Western Reserve University, Cleveland, Ohio, USA, **1970**.
41. Cook, M. J.; McMurdo, J.; Powell, A. K. *J. Chem. Soc., Chem. Commun.* **1993**, 903.
42. Chambrier, I.; Cook, M. J.; Helliwell, M.; Powell, A. K. *J. Chem. Soc., Chem. Commun.* **1992**, 444.
43. Baumann, T. F.; Sibert, J. W.; Olmstead, M. M.; Barrett, A. G. M.; Hoffman, B. M. *J. Amer. Chem. Soc.* **1994**, *116*, 2639.
44. Baumann, T. F.; Nasir, M. S.; Sibert, J. W.; White, A. J. P.; Olmstead, M. M.; Williams, D. J.; Barrett, A. G. M.; Hoffman, B. M. *J. Amer. Chem. Soc.* **1996**, *118*, 10479.
45. Chambrier, I.; Cook, M. J.; Wood, P. T. *Chem. Commun.* **2000**, 2133.
46. Shannon, R. D. *Acta Crystallogr. A* **1976**, *32*, 751.
47. Hoard, J. L. *Ann. N. Y. Acad. Sci.* **1973**, *206*, 18.
48. Berezin, B. D.; Koifman, O. I. *Russ. Chem. Rev.* **1980**, *49*, 1188.
49. Janczak, J.; Kubiak, R. *J. Alloys Comp.* **1993**, *202*, 69.
50. Hoberg, C.; Göldner, M.; Cornelissen, U.; Homborg, H. *Z. Anorg. Allg. Chem.* **2000**, *626*, 2435.
51. Janczak, J.; Kubiak, R.; Richter, J.; Fuess, H. *Polyhedron* **1999**, *18*, 2775.
52. Benihya, K.; Mossoyan-Deneux, M.; Hahn, F.; Boucharat, N.; Terzian, G. *Eur. J. Inorg. Chem.* **2000**, 1771.
53. Janczak, J.; Kubiak, R. *J. Chem. Soc., Dalton Trans.* **1993**, 3809.
54. Buchler, J. W.; Ng, D. K. P. In *The Porphyrin Handbook*. Kadish, K. M., Smith, K. M., Guilard, R., Eds.; Academic Press, **2000**; Vol. 3, p. 245.
55. Strohmaier, R.; Ludwig, C.; Petersen, J.; Gompf, B.; Eisenmenger, W. *J. Vac. Sci. Technol. B* **1996**, *14*, 1079.
56. Walzer, K.; Hietschold, M. *Surf. Sci.* **2001**, *471*, 1.
57. Robertson, J. M. *J. Chem. Soc.* **1935**, 615.
58. Linstead, R. P.; Robertson, J. M. *J. Chem. Soc.* **1936**, 1736.
59. Janczak, J.; Kubiak, R. *Polyhedron* **2001**, *20*, 2901.
60. Mizuguchi, J. *Z. Kristallogr. - NCS* **2001**, *216*, 377.
61. Mizuguchi, J. *J. Phys. Chem. A* **2001**, *105*, 1121.
62. Bluhm, T. L.; Wagner, H. J.; Loutfy, R. O. *J. Mater. Sci. Lett.* **1983**, *2*, 85.
63. Sievertsen, S. *Doctoral thesis*, University of Kiel, Germany, **1999**.
64. Susich, G. *FIAT Final Report* **1948**, *1313*, PB85172, 412.
65. Susich, G. *Anal. Chem.* **1950**, *22*, 425.
66. Barrett, P. A.; Linstead, R. P.; Tuey, G. A. P.; Robertson, J. M. *J. Chem. Soc.* **1939**, 1809.
67. Woodward, I. *J. Chem. Soc.* **1940**, 601.
68. Das, I. M.; Chandhuri, B. *Acta Crystallogr. B* **1972**, *28*, 579.
69. Huggins, M. L. *Nature* **1945**, *155*, 18.
70. Robertson, J. M. In *Organic Crystals and Molecules*. Cornell Univ. Press, Ithaca, N.Y., **1953**; p. 262.
71. Moser, F. H.; Thomas, A. L. In *Phthalocyanine Compounds*. Rheinhold Publ. Corp.: New York, **1963**.
72. Lever, A. B. P. *Adv. Inorg. Chem. Radiochem.* **1965**, *7*, 32.
73. Fielding, P. E.; Stephenson, N. C. *Aust. J. Chem.* **1965**, *18*, 1691.
74. Honigmann, B.; Lenné, H.-U.; Schrödel, R. *Z. Kristallogr.* **1965**, *122*, 185.
75. Suito, E.; Uyeda, N.; Ashida, M. *Senryo to Yakuhin* **1967**, *12*, 41.
76. Brown, C. J. *J. Chem. Soc., (A)* **1968**, 2488.
77. Brown, C. J. *J. Chem. Soc., (A)* **1968**, 2494.
78. Kirk, R. S. *Amer. Cryst. Assoc. Summer Meeting*, **1968**; p. 63 (abstract FF1).
79. Kirk, R. S. *Amer. Cryst. Assoc. Winter Meeting*, **1969**; p. 25 (abstract B6).
80. Bissell, E. C.; Knox, K.; Furnas, T. C.; Beard, D. W. *Amer. Cryst. Assoc. Summer Meeting*, **1968**, p. 63 (abstract FF2).
81. Kirner, J. F.; Dow, W.; Scheidt, W. R. *Inorg. Chem.* **1976**, *15*, 1685.
82. Scheidt, W. R.; Dow, W. *J. Amer. Chem. Soc.* **1977**, *99*, 1101.
83. Figgis, B. N.; Mason, R.; Williams, G. A. *Acta Crystallogr. A* **1978**, *34*, S158.
84. Mason, R.; Williams, G. A.; Fielding, P. E. *J. Chem. Soc., Dalton Trans.* **1979**, 676.
85. Figgis, B. N.; Kucharski, E. S.; Williams, G. A. *J. Chem. Soc., Dalton Trans.* **1980**, 1515.
86. Figgis, B. N.; Mason, R.; Williams, G. A. *Acta Crystallogr. B* **1980**, *36*, 2963.
87. Aldoshin, S. M.; Dyochenko, O. A.; Atovmyan, L. O.; Cheklov, A. N.; Alyanov, M. I. *Koord. Khim.* **1980**, *6*, 936.
88. Kirkland, E. J. *Ultramicroscopy* **1982**, *9*, 45.
89. Saito, Y.; Shiojiri, M. *J. Cryst. Growth* **1984**, *67*, 91.
90. Homborg, H.; Teske, C. L. *Z. Anorg. Allg. Chem.* **1985**, *527*, 45.
91. Sugimoto, H.; Mori, M.; Masuda, H.; Taga, T. *J. Chem. Soc., Chem. Commun.* **1986**, 962.
92. Perovic, A. *J. Mater. Sci.* **1987**, *22*, 835.
93. Yase, K.; Yasuoka, N.; Kobayashi, T.; Uyeda, N. *Acta Crystallogr. C* **1988**, *44*, 514.
94. Figgis, B. N.; Kucharski, E. S.; Reynolds, P. A. *J. Amer. Chem. Soc.* **1989**, *111*, 1683.
95. Reynolds, P. A.; Figgis, B. N.; Kucharski, E. S.; Mason, S. A. *Acta Crystallogr. B* **1991**, *47*, 899.
96. Kubiak, R.; Janczak, J. *J. Alloys Comp.* **1992**, *190*, 117.
97. Janczak, J.; Kubiak, R. *J. Alloys Comp.* **1992**, *190*, 121.
98. Janczak, J. *Pol. J. Chem.* **2000**, *74*, 157.
99. Oka, K.; Okada, O. *J. Imag. Sci. Technol.* **1993**, *37*, 13.
100. Kakuta, A. *Personal communication*, **1994**.
101. Hammond, R. B.; Roberts, K. J.; Docherty, R.; Edmondson, M.; Gairns, R. *J. Chem. Soc., Perkin Trans. 2* **1996**, 1527.
102. Zugenmaier, P.; Bluhm, T. L.; Deslandes, Y.; Orts, W. J.; Hamer, G. K. *J. Mater. Sci.* **1997**, *32*, 5561.
103. Liu, D.-J.; Blumberg Selinger, R. L.; Weeks, J. D. *J. Chem. Phys.* **1996**, *105*, 4751.
104. Morishige, K.; Araki, K. *J. Chem. Soc., Dalton Trans.* **1996**, 4303.
105. Brinkmann, M.; Chaumont, C.; Wachtel, H.; André, J.-J. *Thin Solid Films* **1996**, *283*, 97.
106. Brinkmann, M.; Turek, P.; André, J.-J. *J. Mater. Chem.* **1998**, *8*, 675.
107. Ballirano, P.; Caminiti, R.; Ercolani, C.; Maras, A.; Orru, M. A. *J. Amer. Chem. Soc.* **1998**, *120*, 12798.
108. Homborg, H. *Personal communication*.
109. Miller, J. S.; Epstein, A. J. *Progr. Inorg. Chem.* **1976**, *20*, 1.
110. *Low Dimensional Cooperative Phenomena*. Keller, H. J., Ed.; Plenum Press: New York, **1975**.
111. Schramm, C. J.; Scaringe, R. P.; Stojakovic, D. R.; Hoffman, B. M.; Ibers, J. A.; Marks, T. J. *J. Amer. Chem. Soc.* **1980**, *102*, 6702.
112. Martinsen, J.; Stanton, J. L.; Greene, R. L.; Tanaka, J.; Hoffman, B. M.; Ibers, J. A. *J. Amer. Chem. Soc.* **1985**, *107*, 6915.
113. Ogawa, M. Y.; Martinsen, J.; Palmer, S. M.; Stanton, J. L.; Tanaka, J.; Greene, R. L.; Hoffman, B. M.; Ibers, J. A. *J. Amer. Chem. Soc.* **1987**, *109*, 1115.
114. Gardberg, A. S.; Yang, S.; Hoffman, B. M.; Ibers, J. A. *Inorg. Chem.* **2002**, *41*, 1778.
115. Latte, B.; Aßmann, B.; Homborg, H. *Z. Anorg. Allg. Chem.* **1997**, *623*, 1281.
116. Rogers, D.; Osborn, R. S. *J. Chem. Soc., Chem. Commun.* **1971**, 840.
117. Osborn, R. S. *Doctoral thesis*, University of London, UK, **1972**.
118. Janczak, J.; Kubiak, R. *Pol. J. Chem.* **1999**, *73*, 1587.

119. Ejsmont, K.; Kubiak, R. *Acta Crystallogr. C* **1997**, *53*, 1051.
120. Krämer, T. *Doctoral thesis*, University of Tübingen, Germany, **1985**.
121. Nohr, R. S.; Wynne, K. J. *J. Chem. Soc., Chem. Commun.* **1981**, 1210.
122. Wynne, K. J. *Inorg. Chem.* **1985**, *24*, 1339.
123. Moubaraki, B.; Ley, M.; Benlian, D.; Sorbier, J.-P. *Acta Crystallogr. C* **1990**, *46*, 379.
124. Gingl, F. *Doctoral thesis*, University of Tübingen, Germany, **1990**.
125. Kobayashi, T.; Furukawa, C.; Ogawa, T.; Isoda, S. *J. Porph. Phthalocyan.* **1997**, *1*, 297.
126. Kojima, Y.; Osano, Y. T.; Ohashi, T. *Bull. Chem. Soc. Jpn* **1999**, *72*, 2203.
127. Kojima, Y.; Osano, Y. T.; Ohashi, T. *Bull. Chem. Soc. Jpn* **2000**, *73*, 2469.
128. Silver, J.; Frampton, C. S.; Fern, G.; Davies, D. A.; Miller, J. R.; Sosa-Sanchez, J. L. *Inorg. Chem.* **2001**, *40*, 5434.
129. Küppers, H.; Kalz, W.; Homborg, H. *Acta Crystallogr. C* **1985**, *41*, 1420.
130. Inabe, T. *J. Porph. Phthalocyan.* **2001**, *5*, 3.
131. Weidemann, M. *Doctoral thesis*, University of Kiel, Germany, **1997**; Homborg, H. *Personal communication* **1997**.
132. Iyechika, Y.; Yakushi, K.; Ikemoto, I.; Kuroda, H. *Acta Crystallogr. B* **1982**, *38*, 766.
133. Ukei, K. *Acta Crystallogr. B* **1973**, *29*, 2290.
134. Mossoyan-Deneux, M.; Benlian, D.; Pierrot, M.; Fournel, A.; Sorbier, J.-P. *Inorg. Chem.* **1985**, *24*, 1878.
135. Gacho, E. H.; Naito, T.; Inabe, T.; Fukuda, T.; Kobayashi, N. *Chem. Lett.* **2001**, 260.
136. Hückstädt, H.; Jaouen, C.; Göldner, M.; Cornelissen, U.; Tutaß, A.; Homborg, H. *Z. Anorg. Allg. Chem.* **2000**, *626*, 671.
137. Engel, M. K.; Yao, J.; Maki, H.; Takeuchi, H.; Yonehara, H.; Pac, C. *Kawamura Rikagaku Kenkyusho Hokoku* **1998**, *9 (1997)*, 53; http://www.mkengel.de/phthalo/subpc.htm.
138. Kietaibl, H. *Monatsh. Chem.* **1974**, *105*, 405.
139. Rauschnabel, J.; Hanack, M. *Tetrahedron Lett.* **1995**, *36*, 1629.
140. Kasuga, K.; Idehara, T.; Handa, M.; Ueda, Y.; Fujiwara, T.; Isa, K. *Bull. Chem. Soc. Jpn* **1996**, *69*, 2559.
141. Potz, R.; Göldner, M.; Hückstädt, H.; Cornelissen, U.; Tutaß, A.; Homborg, H. *Z. Anorg. Allg. Chem.* **2000**, *626*, 588.
142. Friedel, M. K.; Hoskins, B. F.; Martin, R. L.; Mason, S. A. *J. Chem. Soc., Chem. Commun.* **1970**, 400.
143. Kubiak, R.; Janczak, J. *J. Alloys Comp.* **1992**, *189*, 107.
144. Kobayashi, T.; Ashida, T.; Uyeda, N.; Suito, E.; Kakudo, M. *Bull. Chem. Soc. Jpn* **1971**, *44*, 2095.
145. Fischer, M. S.; Templeton, D. H.; Zalkin, A.; Calvin, M. *J. Amer. Chem. Soc.* **1971**, *93*, 2622.
146. Engel, M. K.; Saito, T.; Isogai, M.; Iwayanagi, Y. *Unpublished results* **1995**.
147. Suga, T.; Isoda, S.; Kobayashi, T.; Isogai, M.; Engel, M. *Mol. Cryst. Liq. Cryst.* **1999**, *337*, 377.
148. Ukei, K. *J. Phys. Soc. Jpn* **1976**, *40*, 140.
149. Griffiths, C. H.; Walker, M. S.; Goldstein, P. *Mol. Cryst. Liq. Cryst.* **1976**, *33*, 149.
150. Ziolo, R. F.; Griffiths, C. H.; Troup, J. M. *J. Chem. Soc., Dalton Trans.* **1980**, 2300.
151. Przyborowski, F.; Hamann, C. *Cryst. Res. Technol.* **1982**, *17*, 1041.
152. Ukei, K. *Acta Crystallogr. B* **1982**, *38*, 1288.
153. Börschel, V.; Strähle, J. *Z. Naturforsch. B* **1984**, *39*, 1664.
154. Wynne, K. J. *Inorg. Chem.* **1984**, *23*, 4658.
155. Hasegawa, M. *Master thesis*, University of Kyoto, Japan, **1995**.
156. Hasegawa, M.; Sato, N. *Mol. Cryst. Liq. Cryst.* **1997**, *296*, 409.
157. Palmer, S. M.; Stanton, J. L.; Jaggi, N. K.; Hoffman, B. M.; Ibers, J. A.; Schwartz, C. H. *Inorg. Chem.* **1985**, *24*, 2040.
158. Goedken, V. L.; Dessy, G.; Ercolani, C.; Fares, V.; Gastaldi, L. *Inorg. Chem.* **1985**, *24*, 991.
159. DeCian, A.; Moussavi, M.; Fischer, J.; Weiss, R. *Inorg. Chem.* **1985**, *24*, 3162.
160. Gingl, F.; Strähle, J. *Z. Naturforsch. B* **1988**, *43*, 445.
161. Tahiri, M.; Doppelt, P.; Fischer, J.; Weiss, R. *Inorg. Chem.* **1988**, *27*, 2897.
162. Gingl, F.; Strähle, J. *Z. Naturforsch. B* **1989**, *44*, 110.
163. Cellucci, L.; Ercolani, C.; Lukes, P. J.; Chiesi-Villa, A.; Rizzoli, C. *J. Porph. Phthalocyan.* **1998**, *2*, 9.
164. Bluhm, T.; Mayo, J.; Hamer, G.; Martin, T. *Proc. SPIE* **1992**, *1670*, 160.
165. Oka, K.; Okada, O.; Nukada, K. *Jpn. J. Appl. Phys.* **1992**, *31*, 2181.
166. Chen, M. J.; Rathke, J. W.; Huffman, J. C. *Organometallics* **1993**, *12*, 4673.
167. Inami, K.; Daimon, K.; Oka, K. *Japan Hardcopy Proc.*, Tokyo, **1994**; p. 221.
168. Inami, K.; Oka, K.; Daimon, K. *J. Imag. Sci. Technol.* **1995**, *39*, 298.
169. Gorsch, M.; Franken, A.; Sievertsen, S.; Homborg, H. *Z. Anorg. Allg. Chem.* **1995**, *621*, 607.
170. Aßmann, B.; Sievertsen, S.; Homborg, H. *Acta Crystallogr. C* **1996**, *52*, 876.
171. Chen, M. J.; Nunez, L.; Rathke, J. W.; Rogers, R. D. *Organometallics* **1996**, *15*, 2338.
172. Aßmann, B.; Ostendorp, G.; Lehmann, G.; Homborg, H. *Z. Anorg. Allg. Chem.* **1996**, *622*, 1085.
173. Aßmann, B.; Franken, A.; Homborg, H. *Z. Naturforsch. B* **1996**, *51*, 325.
174. Yamasaki, K.; Okada, O.; Inami, K.; Oka, K.; Kotani, M.; Yamada, H. *J. Phys. Chem. B* **1997**, *101*, 13.
175. Desiraju, G. R. *Cryst. Eng. Comm.* **2002**, *4*, 499.
176. Kubiak, R.; Ejsmont, K. *Acta Crystallogr. C* **1998**, *54*, 572.
177. Schweiger, K.; Kienast, A.; Latte, B.; Homborg, H. *Z. Anorg. Allg. Chem.* **1997**, *623*, 973.
178. Schweiger, K.; Hückstädt, H.; Homborg, H. *Z. Anorg. Allg. Chem.* **1998**, *624*, 44.
179. Schweiger, K.; Hückstädt, H.; Homborg, H. *Z. Anorg. Allg. Chem.* **1997**, *623*, 1853.
180. Schweiger, K.; Hückstädt, H.; Homborg, H. *Z. Anorg. Allg. Chem.* **1998**, *624*, 167.
181. Göldner, M.; Hückstädt, H.; Murray, K. S.; Moubaraki, B.; Homborg, H. *Z. Anorg. Allg. Chem.* **1998**, *624*, 288.
182. Schweiger, K.; Hückstädt, H.; Homborg, H. *Z. Anorg. Allg. Chem.* **1998**, *624*, 1298.
183. Kubiak, R.; Razik, M. *Acta Crystallogr. C* **1998**, *54*, 483.
184. Kubiak, R.; Ejsmont, K. *Pol. J. Chem.* **1999**, *73*, 1771.
185. Ejsmont, K.; Kubiak, R. *Acta Crystallogr. C* **1998**, *54*, 1844.
186. Janczak, J.; Kubiak, R. *Inorg. Chem.* **1999**, *38*, 2429.
187. Janczak, J.; Kubiak, R. *Inorg. Chim. Acta* **1999**, *288*, 174.
188. Kubiak, R.; Janczak, J.; Razik, M. *Inorg. Chim. Acta* **1999**, *293*, 155.
189. Janczak, J.; Kubiak, R.; Hahn, F. *Inorg. Chim. Acta* **1999**, *287*, 101.
190. Kubiak, R.; Ejsmont, K. *J. Mol. Struct.* **1999**, *474*, 275.
191. Schweiger, K.; Göldner, M.; Hückstädt, H.; Homborg, H. *Z. Anorg. Allg. Chem.* **1999**, *625*, 1693.
192. Tutaß, A.; Göldner, M.; Hückstädt, H.; Cornelissen, U.; Homborg, H. *Z. Anorg. Allg. Chem.* **2001**, *627*, 2323.
193. Janczak, J.; Kubiak, R. *Acta Crystallogr. C* **2001**, *57*, 55.
194. Janczak, J.; Idemori, Y. M. *Acta Crystallogr. C* **2002**, *58*, m23.
195. Troyanov, S. I.; Lapkina, L. A.; Larchenko, V. E.; Tsivadze, A. Y. *Dokl. Chem.* **1999**, *367*, 192.
196. Darovskikh, A. N.; Tsytsenko, A. K.; Frank-Kamenetskaya, O. V.; Fundamenskii, V. S.; Moskalev, P. N. *Sov. Phys.-Crystallogr.* **1984**, *29*, 273.

197. Koike, N.; Uekusa, H.; Ohashi, Y.; Harnoode, C.; Kitamura, F.; Ohsaka, T.; Tokuda, K. *Inorg. Chem.* **1996**, *35*, 5798.
198. Hückstädt, H.; Tutaß, A.; Göldner, M.; Cornelissen, U.; Homborg, H. *Z. Anorg. Allg. Chem.* **2001**, *627*, 485.
199. Broberg, D. E. *Ph.D. thesis*, University of Iowa, Iowa City, USA, **1967**.
200. Gieren, A.; Hoppe, W. *J. Chem. Soc., Chem. Commun.* **1971**, 413.
201. Bennett, W. E.; Broberg, D. E.; Baenziger, N. C. *Inorg. Chem.* **1973**, *12*, 930.
202. Kirin, I. S.; Kolyadin, A. B.; Lychev, A. A. *Russ. J. Inorg. Chem.* **1973**, *18*, 1215.
203. Kirin, I. S.; Kolyadin, A. B.; Lychev, A. A. *Sov. J. Struct. Chem.* **1974**, *15*, 415.
204. Kobayashi, T. *Bull. Inst. Chem. Res., Kyoto Univ.* **1978**, *56*, 204.
205. Kasuga, K.; Tsutsui, M.; Petterson, R. C.; Tatsumi, K.; Van Opdenbosch, N.; Pepe, G.; Meyer, E. F. *J. Amer. Chem. Soc.* **1980**, *102*, 4835.
206. Kasuga, K. *Personal communication* **1994**.
207. Darovskikh, A. N.; Frank-Kamenetskaya, O. V.; Fundamenskii, V. S.; Golynskaya, O. A. *Sov. Phys.-Crystallogr.* **1985**, *30*, 631.
208. Darovskikh, A. N.; Frank-Kamenetskaya, O. V.; Fundamenskii, V. S.; Golubev, A. M. *Sov. Phys.-Crystallogr.* **1986**, *31*, 165.
209. Darovskikh, A. N.; Frank-Kamenetskaya, O. V.; Fundamenskii, V. S. *Sov. Phys.-Crystallogr.* **1986**, *31*, 534.
210. Jullien, J.; Mossoyan-Deneux, M.; Pierrot, M.; Sorbier, J.-P.; Fournel, A.; Benlian, D. *C.R. Acad. Sci. II* **1986**, *303*, 669.
211. Darovsky, A.; Wu, L. Y.; Lee, P.; Sheu, H. S. *Acta Crystallogr. C* **1991**, *47*, 1836.
212. Moussavi, M.; DeCian, A.; Fischer, J.; Weiss, R. *Inorg. Chem.* **1988**, *27*, 1287.
213. Mossoyan-Deneux, M.; Benlian, D.; Baldy, A.; Pierrot, M. *Mol. Cryst. Liq. Cryst.* **1988**, *156*, 247.
214. Ercolani, C.; Paoletti, A. M.; Pennesi, G.; Rossi, G.; Chiesi-Villa, A.; Rizzoli, C. *J. Chem. Soc., Dalton Trans.* **1990**, 1971.
215. Silver, J.; Lukes, P.; Howe, S. D.; Howlin, B. *J. Mater. Chem.* **1991**, *1*, 29.
216. Paillaud, J. L.; Drillon, M.; DeCian, A.; Fischer, J.; Weiss, R.; Poinsot, R.; Herr, A. *Physica B* **1991**, *175*, 337.
217. Haghighi, M. S.; Teske, C. L.; Homborg, H. *Z. Anorg. Allg. Chem.* **1992**, *608*, 73.
218. Haghighi, M. S.; Rath, M.; Rotter, H. W.; Homborg, H. *Z. Anorg. Allg. Chem.* **1993**, *619*, 1887.
219. Capobianchi, A.; Ercolani, C.; Paoletti, A. M.; Pennesi, G.; Rossi, G.; Chiesi-Villa, A.; Rizzoli, C. *Inorg. Chem.* **1993**, *32*, 4605.
220. Haghighi, M. S.; Franken, A.; Homborg, H. *Z. Naturforsch. B* **1994**, *49*, 812.
221. Homborg, H. *Personal communication* **2001**.
222. Darovsky, A.; Keserashvili, V.; Harlow, R.; Mutikainen, I. *Acta Crystallogr. B* **1994**, *50*, 582.
223. Janczak, J.; Kubiak, R. *J. Alloys Comp.* **1994**, *204*, 5.
224. Janczak, J.; Kubiak, R.; Jezierski, A. *Inorg. Chem.* **1995**, *34*, 3505.
225. Ostendorp, G.; Werner, J.-P.; Homborg, H. *Acta Crystallogr. C* **1995**, *51*, 1125.
226. Ostendorp, G.; Homborg, H. *Z. Naturforsch. B* **1995**, *50*, 1200.
227. Ostendorp, G.; Rotter, H. W.; Homborg, H. *Z. Anorg. Allg. Chem.* **1996**, *622*, 235.
228. Ostendorp, G.; Homborg, H. *Z. Anorg. Allg. Chem.* **1996**, *622*, 873.
229. Ostendorp, G.; Homborg, H. *Z. Anorg. Allg. Chem.* **1996**, *622*, 1358.
230. Ostendorp, G.; Rotter, H. W.; Homborg, H. *Z. Naturforsch. B* **1996**, *51*, 567.
231. Donzello, M. P.; Ercolani, C.; Chiesi-Villa, A.; Rizzoli, C. *Inorg. Chem.* **1998**, *37*, 1347.
232. Janczak, J.; Kubiak, R.; Hahn, F. *Inorg. Chim. Acta* **1998**, *281*, 195.
233. Janczak, J.; Kubiak, R.; Jezierski, A. *Inorg. Chem.* **1999**, *38*, 2043.
234. Janczak, J.; Kubiak, R. *Polyhedron* **1999**, *18*, 1621.
235. Janczak, J.; Kubiak, R.; Svoboda, I.; Jezierski, A.; Fuess, H. *Inorg. Chim. Acta* **2000**, *304*, 150.
236. Rittenberg, D. K.; Sugiura, K.-i.; Sakata, Y.; Arif, A. M.; Miller, J. S. *Inorg. Chem.* **2001**, *40*, 3654.
237. Jiang, J.; Kasuga, K.; Arnold, D. P. In *Supramolecular Photosensitive and Electroactive Materials*. Nalwa, H. S., Ed.; Academic Press: Boston, **2001**; p. 113.
238. McKeown, N. B.; Helliwell, M. *Personal communication* **2002**.
239. Kobayashi, N.; Fukuda, T.; Ueno, K.; Ogino, H. *J. Amer. Chem. Soc.* **2001**, *123*, 10740.
240. Helliwell, M.; Deacon, A.; Moon, K. J.; Powell, A. K.; Cook, M. J. *Acta Crystallogr. B* **1997**, *53*, 231.
241. Kasuga, K.; Kawashima, M.; Asano, K.; Sugimori, T.; Abe, K.; Kikkawa, T.; Fujiwara, T. *Chem. Lett.* **1996**, 867.
242. Kasuga, K.; Asano, K.; Lin, L.; Sugimori, T.; Handa, M.; Abe, K.; Kikkawa, T.; Fujiwara, T. *Bull. Chem. Soc. Jpn* **1997**, *70*, 1859.
243. Ruf, M.; Lawrence, A. M.; Noll, B. C.; Pierpont, C. G. *Inorg. Chem.* **1998**, *37*, 1992.
244. Sugimori, T.; Kasuga, K. *Personal communication* **2002**.
245. Sugimori, T.; Okamoto, S.; Kotoh, N.; Handa, M.; Kasuga, K. *Chem. Lett.* **2000**, 1200.
246. Contakes, S. M.; Beatty, S. T.; Dailey, K. K.; Rauchfuss, T. B.; Fenske, D. *Organometallics* **2000**, *19*, 4767.
247. Gül Gürek, A.; Ahsen, V.; Luneau, D.; Pecaut, J. *Inorg. Chem.* **2001**, *40*, 4793.
248. Liu, W.; Lee, C.-H.; Li, H.-W.; Lam, C.-K.; Wang, J.; Mak, T. C. W.; Ng, D. K. P. *Chem. Commun.* **2002**, 628.
249. Hanack, M.; Renz, G.; Strähle, J.; Schmid, S. *Chem. Ber.* **1988**, *121*, 1479.
250. Hanack, M.; Renz, G.; Strähle, J.; Schmid, S. *J. Org. Chem.* **1991**, *56*, 3501.
251. Matsumura, N.; Fujita, A.; Naito, T.; Inabe, T. *J. Mater. Chem.* **2000**, *10*, 2266.
252. Furuya, F.; Kobayashi, N.; Bian, Y.; Jiang, J. *Chem. Lett.* **2001**, 944.
253. Baumann, F.; Bienert, B.; Rösch, G.; Vollmann, H.; Wolf, W. *Angew. Chem.* **1956**, *68*, 133.
254. Gingl, F.; Strähle, J. *Acta Crystallogr. C* **1990**, *46*, 1841.
255. Janczak, J.; Kubiak, R. *J. Chem. Soc., Dalton Trans.* **1994**, 2539.
256. Janczak, J.; Kubiak, R. *Acta Crystallogr. C* **1995**, *51*, 2039.
257. Janczak, J.; Kubiak, R. *Acta Chem. Scand.* **1995**, *49*, 871.
258. Benihya, K.; Mossoyan-Deneux, M.; Giorgi, M. *Eur. J. Inorg. Chem.* **2001**, 1343.
259. Claessens, C. G.; Gonzalez-Rodriguez, D.; Torres, T. *Chem. Rev.* **2002**, *102*, 835.
260. Day, V. W.; Marks, T. J.; Wachter, W. A. *J. Amer. Chem. Soc.* **1975**, *97*, 4519.
261. Fukuda, T.; Stork, J. R.; Potucek, R. J.; Olmstead, M. M.; Noll, B. C.; Kobayashi, N.; Durfee, W. S. S. *Angew. Chem. Int. Ed. Engl.* **2002**, *114*, 2565.
262. Idehara, T. *Master thesis*, Shimane University, **1992**.
263. Kasuga, K. *Personal communication* **1996**.
264. Stork, J. R.; Potucek, R. J.; Durfee, W. S.; Noll, B. C. *Tetrahedron Lett.* **1999**, *40*, 8055.
265. Claessens, C. G.; Torres, T. *Angew. Chem. Int. Ed. Engl.* **2002**, *114*, 2561.
266. Kopranenkov, V. N.; Luk'yanets, E. A. *Russ. Chem. Bull.* **1995**, *44*, 2216.
267. Stuzhin, P. A.; Khelevina, O. G. *Coord. Chem. Rev.* **1996**, *147*, 41.
268. Andersen, K.; Anderson, M.; Anderson, O. P.; Baum, S.; Baumann, T. F.; Beall, L. S.; Broderick, W. E.; Cook, A. S.;

Eichhorn, D. M.; Goldberg, D.; Hope, H.; Jarrell, W.; Lange, S. J.; McCubbin, Q. J.; Mani, N. S.; Miller, T.; Montalban, A. G.; Rodriguez-Morgade, M. S.; Lee, S.; Nie, H.; Olmstead, M. M.; Sabat, M.; Sibert, J. W.; Stern, C.; White, A. J. P.; Williams, D. B. G.; Williams, D. J.; Barrett, A. G. M.; Hoffman, B. M. *J. Heterocycl. Chem.* **1999**, *35*, 1013.

269. Swift, D. R. *Ph.D. thesis*, Case Western Reserve University, Cleveland, Ohio, USA, **1970**.
270. D'Addario, A. P. *Ph.D. thesis*, Case Western Reserve University, Cleveland, Ohio, USA, **1971**.
271. Miller, J. S. *Personal communication* **1996**.
272. Inabe, T.; Marks, T. J.; Burton, R. L.; Lyding, J. W.; McCarthy, W. J.; Kannewurf, C. R.; Reisner, G. M.; Herbstein, F. H. *Solid State Commun.* **1985**, *54*, 501.
273. Thompson, J. A.; Murata, K.; Miller, D. C.; Stanton, J. L.; Broderick, W. E.; Hoffman, B. M.; Ibers, J. A. *Inorg. Chem.* **1993**, *32*, 3546.
274. Yakushi, K.; Ida, T.; Ugawa, A.; Yamakado, H.; Ishii, H.; Kuroda, H. *J. Phys. Chem.* **1991**, *95*, 7636.
275. Gardberg, A. S.; Sprauve, A. E.; Ibers, J. A. *Inorg. Chim. Acta* **2002**, *328*, 179.
276. Homborg, H.; Kalz, W. *Z. Anorg. Allg. Chem.* **1984**, *514*, 115.
277. Ercolani, C.; Neri, C.; Porta, P. *Inorg. Chim. Acta* **1967**, *1*, 415.
278. Yakushi, K.; Yamakado, H.; Ida, T.; Ugawa, A.; Awaga, K.; Maruyama, Y.; Imaeda, K.; Inokuchi, H. *Synth. Met.* **1993**, *56*, 1699.
279. Hückstädt, H.; Homborg, H. *Z. Anorg. Allg. Chem.* **1998**, *624*, 715.
280. Ding, Y.; Simonyan, M.; Yonehara, Y.; Uruichi, M.; Yakushi, K. *J. Mater. Chem.* **2001**, *11*, 1469.
281. Dorset, D. L. *Acta Crystallogr. A* **1997**, *53*, 356.
282. Moxon, N. T.; Fielding, P. E.; Gregson, A. K. *Aust. J. Chem.* **1981**, *34*, 489.
283. Robinson, M. T.; Klein, G. E. *J. Amer. Chem. Soc.* **1952**, *74*, 6294.
284. Miyahara, T.; Shimizu, M. *J. Cryst. Growth* **2001**, *229*, 553.
285. Miyahara, T.; Shimizu, M. *J. Cryst. Growth* **2001**, *226*, 130.
286. Brown, C. J. *Acta Crystallogr. A* **1966**, *21*, 137.
287. Godfrey, M. R.; Liou, K.; Newcomb, T. P.; Halperin, W. P.; Hoffman, B. M.; Ibers, J. A. *Synth. Met.* **1989**, *29*, F51.
288. Tahiri, M.; Doppelt, P.; Fischer, J.; Weiss, R. *Inorg. Chim. Acta* **1987**, *127*, L1.
289. Aßmann, B.; Franken, A.; Homborg, H. *Z. Anorg. Allg. Chem.* **1995**, *621*, 1715.
290. Latte, B.; Kienast, A.; Bruhn, C.; Loidl, A.; Homborg, H. *J. Porph. Phthalocyan.* **1997**, *1*, 267.
291. Figgis, B. N.; Williams, G. A.; Forsyth, J. B.; Mason, R. *J. Chem. Soc., Dalton Trans.* **1981**, 1837.
292. Yakushi, K.; Sakuda, M.; Hamada, I.; Kuroda, H.; Kawamoto, A.; Tanaka, J.; Sugano, T.; Kinoshita, M. *Synth. Met.* **1987**, *19*, 769.
293. Yakushi, K.; Yamakado, H.; Yoshitake, M.; Kosugi, N.; Kuroda, H.; Sugano, T.; Kinoshita, M.; Kawamoto, A.; Tanaka, J. *Bull. Chem. Soc. Jpn* **1989**, *62*, 687.
294. Yakushi, K.; Yamakado, H.; Yoshitake, M.; Kosugi, N.; Kuroda, H.; Kawamoto, A.; Tanaka, J.; Sugano, T.; Kinoshita, M.; Hino, S. *Synth. Met.* **1989**, *29*, F95.
295. Yakushi, K.; Sakuda, M.; Kuroda, H.; Kawamoto, A.; Tanaka, J. *Chem. Lett.* **1986**, 1161.
296. Inabe, T.; Nakamura, S.; Liang, W.-B.; Marks, T. J.; Burton, R. L.; Kannewurf, C. R.; Imaeda, K.-i. *J. Amer. Chem. Soc.* **1985**, *107*, 7224.
297. Almeida, M.; Kanatzidis, M. G.; Tonge, L. M.; Marks, T. J.; McCarthy, W. J.; Marcy, H. O.; Kannewurf, C. R. *Synth. Met.* **1989**, *29*, F37.
298. Inabe, T.; Gaudiello, J. G.; Moguel, M. K.; Lyding, J. W.; Burton, R. L.; McCarthy, W. J.; Kannewurf, C. R.; Marks, T. J. *J. Amer. Chem. Soc.* **1986**, *108*, 7595.
299. Almeida, M.; Kanatzidis, M. G.; Tonge, L. M.; Marks, T. J.; Marcy, H. O.; McCarthy, W. J.; Kannewurf, C. R. *Solid State Commun.* **1987**, *63*, 457.
300. Palmer, S. M.; Stanton, J. M.; Hoffman, B. M.; Ibers, J. A. *Inorg. Chem.* **1986**, *25*, 2296.
301. Yamakado, H.; Yakushi, K.; Kosugi, N.; Kuroda, H.; Kawamoto, A.; Tanaka, J.; Sugano, T.; Kinoshita, M.; Hino, S. *Bull. Chem. Soc. Jpn* **1989**, *62*, 2267.
302. Gardberg, A. S.; Deng, K.; Ellis, D. E.; Ibers, J. A. *J. Amer. Chem. Soc.* **2002**, *124*, 5476.
303. Janczak, J.; Idemori, Y. M. *Acta Crystallogr. E* **2002**, *58*, m36.
304. Schröder, E. *Doctoral thesis*, University of Kiel, Germany, **1997**.
305. Janczak, J.; Idemori, Y. M. *Acta Crystallogr. C* **2001**, *57*, 924.
306. Gardberg, A. S.; Doan, P. E.; Hoffman, B. M.; Ibers, J. A. *Angew. Chem. Int. Ed. Engl.* **2001**, *40*, 244.
307. Ercolani, C.; Gardini, M.; Monacelli, F.; Pennesi, G.; Rossi, G. *Inorg. Chem.* **1983**, *22*, 2584.
308. Schweiger, K. *Doctoral thesis*, University of Kiel, Germany, **1998**.
309. Mizuguchi, J. *Z. Kristallogr. - NCS* **2002**, *217*, 255.
310. Börschel, V. *Doctoral thesis*, University of Tübingen, Germany, **1985**.
311. Mizuguchi, J.; Mochizuki, M. *Z. Kristallogr.-NCS* **2002**, *217*, 244.
312. Okada, O.; Klein, M. L. *J. Chem. Soc., Faraday Trans.* **1996**, *92*, 2463.
313. Yanagi, H.; Mikami, T.; Tada, H.; Terui, T.; Mashiko, S. *J. Appl. Phys.* **1997**, *81*, 7306.
314. Geiss, A.; Vahrenkamp, H. *Inorg. Chem.* **2000**, *39*, 4029.
315. Mossoyan-Deneux, M.; Benlian, D.; Ley, M.; Pierrot, M.; Sorbier, J. P.; Fournel, A. *Mol. Cryst. Liq. Cryst.* **1985**, *120*, 437.
316. Ziolo, R. F.; Guenther, W. H. H.; Troup, J. M. *J. Amer. Chem. Soc.* **1981**, *103*, 4629.
317. Ziolo, R. F.; Extine, M. *Inorg. Chem.* **1981**, *20*, 2709.
318. Sugimoto, H.; Higashi, T.; Maeda, A.; Hirai, Y.; Teraoka, J.; Mori, M.; Masuda, H.; Taga, T. *J. Less-Common Met.* **1985**, *112*, 387.
319. Sugimoto, H.; Higashi, T.; Maeda, A.; Mori, M.; Masuda, H.; Taga, T. *J. Chem. Soc., Chem. Commun.* **1983**, 1234.
320. Hedtmann-Rein, C. *Doctoral thesis*, University of Tübingen, Germany, **1986**.
321. Orihashi, Y.; Kobayashi, N.; Tsuchida, E.; Matsuda, H.; Nakanishi, H.; Kato, M. *Chem. Lett.* **1985**, 1617.
322. Tsuchida, E.; Orihashi, Y.; Kobayashi, N.; Ohno, H. *Synth. Met.* **1986**, *15*, 201.
323. Orihashi, Y.; Kobayashi, N.; Ohno, H.; Tsuchida, E.; Matsuda, H.; Nakanishi, H.; Kato, M. *Synth. Met.* **1987**, *19*, 751.
324. Aßmann, B.; Homborg, H. *Z. Anorg. Allg. Chem.* **1996**, *622*, 766.
325. Cariati, F.; Morazzoni, F.; Zocchi, M. *Inorg. Chim. Acta* **1975**, *14*, L31.
326. Cariati, F.; Morazzoni, F.; Zocchi, M. *J. Chem. Soc., Dalton Trans.* **1978**, 1018.
327. Inabe, T.; Morimoto, K. *Mol. Cryst. Liq. Cryst.* **1996**, *285*, 107.
328. Morimoto, K.; Inabe, T. *Mol. Cryst. Liq. Cryst.* **1996**, *284*, 291.
329. Morimoto, K.; Inabe, T. *J. Mater. Chem.* **1995**, *5*, 1749.
330. Fujita, A.; Hasegawa, H.; Naito, T.; Inabe, T. *J. Porph. Phthalocyan.* **1999**, *3*, 720.
331. Inabe, T.; Maruyama, Y. *Bull. Chem. Soc. Jpn* **1990**, *63*, 2273.
332. Inabe, T.; Maruyama, Y.; Mitsuhashi, T. *Synth. Met.* **1991**, *41–43*, 2629.
333. Hedtmann-Rein, C.; Hanack, M.; Peters, K.; Peters, E.-M.; von Schnering, H. G. *Inorg. Chem.* **1987**, *26*, 2647.
334. Takano, S.; Naito, T.; Inabe, T. *J. Mater. Chem.* **1998**, *8*, 511.
335. Inabe, T.; Maruyama, Y. *Chem. Lett.* **1989**, 55.

336. Matsuda, M.; Naito, T.; Inabe, T.; Hanasaki, N.; Tajima, H. *J. Mater. Chem.* **2001**, *11*, 2493.
337. Takano, S.; Naito, T.; Inabe, T. *Chem. Lett.* **1998**, 1249.
338. Mitsuya, K.-I.; Inabe, T.; Okuno, T.; Awaga, K. *Mol. Cryst. Liq. Cryst.* **1997**, *296*, 293.
339. Hasegawa, H.; Takano, S.; Miyajima, N.; Inabe, T. *Mol. Cryst. Liq. Cryst.* **1996**, *285*, 113.
340. Hasegawa, H.; Naito, T.; Inabe, T.; Akutagawa, T.; Nakamura, T. *J. Mater. Chem.* **1998**, *8*, 1567.
341. Bench, B. A.; Brennessel, W. W.; Lee, H.-J.; Gorun, S. M. *Angew. Chem. Int. Ed. Engl.* **2002**, *41*, 750.
342. Gorsch, M.; Homborg, H. *Personal communication* **2001**.
343. Sievertsen, S.; Aßmann, B.; Homborg, H. *Z. Anorg. Allg. Chem.* **1996**, *622*, 1685.
344. Gardberg, A. S.; Ibers, J. A. *Acta Crystallogr. C* **2001**, *57*, 528.
345. Kobayashi, T.; Kurokawa, F.; Ashida, T.; Uyeda, N.; Suito, E. *J. Chem. Soc., Chem. Commun.* **1971**, 1631.
346. Calderazzo, F.; Pampaloni, G.; Vitali, D.; Pelizzi, G.; Collamati, I.; Frediani, S.; Serra, A. M. *J. Organomet. Chem.* **1980**, *191*, 217.
347. Calderazzo, F.; Pampaloni, G.; Vitali, D.; Collamati, I.; Dessy, G.; Fares, V. *J. Chem. Soc., Dalton Trans.* **1980**, 1965.
348. Rossi, G.; Goedken, V. L.; Ercolani, C. *J. Chem. Soc., Chem. Commun.* **1988**, 46.
349. Ercolani, C.; Monacelli, F.; Dzugan, S.; Goedken, V. L.; Pennesi, G.; Rossi, G. *J. Chem. Soc., Dalton Trans.* **1991**, 1309.
350. Moubaraki, B.; Benlian, D.; Baldy, A.; Pierrot, M. *Acta Crystallogr. C* **1989**, *45*, 393.
351. Nemykin, V. N.; Kobayashi, N.; Chernii, V. Y.; Belsky, V. K. *Eur. J. Inorg. Chem.* **2001**, 733.
352. Kienast, A.; Bruhn, C.; Homborg, H. *Z. Anorg. Allg. Chem.* **1997**, *623*, 967.
353. Galich, L.; Kienast, A.; Hückstädt, H.; Homborg, H. *Z. Anorg. Allg. Chem.* **1998**, *624*, 1235.
354. Potz, R.; Hückstädt, H.; Homborg, H. *Personal communication* **2001**.
355. Kienast, A.; Homborg, H. *Z. Anorg. Allg. Chem.* **1998**, *624*, 107.
356. Kienast, A.; Homborg, H. *Z. Anorg. Allg. Chem.* **1998**, *624*, 233.
357. Küppers, H.; Kalz, W.; Homborg, H. *Acta Crystallogr. C* **1986**, *42*, 925.
358. Küppers, H.; Eulert, H.-H.; Hesse, K.-F.; Kalz, W.; Homborg, H. *Z. Naturforsch. B* **1986**, *41*, 44.
359. Matsuda, M.; Naito, T.; Inabe, T.; Hanasaki, N.; Tajima, H.; Otsuka, T.; Awaga, K.; Narymbetov, B.; Kobayashi, H. *J. Mater. Chem.* **2000**, *10*, 631.
360. Potz, R.; Hückstädt, H.; Homborg, H. *Z. Anorg. Allg. Chem.* **1998**, *624*, 173.
361. Ercolani, C.; Jubb, J.; Pennesi, G.; Russo, U.; Trigiante, G. *Inorg. Chem.* **1995**, *34*, 2535.
362. Donzello, M. P.; Ercolani, C.; Russo, U.; Chiesi-Villa, A.; Rizzoli, C. *Inorg. Chem.* **2001**, *40*, 2963.
363. Dirk, C. W.; Inabe, T.; Schoch, K. F.; Marks, T. J. *J. Amer. Chem. Soc.* **1983**, *105*, 1539.
364. Wang, H.-Y.; Chiang, C. W.; Lando, J. B. *Thin Solid Films* **1996**, *273*, 90.
365. Diel, C. W.; Inabe, T.; Lyding, J. W.; Schoch, K. F.; Kannewurf, C. R.; Marks, T. J. *J. Amer. Chem. Soc.* **1983**, *105*, 1551.
366. Janczak, J.; Razik, M.; Kubiak, R. *Acta Crystallogr. C* **1999**, *55*, 359.
367. Hückstädt, H.; Homborg, H. *J. Porph. Phthalocyan.* **1998**, *2*, 269.
368. Hückstädt, H.; Homborg, H. *Z. Naturforsch. B* **1997**, *52*, 728.
369. Hückstädt, H.; Homborg, H. *Z. Anorg. Allg. Chem.* **1997**, *623*, 369.
370. Hückstädt, H.; Homborg, H. *Z. Anorg. Allg. Chem.* **1998**, *624*, 980.
371. Hückstädt, H.; Homborg, H. *Z. Naturforsch. B* **1997**, *52*, 1003.
372. Hückstädt, H.; Homborg, H. *J. Porph. Phthalocyan.* **1999**, *3*, 169.
373. Matsumoto, S.; Endo, A.; Mizuguchi, J. *Z. Kristallogr.* **2000**, *215*, 182.
374. Janczak, J.; Kubiak, R. *Polyhedron* **2002**, *21*, 265.
375. Geiss, A.; Keller, M.; Vahrenkamp, H. *J. Organomet. Chem.* **1997**, *541*, 441.
376. Miller, J. S.; Vazquez, C.; Calabrese, J. C.; McLean, R. S.; Epstein, A. J. *Adv. Mater.* **1994**, *6*, 217.
377. Göldner, M.; Geniffke, B.; Franken, A.; Murray, K. S.; Homborg, H. *Z. Anorg. Allg. Chem.* **2001**, *627*, 935.
378. Vogt, L. H.; Zalkin, A.; Templeton, D. H. *Science* **1966**, *151*, 569.
379. Vogt, L. H.; Zalkin, A.; Templeton, D. H. *Inorg. Chem.* **1967**, *6*, 1725.
380. Gorsch, M.; Kienast, A.; Hückstädt, H.; Homborg, H. *Z. Anorg. Allg. Chem.* **1997**, *623*, 1433.
381. Omiya, S.; Tsutsui, M.; Meyer, E. F.; Bernal, I.; Cullen, D. L. *Inorg. Chem.* **1980**, *19*, 134.
382. Schlehahn, H. *Doctoral thesis*, University of Kiel, Germany, **1996**.
383. Göldner, M.; Kienast, A.; Homborg, H. *Z. Anorg. Allg. Chem.* **1998**, *624*, 141.
384. Göldner, M.; Homborg, H. *Z. Anorg. Allg. Chem.* **2000**, *626*, 1803.
385. Göldner, M.; Hückstädt, H.; Homborg, H. *Z. Anorg. Allg. Chem.* **1998**, *624*, 897.
386. Hückstädt, H. *Doctoral thesis*, University of Kiel, Germany, **1997**.
387. Hückstädt, H.; Bruhn, C.; Homborg, H. *J. Porph. Phthalocyan.* **1997**, *1*, 367.
388. Sievertsen, S.; Weidemann, M.; Hückstädt, H.; Homborg, H. *J. Porph. Phthalocyan.* **1997**, *1*, 379.
389. Weidemann, M.; Hückstädt, H.; Homborg, H. *Z. Anorg. Allg. Chem.* **1998**, *624*, 846.
390. Okada, K.; Sasa, N.; Nakamura, K.; Okada, S. *Amer. Cryst. Assoc. Meeting*, **1997**; Abstr. E191.
391. Sasa, N.; Okada, K.; Nakamura, K.; Okada, S. *J. Mol. Struct.* **1998**, *446*, 163.
392. Silver, J.; Sosa-Sanchez, J. L.; Frampton, C. S. *Inorg. Chem.* **1998**, *37*, 411.
393. Ciliberto, E.; Doris, K. A.; Pietro, W. J.; Reisner, G. M.; Ellis, D. E.; Fragala, I.; Herbstein, F. H.; Ratner, M. A.; Marks, T. J. *J. Amer. Chem. Soc.* **1984**, *106*, 7748.
394. Kawase, N.; Kurata, H.; Kubono, K.; Isoda, S.; Kobayashi, T. *J. Polym. Sci. B* **1993**, *31*, 1713.
395. Mooney, J. R.; Choy, C. K.; Knox, K.; Kenney, M. E. *J. Amer. Chem. Soc.* **1975**, *97*, 3033.
396. Brewis, M.; Helliwell, M.; McKeown, N. B.; Reynolds, S.; Shawcross, A. *Tetrahedron Lett.* **2001**, *42*, 813.
397. Brewis, M.; Clarkson, G. J.; Goddard, V.; Helliwell, M.; Holder, A. M.; McKeown, N. B. *Angew. Chem. Int. Ed. Engl.* **1998**, *37*, 1092.
398. Decreau, R.; Julliard, M.; Giorgi, M. *Acta Crystallogr. C* **1999**, *C55*, 1717.
399. Farren, C.; FitzGerald, S.; Bryce, M. R.; Beeby, A.; Batsanov, A. S. *Perkin Trans. 2* **2002**, 59.
400. Tutaß, A.; Klöpfer, M.; Hückstädt, H.; Cornelissen, U.; Homborg, H. *Z. Anorg. Allg. Chem.* **2002**, *628*, 1027.
401. Schweiger, K.; Hückstädt, H.; Homborg, H. *Z. Naturforsch. B* **1999**, *54*, 963.
402. Gingl, F. *Diplomal thesis*, University of Tübingen, Germany, **1986**.
403. Kubiak, R.; Ejsmont, K. *Progr. Coord. Organomet. Chem.* **1997**, 81.
404. Aßmann, B.; Ostendorp, G.; Homborg, H. *Z. Anorg. Allg. Chem.* **1995**, *621*, 1708.
405. Lachkar, M.; DeCian, A.; Fischer, J.; Weiss, R. *New J. Chem.* **1988**, *12*, 729.
406. Chabach, D.; Lachkar, M.; DeCian, A.; Fischer, J.; Weiss, R. *New J. Chem.* **1992**, *16*, 431.
407. Chabach, D.; DeCian, A.; Fischer, J.; Weiss, R.; El Mahouli Bibout, M. *Angew. Chem. Int. Ed. Engl.* **1996**, *35*, 898.

408. Chabach, D.; Tahiri, M.; DeCian, A.; Fischer, J.; Weiss, R.; El Malouli Bibout, M. *J. Amer. Chem. Soc.* **1995**, *117*, 8548.
409. Janczak, J.; Idemori, Y. M. *Inorg. Chim. Acta* **2001**, *325*, 85.
410. Tahiri, M.; Chabach, D.; El Mahouli Bibout, M.; DeCian, A.; Fischer, J.; Weiss, R. *Ann. Chim. France* **1995**, *20*, 81.
411. Kubiak, R.; Janczak, J. *Cryst. Res. Technol.* **2001**, *36*, 1095.
412. Moussavi, M.; DeCian, A.; Fischer, J.; Weiss, R. *Inorg. Chem.* **1986**, *25*, 2107.
413. Lux, F.; Beck, O. F.; Krauss, H.; Brown, D.; Tso, T. C. *Z. Naturforsch. B* **1980**, *35*, 564.
414. Molek, C. D.; Halfen, J. A.; Loe, J. C.; McGaff, R. W. *Chem. Commun.* **2001**, 2644
415. Velazquez, C. S.; Broderick, W. E.; Sabat, M.; Barrett, A. G. M.; Hoffman, B. M. *J. Amer. Chem. Soc.* **1990**, *112*, 7408.
416. Velazquez, C. S.; Fox, G. A.; Broderick, W. E.; Andersen, K. A.; Anderson, O. P.; Barrett, A. G. M.; Hoffman, B. M. *J. Amer. Chem. Soc.* **1992**, *114*, 7416.
417. Fitzgerald, J. P.; Haggerty, B. S.; Rheingold, A. L.; May, L.; Brewer, G. A. *Inorg. Chem.* **1992**, *31*, 2006.
418. Velazquez, C. S.; Baumann, T. F.; Olmstead, M. M.; Hope, H.; Barrett, A. G. M.; Hoffman, B. M. *J. Amer. Chem. Soc.* **1993**, *115*, 9997.
419. Balch, A. L.; Olmstead, M. M.; Safari, N. *Inorg. Chem.* **1993**, *32*, 291.
420. van Nostrum, C. F.; Benneker, F. B. G.; Veldman, N.; Spek, A. L.; Schouten, A.-J.; Nolte, R. J. M. *Recl. Trav. Chim. Pays-Bas* **1994**, *113*, 109.
421. van Nostrum, C. F.; Benneker, F. B. G.; Brussaard, H.; Kooijman, H.; Veldman, N.; Spek, A. L.; Schoonman, J.; Feiters, M. C.; Nolte, R. J. M. *Inorg. Chem.* **1996**, *35*, 959.
422. Ni, Y.; Fitzgerald, J. P.; Carroll, P.; Wayland, B. B. *Inorg. Chem.* **1994**, *33*, 2029.
423. Mani, N. S.; Beall, L. S.; White, A. J. P.; Williams, D. J.; Barrett, A. G. M.; Hoffman, B. M. *J. Chem. Soc., Chem. Commun.* **1994**, 1943.
424. Montalban, A. G.; Lange, S. J.; Beall, L. S.; Mani, N. S.; Williams, D. J.; White, A. J. P.; Barrett, A. G. M.; Hoffman, B. M. *J. Org. Chem.* **1997**, *62*, 9284.
425. Mani, N. S.; Beall, L. S.; Miller, T.; Anderson, O. P.; Hope, H.; Parkin, S. R.; Williams, D. J.; Barrett, A. G. M.; Hoffman, B. M. *J. Chem. Soc., Chem. Commun.* **1994**, 2095.
426. Eichhorn, D. M.; Yang, S.; Jarrell, W.; Baumann, T. F.; Beall, L. S.; White, A. J. P.; Williams, D. J.; Barrett, A. G. M.; Hoffman, B. M. *J. Chem. Soc., Chem. Commun.* **1995**, 1703.
427. Sibert, J. W.; Lange, S. J.; Stern, C. L.; Barrett, A. G. M.; Hoffman, B. M. *Angew. Chem. Int. Ed. Engl.* **1995**, *34*, 2020.
428. Ricciardi, G.; Bavoso, A.; Bencini, A.; Rosa, A.; Lelj, F.; Bonosi, F. *J. Chem. Soc., Dalton Trans.* **1996**, 2799.
429. Ricciardi, G.; Bencini, A.; Bavoso, A.; Rosa, A.; Lelj, F. *J. Chem. Soc., Dalton Trans.* **1996**, 3243.
430. Sibert, J. W.; Baumann, T. F.; Williams, D. J.; White, A. J. P.; Barrett, A. G. M.; Hoffman, B. M. *J. Amer. Chem. Soc.* **1996**, *118*, 10487.
431. Fitzgerald, J. P.; Yap, G. P. A.; Rheingold, A. L.; Brewer, C. T.; May, L.; Brewer, G. A. *J. Chem. Soc., Dalton Trans.* **1996**, 1249.
432. Cook, A. S.; Williams, D. B. G.; White, A. J. P.; Williams, D. J.; Lange, S. J.; Barrett, A. G. M.; Hoffman, B. M. *Angew. Chem. Int. Ed. Engl.* **1997**, *36*, 760.
433. Collman, J. P.; Kendall, J. L.; Chen, J. L.; Eberspacher, T. A.; Moylan, C. R. *Inorg. Chem.* **1997**, *36*, 5603.
434. Lange, S. J.; Nie, H.; Stern, C. L.; Barrett, A. G. M.; Hoffman, B. M. *Inorg. Chem.* **1998**, *37*, 6435.
435. Goldberg, D. P.; Montalban, A. G.; White, A. J. P.; Williams, D. J.; Barrett, A. G. M.; Hoffman, B. M. *Inorg. Chem.* **1998**, *37*, 2873.
436. Goldberg, D. P.; Michel, S. L. J.; White, A. J. P.; Williams, D. J.; Barrett, A. G. M.; Hoffman, B. M. *Inorg. Chem.* **1998**, *37*, 2100.
437. Ricciardi, G.; Bencini, A.; Belviso, S.; Bavoso, A.; Lelj, F. *J. Chem. Soc., Dalton Trans.* **1998**, 1985.
438. Beall, L. S.; Mani, N. S.; White, A. J. P.; Williams, D. J.; Barrett, A. G. M.; Hoffman, B. M. *J. Org. Chem.* **1998**, *63*, 5806.
439. Forsyth, T. P.; Williams, D. B. G.; Montalban, A. G.; Stern, C. L.; Barrett, A. G. M.; Hoffman, B. M. *J. Org. Chem.* **1998**, *63*, 331.
440. Nie, H.; Stern, C. L.; Barrett, A. G. M.; Hoffman, B. M. *Chem. Commun.* **1999**, 703.
441. Ricciardi, G.; Rosa, A.; Ciofini, I.; Bencini, A. *Inorg. Chem.* **1999**, *38*, 1422.
442. Hochmuth, D. H.; Michel, S. L. J.; White, A. J. P.; Williams, D. J.; Barrett, A. G. M.; Hoffman, B. M. *Eur. J. Inorg. Chem.* **2000**, 593.
443. Montalban, A. G.; Jarrell, W.; Riguet, E.; McCubbin, Q. J.; Anderson, M. E.; White, A. J. P.; Williams, D. J.; Barrett, A. G. M.; Hoffman, B. M. *J. Org. Chem.* **2000**, *65*, 2472.
444. Michel, S. L. J.; Goldberg, D. P.; Stern, C.; Barrett, A. G. M.; Hoffman, B. M. *J. Amer. Chem. Soc.* **2001**, *123*, 4741.
445. Kobayashi, N.; Furuya, F. *Personal communication* **2002**.
446. Lee, S.; White, A. J. P.; Williams, D. J.; Barrett, A. G. M.; Hoffman, B. M. *J. Org. Chem.* **2001**, *66*, 461.
447. Montalban, A. G.; Michel, S. L. J.; Baum, S. M.; Vesper, B. J.; White, A. J. P.; Williams, D. J.; Barrett, A. G. M.; Hoffman, B. M. *J. Chem. Soc., Dalton Trans.* **2001**, 3269.
448. Ramdhanie, B.; Stern, C. L.; Goldberg, D. P. *J. Amer. Chem. Soc.* **2001**, *123*, 9447.
449. Knox, K.; D'Addario, A. P. *Amer. Cryst. Assoc. Summer Meeting*, **1967**; p. 61.
450. Kobayashi, N.; Furuya, F.; Yug, G.-C.; Wakita, H.; Yokomizo, M.; Ishikawa, N. *Chem.-Eur. J.* **2002**, *8*, 1479.
451. Bench, B. A.; Beveridge, A.; Sharman, W. M.; Diebold, G. J.; van Lier, J. E.; Gorun, S. M. *Angew. Chem. Int. Ed. Engl.* **2002**, *41*, 748.
452. Idehara, T.; Handa, M.; Kasuga, K. *10th. Symp. Coord. Chem. Jpn*, Kanazawa, Oct. 7–9, **1990**; p. 711 (3AP14).
453. Guilard, R.; Latour, J.-M.; Lecomte, C.; Marchon, J.-C.; Protas, J.; Ripoll, D. *Inorg. Chem.* **1978**, *17*, 1228.
454. Murata, K.; Liou, K.; Thompson, J. A.; McGhee, E. M.; Rende, D. E.; Ellis, D. E.; Musselman, R. L.; Hoffman, B. M.; Ibers, J. A. *Inorg. Chem.* **1997**, *36*, 3363.
455. Liou, K.; Ogawa, M. J.; Newcomb, T. P.; Quirion, G.; Lee, M.; Poirier, M.; Halperin, W. P.; Hoffman, B. M.; Ibers, J. A. *Inorg. Chem.* **1989**, *28*, 3889.
456. Martinsen, J.; Pace, L. J.; Phillips, T. E.; Hoffman, B. M.; Ibers, J. A. *J. Amer. Chem. Soc.* **1982**, *104*, 83.
457. Yamakado, H. *Ph.D. thesis*, **1991**.
458. Kanazawa, D. *Ph.D. thesis*, **1989**.
459. Yakushi, K.; Yoshitake, M.; Kuroda, H.; Kawamoto, A.; Tanaka, J.; Sugano, T.; Kinoshita, M. *Bull. Chem. Soc. Jpn* **1988**, *61*, 1571.
460. Yakushi, K. In *Phthalocyanines - Chemistry & Function*. Shirai, H., Kobayashi, N., Eds.; IPC Publisher: Tokyo, **1997**; p. 85.
461. Raidt, H. *Doctoral thesis*, Humboldt University of Berlin, Germany, **1967**.
462. Suito, E.; Uyeda, N. *Kolloid Z. Z. Polym.* **1963**, *193*, 97.
463. Simon, J.; Bassoul, P. *Design of Molecular Materials: Supramolecular Engineering*. John Wiley & Sons, Inc.: New York, **2000**, p. 121.
464. Lee, S. *Master thesis*, Cleveland State University, Cleveland, Ohio, USA, **1974**.
465. Rose, M. *Master thesis*, Cleveland State University, Cleveland, Ohio, USA, **1977**.
466. Janczak, J.; Idemori, Y. M. *Inorg. Chem.* **2002**, *41*, 5059.

Index to Volume 20

Page numbers in italics, e.g. *48*, refer to figures. Page numbers in bold, e.g. **110**, indicate entries in tables.

E

G

H

I

L

Cumulative Index to Volumes 11-20

Volume numbers are given in Roman numerals followed by a colon and precede the relevant pages. Italics, e.g. *XVIII:290*, refer to figures. Boldface, e.g. **XVI:129**, indicates entries in tables.

B

C

H

I

J

K

L

M

N

O

P

Q

R

ISBN 0-12-393220-3

9 780123 932204